STRUCTURAL GENOMICS
ON
MEMBRANE PROTEINS

STRUCTURAL GENOMICS
ON
MEMBRANE PROTEINS

EDITED BY

KENNETH H. LUNDSTROM

CRC Press
Taylor & Francis Group
Boca Raton London New York

CRC Press is an imprint of the
Taylor & Francis Group, an **informa** business
A TAYLOR & FRANCIS BOOK

CRC Press
Taylor & Francis Group
6000 Broken Sound Parkway NW, Suite 300
Boca Raton, FL 33487-2742

First issued in paperback 2019

ISBN-13: 978-1-57444-526-8 (hbk)
ISBN-13: 978-0-367-39110-2 (pbk)
Library of Congress Card Number 2005019887

Library of Congress Cataloging-in-Publication Data

Structural genomics on membrane proteins / [edited by] Kenneth H. Lundstrom.
     p. cm.
  Includes bibliographical references and index.
  ISBN-13: 978-1-57444-526-8 (alk. paper)
  ISBN-10: 1-57444-526-X (alk. paper)
  1. Membrane proteins--Conformation. 2. Genomics. I. Lundstrom, Kenneth H.
  [DNLM: 1. Membrane Proteins--genetics. 2. Membrane Proteins--ultrastructure. 3. Genomics.
QU 55 S928 2006]

QP552.M44S87 2006
572'.696--dc22   2005019887

Visit the Taylor & Francis Web site at
http://www.taylorandfrancis.com

and the CRC Press Web site at
http://www.crcpress.com

Foreword

In the near future, the unprecedented use of novel technologies for membrane proteins will hopefully aid in conquering the last frontier in structural biology. Although there are more than 30,000 protein structures deposited in the protein data bank (PDB), less than 1% of these represent membrane proteins. In light of the fact that membrane proteins constitute >20% of all proteins, this disparity in the number of membrane proteins and available structures is due to the inherent transmembrane nature of these proteins, which makes their expression, purification, stabilization, and crystallization substantially more difficult than for their soluble counterparts. The significance of structural biology has been recently demonstrated in the field of rational drug design. Unfortunately, from the drug discovery point of view, membrane proteins comprise 60 to 70% of current medicinal targets. In order to improve the success in obtaining high resolution structures on membrane proteins, development of appropriate technology is necessary for all areas including expression, purification, and crystallography. *Structural Genomics on Membrane Proteins* provides an excellent overview on novel research in bioinformatics and modeling on membranes as well as the latest technological developments in recombinant protein expression, refolding of membrane proteins from inclusion bodies, solubilization, and purification methods. Moreover, methods for the application of NMR and miniaturization in structural biology as well as electron and atomic force microscopy on membrane proteins are discussed. It is very helpful and a great read!

Krzysztof Palczewski, Ph.D.,
Chair, Department of Pharmacology
Case Western Reserve University
Cleveland, Ohio, U.S.A.

Preface

Genomics, proteomics, and other types of "omics" approaches have proven most fruitful in both current basic and applied research. There are numerous examples of studies on whole genomes of specific organisms, types of genes/proteins and gene/protein families. In genomics and proteomics strategies, the aims are to study the function of gene activities and to characterize proteins. Structural genomics approaches, again, have the goals set on structure determination of target proteins, which can facilitate rational drug design by speeding up the drug development process and improving the selectivity and efficacy of medicines. In this context, membrane proteins are of great significance as more than 60% of current drugs are based on this group of proteins.

As in many other modern approaches, a thorough knowledge in bioinformatics is a necessity for successful applications in structural genomics. Another cornerstone of any structure determination is the supply of proteins for purification and crystallization attempts. Very few proteins, especially those of therapeutic interest, are available in high quantities in native tissue, and even so, ethical issues prevent proceeding with this approach. The scientific community therefore relies, to a large extent, on recombinant expression technologies. As the requirements for membrane protein expression are more demanding, expression has been evaluated in bacterial, yeast, insect, and animal cells, applying a variety of specifically designed vectors containing various expression-enhancing sequences and tags to facilitate purification. The expression mode dictates the downstream processing requests as recombinant proteins in *Escherichia coli* inclusion bodies need to be subjected to refolding, as those expressed in cell membranes require solubilization procedures. Purification technologies are also of great importance as the presence of detergents makes the procedure more difficult and the target proteins less stable. Furthermore, crystallization of membrane proteins has been more difficult than for their soluble counterparts, clearly indicated by the submission of less than 100 structures in public databases compared with more than 30,000 entries for soluble proteins. However, x-ray crystallography is not the only approach to receive structural information on proteins. Alternative methods include NMR (nuclear magnetic resonance) and EM (electron microscopy) technologies.

I would like to acknowledge the authors of the chapters of this book. The enthusiasm encountered was overwhelming and made the project possible. I am also grateful to CRC Press and Jill Jurgensen, Project Coordinator, and Jay Margolis, Project Editor, for the efficient and professional contribution to get the book published. Sadly, during the preparation of the manuscript, Dr. Helmut Reiländer (Aventis Pharmaceuticals, Frankfurt, previously at Max Planck Institute for Biochemistry, Frankfurt) passed away after a long illness. Helmut, as a colleague and a good friend, made a significant contribution to the field of recombinant membrane proteins, and his support for many structural genomics programs in Europe has been vital. I dedicate this book to his memory.

Kenneth H. Lundstrom, Ph.D.

Editor

Kenneth H. Lundstrom received his Ph.D. in "Overexpression of Viral Membrane Proteins in *Bacillus subtilis*" from the University of Helsinki, Finland. He conducted his postdoctoral research at Cetus Corporation in California on "Antisense Expression and PCR Technologies." Dr. Lundstrom then went back to his native Finland, where he was appointed Senior Scientist at Orion Pharmaceuticals and was involved in cloning, expression, and structural studies on catechol-o-methyltransferases. From 1992 to 1995, he developed Semliki Forest virus vectors for overexpression of receptors (GPCRs) and ion channels at the Glaxo Institute of Molecular Biology, in Geneva, Switzerland. Dr. Lundstrom worked as Principal Biologist at the Glaxo Medicines Research Centre in Stevenage, United Kingdom, from 1995 to 1996. From 1996 to 2001, he was responsible for receptor expression in the CNS Department of F. Hoffmann-La Roche, in Basel, Switzerland. In 2001, Dr. Lundstrom was appointed the Scientific Coordinator of the MePNet program; he became the Chief Scientific Officer of BioXtal in Lausanne, Switzerland, in 2002. He is also part of the Senior Management Team (Vice President, Science & Technology) of Regulon Inc., Mountain View, California, a biotech company involved in cancer therapy. Dr. Lundstrom has published more than 100 scientific papers and reviews in international journals; acts as editor for books in the fields of GPCRs, structural genomics, and gene therapy; and is a frequent speaker at international conferences.

Contributors

Dr. Enrique Abola
The Scripps Research Institute,
La Jolla, California, U.S.A.

Dr. Mark Bacon
University of Leeds
Leeds, United Kingdom

Dr. Monika Bährer
Roche Diagnostics GmbH
Penzberg, Germany

Emma Barksby
University of Leeds
Leeds, United Kingdom

Dr. Sabrina A. Berettta
Abbott Laboratories
Abbott Park, Illinois, U.S.A.

Kim Bettaney
University of Leeds
Leeds, United Kingdom

Dr. Roslyn Bill
Aston University
Birmingham, United Kingdom

Dr. Giel Bosman
Nijmegen Center for Molecular Life
  Sciences
Nijmegen, The Netherlands

Dr. J. Robert Bostwick
AstraZeneca Pharmaceuticals LP
Wilmington, Delaware, U.S.A.

Dr. Bernadette Byrne
Wolfson Laboratories
Imperial College
London, United Kingdom

Dr. Mark Chiu
Abbott Laboratories
Abbott Park, Illinois, U.S.A.

Dr. Joanne Clough
University of Leeds
Leeds, United Kingdom

Dr. Richard Cogdell
Institute of Biomedical & Life Sciences
University of Glasgow
Glasgow, United Kingdom

Dr. Willem de Grip
Nijmegen Center for Molecular Life
  Sciences
Department of Membrane Biochemistry
Nijmegen, The Netherlands

Dr. Andreas Engel
University of Basel
Basel, Switzerland

Dr. Said Eshaghi
Stockholm University
Stockholm, Sweden

Dr. Slawomir Filipek
International Institute of Molecular and
  Cell Biology
Warsaw, Poland

Dr. Alastair Gardiner
Wolfson Laboratories
Imperial College
London, United Kingdom

Dr. Marie Groves
University of Leeds
Leeds, United Kingdom

Dr. Emmanuel G. Guignet
Institute of Biomolecular Sciences
Swiss Federal Institute of Technology
Lausanne, Switzerland

Dr. Frank Gunn-Moore
Imperial College
London, United Kingdom

Dr. Deborah S. Hartman
AstraZeneca Pharmaceuticals LP
Wilmington, Delaware, U.S.A.

Dr. Peter J.F. Henderson
University of Leeds
Leeds, United Kingdom

Dr. Richard Herbert
University of Leeds
Leeds, United Kingdom

Dr. Ruud Hovius
Institute of Biomolecular Sciences
Swiss Federal Institute of Technology
Lausanne, Switzerland

Dr. Mika Jormakka
Wolfson Laboratories
Imperial College
London, United Kingdom

Dr. Hans Kiefer
m-Phasys GmbH
Tuebingen, Germany

Dr. Alla Korepanova
Abbott Laboratories
Abbott Park, Illinois, U.S.A.

Christoph Krettler
Max Planck Institute for Biophysics
Frankfurt, Germany

Dr. Peter Kuhn
The Scripps Research Institute
La Jolla, California, U.S.A.

Dr. Kenneth Lundstrom
BioXtal
Epalinges, Switzerland

Mulugeta Mamo
Abbott Laboratories
Abbott Park, Illinois, U.S.A.

Dr. Johan Meuller
University of Leeds
Leeds, United Kingdom

Dr. Bruno H. Meyer
Institute of Biomolecular Sciences
Swiss Federal Institute of Technology
Lausanne, Switzerland

Anna Modzelewska
International Institute of Molecular &
  Cell Biology
Warsaw, Poland

Dr. Pär Nordlund
Stockholm University
Stockholm, Sweden

Dr. Thomas Ostermann
m-Phasys GmbH
Tuebingen, Germany

Dr. Simon Patching
University of Leeds
Leeds, United Kingdom

Dr. Bengt Persson
Linköping University
Linköping, Sweden

Dr. Mary Phillips-Jones
University of Leeds
Leeds, United Kingdom

John O'Reilly
University of Leeds
Leeds United Kingdom

Dr. Christoph Reinhart
Max Planck Institute for Biophysics
Frankfurt, Germany

Dr. Nick Rutherford
University of Leeds
Leeds United Kingdom

Dr. Massoud Saidijam
University of Leeds
Leeds, United Kingdom

Dr. Keigo Shibayama
University of Leeds
Leeds, United Kingdom

June Southall
University of Glasgow
Glasgow, United Kingdom

Dr. Geoffrey F. Stamper
Abbott Laboratories
Abbott Park, Illinois, U.S.A.

Dr. Raymond C. Stevens
The Scripps Research Institute
La Jolla, California, U.S.A.

Shun'ichi Suzuki
University of Leeds
Leeds, United Kingdom

Dr. Gerda Szakonyi
University of Leeds
Leeds, United Kingdom

Dr. Horst Vogel
Institute of Biomolecular Sciences
Swiss Federal Institute of Technology
Lausanne, Switzerland

Dr. Alison Ward
University of Leeds
Leeds United Kingdom

Dr. Xiang-Qun Xie
University of Houston
Houston, Texas, U.S.A.

Table of Contents

1 Introduction

Kenneth H. Lundstrom

CONTENTS

1.1 SCOPE OF BOOK

The aim of this book is to provide the reader with an overview on structural biology research and recent technology development on integral membrane proteins (IMPs). IMPs represent the last frontier in structural biology that has not been conquered. Paradoxically, although IMPs are the most important drug targets, very few high-resolution structures are available. Among the more than 30,000 structures deposited in public databases, only some 60 are on IMPs.[1] A similar situation was encountered for soluble proteins in the 1970s when structural determination methods were less advanced. Technology development led to an almost exponential increase in the number of resolved structures. However, IMPs are more complicated to handle due to their topology, which affects the inefficient transport and insertion in cellular membranes, the toxic effects of recombinant IMPs on host cells, and the instability of IMPs. As the density of IMPs in native tissue — with the few exceptions of IMPs such as the bacterial rhodopsin in *Halobacterium salinarium*,[2] the bovine rhodopsin in cow retina,[3] and the nicotinic acetylcholine receptor in the electric organ of *Torpedo marmorata*[4] — is insufficient for purification attempts directly from the native tissue, recombinant expression of IMPs is a necessity. Furthermore, large-scale isolation and purification of IMPs from human tissues for structural studies would be ethically unacceptable.

For this reason, much emphasis has been put on the overexpression of IMPs from various expression vectors. Generally, the overexpression of prokaryotic IMPs has been more successful than that of their eukaryotic counterparts.[5] The main reason is that it has been possible to overexpress them in *Escherichia coli* or alternative bacterial organisms. The complexity of prokaryotic IMPs is also lower as most post-translational modifications do exist only in eukaryotes. Chapter 3 describes the **Expression of Bacterial Membrane Proteins**, and Chapter 4 is an overview of **Prokaryotic Membrane Protein Production Strategies** as a high throughput approach. One of the drawbacks of eukaryotic IMP expression in *E. coli* has been the toxicity the foreign IMP has caused when inserted in bacterial membranes. For this reason, an alternative strategy has been to overexpress recombinant IMPs in bacterial inclusion bodies, which has significantly reduced host-cell toxicity and improved IMP yields substantially. However, in this case the drawback has been the requirement of refolding procedures to re-establish the functionality of the IMP, which has been frustratingly inefficient.[6] Recent method development has brought some improvements to the refolding technology as described in Chapter 5 on **Refolding and Purification Technologies**. Chapter 6 is an overview of **Crystallization of Membrane Proteins** with a special emphasis on bacterial IMPs.

Eukaryotic IMPs can be divided into GPCRs, ion channels, transporters and single-transmembrane proteins, which play such important roles in cellular signaling events.[7] A variety of factors such as small molecule ligands, light, odors, ions, changes in cell membrane potential, pressure, and pH can trigger the activation of IMPs.[8] Activation of IMPs results in signal transduction cascades including changes in intracellular calcium levels, protein phosphorylation, transcriptional regulation, proliferation, and cell death. IMPs are therefore directly involved in cardiovascular, metabolic, neurodegenerative, neurological, psychiatric, and viral diseases.[9] Additionally, certain GPCRs and single-transmembrane receptors play a role in cancer development. Chapter 7 has therefore been dedicated to **Signaling through Membrane Proteins** to provide an overview of how IMPs function. Special attention is given to the **Expression of Eukaryotic Membrane Proteins** in various host systems. Chapter 8 deals with **Yeast Expression Vectors**, Chapter 9 with IMP **Expression in Insect Cells**, and Chapter 10 with the application of **Mammalian Cells** for recombinant IMP production. The downstream processing of expressed recombinant IMPs is described in Chapter 11, where methods for **Solubilization and Purification of Membrane Proteins** are outlined. Chapter 12 is dedicated to **Fluorescence Technologies**, which are applied as quality control measurements for the functionality of the overexpressed recombinant IMP. These methods are extremely important and allow fast and reliable optimization of the conditions required for expression and purification.

Chapter 14 describes the **Miniaturization of Structural Biology Technologies**, which provides a means for high-throughput screening of crystallization parameters and conditions and substantially reduces the requirement of precious purified protein. Alternative methods to x-ray crystallography are described in Chapter 13, **Membrane Proteins and NMR**, and in Chapter 16, **Applications of Electron Microscopy Technologies on Membrane Proteins**. As most efforts in structural biology today — especially structural genomics initiatives,[10] where whole gene or protein families or whole genomes are studied — require a major input of expertise in a broad range of areas such as expression, purification and crystallization, large national and international networks have been established as described in Chapter 16.

Chapter 2 provides a description of **Bioinformatics on Membrane Proteins** and *in silico* methods for drug screening. Homology studies on sequences and topologies on membrane proteins are important for a better understanding of their various functions. Chapter 17 discusses in detail the application of **Molecular Modeling** as a prerequisite for drug development programs today. In Chapter 18, **Structure-Based Drug Design**, demonstrates through examples how the drug discovery process can be accelerated and improved with the aim of generating better and safer medicines.

1.2 SUMMARY

The structural biology of IMPs is reaching a critical stage in development. During the past 20 years less than 1% of the accumulated high-resolution structures have been represented by IMPs. This situation is really unsatisfactory as the majority of drug targets today are based on IMPs. Major technology development has taken place in the areas of recombinant protein expression, purification, and crystallization. The advances in molecular and cell biology have also substantially enhanced our understanding of the function of IMPs. However, further technology improvement is necessary. Although tens to hundreds of milligrams of recombinant proteins can be produced from bacterial, yeast, insect, and mammalian cells today to give further insight into the structural characterization of IMPs, obtaining high-resolution structures is a far more commplex process. Although efforts on certain IMPs in individual research teams have generated some success, the trend is now to establish large networks where a large number of targets, whole gene families, or genomes, can be studied in parallel. These efforts seem to be the most efficient way forward to achieve the much awaited breakthrough in structural genomics of IMPs in the near future.

REFERENCES

1. Michel, H. Membrane proteins of known structure. Max Planck Institute of Biophysics, Frankfurt, Germany. http://www.mpibp-frankfurt.mpg.de/michel/public/memprotstruct.html, accessed 2005.
2. Henderson, R., Baldwin, J.M., Ceska, T.A., Zemlin, F., Beckman, E., and Downing, K.H. Model for the structure of bacteriorhodopsin based on high-resolution electron cryomicroscopy. *J. Mol. Biol.,* 213, 899–929, 1990.
3. Palczewski, K., Kumasaka, T., Hori, T., Behnke, C.A., Motoshima, H., Fox, B.A., Le Trong, I., Teller, D.C., Okada, T., Stenkamp, R.E., Yamamoto, M., Miyano, M. Crystal structure of rhodopsin: A G protein-coupled receptor. *Science*, 289, 739–745, 2000.
4. Unwin, N. Structure and action of the nicotinic acetylcholine receptor explored by electron microscopy. *FEBS Lett,* 555, 91–95, 2003.
5. Drew, D., Fröderberg, L., Baars, L., and De Gier, J.W.L. Assembly and overexpression of membrane proteins in *Escherichia coli. Biochim. Biophys. Acta.,* 1610, 3–10, 2003.
6. Kiefer, H. *In vitro* folding of alpha-helical membrane proteins. *Biochim. Biophys. Acta.,* 1610, 57–62, 2003.
7. Pierce, K.L., Premont, R.T., and Lefkowitz, R.J. Seven-transmembrane receptors. *Nat. Rev.* 3, 639–650, 2002.
8. Lundstrom, K. Structural genomics of GPCRs. *Trends Biotechnol.,* 23, 103–108, 2005.
9. Drews, J. Drug discovery. *Science*, 287, 1960–1964, 2000.
10. Lundstrom, K. Structural genomics on membrane proteins: Mini review. *Comb. Chem. High Throughput Screen.,* 7, 431–439, 2004.

2 Bioinformatics in Membrane Protein Analysis

Bengt Persson

CONTENTS

2.1 INTRODUCTION

Membrane proteins are of critical importance for a wide variety of biological processes. They constitute ion channels, transport proteins, receptors for hormones, light, and odorants, just to mention a few examples. Over half of prescription drugs act on G protein-coupled receptors.[1] In the completely sequenced genomes, the proportion of genes coding for membrane proteins is estimated to be about 25%.[2-4] In spite of the biological importance of membrane proteins, there are only a few proteins for which the three-dimensional structures have been solved experimentally due to difficulties in crystallizing these proteins. Currently, only about 1 to 2% of all structures in the PDB[5] are membrane proteins.[6] Thus, there is a large gap to bridge. Bioinformatics can be of great value when it comes to identifying membrane-spanning proteins from the amino acid sequence alone and predicting their topology, that is, delineating the transmembrane segments and the orientation of the protein in the membrane.

This chapter presents a survey of various prediction methods, most of which are available as Web services, allowing for easy and user-friendly access. Several consensus methodologies have also appeared. In addition, the chapter will review evaluations of prediction performance and availability of data sets of experimentally verified membrane proteins.

2.2 PREDICTION OF MEMBRANE-SPANNING REGIONS OF PROTEINS

Among the presently known three-dimensional structures of membrane proteins,[7] most consist of one or several membrane-spanning alpha helices with intervening short or long loops on each side of the membrane (Figure 2.1). There is also an alternative architecture with beta sheets forming a barrel inserted in the membrane,

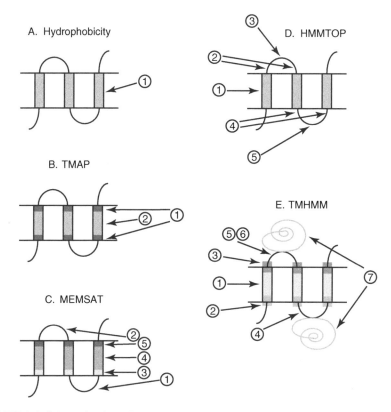

FIGURE 2.1 Schematic view of a transmembrane protein illustrating five different prediction methods. The thick vertical boxes illustrate three transmembrane helices, while the thin lines represent the intervening loops and the N- and C-terminal tails. Encircled numbers with arrows point to the segments used in the predictions (see text for the details).

but only a few proteins of this type are known so far. Most prediction algorithms are developed for alpha-helical membrane proteins, but there are algorithms for beta-barrel proteins as well. The topology prediction methods aim at identifying the membrane-spanning segments and the orientation of the protein in the membrane, that is, if the N terminus is cytosolic or noncytosolic. In the subsequent sections, the different types of prediction algorithms will be described. Most methods are available as Web servers and are listed in Table 2.1.

2.2.1 HYDROPHOBICITY ANALYSIS

One of the first and most basic methods to identify the membrane-spanning regions is to find hydrophobic segments in the protein sequence that are long enough to traverse the membrane (Figure 2.1A). Since the hydrophobic region corresponding to the apolar phospholipid tails of the lipid bilayer is about 30 Å,[8] these segments

TABLE 2.1
Membrane Protein Prediction Methods Available as Web Servers

Method	Web Server	Ref.
A. Alpha-Helical Proteins		
BPROMPT	http://www.jenner.ac.uk/BPROMPT	44
ConPred	http://bioinfo.si.hirosaki-u.ac.jp/~ConPred2/	39
DAS	http://www.sbc.su.se/~miklos/DAS/	17
HMMTOP	http://www.enzim.hu/hmmtop	31
MEMSAT	http://www.psipred.net	28,29
PRED-TMR2	http://o2.db.uoa.gr/PRED-TMR/	46
PRODIV-TMHMM	http://www.sbc.su.se/PRODIV-TMHMM/	34
SOSUI	http://sosui.proteome.bio.tuat.ac.jp/sosui_submit.html	43
SVMtm	http://ccb.imb.uq.edu.au/svmtm/SVMtm_Predictor.shtml	35
THUMBUP	http://phyyz4.med.buffalo.edu/service.htm	15
TMAP	http://www.ifm.liu.se/bioinfo/services	24,25
TMFinder	http://www.bioinformatics-canada.org/TM/	12
TMHMM	http://www.cbs.dtu.dk/services/TMHMM	3
TMpred	http://www.ch.embnet.org/software/TMPRED_form.html	42
TMMOD	http://liao.cis.udel.edu/website/servers/TMMOD/	33
TopPred	http://bioweb.pasteur.fr/seqanal/interfaces/toppred.html	23
B. Beta-Barrel Proteins		
B2TMPRED	http://gpcr.biocomp.unibo.it/	49
Omp_topo_predict	http://strucbio.biologie.uni-konstanz.de/~kay/om_topo_predict.html	48
PRED-TMBB	http://bioinformatics.biol.uoa.gr/PRED-TMBB/	53
PROFtmb	http://cubic.bioc.columbia.edu/services/proftmb	54
TBBPred	http://www.imtech.res.in/raghava/tbbpred/	51
TMBeta-Net	http://psfs.cbrc.jp/tmbeta-net/	50

should be about 20 residues if they form an alpha helix. To identify these segments, a hydropathy plot is made by plotting the hydrophobic character against the residue number. Normally, the values are averaged over a sliding window to smooth the curve.

In 1982, Kyte and Doolittle developed a hydropathy scale for amino acid residues, which they used to calculate a moving average of segment hydrophobicity along the protein chain.[9] They showed that such hydropathy plots could be used to identify membrane-spanning regions and interior regions of globular proteins. Still, the Kyte–Doolittle method is widely used and is considered to be the standard technique for hydrophobicity analysis.

Since then, a number of hydrophobicity scales have been developed.[10] In 1990, Degli-Esposti and co-workers evaluated different hydropathy scales for prediction of membrane proteins.[11] From their comparisons, it can be seen that most hydrophobicity scales correlate quite well, even if some clearly deviate. A new average scale, denoted AMP07, was created based on seven scales.[11]

Optimal window sizes vary among methods and have also been investigated. For the Kyte–Doolittle method, it was found to be five to nine residues,[11] while in the original paper, a 19-residue window was suggested.[9] The drawbacks of long windows are the loss of local information and the risk of an apparent fusion of closely spaced transmembrane segments.[11] It should be remembered that when changing the window size, the threshold values might also have to be changed from those of the original method descriptions.

Prediction accuracy could be improved by selecting a scale developed especially for membrane proteins, which is used in several prediction algorithms (see below), where propensity values are calculated from known membrane proteins.

The method that TMFinder uses is a combination of hydrophobicity and helicity scales for transmembrane protein prediction to reliably distinguish between membrane proteins and globular proteins.[12] It uses a hydrophobicity scale that was derived from experimental properties of transmembrane-mimetic model peptides[13] and a helicity scale derived from the structural properties of these peptides.[14] The principle of TMFinder is that a transmembrane segment must show both hydrophobicity and helicity.

Recently, a scale reflecting burial propensity has been developed from a set of 200 structurally known proteins.[15] The prediction method THUMBUP is based on this scale of burial propensity and the positive inside rule, and the method has been shown to reach similar levels of accuracy as the parametrically more complicated Hidden Markow Model (HMM)-based methods.[15]

Amphiphilicity can be measured using a method developed by Eisenberg and co-workers.[16] The so-called helical hydrophobic moment can be used as a means to distinguish transmembrane helices from those of globular soluble proteins.

A variant of hydrophobicity analysis is the method known as dense alignment surface (DAS).[17] In this method, a number of low-stringency dot plot analyses are performed between the sequence to be predicted and a set of nonhomologous membrane proteins.[17,18] The putative membrane-spanning regions are seen as diagonals reflecting distant similarity.

2.2.2 THE POSITIVE INSIDE RULE

A major breakthrough in the prediction of membrane protein topology was the discovery by von Heijne in 1986 that there is a preponderance of positively charged residues on the cytosolic side of prokaryotic membrane proteins.[19] The rule was later found to be valid also for eukaryotic membrane proteins,[20] even if not to the same extent. Recently, a genome-wide investigation showed that the positive inside rule is detectable in all completely sequenced genomes.[21]

TopPred,[22,23] the first membrane protein topology prediction method, combines hydrophobicity analysis with the positive inside rule. The hydrophobicity analysis was performed using a trapezoid sliding window, giving more weight to the central residues than to the flanking residues. After the identification of putative trans-membrane segments, the positive inside rule was applied to predict the topology. In cases where hydrophobic segments were on the borderline to be judged as transmembranous or not, the positive inside rule was used to distinguish between the alternatives.

2.2.3 USE OF MULTIPLE SEQUENCE ALIGNMENTS

The first method to use the information available from multiple sequence alignments of homologous proteins in membrane protein prediction was TMAP.[24] The idea is that by including information about the amino acid residue variation at each position, the membrane-spanning regions can be identified with higher accuracy than by using only a single sequence, similarly to a strategy applied for secondary structure predictions. The TMAP method also takes into consideration the differences in residue distributions between the central, hydrophobic part of the membrane-spanning helix and the ends, corresponding to the regions interacting with the polar head groups of the lipid bilayer (see Figure 2.1B). Two sets of propensity values are used in the prediction, derived from the statistical analysis of known membrane proteins. The use of two sets of propensity values together with the use of multiple sequences increases the prediction accuracy[24]

The TMAP method has been extended to predict the topology by analyzing the residue distributions of the loops on each side of the membrane.[25] Apart from the prominent Lys/Arg preponderance for the cytosolic side, the distributions of further residue types were also analyzed, which has led to improvements over considering the positive inside rule only.

Another method that also includes knowledge from related sequences is the PHD-htm by Rost and co-workers.[26,27] The sequence to be predicted is first used to search sequence databases for homologues, which subsequently are included in a multiple sequence alignment. From this alignment, a profile reflecting the residue distribution for each alignment position is calculated . The profile is analyzed by a neural network algorithm that estimates, for each residue, the preference to be in a transmembrane helix. After refinements of the results from the neural network, the positive inside rule is applied to predict the orientation of the protein in the membrane.

2.2.4 MODEL-RECOGNITION APPROACHES

In 1994, Jones and co-workers developed a prediction method, MEMSAT, using five sets of propensity values, derived from proteins with known membrane topologies.[28] These values represent the statistical likelihood of each amino acid residue type to be in the structural states: inside loop, outside loop, inside helix end, helix middle, and outside helix end (Figure 2.1C). The helix end regions were set to four residues. For the protein sequence of length n to be predicted, these five scores are assigned to each residue, giving an $n \times 5$ matrix. A dynamic programming algorithm recognizes the optimal membrane topology model, revealed by the path with the highest score through this matrix. In difficult cases, the method might suggest several alternative models. The MEMSAT algorithm has been refined using information from multiple sequence alignments[29]

The modeling concept has been further developed using hidden Markov models (HMMs) in the two prediction methods HMMTOP and TMHMM. HMM is a machine learning technique that can be used for a number of modeling purposes in various fields, and HMMs have become very popular in bioinformatics.[30] When used for membrane protein predictions, an HMM is first trained on a number of known cases, during which phase all parameters of the model are optimized. Subsequently, the HMM is tested and evaluated on a test set that is different from the training set.

HMMTOP, devised by Tusnády and Simon, uses an architecture that models the membrane proteins in five different states: membrane helix, outside tail, outside loop, inside tail, and inside loop[31] (Figure 2.1D). The idea is that the residue distributions in each of these five structural parts better reflects reality than does the mere consideration of the hydrophobic character of the residues. The HMM searches possible topologies and finds the one with the maximum likelihood.

TMHMM, devised by Krogh and co-workers,[3] uses seven different states: helix core, cytosolic cap, noncytosolic cap, cytosolic loop, noncytosolic short loop, noncytosolic long loop, and globular region (see Figure 2.1E). Loops of lengths up to 20 residues are classified as "loop" regions, while longer loops are classified as "globular" regions. The cap regions correspond to the five residues at each end of the transmembrane segments.

One difficulty with transmembrane protein prediction is distinguishing a signal sequence from a transmembrane segment, since both are similar in length and hydrophobic in nature. Käll and co-workers have developed Phobius, an HMM-based predictor that combines transmembrane segment prediction with signal peptide prediction.[32] The method is based on the models in TMHMM and SignalP-HMM. Phobius makes fewer misclassifications between transmembrane segments and signal peptides compared with the TMHMM and SignalP used individually. However, Phobius is less sensitive in signal peptide detection and less accurate when predicting the cleavage site of the signal peptide.

In 2005, a third HMM-based method, TMMOD, was described.[33] This method is based on the ideas of TMHMM but differs in the modeling of loops on both sides of the membrane and in the training procedure of the HMM. The authors report a small improvement compared with the TMHMM method.

Evolutionary information from multiple sequence alignments has long since been shown to improve membrane protein predictions.[24,26] Recently, Viklund and Elofsson showed that the HMM-based predictors could also be significantly improved by inclusion of information from evolutionary related sequences.[34] They have developed a method, PRODIV-TMHMM, which can correctly predict the topology for about two-thirds of all membrane proteins. Sequence profiles are created from the multiple sequence alignments and used as input to the HMMs. The performance is increased by approximately 10 percentage units when multiple sequence information is included; this is in the same range as reported for other methods, including the multiple sequence version of HMMTOP.[34]

2.2.5 SUPPORT VECTOR MACHINES

A support vector machine (SVM) is a machine learning technique suitable for classification purposes. SVMs have also recently been used in membrane protein predictions, as is shown with the method SVMtm.[35] The authors claim a sensitivity of 93.4% and specificity of 92% for the transmembrane helix predictions. The method can distinguish transmembrane proteins from soluble proteins with 99% accuracy. In addition, the method calculates a reliability measure for each transmembrane segment.

2.2.6 CONSENSUS TECHNIQUES

One strategy to improve the accuracy of the predictions is to combine several different methods using the consensus as the result. Such a combination was shown early on to increase the accuracy of secondary structure predictions.[36,37] To determine the consensus, it has to be decided how large the deviations can be between the segments predicted by the different methods. One consensus method uses a minimum overlap of five residues and has shown that there are small differences in the outcome using values between 1 and 10.[38] Another consensus method, in contrast, limits the differences in the distance between the mid-positions of the transmembrane segments to 11 to 15 residues.[39]

In 2000, Nilsson and co-workers reported that by using the consensus of five different prediction methods, the fraction of correctly predicted topologies could be considerably increased.[40] The five methods used were TMHMM, HMMTOP, MEMSAT, TopPred, and PHD. It was shown that when all five methods agreed, the fraction correctly predicted was 100% and that when four methods agreed, the accuracy was over 80%. These numbers can be compared with accuracies of 48 to 73% for each method used individually. However, the number of predictions decreases when applying the stringent criterion of 5:5 or 4:1 majority consensus. For *E. coli* membrane proteins, it was shown that using a threshold of at least four agreeing methods, close to half of the proteins could be predicted with high accuracy.

ConPred is another consensus prediction method[39] that uses nine different methods: KKD,[41] TMpred,[42] TopPred, DAS, TMAP, MEMSAT, SOSUI,[43] TMHMM, and HMMTOP. For topology predictions, only TMpred, TMAP, MEMSAT, TMHMM, and HMMTOP are used. The prediction accuracy reported is almost 100%. However, the prediction coverage is only 20 to 30%. The consensus methodology is calculated to increase accuracy by up to 11 percentage units over the individual methods. The ConPred server reports the predicted model accompanied by graphical representations showing topology, hydropathy plot, and helical wheel diagram.

A third consensus prediction method is BPROMPT, which uses a Bayesian belief network to integrate the results from several Web-based predictors.[44] The methods included are HMMTOP, DAS, SOSUI, TMpred, and TopPred. The BPROMPT method is reported to arrive at a topology prediction accuracy of 70% for prokaryotes and 53% for eukaryotes.

An additional method is CoPreTHi, which combines the output from seven methods (DAS, ISREC-SAPS,[45] PHD, SOSUI, TMpred, TopPred, and PRED-TMR[46]) and considers an amino acid residue to be transmembranous if it is predicted by at least three methods.[47]

2.2.7 PREDICTION OF BETA-BARREL PROTEINS

The beta-barrel membrane proteins have hitherto been characterized in bacteria, where they mediate transport of ions and small molecules. They have also been found in the organelles mitochondria and chloroplasts. The hydrophobic properties of the membrane-spanning beta strands are similar to those of soluble proteins, making predictions difficult. Furthermore, only few proteins are structurally characterized, and therefore only a small training set is available for the development of prediction methods. Consequently, there is a risk of over-training of the algorithms. So far, neural networks and hidden Markov models have been popular in prediction of beta-barrel membrane proteins.

One of the first reports on neural network methods used in topology predictions of beta-strand membrane proteins was published in 1998 by Diederichs and co-workers.[48] This method predicts the residue locations along the z-axis perpendicular to the membrane plane, where low values indicate periplasimc turns, medium values transmembrane beta-strands, and high values extracellular loops. The method was developed based on seven known structures and shown to be able to correctly predict two structures not related to the training set.[48]

Neural networks were combined with dynamic programming in a method developed by Jacoboni and coworkers.[49] They also included evolutionary information in the input to the network and achieved accuracy of 78%. The topology prediction is based on the observation that the longest loops are at the extracellular side of the membrane.

A third example of a neural-network-based method is reported by Gromiha and co-workers.[50] Their method reports probabilities for each beta strand allowing for further interpretations after analysis. Trained on 13 known structures, this method achieved 73% prediction accuracy.

Natt and co-workers have used both neural networks and support vector machines (SVM) to predict the transmembrane regions of beta-barrel proteins.[51] They used a feed-forward neural network with a standard back-propagation training algorithm. By including information from multiple sequence alignments, they could increase the accuracy to 80%, an improvement of the same range as for alpha-helical proteins.[34] They also developed an SVM-based method based on the amino acid sequence together with 36 physicochemical parameters. The accuracy is reported to be 77%. However, by combining the two techniques, the accuracy could be increased to 82%.

In analogy to prediction of alpha-helical membrane proteins, HMMs have also been used for beta-barrel membrane proteins. Martelli and coworkers[52] constructed a predictor using six states — beta-strand transmembrane core (two states), beta strand caps at each side of the membrane, inner loop, outer loop, and globular domain in the middle of each loop. They used input from multiple sequence alignments in the HMM to increase the accuracy, which is reported to be 83%. The information from the multiple sequence alignment is entered into the HMM as vectors, representing the sequence profile. The discriminatory ability between beta-membrane proteins and globular proteins is about 90%.

Another HMM-based method is PRED-TMBB, developed by Bagos and coworkers.[53] This HMM is a cyclic 61-state model, consisting of three submodels, representing the transmembrane strand and the inner and outer loops. The lengths of the transmembrane strands are between 7 and 17 residues. The method that was trained on 14 known proteins and tested using a jack-knife procedure shows 84% accuracy. Furthermore, PRED-TMBB discriminates beta-barrel proteins correctly from water-soluble proteins in 89% of the cases.

Bigelow and co-workers[54] have invented PROFtmb, a profile-based HMM for beta-barrel membrane proteins, with an accuracy of 86%. They have included a new definition of beta-hairpin motifs. This HMM includes 91 states representing the transmembrane beta strand in each direction, beta hairpins, inner loop, and outer loop. The discrimination between membrane proteins and soluble beta proteins is reported to be 100% at 45% coverage. The authors have applied this method on completed genomes from Gram-negative bacteria.

Finally, an alternative approach is used in the beta-barrel finder (BBF) program, which is based on analysis of the secondary structure, hydropathy, and amphipathicity of six outer membrane structures.[55] The authors have used BBF to estimate the proportion of beta-barrel membrane proteins in *E. coli* to be 2.8%. The program is available from the authors.

2.3 PREDICTION CONFIDENCE

When predicting membrane proteins, some regions are correctly predicted, while other regions are wrongly predicted. It would be of great value if the accuracy of the prediction could be estimated using a type of quality measurement, for example, that presented by PHD_htm.[26] Melén and co-workers[56] have developed reliability measures for the transmembrane prediction methods TMHMM, HMMTOP, MEM-SAT, PHD, and TopPred. For TMHMM and MEMSAT, the reliability scores have been shown to correlate with prediction accuracy and will therefore add valuable information to the predictions.

2.3.1 PARTIAL PREDICTIONS WITH HIGH ACCURACY

In cases when it is difficult to get correct predictions of the complete protein, it would still be valuable to get at least a partial prediction, especially if there is additional information available from elsewhere. Many times, it would also be important to know if these partial predictions are of high confidence. It has been shown that using a consensus technique with the criterion that at least four of five methods should agree gives predictions with high accuracy.[40] However, with this strict criterion, only a small number of membrane proteins will be predicted. Thus, in order to increase the number of predicted proteins, a method for prediction of partial topologies was developed using the strict criterion of the consensus methodology.[38] Partial consensus topologies could then be predicted for 60 to 70% of all proteins, on average covering 58% of the sequence length.

2.3.2 COMBINATION OF PREDICTIONS AND EXPERIMENTAL DETERMINATION

Partial predictions can be used in combination with experimental analyses of membrane topology. For instance, the experimental determinations can be directed to those regions for which the predictions are ambiguous. Thus, a limited number of experiments combined with reliable predictions can give the complete picture of a membrane protein topology. It has been shown that a combination of experimental determination of the C-terminal location and consensus predictions can be used to give reliable topology models for *E. coli.*[57,58]

Another example of a successful combination is cases where there are two alternative predicted topologies and one experiment thereby could be sufficient to distinguish between these two models.[57] An experimentally determined C-terminal location can be used as a constraint for TMHMM to improve the outcome of the predictions.[59]

2.4 EVALUATION OF METHODS

It is of importance to try to estimate the accuracy of the available prediction methods, and therefore several evaluations have been reported.

An evaluation of different alpha-helical membrane protein prediction methods has been made by Möller and co-workers.[60] For evaluation, they used a test set of 188 membrane proteins with experimentally verified topology.[61] They measured the reliability of both transmembrane segment predictions and sidedness predictions. Overall, they found the methods TMHMM and MEMSAT to be generally the best performing. HMMTOP was best for sidedness predictions. Interestingly, Kyte–Doolittle-based analyses (KKD)[41] and analysis of hydrophobic moment [16] were quite reliable in identifying membrane-spanning regions, even though the methods lacked specificity for membrane proteins.

TMHMM and SOSUI are most reliable in not making false predictions, that is, predicting transmembrane helices in proteins not bound to the membrane. For signal peptides that often are mispredicted as transmembranous, the methods ALOM,[62] PHD, and TopPred were most successful.

The evaluation also shows that these methods have problems when predicting proteins with four or more transmembrane segments. In proteins with many membrane passages, all transmembrane segments do not need to be hydrophobic, since not all are in direct contact with the lipid bilayer. If the segments are amphiphilic, it is difficult to distinguish them from a helix at the exterior of a globular cytosolic protein. These difficulties are also seen when trying to predict the topology of the G protein-coupled receptors.[63]

In 2002, Ikeda et al. compared 10 transmembrane prediction methods on a test set of 122 experimentally characterized transmembrane topologies.[64] They also reported that methods based on HMMs and other model-based approaches were most successful. Furthermore, they noticed that generally the prediction performance is better for prokaryotic sequences than for eukaryotic ones.

In general, the methods fared less well in these evaluations than in the original reports. One major reason could be that the methods might have been "over-trained" on the proteins available at the time for development. The training set might not have been representative enough due to only a small number of proteins with known topologies being available. Thus, more recently developed techniques would be better, since they are trained on a much larger test set. However, more importantly, significant advances have been made in the recent algorithms. The early techniques only judged single properties, such as hydrophobicity, while recent algorithms have subdivided the membrane protein into several parts, using multiple parameters for the different parts of the protein (see Figure 2.1).

2.5 TEST SETS AND DATABASES OF MEMBRANE PROTEINS

A collection of experimentally characterized membrane proteins has been assembled and made publicly available by Möller and co-workers at ftp://ftp.ebi.ac.uk/databases/testsets/transmembrane.[60] The entries are human curated and annotated, depending on experimental reliability. The top level (A) consists of proteins with known three-dimensional structure, level B of proteins characterized biochemically with at least two complementary methods, followed by level C with proteins for which only basic biochemical characterization has been reported. The database will be continuously updated and is provided in Swissprot format, making it easy to use for development and evaluation of new membrane protein prediction algorithms.

TMPDB is another database of experimentally characterized membrane protein topologies. The release of 2003 contained over 300 proteins, of which the vast majority was of the alpha-helical type.[65] TMPDB is based on information from examination of scientific articles and sequence and structure databases. The data are valuable for all scientists developing and optimizing new methods for transmembrane protein predictions. The database is available at http://bioinfo.si.hirosakiu.ac.jp/~TMPDB/.

PDB_TM is a database of transmembrane proteins with known structures, extracted from PDB. In PDB_TM, the membrane-spanning segments are determined using the TMDET algorithm[66] for calculation of the position of the protein in the lipid bilayer. The PDB_TM is updated weekly and is available at http://www.enzim.hu/PDB_TM.

2.6 CONCLUSION

Even if today's methods for membrane protein prediction are quite accurate and these structural predictions are among the most successful in bioinformatics, there is still much room for improvement. For development and training of the algorithms, the number of experimentally determined structures is still far too low, which might lead to methods that are biased and lack generality. Hopefully, the ongoing structural genomics initiatives worldwide will contribute to a considerable increase in the number of available structures. Also, large-scale experimental topology mappings will add important knowledge regarding membrane protein properties that can be used in new methods.

Hitherto, most methods have been based on neural networks and HMMs, but now methods based on support vector machines have started to appear. It can be anticipated that further sophisticated machine learning techniques will be used in membrane protein predictions. It is also likely that various combinations of these techniques will increase reliability. Thus, more training data together with improved prediction algorithms will hopefully help approach 100% accuracy.

REFERENCES

1. Attwood, T.K. A compendium of specific motifs for diagnosing GPCR subtypes. *Trends Pharmacol. Sci.*, 22, 162–165, 2001.
2. Wallin, E. and von Heijne, G. Genome-wide analysis of integral membrane proteins from eubacterial, archaean, and eukaryotic organisms. *Protein Sci.*, 7, 1029–1038, 1998.
3. Krogh, A., Larsson, B., von Heijne, G., and Sonnhammer, E.L. Predicting transmembrane protein topology with a hidden Markov model: application to complete genomes. *J. Mol. Biol.* 305, 567–580,, 2001.
4. Stevens, T.J. and Arkin, I.T. Do more complex organisms have a greater proportion of membrane proteins in their genomes? *Proteins*, 39, 417–420, 2000.
5. Berman, H.M., Westbrook, J., Feng, Z., et al. The Protein Data Bank. *Nucleic Acids Res.*, 28, 235–242, 2000.
6. Chen, C.P. and Rost, B. State-of-the-art in membrane protein prediction. *Appl. Bioinformatics*, 1, 21–35, 2002.
7. Tusnady, G.E., Dosztanyi, Z., and Simon, I. PDB_TM: selection and membrane localization of transmembrane proteins in the protein data bank. *Nucl. Acids Res.*, 33, D275–D278, 2005.
8. Lewis, B.A. and Engelman, D.M. Lipid bilayer thickness varies linearly with acyl chain length in fluid phosphatidylcholine vesicles. *J. Mol. Biol.*, 166, 211–217, 1983.
9. Kyte, J. and Doolittle, R.F. A simple method for displaying the hydropathic character of a protein. *J. Mol. Biol.*, 157, 105–132, 1982.

10. Cornette, J.L., Cease, K.B., Margalit, H., Spouge, J.L., Berzofsky, J.A., and DeLisi, C. Hydrophobicity scales and computational techniques for detecting amphipathic structures in proteins. *J. Mol. Biol.*, 195, 659–685, 1987.

11. Degli-Esposti, M., Crimi, M., and Venturoli, G. A critical evaluation of the hydropathy profile of membrane proteins. *Eur. J. Biochem.*, 190, 207–219, 1990.

12. Deber, C.M., Wang, C., Liu, L.P., et al. TM Finder: A prediction program for transmembrane protein segments using a combination of hydrophobicity and nonpolar phase helicity scales. *Protein Sci.*, 10, 212–219, 2001.

13. Liu, L.P. and Deber, C.M. Guidelines for membrane protein engineering derived from *de novo* designed model peptides. *Biopolymers*, 47, 41–62, 1998.

14. Liu, L.P. and Deber, C.M.. Uncoupling hydrophobicity and helicity in transmembrane segments. Alpha-helical propensities of the amino acids in non-polar environments. *J. Biol. Chem.*, 273, 23645–23648, 1998.

15. Zhou, H. and Zhou, Y. Predicting the topology of transmembrane helical proteins using mean burial propensity and a hidden-Markov-model-based method. *Protein Sci.*, 12, 1547–1555, 2003.

16. Eisenberg, D., Weiss, R.M., Terwilliger, T.C. The helical hydrophobic moment: a measure of the amphiphilicity of a helix. *Nature*, 299, 371–374, 1982.

17. Cserzo, M., Wallin, E., Simon, I., et al. Prediction of transmembrane alpha-helices in prokaryotic membrane proteins: the dense alignment surface method. New alignment strategy for transmembrane proteins. *Protein Eng.*, 10, 673–676, 1997.

18. Cserzo, M., Bernassau, J.M., Simon, I., and Maigret, B. New alignment strategy for transmembrane proteins. *J. Mol. Biol.*, 243, 388–396, 1994.

19. von Heijne, G. The distribution of positively charged residues in bacterial inner membrane proteins correlates with the trans-membrane topology. *EMBO J.*, 5, 3021–3027, 1986.

20. Sipos, L. and von Heijne, G. Predicting the topology of eukaryotic membrane proteins. *Eur. J. Biochem.*, 213, 1333–1340, 1993.

21. Nilsson, J., Persson, B., and von Heijne, G. Comparative analysis of amino acid distributions in integral membrane proteins from 107 genomes. *Proteins*, 60, 606–616, 2005.

22. von Heijne, G. Membrane protein structure prediction. Hydrophobicity analysis and the positive-inside rule. *J. Mol. Biol.*, 225, 487–494, 1992.

23. Claros, M.G. and von Heijne, G. TopPred II: an improved software for membrane protein structure predictions. *Comput. Appl. Biosci.*, 10, 685–686, 1994.

24. Persson, B. and Argos, P. Prediction of transmembrane segments in proteins utilising multiple sequence alignments. *J. Mol. Biol.*, 237, 182–192, 1994.

25. Persson, B. and Argos, P. Topology prediction of membrane proteins. *Prot. Sci.*, 5, 363–371, 1996.

26. Rost, B., Casadio, R., Fariselli, P., and Sander, C. Transmembrane helices predicted at 95% accuracy. *Prot. Sci.*, 4, 521–533, 1995.

27. Rost, B., Fariselli, P., and Casadio, R. Topology prediction for helical transmembrane proteins at 86% accuracy. *Prot. Sci.*, 5, 1704–1718, 1996.

28. Jones, D.T., Taylor, W.R., and Thornton, J.M. A model recognition approach to the prediction of all-helical membrane protein structure and topology. *Biochemistry*, 33, 3038–3049, 1994.

29. McGuffin, L.J., Bryson, K., and Jones, D.T. The PSIPRED protein structure prediction server. *Bioinformatics*, 16, 404–405, 2000.

30. Krogh, A., Brown, M., Mian, I.S., Sjolander, K., and Haussler, D. Hidden Markov models in computational biology. Applications to protein modeling. *J. Mol. Biol.*, 235, 1501–1531, 1994.
31. Tusnady, G.E. and Simon, I. Principles governing amino acid composition of integral membrane proteins: application to topology prediction. *J. Mol. Biol.*, 283, 489–506, 1998.
32. Kall, L., Krogh, A., and Sonnhammer, E.L. A combined transmembrane topology and signal peptide prediction method. *J. Mol. Biol.*, 338, 1027–1036, 2004.
33. Kahsay, R.Y., Gao, G., and Liao, L. An improved hidden Markov model for transmembrane protein detection and topology prediction and its applications to complete genomes. *Bioinformatics,* 21, 1853–1858, 2005.
34. Viklund, H. and Elofsson, A. Best (alpha)-helical transmembrane protein topology predictions are achieved using hidden Markov models and evolutionary information. *Prot. Sci.,* 13, 1908–1917, 2004.
35. Yuan, Z., Mattick, J.S., and Teasdale, R.D. SVMtm: support vector machines to predict transmembrane segments. *J. Comput. Chem.,* 25, 632–636, 2004.
36. Schulz, G.E., Barry, C.D., Friedman, J., Chou, P.Y., Fasman, G.D., Finkelstein, A.V., Lim, V.I., Ptitsyn, O.B., Kabat, E.A., Wu, T.T., Levitt, M., Robson, B., and Nogano, K. Comparison of predicted and experimentally determined secondary structure of adenyl kinase. *Nature*, 250, 140–142, 1974.
37. Argos, P. and Schwarz, J. An assessment of protein secondary structure prediction methods based on amino acid sequence. *Biochim. Biophys. Acta.,* 439, 261–273, 1976.
38. Nilsson, J., Persson, B., and von Heijne, G. Prediction of partial membrane topologies using a consensus approach. *Prot. Sci.,* 11, 2974–2980, 2002.
39. Arai, M., Mitsuke, H., Ikeda, M., Xia, J.X., Kikuchi, T., Satake, M., and Shimizu, T. ConPred II: a consensus prediction method for obtaining transmembrane topology models with high reliability. *Nucl. Acids Res.,* 32, W390–W393, 2004.
40. Nilsson, J., Persson, B., and von Heijne, G. Consensus predictions of membrane protein topology. *FEBS Lett.,* 486, 267–269, 2004.
41. Klein, P., Kanehisa, M., and DeLisi, C. The detection and classification of membrane-spanning proteins. *Biochim. Biophys. Acta.,* 815, 468–476, 1985.
42. Hofmann, K. and Stoffel, W. TMbase — a database of membrane spanning proteins segments. *Biol. Chem. Hoppe-Seyler,* 347, 166, 1993.
43. Hirokawa, T., Boon-Chieng, S., and Mitaku, S. SOSUI: classification and secondary structure prediction system for membrane proteins. *Bioinformatics*, 14, 378–379, 1998.
44. Taylor, P.D., Attwood, T.K., and Flower, D.R. BPROMPT: a consensus server for membrane protein prediction. *Nucl. Acids Res.,* 31, 3698–3700, 2003.
45. Brendel, V., Bucher, P., Nourbakhsh, I.R., Blaisdell, B.E., and Karlin, S. Methods and algorithms for statistical analysis of protein sequences. *Proc. Natl. Acad. Sci. USA,* 89, 2002–2006, 1992.
46. Pasquier, C., Promponas, V.J., Palaios, G.A., Hamodrakas, J.S., and Hamodrakas, S.J. A novel method for predicting transmembrane segments in proteins based on a statistical analysis of the SwissProt database: the PRED-TMR algorithm. *Prot. Eng.,* 12, 381–385, 1999.
47. Promponas, V.J., Palaios, G.A., Pasquier, C.M., Hamodrakas, J.S., and Hamodrakas, S.J. CoPreTHi: a Web tool which combines transmembrane protein segment prediction methods. *In Silico. Biol.,* 1, 159–162, 1999.

48. Diederichs, K., Freigang, J., Umhau, S., Zeth, K., and Breed, J. Prediction by a neural network of outer membrane (beta)-strand protein topology. *Prot. Sci.,* 7, 2413–2420, 1998.
49. Jacoboni, I., Martelli, P.L., Fariselli, P., De Pinto, V., and Casadio, R. Prediction of the transmembrane regions of (beta)-barrel membrane proteins with a neural network-based predictor. *Prot. Sci.,* 10, 779–787, 2001.
50. Gromiha, M.M., Ahmad, S., and Suwa, M. Neural network-based prediction of transmembrane beta-strand segments in outer membrane proteins. *J. Comput. Chem.,* 25, 762–767, 2004.
51. Natt, N.K., Kaur, H., and Raghava, G.P. Prediction of transmembrane regions of beta-barrel proteins using ANN- and SVM-based methods. *Proteins,* 56, 11–18, 2004.
52. Martelli, P.L., Fariselli, P., Krogh, A., and Casadio, R. A sequence-profile-based HMM for predicting and discriminating beta barrel membrane proteins. *Bioinformatics,* 18 Suppl 1, S46–S53, 2002.
53. Bagos, P.G., Liakopoulos, T.D., Spyropoulos, I.C., and Hamodrakas, S.J. A Hidden Markov Model method, capable of predicting and discriminating beta-barrel outer membrane proteins. *BMC Bioinformatics,* 5, 29, 2004.
54. Bigelow, H.R., Petrey, D.S., Liu, J., Przybylski, D., and Rost, B. Predicting transmembrane beta-barrels in proteomes. *Nucleic Acids Res.,* 32, 2566–2577, 2004.
55. Zhai, Y. and Saier, M.H., Jr. The beta-barrel finder (BBF) program, allowing identification of outer membrane beta-barrel proteins encoded within prokaryotic genomes. *Prot. Sci.* 11, 2196–2207, 2002.
56. Melen, K., Krogh, A., and von Heijne, G. Reliability measures for membrane protein topology prediction algorithms. *J. Mol. Biol.,* 327, 735–744, 2003.
57. Drew, D., Sjostrand, D., Nilsson, J., Urbig, T., Chin, C.N., de Gier, J.W., and von Heijne, G. Rapid topology mapping of *Escherichia coli* inner-membrane proteins by prediction and PhoA/GFP fusion analysis. *Proc. Natl. Acad. Sci. U.S.A.,* 99, 2690–2695, 2002.
58. Daley, D.O., Rapp, M., Granseth, E., Melen, K., Drew, D., and von Heijne, G. Global topology analysis of the *Escherichia coli* inner membrane proteome. *Science,* 308, 1321–1323, 2005.
59. Kim, H., Melen, K., and von Heijne, G. Topology models for 37 *Saccharomyces cerevisiae* membrane proteins based on C-terminal reporter fusions and predictions. *J. Biol. Chem.,* 278, 10208–10213, 2003.
60. Möller, S., Croning, M.D., and Apweiler, R. Evaluation of methods for the prediction of membrane spanning regions. *Bioinformatics,* 17, 646–653, 2001.
61. Möller, S., Kriventseva, E.V., and Apweiler, R. A collection of well characterised integral membrane proteins. *Bioinformatics,* 16, 1159–1160, 2000.
62. Nakai, K. and Kanehisa, M. A knowledge base for predicting protein localization sites in eukaryotic cells. *Genomics,* 14, 897–911, 1992.
63. Ji, T.H., Grossmann, M., and Ji, I. G protein-coupled receptors. I. Diversity of receptor-ligand interactions. *J. Biol. Chem.,* 273, 17299–17302, 1998.
64. Ikeda, M., Arai, M., Lao, D.M., and Shimizu, T. Transmembrane topology prediction methods: a re-assessment and improvement by a consensus method using a dataset of experimentally-characterized transmembrane topologies. *In Silico. Biol.,* 2, 19–33, 2002.
65. Ikeda, M., Arai, M., Okuno, T., and Shimizu, T. TMPDB: a database of experimentally-characterized transmembrane topologies. *Nucl. Acids Res.,* 31, 406–409, 2003.
66. Tusnady, G.E., Dosztanyi, Z., and Simon, I. TMDET: Web server for detecting transmembrane regions of proteins by using their 3D coordinates. *Bioinformatics,* 21, 1276–1277, 2005.

3 Prokaryotic Membrane Transport Proteins: Amplified Expression and Purification

Joanne Clough, Massoud Saidijam, Kim Bettaney, Gerda Szakonyi, Simon Patching, Johan Meuller, Shun'ichi Suzuki, Keigo Shibayama, Mark Bacon, Emma Barksby, Marie Groves, Richard Herbert, Mary Phillips-Jones, Alison Ward, Frank Gunn-Moore, John O'Reilly, Nick Rutherford, Roslyn Bill, and Peter Henderson

CONTENTS

3.1   INTRODUCTION

Membrane transport proteins are involved in nutrient capture, antibiotic efflux, protein secretion, toxin production, photosynthesis, oxidative phosphorylation, environmental sensing, and other vital functions in bacteria (Figure 3.1). Already there is commercial interest in inhibiting the activities of some membrane transport proteins, optimizing the activities of others, employing them as transducers of electrical/chemical/mechanical energy for nanotechnology, and so on. However, membrane proteins are notoriously difficult to study. Owing to their extreme hydrophobicity, they are refractory to direct manipulation and can only be removed from the membrane, and their solubility maintained, in the presence of a detergent.[1] In addition, transport proteins are usually only expressed at low levels and constitute less than 0.1% of total cell protein. Such difficulties help explain why fewer than 100 unique membrane protein structures have been resolved (see relevant examples in References[2,3]), although the structures of over 8000 unique soluble proteins (from almost 30,000 total structures, many not unique) have been solved. In fact, less than 1% of unique structures in the Protein Structures Database are membrane proteins, whereas they account for about 30% of all proteins in the cell.[2,3]

Prokaryote membrane transport proteins fall predominantly into two classes.[4,5] One of these uses adenosine triphosphate (ATP) to energize the transport of substrates across the membrane — the "ATP-Binding Cassette" (ABC) superfamily[6,7]

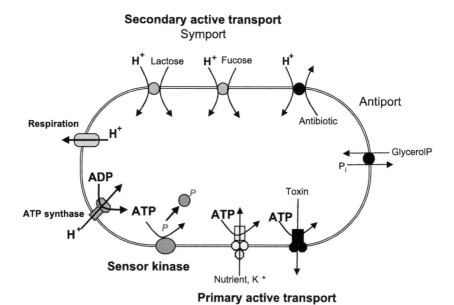

FIGURE 3.1 Active transport systems in bacteria. The large oval represents the cytoplasmic membrane of the microorganism. A transmembrane electrochemical gradient of protons is generated by respiration or ATP hydrolysis, shown on the left. The proton gradient may be used to drive ATP synthesis and the proton–nutrient symport and proton–substrate antiport secondary active transport systems shown along the top; alternatively, sodium (not shown) or phosphate (right side) may be the accompanying ions. Each is generally a single protein, usually of the 12-helix type. Along the bottom are illustrated primary active transport systems energized directly by ATP and a sensor kinase system.

— and the second, the "Major Facilitator Superfamily" (MFS)[8,9] is usually energized by the electrochemical gradient of protons (Figure 3.1) or sometimes sodium or phosphate ions (Figure 3.1). Members of the ABC superfamily and other types of transport systems[5] are not considered further in this chapter, which focuses mainly on the MFS transport proteins. These are found in nearly all organisms, from cyanobacteria to humans.[10,11]

In bacteria, individual MFS proteins may accomplish: the active accumulation of nutrients by a cation–substrate symport mechanism (Figure 3.1); the active efflux of compounds such as antibiotics, antibacterials, or toxins by a cation–substrate antiport mechanism (Figure 3.1); or substrate/substrate antiport reactions (Figure 3.1). They are thought to be single polypeptides comprising 10 to 14 (usually 12) transmembrane alpha helices,[12] illustrated for the *Escherichia coli* "FucP" protein in Figure 3.2. This conclusion is usually based on analysis of the hydropathic profile of the amino acid sequence of each protein predicted from its DNA sequence. In a few cases, the prediction is reinforced by genetic, immuno-chemical or other types of topological experiments.[13] There is structural information consistent with the 12-helix composition from electron diffraction analyses of two-dimensional protein crystals.[14–16] A spectacular confirmation came from x-ray diffraction analysis of three-dimensional crystals of three MFS proteins from *E. coli:* the lactose-H^+ symporter LacY[17], the glycerol-P/Pi antiporter GlpT[18] (Figure 3.1), and the Na^+/H^+ antiporter NhaA.[19]

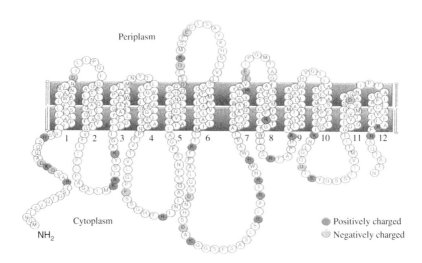

FIGURE 3.2 (See color insert following page 272.) A two-dimensional model for the folding of the FucP protein of *Escherichia coli* in the cell membrane based on hydropathy plot, positive inside rule, and beta-lactamase fusions. Positive and negative residues are highlighted in gray and black, respectively. From Gunn, F., Tate, C.G., Sansom, C.E., Henderson, P.J.F. *Molec. Microbiol.*, 15, 771–783, 1995; and Clough, J.L. Ph.D. Thesis, University of Leeds, 2001. With permission.

To enable the determination of structures of membrane transport proteins, a continuing supply of milligram quantities of protein is required. As native expression levels are usually less than 0.1% of total cell protein, heterologous gene expression is a necessity. Even if this approach is successful, a suitable detergent must be found for purification. The protein may also require "conformation locking" to overcome the probable flexibility of transport proteins, which is invoked to account for their ability to bind substrate in or on one side of the membrane, and effect its translocation. The strategy needed to purify sufficient active protein is therefore complex.

A general approach has been devised for the amplified expression, purification, and characterization of bacterial membrane transport proteins (Table 3.1) in E. coli. The strategy is described in this chapter, using the L-fucose-H[+] symport protein, FucP[20] (Figure 3.2) as an example to facilitate future examination of the large number of transport proteins arising from genome analyses; these proteins have potential for development of novel antibacterials and perhaps applications in biotechnology. L-Fucose — 6-deoxy-L-galactose — is reasonably abundant in nature as the breakdown product of plant cell wall polysaccharides and is used as a carbon source by free-living bacteria.[20] So far, the strategy has been successful for over 30 prokaryotic transporters, including MFS transport proteins from E. coli, Brucella abortus, B. melitensis, Helicobacter pylori, Microbacterium liquefaciens, Enterococcus faecalis, Bacillus subtilis, Staphylococcus aureus, Campylobacter jejuni, and Neisseria meningitides (Table 3.1). Proteins produced in this way have been tested in crystallization trials, and so far, two have yielded diffracting crystals.

Other organisms that appear to be successful for the propagation of vectors and expression of heterologous prokaryotic membrane proteins are Lactococcus lactis, Streptococcus thermophilus,[21] and Halobacterium salinarum.[22]

3.2   E. COLI EXPRESSION SYSTEMS

To facilitate the study of prokaryotic membrane proteins, numerous E. coli expression systems have been used,[23,24] with levels of expression as high as 50% of inner membrane protein[24] and 80% of outer membrane protein.[25]

The pET system (Novagen), which is widely used for the expression of soluble proteins, has also led to the amplified production of membrane proteins.[26–30] In pET vectors, the strong bacteriophage T7 promoter is recognized by T7 but not by E. coli RNA polymerase. Expression from pET vectors is achieved by transforming the recombinant plasmid into a host strain that carries a chromosomal copy of T7 RNA polymerase. For the usual E. coli (DE3) lysogen host strains, the T7 RNA polymerase is under the control of the lacUV5 promoter,[30] thereby allowing some expression of T7 RNA polymerase even in the absence of an inducer. In the case of the fucP gene, a construct exploiting the T7 system (Figure 3.3A) did yield expression but not amplification of the protein.[31] However, constructs exploiting transcription from the lambda leftward promoter (plasmid AD5827)[31–33] (Figure 3.3A) or the tac promoter (plasmid pTTQ18)[34] (Figure 3.3A) — which in the absence of isopropyl-β-D-thiogalactoside (IPTG) is repressed by the plasmid-encoded lac repressor (Figures 3A,B),[34–35] — were successful.[31,36] We have now successfully overexpressed more than 30 prokaryotic membrane transport proteins in E. coli using the plasmid pTTQ18

TABLE 3.1
Examples of 34 Overexpressed Proteins and Their 12 Organisms of Origin

Protein	Organism	Major Substrate
KgtP (JHP0334)	*Helicobacter pylori*	Ketoglutarate
"ProP"	*Helicobacter pylori*	Not known
GluP	*Helicobacter pylori*	Glucose
NupC	*Helicobacter pylori*	Not known
PutP	*Helicobacter pylori*	Proline–Na$^+$
NixA	*Helicobacter pylori*	Nickel
Hp1092	*Helicobacter pylori*	Multidrugs
Hp1181	*Helicobacter pylori*	Multidrugs
AraE	*Escherichia coli*	Arabinose–H$^+$
XylE	*Escherichia coli*	Xylose–H$^+$
GalP	*Escherichia coli*	Galactose–H$^+$
ProP	*Escherichia coli*	Proline-H$^+$
PutP	*Escherichia coli*	Proline–Na$^+$
Bcr	*Escherichia coli*	Bicyclomycin
FucP	*Escherichia coli*	Fucose–H$^+$
GusB	*Escherichia coli*	Glucuronide–H$^+$
"ProP" (Cj0250c)	*Campylobacter jejuni*	Not known
FucP (Cj0486)	*Campylobacter jejuni*	Fucose
Nma2100	*Neisseria meningitidis*	Sugar?
GluP (Nma0714)	*Neisseria meningitidis*	Glucose
"Bcr" (Nma2040)	*Neisseria meningitidis*	Multidrugs?
GluP	*Brucella abortus*	Glucose
KgtP	*Brucella melitensis*	Ketoglutarate
LmrB	*Brucella melitensis*	Lincomycin
NorA	*Staphylococcus aureus*	Multidrugs
Mj 1560	*Methanococcus janaschii*	Multidrugs?
Mhp1	*Microbacterium liquefaciens*	Hydantoins
AraE	*Bacillus subtilis*	Arabinose
Bmr	*Bacillus subtilis*	Multidrugs
Blt	*Bacillus subtilis*	Multidrugs

(Table 3.1 and Figure 3.3B) and have achieved expression levels ranging from 5 to 50% of inner membrane protein. Since plasmid pTTQ18 is not host-specific, unlike plasmid AD5827, we now generally ligate the gene for each membrane protein downstream of the pTTQ18 *tac* promoter (Figure 3.3B); an oligonucleotide encoding six histidines and usually an epitope tag are also added in-frame to the C terminus (amino acid sequence -G-G-R-G-S-H-H-H-H-H-H)[33, 35–38] (see below). The N-terminal amino acid sequence of the membrane protein may be slightly modified by a few amino acids from the alpha peptide of the LacZ protein, which is in the multicloning site, but this does not prevent expression or attenuate activity.[35–40]

The putative FucP(His)$_6$ protein resulting from growth of *E. coli* cells transformed with the appropriate construct and induced with IPTG is shown in Figure 3.4.

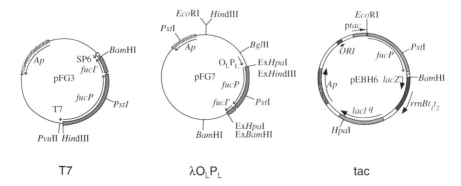

Figure 3.3A. Three plasmid constructs prepared to test for amplified expression of the cloned *fucP* gene using the indicated promoters. From Gunn, F.J., Tate, C.G., Henderson, P.J.F.; *Mol. Microbiol.*, 12, 799–809, 1994; and Clough, J.L. Ph.D. Thesis, University of Leeds, 2001. With permission.

3.3 OPTIMIZATION OF CONDITIONS FOR EXPRESSION

Growth conditions are optimized in 50- to 400-ml cultures of *E. coli* host strains testing both minimal M9 and complex LB media to maximize expression of each protein[35–36] (see Figure 3.4). The concentration of IPTG required is tested (Figure 3.5), and the period of growth before induction is varied to obtain as high a cell density as possible commensurate with optimal protein expression (growth is often diminished, or abolished, after induction) (data not shown). Similarly, the period of exposure to IPTG (2 to 24 hours) is investigated in order to promote maximum expression and minimize the problems described below.

Perhaps the most common difficulties associated with membrane protein expression are toxicity and a reduction in host-cell viability following induction. For example, overexpression of the *E. coli* sugar transporters such as FucP(His)$_6$ slows cell growth (Figure 3.6), and the expression in *E. coli* of multidrug resistance (Mdr) proteins from *B. subtilis* (Bmr), *S. aureus* (NorA), or *H. pylori* (JM1181 or JM1092) results in a rapid attenuation of cell growth[24,37,39–41] and ultimately cell death. Cells expressing Bmr can lyze after only a few hours (Ward and Hoyle, unpublished), whereas cells overexpressing NorA have impaired cell division and elongate,[37] and cells overexpressing TetA undergo a rapid loss of membrane potential, similar to that observed on the addition of uncoupling agents.[42] These problems are often more evident with Mdr proteins than with other members of the MFS family.[37, 40–42] In such cases, overexpression can still be achieved by the careful choice of media, host strain, growth temperature, titration of inducer, and time period of induction.[37,41,43] Miroux and Walker[29] have described mutants of BL21(DE3) host strains for the successful expression of membrane proteins, with avoidance of toxicity. In some cases, the protein of interest accumulated as inclusion bodies. Inclusion bodies result from generation of large quantities of protein (up to grams of recombinant protein per liter of culture) but in an insoluble form. While there are no general methods

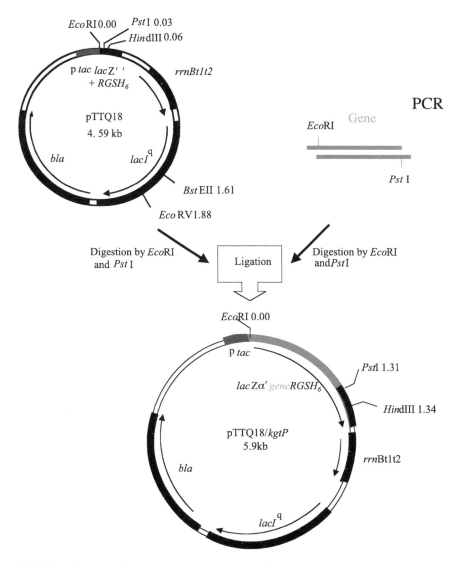

FIGURE 3.3B. The pTTQ18 plasmid vector applied for membrane transport protein expression in *E. coli*. The plasmid comprises the synthetic tac promoter (*ptac*), the beta galactosidase alpha fragment (*lac*Zalpha'), the *E. coli* *rrn*B operon transcription terminator (*rrnBtlt2*), the gene encoding a de-repressed LacI repressor protein (lacI^q), pUC18 origin of replication (not shown), and the gene encoding beta lactamase and ampicillin resistance (*bla*). Relevant restriction enzyme sites, especially in the multiple cloning site, are illustrated. There is also an oligonucleotide sequence encoding GGRGSH6, which becomes fused in-frame to the ligated gene (not shown). From Clough, J.L. Ph.D. Thesis, University of Leeds, 2001 and Stark, M.J.R. *Gene*, 51, 255–267, 1987. With permission.

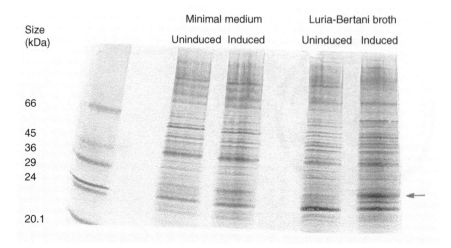

FIGURE 3.4 Detection of expressed FucP(His)$_6$ protein. *E. coli* BLR cells harboring the *fucP* gene-containing pTTQ18 plasmid were grown on Minimal salts or LB medium and harvested 12 hours after induction by 0.2 mM IPTG. Uninduced cells were cultured in parallel. Membrane preparations were made by the water-lysis method analyzed by SDS-PAGE and stained with Coomassie blue. Scanning densitometry indicated that the FucP(His)$_6$ protein comprised 8.2% of the preparation in minimal medium and 16.2% in rich medium. From Ward, A., Sanderson, N.M., O'Reilly, J., Rutherford, N.G., Poolman, B., Henderson, P.J.F., in Baldwin, S.A., ed. *Membrane Transport — a Practical Approach.* Blackwell Press, Oxford:, 2000, pp.141–166; and Clough, J.L. Ph.D. Thesis, University of Leeds, 2001. With permission.

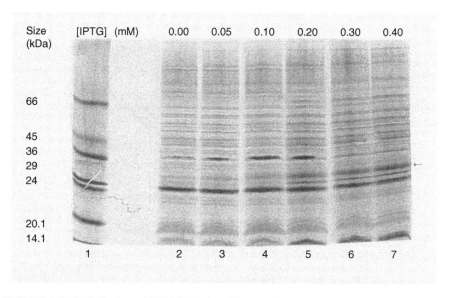

FIGURE 3.5 Optimization of IPTG induction. The experimental culture conditions were the same as in Figure 3.4 at different indicated concentrations of IPTG. The cells were harvested one hour after the induction.

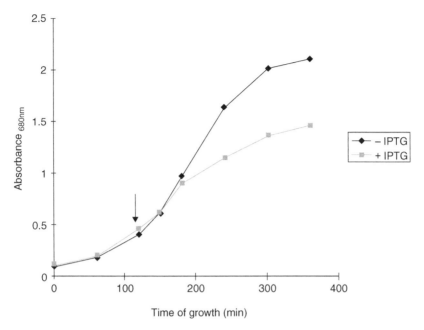

FIGURE 3.6 Growth inhibition of host strain after IPTG induction. *E. coli* strain BLR harboring the *fucP* gene-containing plasmid was cultured in LB medium. Induction took place at mid-log phase (A_{680nm} = 0.4 to 0.6) with 0.2 mM IPTG (indicated by arrow). From Clough, J.L. Ph.D. Thesis, University of Leeds, 2001. With permission.

available for the refolding of polytopic membrane protein inclusion bodies, recent work has demonstrated the facilitation of refolding and membrane insertion of hexa-histidine tagged membrane proteins.[44–48]

In many cases, a 30-L fermenter can conveniently be used without compromising expression, but for some proteins, the level of expression is always higher in batch cultures of 500 to 800 ml in 2-L baffled flasks. Further examination of the parameters regulating growth and protein production in these conditions may enhance our understanding of expression.

It is possible that the cellular machinery for the folding and insertion of transport proteins into the membrane, which require multiple factors in *E. coli*, may limit overexpression. Studies with the *E. coli* lactose transporter have shown that the folding and insertion of this protein into the membrane require phosphatidyl-ethanolamine,[46] the chaperonin GroEL,[45] and the bacterial signal recognition particle.[49] Growth at lower temperatures, or using lower IPTG to slow expression, may alleviate these effects.

The *tac* promoter has also been successfully employed for the overexpression in *E. coli* of the enzyme II mannitol transport protein from the thermophilic *Bacillus stearothermophilis*[50] and the alanine carrier from the thermophilic bacterium PS3[43] (as an N-terminal maltose-binding protein fusion) and the *E. coli* melibiose transporter.[27]

3.4 IDENTIFICATION OF OVEREXPRESSED MEMBRANE PROTEINS

The success of the amplified expression is usually first tested by measuring the appearance of IPTG-inducible transport activity in the host strain. In this case, uptake of 1-[^{3}H]-L-fucose into the BL219(DE3) host was measured, and, depending on the time of induction and the culture medium, fucose transport activity was 4 to 14 times higher in the induced than in the uninduced cultures[36] (data not shown). Then, an extra protein is sought in preparations of inner membranes in Coomassie Blue (shown for FucP(His)$_6$ in Figure 3.4 and Figure 3.5) or silver-stained SDS-PAGE (sodium dodecyl sulfate–polyacrylamide gel electrophoresis) gels. It is very important to note two things: membrane proteins rarely solubilize well when boiled in SDS, unlike soluble proteins; also, all MFS proteins, in our experience, migrate at apparent Mr values below those predicted from their amino acid sequences. Therefore, lower temperatures must be used for solubilization.[35] It is then highly advisable to confirm the identity, as well as the absence of post-translational modification, by eluting the protein and carrying out N-terminal amino acid sequencing (see below) and by using Western blotting to ensure that the C-terminal His-tag is present (shown for FucP(His)$_6$ in Figure 3.7). A variable phenomenon, in our experience, is the appearance of apparently higher-molecular-weight forms of the same protein (Figure 3.7) in stained SDS-PAGE gels and Western blots. If this represents some aggregation of the protein, additional steps may be needed to eliminate it before crystallization trials. Alternatively, it may represent different conformers of the fully or partially unfolded protein, even in SDS-PAGE.

Recently, protocols have been devised to enable MFS proteins to be analyzed in mass spectrometers to determine the precise Mr of purified material.[35,51–53] Partial proteolytic digestion followed by tandem MS amino acid sequencing can then reinforce identification.

3.5 PURIFICATION OF MEMBRANE PROTEINS FACILITATED BY THE ADDITION OF AFFINITY TAGS

While many different affinity fusions[54] are available to be added to membrane proteins to facilitate their purification, perhaps the most widely used is hexa-histidine. The addition of a hexa-histidine tag to the membrane protein of interest allows its rapid purification using nickel chelate affinity chromatography.[27,55–57] This tag is small and, in our experience, does not decrease expression levels or have a deleterious effect on protein folding or insertion into the membrane. Experience indicates that the hexa-histidine tag is in itself not a barrier to crystallization,[17,18,58] but if it is desirable to remove the hexa-histidine tag after protein purification, the genetic insertion of a suitable protease cleavage site such as thrombin readily allows this. An additional advantage to the use of such tags is that often antibodies to the tags are commercially available, which facilitates Western blotting for protein detection (e.g., Figure 3.7) without the need to raise antibodies against the specific protein of interest. Using nickel chelate affinity chromatography, we have routinely purified

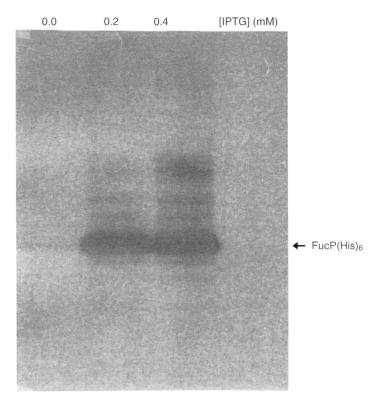

FIGURE 3.7 Identification of the FucP(His)$_6$ protein by Western blot analysis. Membranes were prepared from *E. coli* BLR harboring the *fucP*-containing plasmid grown on LB medium and harvested one hour after inducton with 0.0 mM (lane 1), 0.2 mM (lane 2), and 0.4 mM (lane3) IPTG, respectively. The position of the proposed FucP(His)$_6$ protein is indicated by the arrow. Molecular weight markers are not shown, but the intense bands shown superimpose on the putative FucP(His)$_6$ protein migrating at approximately 32 kDa (kilodaltons) in Coomassie- and silver-stained SDS-PAGE gels (see Figure 3.4, Figure 3.5, Figure 3.7, and Figure 3.8).

proteins to greater than 90% homogeneity,[35–41] as judged by densitometry and shown for FucP(His)$_6$ in Figure 3.8.

In order to aid crystallization, we have also fused the genes encoding FucP(His)$_6$, GusB(His)$_6$, Bmr(His)$_6$, or NorA(His)$_6$ in-frame to "green fluorescent protein" (GFP). In these four examples, a chimeric protein was successfully produced, but the expression levels were severely attenuated. When the fusion of MalE (maltose binding protein [MBP] lacking a signal sequence for export) was made to the N terminus of FucP, GusB, or Bmr, respectively, the chimeric proteins were expressed at levels sufficient for structural studies. In all cases, the attached hexa-histidine tag was used to effect purification, though the additional MBP protein also allows purification using amylose resin. This may require exchange of the detergent, from, for example, dodecyl-beta-D-maltoside to Triton X-100.

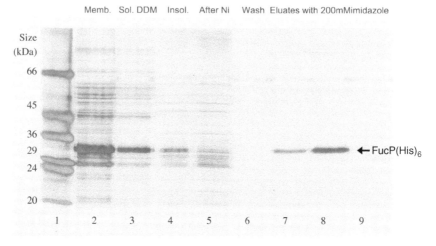

FIGURE 3.8 The purification of FucP(His)$_6$ from inner membranes of *E. coli* using Ni-NTA affinity chromatography. Silver stained 15% SDS-PAGE gel. **Lane 1:** molecular weight markers ("Mr"). **Lane 2:** inner membranes ("IM"), 20 mg solubilized in 100 mM Hepes buffer, pH 7.9, 20% glycerol, 1% dodecyl-beta-D-maltoside (DDM), 20 mM imidazole, while stirring on ice for 30 min. **Lane 3:** the treated membranes were centrifuged at 100,000 g for 30 min. at 4°C to remove unsolubilized material, and the supernatant ("+Sol DDM") was mixed with 1 ml Ni-NTA resin after equilibration in buffer E, and incubated with gentle shaking overnight (12 to 16 h) at 4°C. **Lane 4:** the resin was collected by centrifugation, 198 g 1 min. at room temperature and the supernatant retained ("Unbound"). **Lanes 5 and 6:** the protein–resin complex was packed into a disposable column, and the unbound protein was further removed by washing the resin with 25 ml buffer E containing 0.05% DDM, pH 8 ("After Ni" lane 5, and "Wash" lane 6). **Lanes 7–9:** finally, FucP(His)$_6$ was eluted from the column with 5 ml buffer E containing 0.05% DDM and 200 mM imidazole pH 8, and collected in fractions; those fractions containing protein were concentrated ("Eluates"). The tendency of some of the transport proteins to reveal an apparently higher Mr form (oligomer/conformer) is illustrated in this case.

3.6 CHOICE OF DETERGENT AND SOLUBILIZATION CONDITIONS

Integral membrane proteins are removed from lipid bilayers by the action of detergents. Detergents are amphipathic molecules comprising a polar head group and a hydrocarbon tail. At a determined concentration, referred to as the critical micelle concentration (CMC), detergent monomers aggregate to form ordered structures into which membrane proteins can insert.[1,59,60] Detergents are classified into three broad categories:

- Ionic, which carry a net charge associated with their head group, for example, the anionic sodium dodecyl sulfate.
- Nonionic, which have uncharged hydrophilic head groups, for example, dodecyl-beta-D-maltoside.
- Zwitterionic, which have both positive and negative charges but carry no net charge, for example, CHAPS.

Detergents used in the solubilization and purification of membrane proteins must maintain the structural integrity of the protein and its activity on reconstitution. In many cases, it may not be possible to select a detergent suitable for both solubilization and purification. In such cases, detergent exchange may be carried out.[35] The solubilization of membrane proteins is a multistep process. The CMC detergent monomers partition into the lipid bilayer as described below. As the concentration of the detergent increases to CMC and above, the membrane breaks down to generate mixed micelles of protein–detergent, protein–detergent–lipid, lipid–detergent, and detergent alone. Alternatively, a mixture of detergents used below their CMC values may be effective.[60, 61]

Membrane protein solubilization trials should be carried out using a wide range of detergents ranging from CMC to protein concentrations from 1 to 10 mg/ml and in the presence of stabilizing additives such as glycerol and sodium chloride (NaCl). The solubilization mix is incubated (usually on ice) before the unsolubilized material is removed by ultracentrifugation. Individual membrane proteins will show different solubilization requirements, especially of pH and salt concentration. This property can sometimes be used to pre-extract unwanted membrane proteins with one detergent and subsequently solubilize the protein of interest with a second detergent.[62] The process of solubilization can be monitored using SDS-PAGE and Western blotting.

Detailed protocols for optimizing solubilization of membrane proteins are given elsewhere.[35] For the FucP(His)$_6$ and many other MFS proteins, dodecyl-beta-D-maltoside (DDM) or cyclohexyl-n-hexyl-beta-D-maltoside have proven to be the detergent of choice for solubilization and purification. However, the best concentration of detergent and stabilizing additives (e.g., benzamidine, glycerol, sulfobetaine) still needs to be determined for each protein.

3.7 RECONSTITUTION AND ASSAY OF MEMBRANE PROTEIN ACTIVITY

There are two basic methods for the reconstitution of detergent-solubilized membrane symporters. In the first method, the membrane protein is mixed with detergent-destabilized liposomes, rapidly diluted (to below the CMC of the detergent), and the proteoliposomes recovered by ultracentrifugation.[63] In the second method, the protein is mixed with detergent-destabilized liposomes and the detergent removed by successive additions of polystyrene beads[64] (Biobeads™, Biorad): one or more dialysis steps may also be required.[65] These methods are reviewed by Rigaud and Levy.[66]

The activity of the reconstituted protein can be assayed using artificial ion gradient-driven uptake,[67] enzyme-linked spectroscopic assays,[68] or possibly ligand-binding assays. For artificial ion gradient-driven studies, proteoliposomes can be preloaded with 100 mM potassium acetate and 25 mM potassium phosphate (at pH 7). The reaction is initiated by dilution of the proteoliposomes into 125 mM sodium phosphate plus the potassium ionophore valinomycin and radioactive substrate. In this system, acetic acid passively diffuses across the proteoliposome membrane and then dissociates, thereby removing protons from the vesicle lumen and increasing the internal pH and transmembrane Δ pH. The addition of valinomycin causes a flux of potassium, from the inside to the outside of the proteoliposome, thereby generating a transmembrane electrical potential.[65] For Mdr antiporters, similar methods have been employed.[69]

3.8 PHYSICAL MEASUREMENTS ON DETERGENT-SOLUBILIZED AND RECONSTITUTED MEMBRANE PROTEIN

There are two methods that give an indication of a protein's secondary structure content: circular dichroism (CD) and Fourier transform infra-red (FTIR) spectroscopy. Both methods can be applied to both detergent-solubilized and membrane-reconstituted protein. For FTIR spectroscopy, exchange of H_2O by D_2O is usually carried out to minimize errors caused by imprecise correction for infra-red absorbance by H_2O.[70]

CD signals are generated when a molecule differentially absorbs left- and right-handed circularly polarized light. Characteristic spectra occur for each type of secondary structure, for example, alpha helix has a positive peak at 195 nm and two negative peaks at 208 and 222 nm, with an indent between these two negative peaks at 215 nm.[71] CD can be affected by light scattering, especially from proteoliposome solutions,[72,73] and interpretation is dependent on an accurate estimate of protein concentration. Algorithms for fitting CD data tend to be heavily biased toward soluble protein structure, and so anomalous fitting results can occur.

FTIR spectra of proteins arise from peptide C = O stretching vibrations. Two bands are evident, amide I and amide II.[74,75] The deconvolution of amide I separates the individual subcomponents of the peak, and from the wave numbers and intensities of these deconvoluted peaks, secondary structure assignments can be made. For example, the peaks for alpha helices are found in the region of 1650 to 1658 cm^{-1}. Unlike CD, this method is not affected by light scattering and is independent of protein concentration. With the FucP(His)$_6$ protein (Figure 3.9) and all the MFS proteins we have isolated so far (Table 3.2), both CD and FTIR measurements indicate a high content of alpha helices (50 to 90%), which is tenaciously retained in the detergent-solubilized state.

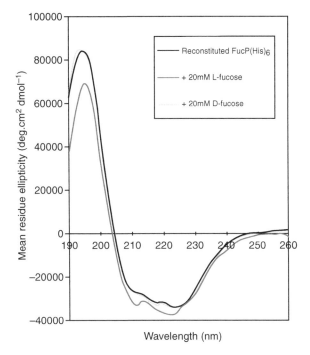

FIGURE 3.9 The FucP(His)$_6$ protein has predominantly alpha-helical secondary structure. The CD spectrum of the FucP(His)$_6$ protein was determined in 10 mM KPi, pH 7.6, after reconstitution into liposomes of *E. coli* lipid. The spectra were obtained in a JascoJ-715 spectropolarimeter at 20°C with constant nitrogen flushing. Each curve is the average of 20 scans. The spectra of a suspension of liposomes without protein in the same buffer was subtracted as background; this included 20 mM sugar when appropriate; and so the observed effect was not caused by the addition of the sugar but by a perturbation in the structure of the protein. From Clough, J.L. Ph.D. Thesis, University of Leeds, 2001. With permission.

3.9 CRYSTALLIZATION OF PURIFIED MEMBRANE TRANSPORT PROTEINS

To gain structural information at atomic resolution, electron or x-ray crystallography of two- or three-dimensional crystals, respectively, is required. Two-dimensional crystals are obtained when detergent-solubilized protein is mixed with lipid-detergent mixed micelles,[76] after which the detergent is removed by dialysis or adsorption onto polystyrene beads (Biobeads™, Biorad).[76,77] Lipid-to-protein ratios (weight:weight) are restricted so as to encourage close packing of the protein into lipid bilayers when the detergent is removed. The successes with the sodium–H$^+$ antiporter,[14] the tetracycline antiporter,[28] the lactose–H$^+$ symporter,[78] the erythromycin efflux protein,[79] the melibiose–Na$^+$ symporter,[15] and the oxalate transporter[16] generate cautious optimism for this methodology.

However, there have now been several ground-breaking successes where diffracting three-dimensional crystals of transport proteins have been obtained at high enough resolution to yield molecular models of protein structure. Two of these, the lactose–H[+17] and the glycerol-3-P/Pi antiporter,[18] have structures of recognizable similarity, which are probably typical of the MFS transporters that are the subjects of the investigations described here. Three other transporters, the ammonium transporter[80] from *E. coli*, the glutamate transporter[81] from *Pyrococcus horikoshii,* and the leucine transporter from *Aquifex aeolicus*[82] have different structures; these are particularly important because they serve as models for proteins found in mammals, including man.[80–82] Additionally, the integral membrane component AcrB[83] of an *E. coli* multidrug efflux system[62] found widely in bacteria, including pathogens, is a potential target for new antibacterials.

Eighteen of the transport proteins listed in Table 3.1 have been purified using the NiNTA IMAC strategy, and two transport proteins have been subjected to Strep-tag purification. Six proteins have entered preliminary crystallization screens, and two have yielded diffracting three-dimensional crystals (Shimamura, Ferrandon, Yajima, Mirza, Suzuki, Rutherford, Henderson, Byrne, and Iwata, unpublished results).

3.10 NUCLEAR MAGNETIC RESONANCE (NMR) APPROACHES

The levels of transport protein yields obtained in our vector–host strains are sufficient to enable solid-state NMR Magic Angle Spinning Sample experiments to investigate the binding of ligands labeled with ^{13}C and ^{15}N, not only on the FucP protein[84] but also on several membrane transport proteins.[85,86] Furthermore, if the host cells grow on minimal medium and maintain the amplified level of expression of the cloned protein, the proteins themselves can be labeled[87] in order to implement future NMR studies on their structures.

3.11 CONCLUSION

The strategies described above have generated amplified expression of active prokaryotic MFS transport proteins. Furthermore, the levels reached are sufficient for purification, characterization, and crystallization trials. The majority of successes have been achieved by using the *tac* promoter for induction of protein expression with IPTG in various *E. coli* host strains. However, other plasmids such as the pET vectors based on the T7 promoter, the AD5827 plasmid based on the lambda leftward promoter, and the *galP* promoter in the plasmid pBR322 have been successful in some instances. Expression of membrane transport proteins as inclusion bodies is to be avoided but may prove a viable option as improved protocols for refolding of membrane protein from inclusion bodies become available.[44–48]

The most commonly used tag for the affinity purification of membrane proteins involves the addition of hexa-histidine to the C terminus of the protein to facilitate purification using nickel (or cobalt) chelate affinity chromatography. This tag is small and has been shown generally not to interfere with levels of overexpression achieved or the targeting or folding of the protein. Importantly, this tag rarely affects the transport activity of any of the transport proteins we have investigated, and when it does, it has only a mildly deleterious effect. Moreover, the successful crystallization and structure determination of solubilized hexa-histidine tagged transport proteins[17,18,82,83] suggests that the histidine tag in itself is not generally a barrier to crystallization, at least for transport proteins;[88] neither does it prevent ligand binding in NMR experiments.[84–87]

In conclusion, with the careful choice of vector, *E. coli* host strain, and expression conditions, it is now possible to overexpress prokaryotic MFS membrane transport proteins to produce milligram quantities of pure protein per liter of culture. This paves the way for the elucidation of increasing numbers of prokaryotic membrane transport protein structures and ultimately the determination of their modes of action at the molecular level. Some of these proteins will lead to increased understanding of their mammalian homologues,[81,82,89] and others are potential targets for new generations of antibacterials.[40,83]

ACKNOWLEDGMENTS

We are grateful to the BBSRC, EU, Ajinomoto Co., and Wellcome Trust for financial support. JC acknowledges a Ph.D. studentship from the BBSRC, and MS thanks the Iranian Government for support. Equipment purchases were funded by the BBSRC and the Wellcome Trust.

REFERENCES

1. Byrne, B. and Jormakka, M. Solubilisation and purification of membrane proteins, in *Structural Genomics on Membrane Proteins,* Lundstrom, K., ed. CRC Press, Boca Raton, 2006.
2. Michel, H. http://www.mpibp-frankfurt.mpg.de/michel/public/memprotstruct.html, 2005.
3. White, S.H. http://blanco.biomol.uci.edu/Membrane_Proteins_xtal.html, 2005.
4. Saier, M.H. Jr. Molecular phylogeny as a basis for the classification of transport proteins from bacteria, archaea, and eukarya. *Adv. Microb. Physiol.,* 40, 81–136, 1998.
5. Saier, M.H. Jr. A functional-phylogenetic classification system for transmembrane solute transporters. *Microbiol. Mol. Biol. Rev.,* 64, 354–411, 2000.
6. Higgins, C.F. ABC Transporters: from microorganism to man. *Annu. Rev. Cell Biol.,* 8, 67–113, 1992.
7. Schmitt, L. and Tampe, R. Structure and mechanism of ABC transporters. *Curr. Opin. Struct. Biol.,* 12, 754–760, 2002.
8. Pao, S.S., Paulsen, I.T., and Saier, M.H. Jr. Major facilitator superfamily. *Microbiol. Mol. Biol. Rev.,* 62, 1–34, 1998.
9. Saier, M.H., Beatty, J.T., Goffeau A., Harley, K.T., Heijne, W.H.M., Huang, S.C., Jack D.L., Jahn, P.S., Lew, K., Liu, J., Pao, S.S., Paulsen, I.T., Tseng, T.T., and Virk, P.S. The major facilitator superfamily. *J. Mol. Microbiol. Biotechnol.* 1, 257–280, 2000.
10. Paulsen, I.T. and Lewis, K. Microbial multidrug efflux. Horizon Press, Norwich, 2002.

11. Ren, Q., Kang, K.H., and Paulsen, I.T. Transport DB: a relational database of cellular membrane transport systems. *Nucleic Acids Res.,* 32, D284–D288, 2004.

12. Henderson, P.J.F. Function and structure of membrane transport proteins, in Griffith, J.K., Sansom, C., eds. *The Transporter Factsbook.* Academic Press, Oxford, 1998, pp. 3–29.

13. Gunn, F., Tate, C.G., Sansom, C.E., and Henderson, P.J.F. Topological analysis of the L-fucose-proton symport protein, FucP, of *Escherichia coli. Molec. Microbiol.,* 15, 771–783, 1995.

14. Williams, K.A. Three-dimensional structure of the ion-coupled transport protein NhaA. *Nature,* 403, 112–115, 2000.

15. Hacksell, I., Rigaud, J.L., Purhonen, P., Pourcher, T., Hebert, H., and Leblanc, G. The melibiose-Na^+ membrane transport protein. *EMBO J.* 21, 3569–3574, 2002.

16. Hirai, T., Heymann, J.A.W., Shi, D., Sarker, R., Maloney, P.C., and Subramanian, S. Three-dimensional structure of a bacterial oxalate transporter. *Nat. Struct, Biol.,* 9, 597–600, 2002.

17. Abramson, J., Smirnova, I., Kasho, V., Verner, G., Kaback, H.R., and Iwata, S. Structure and mechanism of the lactose permease of *Escherichia coli. Science,* 301, 610–615, 2003.

18. Huang, Y., Lemieux, M.J., Song, J., Auer, M., and Wang, D.N. Structure and mechanism of the glycerol-3-phosphate transporter from *Escherichia coli. Science,* 301, 616–620, 2003.

19. Hunte, C., Screpanti, E., Venturi, M., Rimon, A., Padan, E. and Michel, H. Structure of a Na^+/H^+ antiporter and insights into mechanism of action and regulation by pH. *Nature,* 435, 1197–1202, 2005.

20. Bradley, S.A., Tinsley, C.H., Muiry, J.A.R, and Henderson, P.J.F. Proton-linked L-fucose transport in *Escherichia coli. Biochem. J.,* 248, 495–500, 1987.

21. Kunji, E.R.S., Slotboom, D-J., and Poolman, B. *Lactococcus lactis* as host for over-production of membrane proteins. *Biochim. Biophys. Acta.,* 1610, 97–108, 2003.

22. Turner, G.J., Reusch, R., Winter-Vann, A.M., Martinez, L., and Betlach, M.C. Heterologous gene expression in a membrane-protein-specific system. *Prot. Expr. Purif.,* 17, 312–323, 1999.

23. Grisshammer, R., and Tate, C.G. Overexpression of integral membrane-proteins for structural studies. *Q. Rev. Biophys.,* 28, 315–422, 1995.

24. Ward, A., O'Reilly, J., Rutherford, N.G., Ferguson, S.M., Hoyle, C.K., Palmer, S.A., Clough, J.L., Venter, H., Xie, H., Litherland, G.J., Martin, G.E.M., Wood, J.M., Roberts, P.E., Groves, M.A.T., Liang, W.J., Steel, A., McKeown, B.J., and Henderson, P.J.F. Expression of prokaryotic membrane transport systems in *Escherichia coli. Biochem. Soc. Trans.,* 27, 893–899, 1999.

25. Ghosh, R., Steiert, M., Hardmeyer, A. Wang, Y.F., and Rosenbusch, J.P. Overexpression of outer membrane porins in *E. coli* using pBluescript-derived vectors. *Gene Expression,* 7, 149–161, 1998.

26. Pourcher, T., Bassilana, M., Sarkar, H.K., Kaback, H.R., and Leblanc, G. Melibiose permease and alpha-galactosidase of *Escherichia coli* — identification by selective labeling using A T7 RNA-polymerase promoter expression system. *Biochemistry,* 29, 690–696, 1990.

27. Pourcher, T., Leclercq, S., Brandolin, G., and Leblanc, G. Melibiose permease of *Escherichia coli* — large-scale purification and evidence that H^+, Na^+, and Li^+ sugar symport is catalyzed by a single polypeptide. *Biochemistry,* 34, 4412–4420, 1995.

28. Aldema, M.L., McMurry, L.M., Walmsley, A.R., and Levy, S.B. Purification of the Tn 10-specified tetracycline efflux antiporter TetA in a native state as a polyhistidine fusion protein. *Mol. Microbiol.*, 19, 187–195, 1996.

29. Miroux, B. and Walker, J.E. Over-production of proteins in *Escherichia coli*: mutant hosts that allow synthesis of some membrane proteins and globular proteins at high levels. *J. Mol. Biol.*, 260, 289–298, 1996.

30. Studier, F.W., Rosenberg, A.H., Dunn, J.J., and Dubendorff, J.W. Use of T7 RNA-polymerase to direct expression of cloned genes. *Meth. Enzymol.*, 185, 60–89 , 1990.

31. Gunn, F.J., Tate, C.G., and Henderson, P.J.F. Identification of a novel sugar-H+ symport protein, FucP, for transport of L-fucose into *Escherichia coli*. *Mol. Microbiol.*, 12, 799–809, 1994.

32. Maiden, M.C.J., Jones-Mortimer, M.C., and Henderson, P.J.F. The cloning, DNA sequence and amplified expression of the gene *araE* for arabinose-proton symport in *Escherichia coli*. *J. Biol. Chem.*, 263, 8003–8010, 1988.

33. Liang, W.J., Xie, H., Wilson, K., Suzuki, S., Rutherford, N.G., Henderson, P.J.F., and Jefferson, R. The *gusBC* genes of *Escherichia coli* encode a glucuronide transport system. *J. Bacteriol.*, 187, 2377–2385, 2005.

34. Stark, M.J.R. Multicopy expression vectors carrying the lac repressor gene for regulated high-level expression of genes in *Escherichia coli*. *Gene*, 51, 255–267, 1987.

35. Ward, A., Sanderson, N.M., O'Reilly, J., Rutherford, N.G., Poolman, B., and Henderson, P.J.F. The amplified expression, identification, purification, assay and properties of histidine-tagged bacterial membrane transport proteins, in Baldwin, S.A., ed. *Membrane Transport — a Practical Approach*. Blackwell Press, Oxford, 2000, pp.141–166.

36. Clough, J.L. Structure-function relationships of the L-fucose-H+ symport protein (FucP) of *Escherichia coli*, and the homologous D-glucose transport protein (GluP) of *Brucella abortus*. Ph.D. Thesis, University of Leeds, 2001.

37. Hoyle, C.J. Recombinant expression, purification, and characterisation of bacterial multidrug efflux proteins. Ph.D. Thesis, University of Leeds, 2000.

38. Saidijam, M., Psakis, G., Clough, J.L., Meuller, J., Suzuki, S., Hoyle, C.J., Palmer, S.L., Morrison, S.M., Pos, M.K., Essenberg, R.C., Maiden, M.C., Abu-bakr, A., Baumberg, S.G., Neyfakh, A.A., Griffith, J.K., Stark, M.J., Ward, A., O'Reilly, J., Rutherford, N.G., Phillips-Jones, M.K., and Henderson, P.J.F. Collection and characterisation of bacterial membrane proteins. *FEBS Lett.*, 555, 170–175, 2003.

39. Morrison, S., Ward, A., Hoyle, C.J., and Henderson, P.J.F. Cloning, expression, purification and properties of a putative multidrug resistance efflux protein from *Helicobacter pylori*. *Int.. J. Antimicrob. Agents*, 22/3, 242–249, 2003.

40. Ward, A., O'Reilly, J., Rutherford, N.G., Ferguson, S.M., Hoyle, C.K., Palmer, S.A., Clough, J.L., Venter, H., Xie, H., Litherland, G.J., Martin, G.E.M., Wood, J.M., Roberts, P.E., Groves, M.A.T., Liang, W.J., Steel, A., McKeown, B.J., and Henderson, P.J.F. Expression of prokaryotic membrane transport systems in *Escherichia coli*. *Biochem. Soc. Trans.*, 27, 893–899, 1999.

41. Ward, A., Hoyle, C.J., Palmer, S.E., O'Reilly, J., Griffith, J.K., Pos, K.M., Morrison, S.J., Poolman, B., Gwynne, M., and Henderson, P.J.F. Prokaryote multidrug efflux proteins of the major facilitator superfamily: amplified expression, purification and characterisation. *Mol. Microbiol. Biotech.*, 3, 193–200, 2001.

42. Eckert, B. and Beck, C.F. Overproduction of transposon Tn10-encoded tetracycline resistance protein results in cell-death and loss of membrane-potential. *J. Bacteriol.*, 171, 3557–3559, 1989.

43. Saidijam, M., Benedetti, G., Ren, Q., Xu, Z., Hoyle, C.J., Palmer, S.L., Ward, A., Bettaney, K.E., Szakonyi, G., Meuller, J., Morrison, S., Pos, M.K., Butaye, P., Walravens-Langton, K., Herbert, R.B., Brown, M.H., Skurray, R.A., Paulsen, I.T., O'Reilly, J., Rutherford, N.G. Bill, R.M., and Henderson, P.J.F. Microbial drug efflux proteins of the Major Facilitator Superfamily. *Curr. Drug Targ.* 2006.

44. Rogl, H., Kosemund, K., Kuhlbrandt, W., and Collinson, I. Refolding of *Escherichia coli* produced membrane protein inclusion bodies immobilised by nickel chelating chromatography. *FEBS Lett.*, 432, 21–26, 1998.

45. Bochkareva, E., Seluanov, A., Bibi, E., and Girshovich, A. Chaperonin-promoted post-translational membrane insertion of a multispanning membrane protein lactose permease. *J. Biol. Chem.*, 271, 22256–22261, 1996.

46. Bogdanov, M. and Dowhan, W. Phospholipid-assisted protein folding: phosphatidylethanolamine is required at a late step of the conformational maturation of the polytopic membrane protein lactose permease. *EMBO J.*, 17, 5255–5264, 1998.

47. Buchanan, S.K. Beta-barrel proteins from bacterial outer membranes: structure, function, and refolding. *Curr. Opin. Struct. Biol.*, 9, 455–461, 1999.

48. Anderson, M., Blowers, D., Hewitt, N., Hedge, P., Breeze, A., Hampton, I., and Taylor, L. Refolding, purification, and characterization of a loop deletion mutant of human Bcl-2 from bacterial inclusion bodies. *Prot. Expr. Purif.*, 15, 162–170, 1999.

49. Macfarlane, J. and Muller, M. The functional integration of a polytopic membrane-protein of *Escherichia coli* is dependent on the bacterial signal-recognition particle. *Eur. J. Biochem.*, 233, 766–771, 1995.

50. Henstra, S.A., Tolner, B., Duurkens, R.H.T., Konings, W.N., and Robillard, G.T. Cloning, expression, and isolation of the mannitol transport protein from the thermophilic bacterium *Bacillus stearothermophilus*. *J. Bacteriol.*, 178, 5586–5591, 1996.

51. Venter, H., Ashcroft, A.E., Keen, J.N., Henderson, P.J.F., and Herbert, R.B. Molecular dissection of membrane transport proteins: mass spectrometry and sequence determination of the galactose-H$^+$ symport protein, GalP, of *Escherichia coli* and quantitative assay of the incorporation of [ring-2-^{13}C]histidine and ^{15}NH$_3$. *Biochem. J.*, 363, 243–252, 2002.

52. Le Coutre, J., Whitelegge, J.P., Gross, A., Turk, E., Wright, E.M., Kaback, H.R., and Faull, K.F. Proteomics on full-length membrane proteins using mass spectrometry. *Biochemistry*, 39, 4237–4242, 2000.

53. Weinglass, A.B., Whitelegge, J.P., Hu, Y., Verner, G.E., Faull, K.F., and Kaback, H.R. Elucidation of substrate binding interactions in a membrane transport protein by mass spectrometry. *EMBO J.*, 22,1467–1477, 2003.

54. Nilsson, J., Stahl, S., Lundeberg, J., Uhlen, M., and Nygren, P.A. Affinity fusion strategies for detection, purification, and immobilization of recombinant proteins. *Prot. Expr. Purif.*, 1, 11–16, 1997.

55. Hochuli, E., Bannwarth, W., Dobeli, H., Gentz, R., and Stuber, D. Genetic approach to facilitate purification of recombinant proteins with a novel metal chelate adsorbent. *Biotechnology*, 6, 1321–1325, 1988.

56. Hochuli, E., Dobeli, H., and Schacher, A. New metal chelate adsorbent selective for proteins and peptides containing neighbouring histidine residues. *J. Chromatogr.*, 411, 177–184, 1987.

57. Pos, K.M., Bott, M., and Dimroth, P. Purification of 2 active fusion proteins of the Na⁺-dependent citrate carrier of *Klebsiella pneumoniae*. *FEBS Lett.*, 347, 37–41, 1994.

58. Zhang, F.M., Strand, A., Robbins, D., Cobb, M.H., and Goldsmith, E.J. Atomic-structure of the map Kinase Erk2 At 2.3-Angstrom resolution. *Nature*, 367, 704–711, 1994.

59. Howard, T.D., McAuley-Hecht, K.E., and Cogdell, R.J. Crystallization of membrane proteins, in Baldwin, S.A., ed. *Membrane Transport — a Practical Approach*. Blackwell Press, Oxford, 2000, pp. 269–302.

60. Keyes, M.H., Gray, D.N., Kreh, K.E., and Sanders, C.R. Solubility detergents for membrane proteins. in Iwata, S., ed. *Methods and Results in Crystallisation of Membrane Proteins*. International University Line, La Jolla, CA, 2003, pp.15–38.

61. Koronakis, V., Sharff, A., Koronakis, E., Luisi, B., and Hughes, C. Crystal structure of the bacterial outer membrane protein TolC central to multidrug efflux and protein export. *Nature*, 405, 914–919, 2000.

62. Fang, G., Friesen, R., Lanfermeijer, F., Hagting, A., Poolman, B., and Konings, W.N. Manipulation of activity and orientation of membrane-reconstituted di-tripeptide transport protein DtpT of *Lactococcus lactis*. *Mol. Memb. Biol.*, 16, 297–304, 1999.

63. Henderson, P.J., Kagawa, Y., and Hirata, H. Reconstitution of the GalP galactose transport activity of *Escherichia coli* into liposomes made from soybean phospholipids. *Biochim. Biophys. Acta.*, 732, 204–209, 1983.

64. Knol, J., Veenhoff, L., Liang, W.J., Henderson, P.J.F., Leblanc, G., and Poolman, B. Unidirectional reconstitution into detergent-destabilized liposomes of the purified lactose transport system of *Streptococcus thermophilus*. *J. Biol. Chem.*, 271, 15358–15366, 1996.

65. Racher, K.I., Voegele, R.T., Marshall, E.V., Culham, D.E., Wood, J.M., Jung, H., Bacon, M., Cairns, M.T., Ferguson, S.M., Liang, W.J., Henderson, P.J.F., White, G., and Hallett, F.R. Purification and reconstitution of an osmosensor: transporter ProP of *Escherichia coli* senses and responds to osmotic shifts. *Biochemistry*, 38, 1676–1684, 1999.

66. Rigaud, J.L. and Levy, D. Reconstitution of membrane proteins into liposomes. *Meth, Enzymol.*, 372, 65–86, 2003.

67. Jung, H., Tebbe, S., Schmid, R., and Jung, K. Unidirectional reconstitution and characterization of purified Na⁺/proline transporter of *Escherichia coli*. *Biochemistry*, 37, 11083–11088, 1998.

68. Heuberger, E.H.M.L. and Poolman, B., A spectroscopic assay for the analysis of carbohydrate transport reactions. *Eur. J. Biochem.*, 267, 228–234, 2000.

69. Putman, M., van Veen, H.W., Poolman, B., and Konings, W.N. Restrictive use of detergents in the functional reconstitution of the secondary multidrug transporter LmrP. *Biochemistry*, 38, 1002–1008, 1999.

70. Patzlaff, J.S., Moeller, J.A., Barry, B.A., and Brooker, R.J. Fourier transform infrared analysis of purified lactose permease: a monodisperse lactose permease preparation is stably folded, alpha-helical, and highly accessible to deuterium exchange. *Biochemistry*, 37, 15363–15375, 1998.

71. Wallace, B.A. and Teeters, C.L. Differential absorbance optical flattening effects are significant in the circular dichroism spectra of large membrane fragments. *Biochemistry*, 26, 65–70, 1987.

72. Chin, J.J., Jung, K.Y., and Jung, C.Y. Structural basis of human erythrocyte glucose transporter function in reconstituted vesicles — alpha-helix orientation. *J. Biol. Chem.*, 261, 7101–7104, 1986.

73. Wallace, B.A., Lees, J.G., Orry, A.J., Lobley, A., and Janes, R.W. Analyses of circular dichroism spectra of membrane proteins. *Prot. Sci.*, 12, 875–884, 2003.

74. Patzlaff, J.S., Moeller, J.A., Barry, B.A., and Brooker, R.J. Fourier transform infrared analysis of purified lactose permease: a monodisperse lactose permease preparation is stably folded, alpha-helical, and highly accessible to deuterium exchange. *Biochemistry*, 37, 15363–15375, 1998.

75. Haris, P.I. and Chapman, D. The conformational analysis of peptides using Fourier-transform R spectroscopy. *Biopolymers*, 37, 251–263, 1995.

76. Ringler, P., Heymann, B., and Engel, A. Two-dimensional crystallisation of membrane proteins, in Baldwin, S.A., ed. *Membrane Transport — a Practical Approach*. Blackwell Press, Oxford, 2000, pp. 29–268 .

77. Rigaud, J.L., Mosser, G., Lacapere, J.J., Olofsson, A., Levy, D., and Ranck, J.L. Biobeads: an efficient strategy for two-dimensional crystallization of membrane proteins. *J. Struc. Biol.*, 118, 226–235, 1997.

78. Zhuang, J.P., Prive, G.G., Werner, G.E., Ringler, P., Kaback, H.R., and Engel, A. Two-dimensional crystallization of *Escherichia coli* lactose permease. *J. Struc. Biol.*, 125, 63–75, 1999.

79. Ubarretxena-Belandia, I., and Tate, C.G. New insights into the structure and oligomeric state of the bacterial multidrug transporter EmrE: an unusual asymmetric homodimer. *FEBS Lett.*,. 564, 234–238, 2004.

80. Khademi, S., O'Connell, J., Remis, J., Robles-Colmenares, Y., Miercke, L.J.W., and Stroud, R. Mechanism of ammonia transport by Amt/MEP/Rh. *Science*, 305, 1587–1594, 2004.

81. Yernool, D., Boudker, O., Jin, Y., and Gouaux, E. Structure of a glutamate transporter homologue from *Pyrococcus horikoshii*. *Nature*, 431, 811–818, 2004.

82. Yamashita, A., Singh, S.K., Kawate, T., Jin, Y., and Gouaux, E. Crystal structure of Na$^+$/Cl$^-$ -dependent neurotransmitter transporters. *Nature*, 437, 215–223, 2005.

83. Murakami, S., Nakashima, R., Yamashita, E., and Yamaguchi, A. Crystal structure of bacterial multidrug efflux transporter AcrB. *Nature*, 419, 587–593, 2002.

84. Spooner, P.J.R., O'Reilly, W.J., Homans, S.W., Rutherford, N.G., Henderson, P.J.F., and Watts, A. Weak substrate binding to transport proteins studied by NMR. *Biophys. J.*, 75, 2794–2800, 1998.

85. Appleyard, A., Herbert, R.B., Henderson, P.J.F., Watts, A., and Spooner, P.J.R. Selective NMR observation of inhibitor and sugar binding to the galactose-H$^+$ symport protein, GalP, of *Escherichia coli*. *Biochim. Biophys. Acta.*, 509, 55–64, 2000.

86. Patching, S.G., Brough, A.R., Herbert, R.B., Rajakarier, J.A., Henderson, P.J.F., and Middleton, D.A. Substrate affinities for membrane transport proteins determined by ^{13}C cross-polarisation magic-angle spinning NMR. *J. Am. Chem. Soc.*, 126, 3072–3080, 2004.

87. Patching, S.G., Herbert, R.B., O'Reilly, J., Brough, A.R., and Henderson, P.J.F. Low ^{13}C- background for NMR-based studies of ligand binding using ^{13}C-depleted glucose as carbon source for microbial growth: ^{13}C-labelled glucose and ^{13}C-forskolin binding to the galactose-proton symport protein GalP in *Escherichia coli*. *Am. Chem. Soc.*, 126, 86–87, 2004.

88. Mohanty, A.K. and Wiener, M.C. Membrane protein expression and production: effects of polyhistidine tag length and position. *Prot. Expr. Purif.*, 33, 311–325, 2004.

89. Vardy, E., Arkin, I.T., Gottschalk, K.E., Kaback, H.R., and Schuldiner, S. Structural conservation in the Major Facilitator Superfamily as revealed by comparative modelling. *J. Prot. Sci.*, 13, 1832–1840, 2004.

4 Membrane Protein Production Strategies for Structural Genomics

Said Eshaghi and Pär Nordlund

CONTENTS

4.1 INTRODUCTION

High-resolution structural data play an important role in the process of developing new and efficient drugs, as well as in providing understanding of the structural basis for the function of proteins. Today, more than half of current drug targets are integral membrane proteins (IMPs). Moreover, 20 to 25% of all proteins in a typical cell are IMPs. In contrast, IMPs represent only a very small fraction (< 0.5%) of about 30,000 proteins with known structures present in Public Databases (PDBs).[1] Furthermore, the majority of the successfully determined IMPs are of bacterial origin, and only less than 10 unique eukaryotic IMPs have been structurally characterized to date.[2]

To allow structural studies, tens or even hundreds of milligrams of highly purified protein might be needed. Since most IMPs are present at very low concentration in their natural environments, an efficient recombinant overexpression system for membrane proteins is, in most cases, a requirement. This constitutes the first hurdle in the process of determining an IMP structure: the recombinant membrane protein may not be expressed in sufficient amounts needed for serious purification efforts; the expressed protein may not be inserted properly into the membrane, or it may end up in inclusion bodies.[3] There are potentials for solving these problems by

43

investigating the effects of any of the very large number of expression tools and parameters available, such as expression vectors, strains, and physical conditions. However, current understanding of how to prioritize among these parameters is poor.

Another critical factor is the choice of extracting and stabilizing detergents. Vast arrays of different detergents are potentially available for IMP work: those that are commonly used and those that are less characterized or under development. The detergent most useful for extracting the IMP from the membrane might not be the most useful for stabilization or crystallization of the protein. Moreover, mixtures of detergents during purification and crystallization have been shown to improve the quality of crystals.[4,5] Altogether, the size of the detergent parameter space becomes very large.

Still, the general view among biochemists is that IMPs are significantly harder to produce than are soluble proteins, at levels useful for structural studies. Ongoing high-throughput (HTP) structural genomics efforts have emerged to explore and implement the essential technologies for a gene-to-structure pipeline for soluble proteins, where the implementation of parallel and streamlined methodologies for production of crystallizable proteins is a key component.[6–12] Similar initiatives for membrane proteins are now slowly emerging. An example is the *Mycobacterium tuberculosis* Membrane Protein Structural Genomics initiative.[13] MePNet and E-MeP programs are other examples of technology-oriented structural genomics programs, engaging a large number of membrane protein biochemistry laboratories and structural biology laboratories across Europe.[14–16] The structural genomics consortium in Oxford has also added membrane proteins to its target list.[12] JCSG Center for Innovative Membrane Protein Technologies (JCIMPT) is a recently funded consortium with the mission to develop and disseminate methods and technologies for structure-grade production of integral membrane proteins.[17] Most of these initiatives are, however, currently engaged in implementing the essential technologies for work on larger numbers of proteins, and as yet, the literature providing data relevant to efforts in establishing a structural genomics pipeline for IMP production is still quite sparse. Structural genomics initiatives are described in more detail in Chapter 16.

4.2 GENERAL ASPECTS OF MEMBRANE PROTEIN PRODUCTION TECHNOLOGIES

To increase the number of IMPs that can be processed and to improve the quality of the proteins, making them more useful for structural studies, efficient HTP strategies and technologies should preferably be established for all stages — from expression and membrane extraction to purification and crystallization. These technologies should be equally robust as current "one protein at a time" strategies and involve parallel processing of samples. Such methods have been implemented for soluble proteins for most steps of the process, from cloning to crystallization.[18] Some of these implementations such as cloning and cell growth strategies can be directly applied in the IMP production pipeline. However, the experimental handling of membrane proteins is generally more difficult than that of soluble proteins;

IMPs are, for example, relatively easily degraded during the purification or analysis step due to their high hydrophobicity. Therefore, the implementation of HTP methods for membrane proteins requires particularly careful considerations and evaluations.

General aspects of membrane proteins expression based on one or a few IMPs have been reviewed elsewhere.[19–21] At present, only three groups have presented more extensive expression studies on a larger number of proteins, providing statistically significant data for estimating the success rates of IMP production strategies. These studies are from the *M. tuberculosis* project, where some 80 genes were attributed to expression-screening vectors;[22] from our own laboratory, where 49 *Escherichia coli* (*E. coli*) integral membrane proteins were processed, and effects of different vectors, strains, and detergents were benchmarked;[23] and finally from accumulated work on bacterial transporters from Peter Henderson's group (reviewed in Chapter 3 in this volume).

4.3 *E. COLI* EXPRESSION STRATEGIES

In general, *E. coli* is the species of choice for producing soluble proteins for structural studies. The advantages are ease of manipulation, low cost, and rapid production of high levels of recombinant proteins. An additional important factor is that proteins produced in *E. coli* rarely carry covalent modifications, while proteins produced in eukaryotic hosts often have inhomogeneous covalent modifications, which makes them less likely to crystallize. At present, *E. coli* is, in fact, the exclusive host used for protein production for most structural genomics initiatives working on soluble proteins. As will be discussed below, *E. coli* is likely to provide a sufficiently good solution for producing many bacterial integral membrane proteins.

However, it is clear that the production of eukaryotic IMPs in *E. coli* is significantly more challenging than, for example, the bacterial production of intracellular eukaryotic soluble proteins. Eukaryotic hosts will therefore be very important for expressing eukaryotic IMPs, and these are reviewed in Chapter 8 (yeast), Chapter 9 (insect cells), and Chapter 10 (mammalian cells) in this volume. However, one should keep in mind that even though a significant number of eukaryotic IMPs have been recombinantly expressed, high-resolution structure determination has proven very difficult. Thus, the true usefulness of eukaryotic expression systems for producing proteins for structural studies at large still remains to be demonstrated.

Cell-free production systems and the refolding of protein obtained in inclusion bodies[24,25] are other possibilities for IMP production, but again, the broader usefulness of these potentially important methods still remains to be demonstrated.

4.4 SMALL-SCALE PARALLEL EXPRESSION-SCREENING STRATEGIES

In order to isolate proteins that have been expressed in the membranes, the cell lysate is generally subjected to ultracentrifugation to separate membrane fragments and vesicles from the soluble compartment. Although this step is an efficient purification step, it is also rather tedious, and that is why some labs have excluded this step during the purification of their membrane proteins.[26] To establish a parallel platform for expression screening of soluble recombinant proteins, Knaust and Nordlund designed an efficient platform combining 96-well plates and dot blots.[27] Following the lysis, a filtration step efficiently removed the inclusion bodies and nonbroken cells from the soluble moiety. Furthermore, to allow simultaneous detection of all proteins, dot blot analysis was performed, rather than SDS-PAGE (sodium dodecyl sulfate–polyacrylamide gel electropheresis) analysis, since the latter is less efficient as an initial detection method. In order to extend this platform for expression screening to membrane proteins, it was modified by introducing a detergent extraction step.[23] Hence, the solubilized material was successfully cleared from cell debris using 96-well filter plates. To allow solubilized membrane proteins to be distinguished from incompletely solubilized membrane fragments or vesicles in the filtrate, a second step involving His-tag purification in 96-well filter plates was introduced. This step is efficient in distinguishing IMPs in vesicles and IMPs solubilized in detergent, and therefore no ultracentrifugation is required in the process. The expressed IMPs were then analyzed by SDS-PAGE.

Cross and colleagues have processed some 99 *M. tuberculosis* IMPs for expression in *E. coli*.[22] In this study, traditional IMP expression analysis strategies were used, including differential centrifugation technologies, by which soluble fraction, membrane fraction, and inclusion bodies were obtained. However, no affinity purification step was used in this study.

For detection of proteins, both SDS-PAGE with subsequent Coomassie staining, and immunoblotting were used. Here, relatively few proteins (~ 10%) were found to be expressing in the membrane at sufficient levels for detection with SDS-PAGE, while the majority of proteins (~ 80%) were detectable in the membrane fraction when performing Western blot. Considering the very high sensitivity of Western blot, it is unfortunately not clear whether the proteins detected only with this method are, in fact, expressing at levels useful for structural studies. A further complicating factor in this study is that for a number of proteins, the expressed material was distributed not only in the membranes and inclusion bodies but also in the soluble fraction. The latter was isolated by an extensive sonication step prior to an unusually short ultracentrifugation (30 minutes at 90,000 g instead of 60 minutes at 100,000 g, which is required for total membrane isolation). It is plausible that the soluble fraction obtained in such a procedure would contain some membrane fragments. Thus, the proteins appearing as soluble in this study are most likely associated with these small membrane fragments.

4.5 BENCHMARKING OF EXPRESSION STRATEGIES

A large number of expression parameters such as vectors, strains, and physical parameters can be changed in an expression experiment and often have significant effects on the expression levels. The potential "parameter space" to be tested is very large, and an improved understanding of which of these parameters are most useful will be important to limit the number of experiments to be done when applied to many proteins.

All current structural genomics efforts depend on the use of affinity tags for purification, most often 6-His tags. The presence of affinity tags facilitates the purification and detection of the target proteins. However, the presence of tags can affect expression levels. These effects have been examined for soluble proteins.[28–30] In short, fusion tags and proteins other than the His tags often improved the apparent solubility of the fusion constructs. However, it is not yet established that other fusion tags, including protein tags, which have to be cleaved off before crystallizations, and tags that cannot be efficiently used for purification in a HTP setting, are more efficient in structural genomics projects.

Specific effects of vector elements such as promoters and transcriptional initiation and termination elements can be significant. In particular, the strength of the promoters has been implied to be important for IMP expression, where strong promoters have been suggested to be problematic for integral membrane protein expression.[3] Nevertheless, the strong T7 promoter is now turning out to be quite successful for IMP production.

Growth conditions such as media and temperature also affect expression levels. For example, it is generally anticipated that low temperature is more advantageous for native protein expression.[20,21,23] A complicating factor when considering expression data is that changing a specific expression parameter may also change the metabolic or stress state of the host cell, thereby affecting the expression of, for example, chaperones and proteases. This potentially constitutes a mean for modulating the production efficiency of recombinant proteins but also makes it difficult to interpret potential mechanism of changes in expression efficiency.

In order to estimate the probability of success using different expression parameters, the use of a reasonably large and rather diverse target protein set is desired. Within the *M. tuberculosis* project, a subset of 143 genes was selected for processing.[22] These proteins contained between 1 and 10 transmembrane helices; 52 contained only one predicted transmembrane helix. Ninety-nine proteins were successfully expressed. The expression screening of these proteins was initially done in BL21, and later four more proteins were rescued using the C43 strain. In the expression screening, 37 proteins out of the 75 tested were visible by Western blot, while only 9 of 99 tested were detected by Coomassie staining. Potentially, many of the Western blot-detected proteins might be expressing at levels useful for structural studies, but further studies are needed to determine the exact number.

In our own study focused on the expression of *E. coli* IMPs with human homologues, 49 targets from 39 different families were selected.[23] The overall procedure is outlined in Figure 4.1. The selected genes were cloned into three recombination-based Gateway expression vectors. The major advantage of this system is that once an entry clone has been generated, the gene can easily be recombined into a variety of expression vectors. The efficiency of the subcloning process is close to 100%, and the system is therefore particularly useful for comparative benchmarking of expression vectors. In our study, the genes were subcloned into constructs with 6-His-, FLAG-, or MBP-tags at the N terminus. All the constructs also contained a 6-His tag at the C terminus to allow a common detection strategy using dot-blots. The vectors have a T7 promoter with additional repressor sites minimizing preinduced

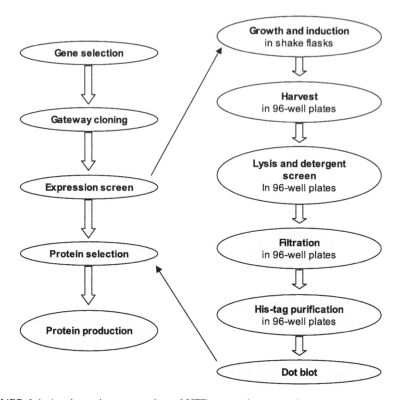

FIGURE 4.1 A schematic presentation of HTP expression screening.

leaky expression.[31] Of the 49 initial genes, only 2 failed cloning into the entry vectors, and all the 47 remaining genes were successfully recombined into the expression vectors.

The comparative expression results are shown in Table 4.1. The efficiency of each of the three tags is in a similar range of ~ 50%, but the accumulated success rate is 63% when using all three vectors. A number of these proteins have been subjected to scale up production; the expression levels in most cases correlated well with the small-scale data (Niegowski and Eshaghi, unpublished results).

Selection of expression strains is known to be important for the optimization of recombinant protein expression. However, there is little knowledge about how the use of multiple strains improves the success rates of obtaining useful proteins for structural studies. In our study on *E. coli* proteins, the expression strain BL21(DE3) and its derivatives C41(DE3) and C43(DE3) were benchmarked (hereafter referred to as BL21, C41, and C43, respectively). C41 is an evolved variant of BL21 and is suggested to be more resistant than BL21 to the toxic effects of the expressed proteins. C43, a further engineered variant of BL21, has a slower onset of transcription than has BL21 after induction.[3] Hence, in C43 cells, slower protein expression can potentially allow incorporation of an overexpressed protein into the membrane rather than into inclusion bodies.[3] Later, overproduction of subunit b of *E. coli* F-ATP synthase in C43 cells resulted in maximal proliferation of intracellular mem-

TABLE 4.1
Summary of Results for HTP Membrane Protein Expression

	Number	Success Rate
Targets	49	100%
Positive clones	47	96%
Expression: C41–HIS	26	53%
Expression: C41–MBP	23	47%
Expression: C41–FLAG	24	49%
Expression: C43–FLAG	22	45%
Expression: BL21–FLAG	25	51%
Expression: C41 ∞ 3 vectors	31	63%
Expression: FLAG ∞ 3 strains	29	59%
Expression: 3 vectors and 3 strains	35	71%

branes containing the overexpressed protein.[32] Therefore, C41 and especially C43 have been implied to be superior to BL21 for overexpression of both soluble and membrane proteins. Nevertheless, these assumptions originate from studies done on a few membrane and soluble proteins, most of them being the subunits of the F-ATPase complex.

In our parallel expression study, all 47 constructs with the N-terminal FLAG-tag were successfully transformed into BL21, C41, and C43 strains. The expression studies revealed that there was no significant superiority of C41 and C43 over BL21, but in fact, more proteins could be produced in BL21 (Table 4.1). However, the cumulative success rate when using all three strains in parallel improved the success rates significantly to 59%. The cumulative success rate for the three strains and three vectors (235 expression experiments) was 71%.

A recent toxicity study also suggested the C41 and C43 strains to be superior to BL21.[33] However, according to this study, the transformation step is the problem, and many plasmids failed to transform BL21 due to toxicity. On the other hand, when the transformation was successful, the expression success rate was slightly better in BL21 than in C41 and especially C43, in agreement with our results.

4.6 PARALLEL DETERGENT SCREENING

Choosing the right detergent is one of the most critical steps in membrane protein purification directed toward structural determination. The detergent determines the efficiency of extracting IMPs from the membrane, as well as the stability, solubility, and homogeneity of the purified proteins and thereby the probability of obtaining useful crystals. At the same time, the large number of detergents available on the market makes the choice rather difficult. Still, in practice, researchers in the field have to limit their experiments to the more common detergents. Technology development simplifying the procedure for detergent screening on a large number of proteins will therefore be very useful. We took the initiative toward this goal using the technologies described above. Twelve proteins were selected for screening against 25 detergents, chosen from nine different families, using the same platform as for the expression screening as described above. Subsequently, the efficiency of different detergents to extract each protein could be compared, leading to the identification of an initial scoreboard for these detergents to extract IMPs from the *E. coli* membrane.

A valuable outcome of this study was also the extraction efficiencies of different detergents versus their working concentrations in relation to their CMCs. For example, in the case of AmtB (signed as EM03 in our target list), HEGA-10 and Fos-Choline-12 (FC12) gave the highest yields (Figure 4.2). Two percent of HEGA-10 was used for the extraction, which is 8 times its critical micelle concentration (CMC). FC12 was used at a concentration 21 times its CMC. On the other hand, TRITON X-100 (TX100) at a concentration 65 times its CMC yielded less than half of the protein amount achieved by FC12 or HEGA-10. TX100 at this concentration showed to be as efficient as octyl glucoside (OG), which was used at only 4 times its CMC. If the efficiency of a detergent was simply defined by its CMC, TX100 must have been much more efficient than HEGA-10, FC12, and OG. However, both HEGA-10 and FC12 were much more efficient than TX100. Therefore, the concept that detergent efficiency is mainly defined by its CMC does not appear to be true.

Using similar small-scale detergent exchange methods and other detection methods, it should be possible to investigate the effects of detergents on the solubility, homogeneity, and long-term stability of proteins (work in progress). However, the

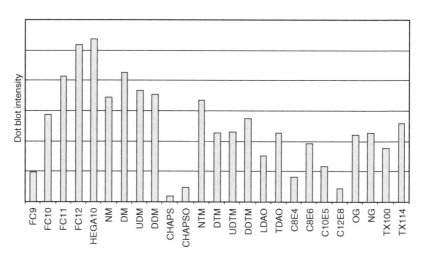

FIGURE 4.2 Detergent screening of EM03, from *E. coli* target list. The concentration of detergents during the extraction were: FC9 2%, FC10 2%, FC11 1%, FC12 1%, HEGA10 2%, NM 2%, DM 1%, UDM 1%, DDM 1%, CHAPS 2%, CHAPSO 2%, NTM 1.5%, DTM 1%, UDTM 1%, DDTM 1%, LDAO 1%, TDAO 1%, C8E4 2%, C8E6 2%, C10E5 1%, C12E8 1%, OG 2%, NG 2%, TX100 1%, TX114 2%. The dot blot signals were detected with FluorSmultiImager (BioRad) and quantified using the Quantity One software (BioRad).

establishment of true HTP methods for IMP analysis is significantly more difficult than for the corresponding methods for soluble proteins, and developments of such analysis methods correlating with crystallizability for IMPs will be of great importance.

The scale-up process of IMPs is likely to follow similar processes as for soluble proteins, except for an extra membrane isolation step. The majority of the *E. coli* proteins in our target list can be scaled up; however, a few are lost on concentration.

4.7 FUTURE PROSPECTS

The production of high quantities of pure and stable IMPs has been a major obstacle for structure determination. Although a number of bacterial proteins have previously been expressed in *E. coli*, the rates of failure have not been documented, and therefore expected success rates have remained unknown. Furthermore, the yields of the expressed proteins are often relatively low, necessitating mega-production in 25- to 50-L fermentors,[19,26] which is not a preferred strategy in structural genomics projects. Therefore, it is likely that the implementation of generic expression optimization strategies, as described above, will be very important for IMP production. In our *E. coli* IMP study, we showed that such technologies can be established, by which it was possible to express two-thirds of *E. coli* IMPs at levels allowing scale-up production (0.5 to 5 mg/L shake-flask culture) for crystallization trials. The success rates are, in fact, comparable with those for producing bacterial soluble proteins.[6] One should, however, note that in our IMP study, the expression levels are 10 to 30% of what is normally considered acceptable for soluble proteins in structural genomics initiatives. However, when IMPs in the scale-up process allow an extra generic purification step, through membrane vesicle isolation, the protein purity of the IMPs compared with that of the soluble proteins is still in a similar range after the affinity purification step. The reasons for the high success rate in *E. coli* are likely to be a combination of different factors. The small-scale technologies implemented in the study are apparently very robust. Furthermore, the low noninduced background expression of the vectors may also have contributed. Additionally, the use of tags both at the N terminus and the C terminus might have some protective properties. The *att*-sites introduced by the Gateway cloning clearly did not cause any problems.

An important factor to get a high cumulative success rate is clearly the multiparameter approach adapted in our study. Such strategies are normally not used in structural genomics initiatives. Comparing the success rates of our "worst single" parameter combination (N-terminal FLAG-tag in C43) and the accumulated success rates for all vectors and strains, it is obvious that nearly twice as many proteins can be produced by the latter approach.

A major point is whether the high success rates seen for the *E. coli* proteins can be extrapolated to other bacterial proteins. The *M. tuberculosis* study, we feel, is not conclusive in this respect due the limited experimental approach (as discussed above). Rather, the studies by Peter Henderson's group on bacterial transporters[19,34] imply that non-*E. coli* IMPs have a similar likelihood to be expressed in *E. coli*, at least for these structural families.

The selection of the most efficient detergent providing proteins at high yield is a critical factor. Thus far, reports of detergent screening have mainly been performed only in one-protein projects[5,35,36] In contrast, when working with several proteins, a single favorite detergent is used for all the proteins.[19,26,34] In the report from the *M. tuberculosis* project, it can be seen that no detergent extraction step was performed at all.[22] Standard methods for detergent screening, involving membrane preparation and ultracentrifugation, are not only time consuming but also expensive due to the high costs of detergents. Therefore, introduction of a robust 96-well His-tag purification method should be helpful in increasing the throughput in screening the solubilizing detergents of IMPs.

Another crucial step in choosing the right detergent is prior to crystallization. In one study, size exclusion chromatography was used to screen 15 detergents for the glycerol-3-phosphate transporter in *E. coli*.[36] Later, it was shown that a detergent mixture was the key to growing high-quality crystals of this protein.[5,37] In order to solve the structure of MsbA from *E. coli*, Chang and Roth screened about 20 detergents in their crystallization setups to finally obtain 35 protein crystals.[26] Thus, this is a time-consuming and elaborate step, and any improvement simplifying this process is greatly appreciated.

In conclusion, an expression screen should be combined with a detergent screen in order to efficiently reach the highest success rates in expression and the highest yields in protein production. Such a route has been suggested in Figure 4.3. Whether

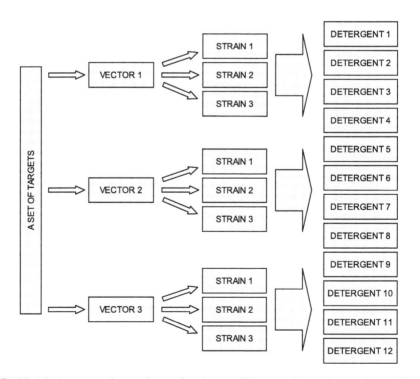

FIGURE 4.3 A suggested route for performing parallel expression and screening studies.

these strategies based on *E. coli* proteins will be useful for non-*E. coli* bacterial proteins, or even eukaryotic proteins, remains to be demonstrated. In any case, considering that about 20% of the 4,356 proteins in *E. coli* are integral membrane proteins, the present study indicates that some 500 *E. coli* integral membrane proteins could be produced for biochemical and structural studies.

ACKNOWLEDGMENTS

We would like to acknowledge the Swedish Research Council, the Wallenberg Consortium North, the Göran Gustafsson Foundation for Research in Natural Science and Medicine, the European Community supported program Structural Proteomics in Europe (SPINE), and European Membrane Proteins Consortium (E-MeP) for financial support.

REFERENCES

1. Protein Data Bank. (http://www.rcsb.org/pdb).
2. Membrane proteins of known structure. (http://www.mpibp-frankfurt.mpg.de/michel/public/memprotstruct.html), accessed in 2005.
3. Miroux, B. and Walker, J.E. Over-production of proteins in *Escherichia coli*: mutant hosts that allow synthesis of some membrane proteins and globular proteins at high levels. *J. Mol. Biol.*, 260, 289–298, 1996.
4. Koronakis, V., Sharff, A., Koronakis, E., Luisi, B., and Hughes, C. Crystal structure of the bacterial membrane protein TolC central to multidrug efflux and protein export. *Nature*, 405, 914–919, 2000.
5. Lemieux, M.J., Song, J., Kim, M.J., Huang, Y., Villa, A., Auer, M., Li, X-D., and Wang, D.N. Three-dimensional crystallization of the *Escherichia coli* glycerol-3-phosphate transporter: a member of the major facilitator superfamily. *Prot. Sci.*, 12, 2748–2756, 2003.
6. Lesley, S.A., Kuhn, P., Godzik, A., Deacon, A.M., Mathews, I., Kreusch, A., Spraggon, G., Klock, H.E., McMullan, D., Shin, T., Vincent, J., Robb, A., Brinen, L.S., Miller, M.D., McPhillips, T.M., Miller, M.A., Scheibe, D., Canaves, J.M., Guda, C., Jaroszewski, L., Selby, T.L., Elsliger, M.A., Wooley, J., Taylor, S.S., Hodgson, K.O., Wilson, I.A., Schultz, P.G., and Stevens, R.C.. Structural genomics of the *Thermotoga* maritime proteome implemented in a high-throughput structure determination pipline. *Proc. Natl. Acad. Sci. USA*, 99, 11664–11669, 2002.
7. Kuhn, P., Wilson, K., Patch, M.G., and Stevens, R.C. The genesis of high-throughput structure-based drug discovery using protein crystallography. *Curr. Opin. Chem. Biol.*, 6, 704–710, 2002.
8. Stewart, L., Clark, R., and Behnke, C. High-throughput crystallization and structure determination in drug discovery. *Drug Disc.*, 7, 187–196, 2002.
9. Hui, R. and Edwards, A. High-throughput protein crystallization. *J. Struct. Biol.*, 142, 154–161, 2003.

10. Christendat, D., Yee, A., Dharamsi, A., Kluger, Y., Savchenko, A., Cort, J.R., Booth, V., Mackereth, C.D., Saridakis, V., Ekiel, I., Kozlov, G., Maxwell, K.L., Wu, N., McIntosh, L.P., Gehring, K., Kennedy, M.A., Davidson, A.R., Pai, E.F., Gerstein, M., Edwards, A.M., and Arrowsmith, C.H. Structural proteomics of an archaeon. *Nature*, 7, 903–909, 2000.

11. Structural Genomics consortium, Toronto. (http://www.sgc.utoronto.ca/), accessed in 2005.

12. Structural Genomics Consortium, Oxford. (http://www.sgc.ox.ac.uk/), accessed in 2005.

13. *Mycobacterium tuberculosis* Membrane Protein Collaboratory. (http://www. membraneprotein.magnet.fsu.edu/), accessed in 2005.

14. Lundstrom, K. Structural genomics on membrane proteins: The MePNet approach. *Curr. Opin. Drug Discov. Dev.*, 7, 342–346, 2004.

15. Lundstrom, K. Structural genomics on membrane proteins: Mini review. *Comb. Chem. High Throughput Screen.*, 7, 431–439, 2004.

16. The European Membrane Protein Consortium. (http://www.e-mep.org/), accessed in 2005.

17. JCSG Center for Innovative Membrane Protein Technologies. (http://stevens. scripps.edu/webpage/htsb/htmembr.html), accessed in 2005.

18. Braun, P. and LaBaer, J. High throughput protein production for functional proteomics. *Trends Biotechnol.*, 21, 383–388, 2003.

19. Henderson, P.J.F., Hoyle, C.K., and Ward, A. Expression, purification, and properties of multidrug efflux proteins. *Biochem. Soc. Trans.*, 28, 513–517, 2000.

20. Wang, D.N., Safferling, M., Lemieux, M.J., Griffith, H., Chen, Y., and Li, X.D. Practical aspects of overexpressing bacterial secondary membrane transporters for structural studies. *Biochim. Biophys. Acta.*, 1610, 23–36, 2003.

21. Walian, P., Cross, T.A., and Jap, B.K. Structural genomics of membrane proteins. *Genome Biol.*, 5, 215, 2004.

22. Korepanova, A., Gao, F.P., Hua, Y., Qin, H., Nakamoto, R.K., and Cross, T.A. Cloning and expression of multiple integral membrane proteins from *Mycobacterium tuberculosis* in *Escherichia coli. Prot. Sci.*, 14, 148–158, 2005.

23. Eshaghi, S., Hedrén, M., Abdel Nasser, M.I., Hammarberg, T., Thornell, A., and Nordlund, P. An efficient strategy for high-throughput expression screening of recombinant integral membrane proteins. *Prot. Sci.*, 14, 676–683, 2005.

24. Vincentelli, R., Canaan, S., Campanacci, V., Valencia, C., Maurin, D., Frassinetti, F., Scappucini-Calvo, L., Bourne, Y., Cambillau, C., and Bignon, C. High-throughput automated refolding screening of inclusion bodies. *Prot. Sci.*, 13, 2782–2792, 2004.

25. Kiefer, H. *In vitro* folding of alpha-helical membrane proteins. *Biochim. Biophys. Acta.*, 1610, 57–62, 2003.

26. Chang, G. and Roth, C.B. Structure of MsbA from *E. coli*: a homolog of the multidrug resistance ATP binding cassette (ABC) transporters. *Science*, 293, 1793–1800, 2001.

27. Knaust, R.K. and Nordlund, P. Screening for soluble expression of recombinant proteins in a 96-well format. *Anal. Biochem.*, 297, 79–85, 2001.

28. Hammarström, M., Hellgren, N., van Den Berg, S., Berglund, H., and Hard, T. Rapid screening for improved solubility of small human proteins produced as fusion proteins in *Escherichia coli. Prot. Sci.*, 11, 313–321, 2002.

29. Zhou, P., Lugovskoy, A.A., and Wagner, G. A solubility-enhancement tag (SET) for NMR studies of poorly behaving proteins. *J. Biomol. NMR.*, 20, 11–14, 2001.

30. Dyson, M.R., Shadbolt, S.P., Vincent, K.J., Perera, R.L., and McCafferty, J. Production of soluble mammalian proteins in *Escherichia coli*: identification of protein features that correlate with successful expression. *BMC Biotech.*, 4, 32, 2004.

31. Tobbell, D.A., Middleton, B.J., Raines, S., Needham, M.R.C., Taylor, I.W.F., Beveridge, J.Y., and Abbott, W.M. Identification of *in vitro* folding conditions for procathepsin S and cathepsin S using fractional factorial screens. *Prot. Expr. Purif.*, 24, 242–254, 2002.

32. Arechaga, I., Miroux, B., Karrasch, S., Huijbregts, R., den Kruijff, B., Runswick, M.J., and Walker, J.E. Characterisation of new intracellular membranes in *Escherichia coli* accompanying large scale over-production of the b subunit of F1Fo ATP synthase. *FEBS Lett.*, 482, 215–219, 2000.

33. Dumon-Seignovert, L., Cariot, G., and Vuillard, L. The toxicity of recombinant proteins in *Escherichia coli*: a comparison of overexpression in BL21(DE3), C41(DE3), and C43(DE3). *Prot. Expr. Purif.*, 37, 203–206, 2004.

34. Saidijam, M., Psakis, G., Clough, J.L., Meuller, J., Suzuki, S., Hoyle, C.J., Palmer, S.L., Morrison, S.M., Pos, M.K., Essenberg, R.C., Maiden, M.C.J., Abu-bakr, A., Baumberg, S.G., Neyfakh, A.A., Griffith, J.K., Stark, M.J., Ward, A., O'Reilly, J., Rutherford, N.G., Phillips-Jones, M.K., and Henderson, P.J.F. Collection and characterisation of bacterial membrane proteins. *FEBS Lett.*, 555, 170–175, 2004.

35. Lemieux, M.J., Reithmeier, R., and Wang, D.N. Importance of detergent and phospholipids in the crystallization of the human erythrocyte anion-exchanger membrane domain. *J. Struct. Biol.*, 137, 322–332, 2002.

36. Auer, M., Kim, M.J., Lemieux, M.J., Villa, A., Song, J., Li, X.D., and Wang, D.N. High-yield expression and fuctional analysis of *Escherichia coli* glycerol-3-phosphate transporter. *Biochemistry*, 40, 6628–6635, 2001.

37. Huang, Y., Lemieux, M.J., Song, J., Auer, M., and Wang, D.N. Structure and mechanism of the glycerol-3-phosphate transporter from *Escherichia coli*. *Science*, 301, 616–620, 2003.

5 Refolding of Membrane Proteins for Large-Scale Production

*Hans Kiefer, Thomas Ostermann, and
Monika Bähner*

CONTENTS

5.1 INTRODUCTION

Of the more than 30,000 protein structures obtained by x-ray crystallography and deposited in the protein databank (PDB),[1] less than 2% represent membrane proteins,[2] (161 entries as of February 2005). The vast majority of these depict bacterial, mitochondrial, and chloroplast proteins. A closer look reveals that only a handful of those structures are derived from proteins produced in recombinant form.[3] This is in sharp contrast to soluble protein structures, where recombinant production is the standard. In fact, not a single cell-surface protein structure of eukaryotic, let alone human, cells has been obtained through recombinant production. This shortage of structures is not only a concern to structural biologists but even more to medicinal chemists who would acclaim experimental structures of human membrane proteins for rational drug design.

Recombinant production of eukaryotic membrane proteins is a well-known issue, and multiple systems have been developed to solve it.[4,5] However, although sufficient amounts of protein for crystallization experiments can be purified in many cases, the first report of a structure derived from a recombinant eukaryotic plasma membrane protein is still being awaited. Why is that? In many cases, the quality of the purified protein is not good enough to allow for the formation of well-ordered crystals. Quality means (1) chemical homogeneity, that is, the absence of contaminating proteins, as well as heterogenous post-translational modifications or proteolysis; (2) physical homogeneity, that is, the protein should neither aggregate nor should it be present in a mixture of different folding states; and (3) stability, that is, the protein has to survive several days to weeks of crystal growth.

It is interesting to note that a large proportion of the currently published protein structures have been solved from proteins produced in *Escherichia coli*: 61% of the PDB protein entries contain the name "*Escherichia.*" This value increases to 71% if only files deposited after 2000 are included in the search and to 76% for structures deposited in 2004. The advantages of this expression system are mainly speed, high yields at low cost, and a wealth of know-how in scale-up and expression optimization. Multiple variants of a protein, often required in order to select mutants that are more stable or more likely to crystallize, can easily be produced in *E. coli*. The same is true for selenomethionine derivatives used to obtain phase information in crystallography and for isotope-labeled proteins required in nuclear magnetic resonance (NMR) spectroscopy. Moreover, since bacteria do not recognize most eukaryote-specific signals of post-translational modifications, the recombinant protein may be more homogenous than if expressed in eukaryotic systems. This is obviously only an advantage for proteins that do not require post-translational modifications for function. Luckily, most ion channels and G protein-coupled receptors (GPCRs) belong to this group. Finally, it is not unlikely that the outstanding success of *E. coli* in this field mainly reflects the fact that once expression is straightforward, one can focus on improving the protein quality and find suitable crystallization conditions.

Is *E. coli* also the host of choice for the production of recombinant eukaryotic membrane proteins? The immediate answer is no, since in most cases, expression of these proteins in the bacterial membrane is toxic and results in low to moderate production levels.[4] However, in some cases, careful optimization has resulted in appreciable expression levels, and the purified protein could be used in crystallization attempts.[6,7] A completely different approach prevents membrane insertion and the resulting toxic effect by deliberately directing the protein to inclusion bodies.[8,9] This results in high levels of recombinant protein that can be easily purified but requires an efficient *in vitro* folding method to meet the requirement of physical homogeneity.

In this chapter, we will briefly review the literature that addresses the driving forces and energetics of membrane protein folding both theoretically and experimentally. We will then discuss in more detail the reports on membrane proteins that have been successfully refolded from a denatured state. Finally, we will attempt to extract from the literature a strategy that can be used to identify folding conditions from scratch. The focus of this chapter is clearly on the practical side: refolding is seen as a method to obtain a supply of pure, functional membrane proteins for biophysical and structural analyses.

Although the term "refolding" is common for all procedures that end with folded protein, irrespective of the starting conditions, we will distinguish here between "folding" and "refolding" experiments. We will call a "folding" procedure any method that describes the transfer of a protein from an *a priori* unfolded state — such as inclusion bodies, aggregated protein produced in cell-free synthesis, or chemically synthesized polypeptides — into the native conformation. "Refolding" will be reserved for procedures that start by unfolding an originally native protein to study the kinetics and thermodynamics of the unfolding and refolding processes.

5.2 THEORETICAL CONSIDERATIONS

In contrast to random peptide sequences, biological proteins have been optimized by evolution to fold into a single, functional state. The folding process is dictated by the energy landscape of the accessible protein conformations. Efficient folding requires (1) a steep energy minimum to ensure that most of the protein molecules remain in the native conformation, and (2) a funnel-like shape of the energy landscape that makes the energy minimum accessible from any "unfolded" starting conformation.[10] Any local minima or barriers would (and will in practice) slow down the folding kinetics, which is energy-costly and therefore disfavored by evolution.

Obviously, the shape of the energy landscape depends not only on the polypeptide but also on its environment, that is, the solvent. However, abundant literature describing successful *in vitro* folding of soluble proteins shows that in many cases, the native conformation can be accessed in a wide range of buffer compositions, sometimes far from the natural environment.[11] The key point here is whether this also holds for membrane proteins.

The lipid bilayer strongly limits the range of possible structures. Therefore, the folding of transmembrane (TM) proteins may be more easily understood than that of soluble proteins as a consequence of environmental constraints. Indeed, the two-stage model of membrane protein folding, introduced by Popot and Engelman in 1990,[12] explains many experimental findings, and topology predictions of membrane proteins can be done with a high degree of confidence.[13] The latter is not the case for soluble proteins.

From a practical point of view, these conformational restraints do not make *in vitro* folding of membrane proteins any easier. This is because in a cell, vectorial insertion of the TM helices into the lipid bilayer is catalyzed by the translocation machinery, something that cannot be mimicked *in vitro* for large-scale production. Instead, the most common approach to *in vitro* refolding is to transfer unfolded proteins into a micellar environment composed of detergents and lipids, where helices are free to move and adopt orientations that would be inaccessible in a membrane. Whether this leads to the native conformation is closely related to the question whether the native fold of a membrane protein corresponds only to a local minimum or to the overall free energy minimum. From a theoretical standpoint, it is conceivable that in a detergent micelle, non-native low energy conformations will trap the polypeptide chain and prevent its folding into the functional state. Whether this happens, and how often, will remain an open question here because the available data are too scarce to give a definite answer.

5.2.1 THE TWO-STAGE MODEL OF MEMBRANE PROTEIN FOLDING

According to this model,[12] membrane protein folding takes place in two consecutive steps. First, the individual TM regions insert into the membrane and adopt their alpha-helical secondary structure. Second, contacts between the helix interfaces form and guide the protein into its native conformation. Abundant experimental evidence supports this model. Stage 1 is demonstrated by the observation that many TM regions fold into stable helices when synthesized individually and inserted into the bilayer.[14] This is by no means self-evident, since in the final structure, the helices will mainly be in contact with other parts of the protein and much less with lipid. Stage 2 is best supported by experiments where membrane proteins have been split into two or more fragments and assemble spontaneously to the native conformation when co-reconstituted into bilayers.[15] As discussed above, the conformational free-dom of individual helices in the membrane is very much restricted, and the second stage consists mainly of a combination of lateral diffusion and rotation of the helices about the membrane normal. The energetics of membrane insertion and helix asso-ciation have first been discussed in detail by Jähnig[16] and later by many others. The dominant free energy contribution in stage 1 is the transfer of hydrophobic side chains from water to lipid, estimated to be ~ –35 kcal/mole for a 20-residue TM segment. Stage 2 is driven mainly by Van-der-Waals interactions and, to some extent, by inter-helical hydrogen bonds and salt bridges. The free energy of this process has been estimated from an alanine scanning mutagenesis study to be –1 kcal/mole per 38 $Å^2$ buried surface area,[17] a value surprisingly close to that found for soluble proteins.

Despite its popularity, the two-stage model has limitations. Of the seven helices of bacteriorhodopsin, only five insert stably if synthesized individually.[14] The two remaining helices require additional portions of the protein for stable insertion. This and similar findings with other proteins have recently been accounted for by referring to all folding events that take place after assembly of a helix core as "stage 3." These include binding of prosthetic groups, folding of loop regions or additional helices, and formation of quarternary structure.[18]

Does the "two plus"-stage model tell us how to fold a membrane protein *in vitro*? Not immediately, as addition of a multispanning membrane protein to lipid bilayers leads to aggregation rather than to correct insertion. This is because the translocation of hydrophilic loops across the membrane faces a high energy barrier, and the translocon that normally catalyzes this process in the cell is absent in a liposome system. Bilayer destabilization by inducing extreme curvature or by the addition of detergents lowers this energy barrier and allows spontaneous insertion in some cases.[19] Even then, proteins might not insert properly because different helices might have a propensity to insert with incompatible orientations, leading to a "frustrated" topology. For this reason, protocols for membrane protein refolding often use detergents or mixed detergent–lipid micelles instead of preformed bilayer membranes to induce folding. Reconstitution into the lipid bilayer is done as a second, optional step, often by removing the detergent from mixed micelles according to methods used for native membrane proteins. Stage 1 is here replaced by the insertion into a mixed micelle and offers the helices the opportunity to flip their orientation. Formation of a helical secondary structure occurs spontaneously in most detergents and is therefore not an issue we will further discuss. Stage 2, that is, formation of helix–helix contacts, is much more sensitive to the environment and requires that the micellar environment is not disruptive and is similar enough to the lipid membrane to allow interhelical contacts to form. This normally requires fine-tuning of the detergent–lipid composition, which can be different for every new membrane protein studied.

5.2.2 STABILITY OF MEMBRANE PROTEINS

If inserted into the bilayer, membrane proteins can be very stable. Purple membranes, which are composed of lipid and bacteriorhodopsin, can be stored at room temperature for months without losing activity. Once removed from the bilayer and transferred into a detergent, however, the half-life of a membrane protein can be reduced to a few minutes. This is probably related to the increased flexibility[20] and depends very much on the nature of the detergents used. Although high flexibility is a nuisance from a practical aspect, it is probably a result of evolutionary fine-tuning and required for activity. Many membrane proteins such as receptors and transporters undergo conformational changes as part of signaling or substrate transport. The same is true for ion channel gating and for certain enzymatic activities. As a consequence, many natural membrane proteins are not optimized for rigidity and stability but, instead, are tuned to a certain degree of flexibility. This was directly demonstrated by introducing single Ala mutations at each position into helix B of bacteriorhodopsin and measuring the stability of the mutants: Mutants with higher stability than the wild-type were obtained surprisingly often.[17]

The folding units or domains of soluble and membrane proteins differ fundamentally in several respects: first, domains have different average sizes. Individual TM domains contain only ~ 25 residues, whereas the smallest autonomous folding units of soluble proteins are about three times as large. Second, TM domains are more stable than the typical soluble domains: a denaturant like 8 M urea will not be able to transform a membrane-inserted TM domain into a random coil. This effect is exploited when membranes are washed with urea to separate peripheral membrane proteins from integral membrane proteins. Third, the completely unfolded state of a TM domain, that is, a random coil, can only be achieved under extremely harsh conditions such as in organic solvents.[21] Refolding from this state always requires an intermediary step that ensures formation of alpha-helical secondary structure in the TM domains by adding detergent or lipid. Once removed from the hydrophobic environment, TM domains do not form alpha-helical structure and aggregate.

Fourth, conformational flexibility of a protein in the membrane is restricted. Correctly inserted TM domains in a membrane are confined to two-dimensional diffusion and rotation about the membrane's normal axis. The free energy of transfer from an aqueous environment to a membrane is 1.5 kcal/mole per residue, excluding the possibility of thermal unfolding of a membrane-inserted TM segment. In a detergent micelle, the situation is quite different, since the protein, as well as parts of it, is free to move in three dimensions, which might decrease the stability considerably. It is therefore possible that some membrane proteins require insertion into a lipid bilayer to fold correctly.

5.2.3 Is There a Rational Approach to Membrane Protein Refolding?

Since, according to the two-stage model, folding is determined exclusively by the portions of the protein that are inserted into the membrane, it would be sufficient to ensure that the protein is inserted into membranes composed of the natural lipid with the right topology. The proper tertiary contacts should then form automatically. Unfortunately, the vectorial insertion process in the cell cannot be replaced by *in vitro* methods that result in a defined topology. Instead, refolding protocols are designed to provide a membrane-like environment and at the same time allow for enough flexibility for the protein to end up in the native conformation. This approach assumes that the native conformation is not only a local minimum of orientationally constrained helices, but a global minimum, as in the case of soluble proteins. As there are no theoretical reasons for or against this assumption, it is instructive to analyze the cases where membrane proteins have been refolded successfully. Any such report is a strong indication that for the respective protein, the native conformation is at the global energy minimum not only in the lipid bilayer but also under the artificial conditions of the refolding protocol. Another important aspect is that any refolding protocol has to avoid off-pathway reactions that trap the protein in a non-native conformation. Aggregation is the most common reaction competing with folding. This reaction is very much determined by the type of detergent, as the detergents control the interaction between micelles and, as a consequence, the interaction between protein molecules contained in the micelles. Unfortunately, those detergents that are able to prevent aggregation are normally the same that denature proteins. The reason is that strong detergents with high solubilization power are electrically charged, leading to repulsion between the micelles and thereby preventing aggregation. However, the headgroups of these same detergents such as sodium dodecyl sulfate (SDS) will also interact electrostatically with parts of the protein and interfere with intramolecular contacts required to stabilize the native structure.

Another important aspect is the requirement of specific lipids for the function of certain membrane proteins.[22] Lipids can be supplied as mixed micelles in the folding procedure, but the question is which lipids should be used. A good starting point is a mixture that closely mimics the native bilayer of the respective protein. If activity is recovered, the natural lipid mixtures can then be successively replaced by synthetic lipids, possibly leading to a more homogenous preparation.

5.3 MEMBRANE PROTEIN REFOLDING IN PRACTICE

There are two main classes of membrane protein structures: alpha-helical bundles and beta barrels. The latter occur only in the outer membranes of bacteria, mitochondria, and chloroplasts but not on the surface of eukaryotic cells. Beta-barrel proteins have been unfolded and refolded very efficiently, as reviewed by Buchanan.[23] It is noteworthy that for this class of membrane proteins, many structures have been solved from proteins produced in inclusion bodies and subjected to *in vitro* folding. The average hydrophobicity of beta-barrel proteins is close to that of soluble proteins, which makes them soluble in denaturants such as urea and guanidinium chloride. Folding *in vivo* does not occur according to the two-stage model, as individual beta strands are not stable in the bilayer. As a consequence, refolding *in vitro* follows protocols that are totally unrelated to those used for alpha-helical bundles. We will not review these methods here and refer to the cited literature instead.

The following alpha-helical proteins have been successfully (re)folded: bacteriorhodopsin (bR), light-harvesting complex (LHC2), diacyl glycerol kinase (DAGK), several mitochondrial transporters, the rat olfactory receptor OR5, the potassium channel KcsA, the human leukotriene receptor BLT1, and the disulfide bond–reducing protein B (DsbB). Some of these (bR, LHCP, DAGK, DsbB, KcsA) have been deliberately unfolded from the native state to study the refolding behavior. In most cases, however, the cited literature deals with folding from inclusion bodies with the goal to obtain large amounts of functional protein.

5.3.1 BACTERIORHODOPSIN

bR, a 25-kDa (248 residues, 7 TM domains) photoactive monomer, is the best studied membrane protein, mainly because of its unusually high stability and because it can easily be prepared from bacterial purple membrane in large quantities. It was the first membrane protein to be completely unfolded in organic solvent (trifluoroacetic acid, TFA) and refolded to the native state.[21] While many variants of the initial refolding protocol have been published since, a common intermediate step is the transfer into SDS, where bR contains about half of the alpha-helical secondary structure present in the native state.[24] In SDS, the chromophore is exposed to the solvent as indicated by its yellow color. Refolding from SDS can therefore be studied by spectroscopy. A widely used refolding protocol[25] decreases the SDS concentration below the critical micellar concentration (CMC) by dilution into a reconstitution buffer containing a neutral detergent (CHAPS) above the CMC, lipids, and the chromophore all-trans retinal. In the presence of retinal, the protein folds and becomes purple. Pompejus and co-workers carried out their folding protocol with bR from inclusion bodies (ebR) and with denatured bR prepared from halobacterial purple membrane (hbR). Interestingly, reconstitution yields were approximately 50% for hbR and 25% for ebR.[25]

In contrast to many other membrane proteins, bR can also be folded by adding SDS-solubilized protein directly to phosphatidyl choline (PC) vesicles, showing that all helices insert with native topology. The effect on the folding kinetics of different lipid compositions has been studied in detail. Lipids conferring high lateral tension will slow down the folding kinetics, that is, bR folds more slowly in membranes with high phosphatidyl ethanolamine (PE) content than in membranes composed of PC.[26]

A cell-free synthesis approach for the production of bacterioopsin has been described by Rothschild and co-workers.[27] The authors used a wheat germ extract and folded the aggregated material as described above with the exception that the bR-peptide was dissolved in SDS after organic extraction and folded into the polar lipids of *Halobacterium halobium* with additional all-trans-retinal present. SDS was then removed by potassium precipitation and dialysis. The overall yield of this procedure was 24 µg of functional bR from a 1 ml reaction.

5.3.2 LIGHT-HARVESTING COMPLEX

Light-harvesting complex II (LHC2) of plant chloroplasts is a 25-kDa protein (3 TM regions per protomer) that forms trimers and binds chlorophylls and carotenoids in the native conformation. The structure is unusual in the sense that most of the helix–helix contacts are mediated by intercalating pigments. Only two helices per monomer are in direct contact.[28] The recombinant monomer can be reconstituted with lipids and pigments, yielding active trimeric protein.[29] Another route starts with the solubilization of inclusion bodies in urea.[30] Purifying and folding was done in one step. To this purpose, the protein was coupled to a nickel column in urea and sequentially washed with lithium dodecyl sulfate (LDS) at pH 9, octyl-ß-D-glucoside (OG) at pH 9, and finally Triton X-100–dipalmitoyl phosphoryl choline (DPPC) mixed micelles at pH 7.5. In the OG step, chlorophylls and carotenoids were added to yield properly folded monomers. The latter was concluded from the pale green color indicative of native protein that was stably associated with the column. The authors assume that trimerization occurred upon elution from the column. The overall yield of this column-based procedure was comparable with a batch method, where LHC2 was folded in analogy to the bR refolding protocol by Pompejus and co-workers.[25]

5.3.3 DIACYLGLYCEROL KINASE

DAGK is a small homotrimeric alpha-helical protein with 121 amino acids per monomer. It is localized in the cytoplasmic membrane of *E. coli* and requires trimerization for function. The folding of DAGK has been studied extensively, and various protocols have been compared for efficiency. The monomer is soluble in urea or guanidinium chloride (GdmCl) and can be refolded by dilution in the presence of preformed liposomes.[31] However, it was observed that low concentrations of detergents were able to enhance the lipid bilayer insertion rate and the recovered activity. Overproduction of DAGK in *E. coli* often results in misfolding. It is unclear whether DAGK is deposited in inclusion bodies or whether it assembles in nonclassical membrane structures of unusually high density[32]. In a "reconstitutive refolding" approach, the protein was solubilized in lauryl-N,N-dimethylglycine (Empigen). The protein was purified, and the detergent was exchanged for decyl-ß-D-maltoside (DM) and finally for dodecyl phosphocholine (DPC) on a nickel column. The eluted protein was then diluted into a refolding solution containing 4 M urea, 1% Triton X-100, 200 mM imidazole, and 0.25 mM TCEP at pH 8, where it was incubated for 1 to 24 h at room temperature or 37°C. This was followed by an additional dilution into mixed micelles composed of DPC and 1-palmitoyl-2-oleoyl-phosphatidylcholine (POPC) to complete the refolding process. The detergent was then removed by extensive dialysis. Other detergents tested, instead of DPC, did not result in appreciable activity levels. Refolding, as monitored by activity measurement, took place at a late stage of detergent removal: activity was not recovered before the solution became opaque, indicative of vesicle formation. NMR analysis confirmed that at this point, the DPC concentration in the proteoliposomes was below the CMC. No refolding was observed when dipalmitoyl phosphatidylcholine (DPPC), instead of POPC, was used or when residual DM was present before reconstitution. In a later report, Sanders and co-workers used an SDS-stable triple mutant to study conformationally specific misfolding of DAGK.[33] This mutant, which was not impaired in catalytic activity, allowed analysis of the trimerization state by SDS-PAGE (polyacrylamide gel electrophoresis). An extensive cysteine scanning mutagenesis study, starting from the SDS-stable mutant, revealed which residues are important for DAGK trimer stability. In another study, the stability of DAGK was assessed directly by measuring reversible unfolding in dodecyl-ß-D-maltoside (DDM)–SDS mixed micelles.[34] Several amino acids in the TM region were mutated to cysteine, and the stability was compared with the wild-type. As for bR, several stabilizing mutants were found.

5.3.4 DISULFIDE BOND–REDUCING PROTEIN B

DsbB (20 kDa, 176 amino acids, 4 TM domains) is a bacterial protein that can be readily purified from *E. coli* in the native state.[35] The aim of the cited work was to study the unfolding and refolding processes and to measure their kinetics and thermodynamics. As for DAGK in the previous section, the protein solubilized in DDM was denatured reversibly by the addition of increasing amounts of SDS. Otzen could show by stopped-flow measurement that the kinetics of refolding was monophasic. No refolding could be observed when DDM was replaced by OG or Triton X-100. These results were compatible with the observation that DsbB shows no enzymatic activity in these detergents.

5.3.5 MITOCHONDRIAL TRANSPORT PROTEINS

Several mitochondrial transporters have been expressed in *E. coli* inclusion bodies. Here, we focus on the 2-oxoglutarate carrier (OGC), a 35-kDa protein composed of six TM segments that is thought to form dimers. After solubilization of the inclusion bodies using the detergent N-lauroyl sarcosine (sarcosyl), the protein was reconstituted into soybean PC lipid vesicles. The detergent was removed by adsorption to a hydrophobic polystyrene matrix, leading to proteoliposomes that could be used in transport assays. However, in a recent publication, the authors describe a protocol for OGC folding[36] that was modified because the levels of renatured protein obtained by the previous method were apparently insufficient for structural characterization. In the modified protocol, the authors solubilized the inclusion bodies in guanidinium chloride (GdmCl) and dithiothreitol (DTT) at room temperature. The purification and folding of the transporter was done by nickel-chelating chromatography. OGC was coupled to the column and GdmCl exchanged for urea with 1% Triton X-100. The column was then washed with 0.1% Triton X-100 and the temperature was reduced to 4°C. The overall yield of this method was about 10% with respect to the inclusion body fraction.

5.3.6 POTASSIUM CHANNEL KcsA

KcsA is a small tetrameric 60-kDa potassium channel with 160 amino acids and 2 TM domains per subunit. Valiyaveetil and co-workers[37] chemically ligated a synthetic C-terminal peptide to the remaining N-terminal portion produced in inclusion bodies. The solubilization and chemical ligation of the purified fragments was done in a buffer containing SDS. After the ligation step, the tetrameric state was obtained by reconstituting the monomers into lipid vesicles through dilution. Folding was not observed at SDS concentrations above 0.1%. Toxin binding showed that the refolded protein had adopted a native conformation. The same protocol could be applied to native KcsA that had been denatured by boiling in SDS for 30 min.[38] The authors tested many refolding conditions and found that the reconstitution step and the complete removal of the detergent were essential to obtain tetrameric KcsA. As the crystal structure of KcsA reveals a single tightly bound phosphatidyl glycerol (PG) per monomer, the authors tested whether refolding would require specific lipids. Surprisingly, tetramer formation occurred in different lipids and required neither PG nor any negatively charged lipid headgroups at all. However, for proper ion channel function, negative lipids were mandatory.

5.3.7 OLFACTORY RECEPTOR OR5

The rat olfactory receptor OR5 (36 kDa, 314 amino acid residues, 7 TM domains) was the first GPCR reported to have been refolded from inclusion bodies.[8] The protein was solubilized in sarcosyl and purified on an Ni-NTA agarose column. Detergent exchange from sarcosyl to digitonin, a detergent widely used to solubilize native GPCRs,[39] was also carried out on an Ni-NTA agarose column. The digitonin concentration was 0.1%, which is close to the CMC of the detergent. After elution, the protein was dialyzed to remove imidazol and added to a lipid–detergent film containing egg PC, egg PG, and DDM in order to form mixed protein–lipid–detergent micelles. Proteoliposomes were obtained by removal of the detergent with an Extractigel column. Alternative attempts to fold OR5 using DDM, Triton X-100, cholate, deoxycholate, or CHAPS, instead of digitonin, led to aggregation of the protein.

Ligand binding was measured both with fluorescence and radioligand competition assays, showing that the protein had been refolded in the detergent micelle. The protocol has since been applied in modified forms to several other GPCRs (our unpublished results).

5.3.8 LEUKOTRIENE RECEPTOR BLT1

The leukotriene receptor BLT1 (43 kDa, 352 residues, 7 TM segments) was folded from inclusion bodies using a protocol reminiscent of the methods used for LHC2 and DAGK. Urea was used as solubilizing agent.[40] The protein was then bound to a nickel column, a neutral detergent (LDAO or DDM) was added, and urea was removed in a shallow gradient. The folding yield of fully functional receptor was about 20% of the purified, urea denatured BLT1. Three important findings should be emphasized: First, the protein-to-resin ratio in the column-folding step had to be 0.6 mg of protein per milliliter of hydrated Ni-NTA agarose, which is far below column saturation. Second, the stability of the purified receptor in the detergent micelles was strongly dependent on the alkyl chain length, with a preference for longer chains as in LDAO or DDM. Third, to recover activity, it was necessary to work only slightly above the CMC of the mild detergents applied.

The eluted protein was only partly soluble, but the soluble fraction had a ligand-binding activity of near 100%. Impressively, the refolded protein could be assembled quantitatively with G protein to form a pentameric complex with a stoichiometry of two GPCR molecules and one G-protein trimer. The high efficiency of this procedure makes this system ideal for any kind of spectroscopic studies. However, it is not clear whether the method can be applied to other GPCRs, since the solubility of BLT1 in urea without the addition of detergents might be the exception rather than the rule. In a recent publication,[41] the authors used sarcosyl in addition to urea to increase the solubilization yield.

5.4 STRATEGIES TO IDENTIFY (RE-)FOLDING CONDITIONS

A common situation is that a membrane protein supposed to be expressed in the bacterial membrane does not insert and forms inclusion bodies instead. Alternatively, a non-native protein may have been obtained from a cell-free protein synthesis system or by chemical synthesis. How should one proceed to identify folding conditions from scratch?

The general procedure is outlined in the refolding scheme (Figure 5.1). The first step would be to search for a solubilizing agent. The choice is among harsh detergents such as SDS, sarcosyl, DPC, and chaotropes such as urea or guanidinium chloride. SDS will probably always work but is also difficult to remove quantitatively in subsequent steps because of its low CMC. Residual SDS, however, is likely to interfere with the function of many proteins. Therefore, any of the other detergents or denaturants would be a better choice if they work. It is important already at this stage not only to screen different concentrations of the denaturants but also to test at least two pH values and ideally a low- and a high-salt concentration. This is because a protein at a pH close to its isoelectric point might be very difficult to solubilize.

Once solubilized, the protein can be subjected to a first purification step prior to folding. Using a His-tag is the method of choice, since it works equally well in the presence of most detergents and denaturants. Classical chromatography such as ion exchange chromatography is of little use if the detergent is charged or if high salt is present. The next decision is whether to elute the protein under denaturing conditions or fold it while it is bound to the column. In the latter procedure, using mixed micelles of neutral detergent in combination with phospholipids is a method

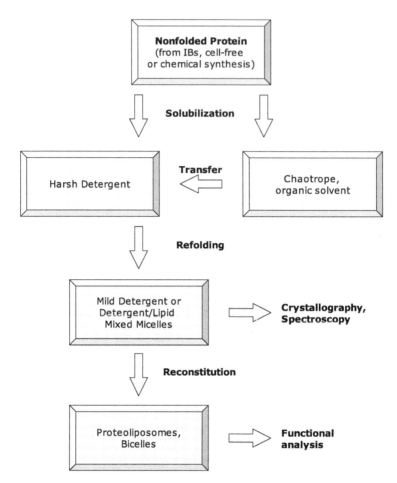

FIGURE 5.1 General refolding scheme for membrane proteins. The nonfolded protein from inclusion bodies or other sources is solubilized in a harsh detergent, in a chaotrope, or in organic solvent. The protein may be purified while denatured. Refolding normally starts from the "harsh detergent" solubilized state and is induced by adding milder detergents, often in combination with lipids, and removing the harsh detergent. The refolded, detergent-solubilized protein can be directly subjected to crystallization trials or spectroscopy. For functional experiments that require a lipid membrane, for example, measurement of ion channel current, an additional reconstitution step where the protein is transferred into a lipid bilayer is required.

that has been proven successful in many cases. Since the specific lipid requirement is often unknown, a mixture that corresponds to the natural environment of the respective protein will be a good starting point. Once activity is recovered, one can always attempt to reduce the complexity of the lipid mixture later. In any case, the eluted protein is now either still solubilized in a harsh detergent or chaotrope or in an environment that induces refolding. Ideally, an activity assay that can be carried out in detergent should be available at this stage. If the native protein, not necessarily in pure form, is available as a positive control, the compatibility of the detergents used for refolding with the activity assay should be tested in advance. On the other hand, reconstitution is required if the activity assay requires a lipid bilayer such as a conductivity or transport measurement. The same is true if the activity assay in a detergent is always negative. In the latter case, the situation could be such as that for KcsA, where detergents apparently interfere with folding and complete removal of detergents is mandatory.

Finding proper folding conditions is a screening process during which parameter setting (type of detergent and lipid, pH, additives, and so on) are optimized. As such, it is helpful to apply experimental design procedures in order to extract as much information as possible from a limited set of experiments. Factorial and sparse matrix screens have been published to identify folding conditions for soluble proteins.[42,43] A similar approach might help in the future in defining the best protocols for membrane protein folding.

5.5 FUTURE PROSPECTS

In summary, efficient folding, once established as a routine method, can considerably speed up the production of highly purified membrane proteins. *E. coli* expression, if available for a specific protein, is likely to outpace any competing expression system in structural biology for reasons of speed, cost, and versatility. However, the race for the first recombinant human membrane protein structure has now been on for more than 20 years, and there is still no winner. Here, we champion the use of *E. coli* expression in combination with *in vitro* folding, but it remains to be demonstrated whether along this route, structures can be solved in the future. The history of other expressions systems such as yeast, insect, or mammalian cells, teaches that the availability of a purified membrane protein, although certainly necessary, is far from being a sufficient requirement to solve the structure.

Membrane protein structural analysis is a very challenging, though potentially very rewarding, task, especially for those proteins that function as human drug targets. Whatever seems to work should be seriously pursued, and may the best expression host win!

REFERENCES

1. Berman, H.M., Westbrook J., Feng, Z., Gilliland, G., Bhat, T.N., Weissig, H., Shindyalov, I.N., and Bourne P.E. The Protein Data Bank. *Nucleic Acids Res.*, 28, 235–242, 2000.
2. http://blanco.biomol.uci.edu/Membrane_Proteins_xtal.html (accessed February 2005).
3. Kiefer, H. *In vitro* folding of alpha-helical membrane proteins. *Biochim. Biophys. Acta.*, 1610, 57–62, 2003.
4. Grisshammer, R., and Tate, C.G. Overexpression of integral membrane proteins for structural studies. *Q. Rev. Biophys.*, 28, 315–422, 1995.
5. Sarramegna, V., Talmont, F., Demange, P., and Milon, A. Heterologous expression of G-protein-coupled receptors: comparison of expression systems from the standpoint of large-scale production and purification. *Cell Mol. Life Sci.*, 60, 1529–1546, 2003.

6. White, J.F., Trinh, L.B., Shiloach, J., and Grisshammer, R. Automated large-scale purification of a G protein-coupled receptor for neurotensin. *FEBS Lett.*, 564, 289–293, 2004.
7. Weiss, H.M. and Grisshammer, R. Purification and characterization of the human adenosine a(2a) receptor functionally expressed in *Escherichia coli*. *Eur. J. Biochem.*, 269, 82–92, 2002.
8. Kiefer, H., Krieger, J., Olszewski, J.D., Von Heijne, G., Prestwich, G.D., and Breer H. Expression of an olfactory receptor in *Escherichia coli*: purification, reconstitution, and ligand binding. *Biochemistry*, 35, 16077–16084, 1996.
9. Kiefer, H., Vogel, R., and Maier, K. Bacterial expression of G protein-coupled receptors: prediction of expression levels from sequence. *Receptors Channels*, 7, 109–119, 2000.
10. Dill, K.A. and Chan, H.S. From levinthal to pathways to funnels. *Nat. Struct. Biol.*, 4, 10–19, 1997.
11. Lilie, H., Schwarz, E., and Rudolph, R. Advances in refolding of proteins produced in *E.coli*. *Curr. Opin. Biotechnol.*, 9, 497–501, 1998.
12. Popot, J.L. and Engelman, D.M. Membrane protein folding and oligomerization: the two-stage model. *Biochemistry*, 29, 4031–4037, 1990.
13. von Heijne, G. Principles of membrane protein assembly and structure. *Prog. Biophys. Mol. Biol.*, 66, 113–139, 1996.
14. Hunt, J.F., Earnest, T.N., Bousche, O., Kalghatgi, K., Reilly, K., Horvath, C., Rothschild, K.J., and Engelman, D.M. A biophysical study of integral membrane protein folding. *Biochemistry*, 36, 15156–15176, 1997.
15. Popot, J.L. and Engelman, D.M. Helical membrane protein folding, stability, and evolution. *Annu. Rev. Biochem.*, 69, 881–922, 2000.
16. Jähnig, F. Thermodynamics and kinetics of protein incorporation into membranes. *Proc. Natl. Acad. Sci. USA*, 8, 3691–3695, 1983.
17. Faham, S., Yang, D., Bare, E., Yohannan, S., Whitelegge, J.P., and Bowie, J.U. Side-chain contributions to membrane protein structure and stability. *J. Mol. Biol.*, 335, 297–305, 2004.
18. Engelman, D.M., Chen, Y., Chin, C.N., Curran, A.R., Dixon, A.M., Dupuy, A.D., Lee, A.S., Lehnert, U., Matthews, E.E., Reshetnyak, Y.K., Senes, A., and Popot, J.L. Membrane protein folding: beyond the two stage model. *FEBS Lett.*, 555, 122–125, 2003.
19. Levy, D., Bluzat, A., Seigneuret, M., and Rigaud, J.L. A systematic study of liposome and proteoliposome reconstitution involving bio-bead-mediated Triton X-100 removal. *Biochim. Biophys. Acta.*, 1025, 179–190, 1990.
20. Bowie, J.U. Stabilizing membrane proteins. *Curr. Opin. Struct. Biol.*, 11, 397–402, 2001.
21. Huang, K.S., Bayley, H., Liao, M.J., London, E., and Khorana, H.G. Refolding of an integral membrane protein: denaturation, renaturation, and reconstitution of intact bacteriorhodopsin and two proteolytic fragments. *J. Biol. Chem.*, 256, 3802–3809, 1981.
22. Opekarova, M. and Tanner, W. Specific lipid requirements of membrane proteins — a putative bottleneck in heterologous expression. *Biochim. Biophys. Acta.*, 1610, 11–22, 2003.
23. Buchanan, S.K. Beta-barrel proteins from bacterial outer membranes: structure, function and refolding. *Curr. Opin. Struct. Biol.*, 9, 455–461, 1999.

24. Allen, S.J., Curran, A.R., Templer, R.H., Meijberg, W., and Booth, P.J. Folding kinetics of an alpha helical membrane protein in phospholipid bilayer vesicles. *J. Mol. Biol.*, 342, 1279-1291, 2004.

25. Pompejus, M., Friedrich, K., Teufel, M., and Fritz, H.J. High-yield production of bacteriorhodopsin via expression of a synthetic gene in *Escherichia coli. Eur. J. Biochem.*, 211, 27–35, 1993.

26. Allen, S.J., Curran, A.R., Templer, R.H., Meijberg, W., and Booth, P.J. Controlling the folding efficiency of an integral membrane protein. *J. Mol. Biol.*, 342, 1293–1304, 2004.

27. Sonar, S., Patel, N., Fischer, W., and Rothschild, K.J. Cell-free synthesis, functional refolding, and spectroscopic characterization of bacteriorhodopsin, an integral membrane protein. *Biochemistry*, 32, 13777–13781, 1993.

28. Liu, Z., Yan, H., Wang, K., Kuang, T., Zhang, J., Gui, L., An, X., and Chang, W. Crystal structure of spinach major light-harvesting complex at a 2.72 resolution. *Nature*, 428, 287-292, 2004.

29. Hobe, S., Prytulla, S., Kuhlbrandt, W., and Paulsen, H. Trimerization and crystallization of reconstituted light-harvesting chlorophyll a/B complex. *EMBO J.*, 13, 3423–3429, 1994.

30. Rogl, H., Kosemund, K., Kuhlbrandt, W., and Collinson, I. Refolding of *Escherichia coli* produced membrane protein inclusion bodies immobilised by nickel chelating chromatography. *FEBS Lett.*, 432, 21-26, 1998.

31. Nagy, J.K., Lonzer, W.L., and Sanders, C.R. Kinetic study of folding and misfolding of diacylglycerol kinase in model membranes. *Biochemistry*, 40, 8971-8980, 2001.

32. Gorzelle, B.M., Nagy, J.K., Oxenoid, K., Lonzer, W.L., Cafiso, D.S., and Sanders, C.R. Reconstitutive refolding of diacylglycerol kinase, an integral membrane protein. *Biochemistry*, 38, 16373–16382, 1999.

33. Oxenoid, K., Sonnichsen, F.D., and Sanders, C.R. Conformationally specific misfolding of an integral membrane protein. *Biochemistry*, 40, 5111–5118, 2001.

34. Lau, F.W., Nauli, S., Zhou, Y., and Bowie, J.U. Changing single side-chains can greatly enhance the resistance of a membrane protein to irreversible inactivation. *J. Mol. Biol.*, 290, 559–564, 1999.

35. Otzen, D.E. Folding of Dsbb in mixed micelles: a kinetic analysis of the stability of a bacterial membrane protein. *J. Mol. Biol.*, 330, 641–649, 2003.

36. Smith, V.R. and Walker, J.E. Purification and folding of recombinant bovine oxoglutarate/malate carrier by immobilized metal-ion affinity chromatography. *Prot. Expr. Purif.*, 29, 209–216, 2003.

37. Valiyaveetil, F.I., MacKinnon, R., and Muir, T.W. Semisynthesis and folding of the potassium channel Kcsa. *J. Am. Chem. Soc.*, 124, 9113–9120, 2002.

38. Valiyaveetil, F.I., Zhou, Y., and MacKinnon, R. Lipids in the structure, folding, and function of the Kcsa K+ channel. *Biochemistry*, 41,10771–10777, 2002.

39. E.C. Hulme, Ed., *Receptor Biochemistry: A Practical Approach*. Practical Approach Series, Oxford University Press, New York, 1990.

40. Baneres, J.L., Martin, A., Hullot, P., Girard, J.P., Rossi, J.C., and Parello, J. Structure-based analysis of Gpcr function: conformational adaptation of both agonist and receptor upon leukotriene B4 binding to recombinant Blt1. *J. Mol. Biol.*, 329, 801–814, 2003.

41. Mesnier, D. and Baneres, J.L. Cooperative conformational changes in a GPCR dimer, the leukotriene B4 receptor BLT1. *J. Biol. Chem.*, 279, 49664–49670, 2004.

42. Armstrong, N., de Lencastre, A., and Gouaux, E. A new protein folding screen: application to the ligand binding domains of a glutamate and kainate receptor and to lysozyme and carbonic anhydrase. *Prot. Sci.*, 8, 1475–1483, 1999.
43. Vincentelli, R., Canaan, S., Campanacci, V., Valencia, C., Maurin, D., Frassinetti, F., Scappucini-Calvo, L., Bourne, Y., Cambillau, C., and Bignon, C. High-throughput automated refolding screening of inclusion bodies. *Prot. Sci.*, 13, 2782–2792, 2004.

6 Crystallization of Membrane Proteins

Alastair T. Gardiner, June Southall, and Richard J. Cogdell

CONTENTS

6.1 INTRODUCTION

In 1990, Hartmut Michel edited a book that serves still today as a useful reference for methodology and background when attempting to crystallize membrane proteins.[1] At that time, the structure of 400 proteins was known at, or nearly at, atomic resolution. However, only two of these were membrane proteins: the bacterial reaction centre from *Rhodopseudomonas viridis*[2] and bacteriorhodopsin from *Halobacterium salinarum*.[3] At the time of writing, the number of structures deposited in the Research Collaboratory for Structural Bioinformatics (RCSB) Data Bank had grown to 32,600, and yet the number of known membrane proteins amounts to no more than 58, of which only 26 can be classed as unrelated. A thorough and regularly updated list of all membrane protein structures, along with

73

a tabulation of the crystallization conditions, is maintained at the following Web site: www.mpibp-frankfurt.mpg. de/michel/public/memprotstruct.html.

Why is there this imbalance between water-soluble protein structures and membrane protein structures? Undoubtedly, the biggest barrier is obtaining sufficient quantities of pure, active membrane protein. It is not surprising that most of the membrane proteins for which there are high-resolution crystal structures are both naturally overexpressed and pigmented. Only relatively recently have structures appeared for membrane proteins that have been heterologously overexpressed.[4] Strikingly, even in these few cases, success has only been achieved with prokaryotic membrane proteins. These issues are covered elsewhere in this book, and this chapter concentrates on the crystallization methodologies, assuming a sufficient amount of pure, active membrane protein.

Integral membrane proteins are, by definition, amphipathic molecules. That is, they are composed of both hydrophilic and hydrophobic regions. The hydrophilic domain(s) are present on either side of the membrane and interact with the aqueous solvent and the polar head groups of the constituent lipids. The hydrophobic portion of the protein that spans the membrane is intimately associated with the lipid alkyl chains and must be shielded from the aqueous phase at all times during purification if the stability, integrity, and functionality of the protein are to be maintained. Obviously, these properties are prerequisites before crystallization trials of any given protein can even be attempted.

In principle, membrane proteins can form at least three types of three-dimensional (3D) crystals. These are reproduced schematically in Figure 6.1. Type I crystals are layered two-dimensional (2D) crystals that are formed in the plane of a reconstituted membrane and are then induced to stack on top of each other to form the 3D lattice. Type I crystals are stabilized by both hydrophobic and hydrophilic interactions. Polar interactions arise from contact between the stacked hydrophilic domains of adjacent layers within the lattice and are hence perpendicular to the membrane. Interactions between the polar head groups of the lipids may possibly also play a role in stabilizing the lattice in the plane of the membrane. Hydrophobic interactions among the alkyl chains of the lipid help stabilize the lattice in the plane of the membrane. For many years, it was thought that the prospect of obtaining large, well-ordered type I crystals suitable for x-ray diffraction studies was rather slim, at least by conventional crystallization methodology. However, recently, the lipidic cubic phase concept of crystallizing membrane proteins has emerged.[5] This produces type I crystals by generating a continuous lipid phase that acts as a solvent for membrane proteins. An attractive feature of crystals produced in lipidic cubic phases is that they contain a lower percentage of the solvent. This means even the microcrystals that are generally produced by this method yield diffraction patterns with enough intensity to produce high-resolution crystal structures.

The second strategy, and until now the most successful in terms of the number of structures solved, is to produce type II crystals. These are grown using the conventional crystallization methodology of inducing supersaturation, except that the protein to be crystallized has been solubilized by a detergent. A suitable detergent, in essence, mimics and replaces the natural hydrophobic environment provided by the natural lipid alkyl chains within the lipid bilayer to form a protein-detergent

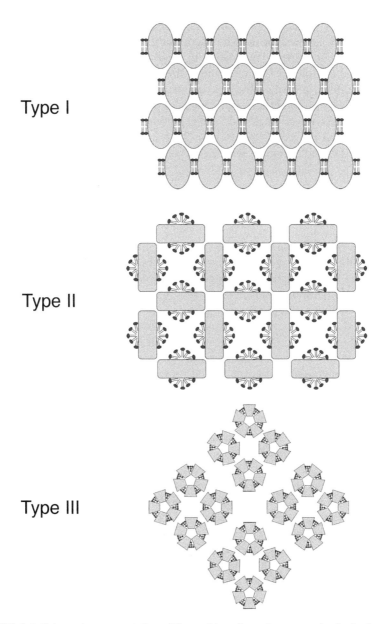

FIGURE 6.1 Schematic representation of the packing of membrane proteins in the three main crystal types so far identified.

Type I: This type of crystal is formed by the superposition of 2D crystals.

Type II: This type of crystal is equivalent to the types of crystals formed from water-soluble proteins. Most of the currently solved structures involve type II crystals.

Type III: This type of crystal resembles the crystals formed when icosehedral viruses crystallize. The membrane protein is ordered within a microliposome, and then these vesicles crystallize to form the macrocrystal lattice.

micelle, thereby protecting the hydrophobic region of the protein and shielding it from the aqueous phase. The main interactions that stabilize the crystal lattice involve the hydrophilic domains of the protein. Any method that increases or promotes these polar interactions should facilitate type II crystal growth.

Very recently, a totally new type of membrane protein crystal (type III) was described.[6] The major light-harvesting pigment-protein complex (LHC-II) from spinach chloroplasts was crystallized in the presence of a nonbilayer forming lipid digalactosyl diacylglycerol (DGDG). So far, there is only this single example of a type III crystal, and only time will tell if this approach will be generally useful.

6.2 PRACTICAL CONSIDERATIONS

6.2.1 Detergent Choice for Sample Preparation

In general, crystallization of any protein requires milligram quantities of pure, active protein. In this chapter, we discuss our experiences working with several membrane proteins and how to approach the general problem of crystallizing membrane proteins. The first question to be answered is: which are the suitable detergents for the initial solubilization and purification process?

There are no rules to guide this choice, and it can only be determined empirically. A good detergent for protein purification should have the following properties in addition to stabilizing the membrane protein of interest: it should have a low critical micellar concentration (CMC); it should be inexpensive to purchase in relatively large quantities; and it should be free from contaminants. In our laboratory, N,N-dimethyldodecylamine-N-oxide (LDAO) is routinely used. This detergent has proven to be excellent for use with the membrane proteins from purple, photosynthetic bacteria that we mainly study. It is cheap, effective in solubilizing reaction centers (RC) and light-harvesting (LH) antenna complexes, and able to maintain these proteins in their native state for a relatively prolonged period, that is, weeks. Moreover, LDAO is easy to remove by dialysis, and therefore proteins purified in LDAO can be readily exchanged into other detergents for crystallization trials. Before embarking on large-scale protein purification of the membrane protein of choice, it is desirable to determine from minipreps the optimal detergent for the protein in question. The common laboratory detergent Triton X-100 should be avoided because it is both a complex mixture of different molecules and difficult to exchange. This, in turn, could compromise any subsequent chance of successful crystallization.

One factor that is usually neglected and that can potentially introduce an unknown degree of heterogeneity is the lipid composition of the purified membrane protein. Clearly, it is easiest to disregard the lipid composition when beginning crystallization trials of the pure membrane protein of choice. However, if crystals that do not diffract x-rays to sufficient resolution are obtained, then it may be necessary to assay the lipid content of the pure protein and to tackle this problem as part of the crystal optimization procedure. Unfortunately, most methods of lipid analysis are not very sensitive, and so one may be faced with the prospect of sacrificing the entire protein preparation to ensure enough material for the lipid analysis. Furthermore, if the lipid composition varies between preparations, this can be a serious hurdle. This means that it is very important to rigorously standardize the protein preparation protocols. It could also be worthwhile, within this context, to experiment with different purification strategies, even when one has a pure preparation. Even small variations in the initial solubilization, for example, the temperature at which it is performed as well as changes in the chromatographic material used, can affect the final lipid composition of the purified membrane protein. Hopefully, mass-spectroscopic techniques will soon be developed, which will allow the routine analysis of the lipid content of membrane protein preparations that will only require the use of a small fraction of the total preparation.

6.2.2 CHOICE OF DETERGENTS FOR CRYSTALLIZATION TRIALS

The particular detergent used initially to isolate the membrane protein of interest often proves inappropriate for crystallization trials. This is because requirements for crystal growth are usually different from those for isolation. The choice of detergent is often the single most important parameter that must be successfully elucidated in the quest to obtain crystals suitable for structural analysis. This is understandable, since the detergent micelle has to fit optimally into the crystal lattice: the length of the detergent alkyl chain must completely protect the hydrophobic transmembranous portion of the protein and yet not be so large that the detergent belt around the protein sterically hinders crystal growth. Similarly, the role of the detergent head group is also important. It must successfully mimic the lipid head group by providing sufficient polarity to be stable in its interaction with the aqueous phase without increasing the micelle size. The headgroup itself should neither be excessively large — which would impinge on the hydrophilic regions of the protein that are crucial for forming protein–protein contacts within the crystal lattice — nor too small, since such detergents tend to be stronger denaturants. The optimal detergent will strike a fine balance between these conflicting properties, and only methodical screening can discover, for any given membrane protein, which detergent this is. As a rule, ionic detergents are good at breaking protein–protein interactions and so are less suitable for crystallization trials. Milder, nonionic detergents are able to disrupt lipid–lipid interactions and are better suited for this purpose. It has been found in our laboratory that in general, the zwitterionic (such as LDAO) and nonionic sugar-based detergents have been the most useful (such as octyl-β-D-glucopyranoside or dodecyl-β-D-maltoside). Usually, crystallization trials begin with detergent concentrations just above the CMC. Table 6.1[7–18] lists the CMCs of some detergents that have proven most useful for membrane protein crystallization so far. We should, however, always be on the lookout for novel detergents that can be tested in membrane protein crystallization screens. Moreover, we should not be afraid to try detergent mixtures in order to manipulate the micelle size.[19]

6.2.3 CHOICE OF PRECIPITANT

Protein crystallization requires the solubility of the protein to be reduced to the point of precipitation. This can be achieved by two extreme approaches. Some proteins such as the cyanobacterial photosystem I complex (PSI) precipitates in pure water.[20]

TABLE 6.1
Useful Properties of Some of the Main Detergents That Have So Far Proven Successful in Membrane Protein Crystallization Trials

	CMC (mM)	CMC % (w/v)	Mol. Wt. (anhyd.)	Reference
Nonionic Alkyl Glycosides				
Nonyl-β-ᴅ-glucopyranoside (NG)	6.5	0.2	306.4	7
Octyl-β-ᴅ-glucopyranoside (BOG)	14.0–30.0	0.4–0.9	292.4	8–11
Heptyl-β-ᴅ-thioglucopyranoside (HTG)	30.0	0.9	294.3	12
Dodecyl-β-ᴅ-maltoside (DDM or LM)	0.1–0.6	0.005–0.003	510.6	11
Decyl-β-ᴅ-maltoside (DM)	1.5–2.2	0.075–0.1	482.6	13
6-O-(N-heptylcarbamoyl) methyl-α-ᴅ-glucoside (HECAMEG)	19.5	0.65	335.4	14
Nonionic Bile Salt Derivative				
Sucrose monocholate (SMC)	4.7	0.35	732.8	Dojondo, personal comm.
Zwitterionic				
N,N-dimethyldodecylamine-ɴ-oxide(LDAO)	0.4–2.0	0.01–0.07	229.4	12,15–18

However, it is more common to reduce the protein's solubility by the addition of precipitating agents. These can be ionic such as ammonium sulfate or potassium phosphate or nonionic such as polyethylene glycol. There are no rules to determine which type of precipitant is better for which type of protein. As many alternatives as possible should be tried. To aid this process, there are now several sparse matrix screens commercially available, and their use greatly speeds up the initial screening process. Interested readers are directed to the book by Iwata,[21] where some of these screens and their underlying methodologies are described.

As a general starting point, when deciding which precipitant to use, it has been observed that salts such as ammonium sulfate, sodium citrate, or potassium phosphate are more suitable for use with ionic or zwitterionic detergents. Long-chain polymers (PEG 400–4000, monomethylether PEG, Jeffamine) tend to be more successful when coupled with nonionic detergents.

6.2.4 PHASE SEPARATION AND THE SMALL AMPHIPHILE CONCEPT

One of the most intractable problems encountered by early researchers trying to crystallize membrane proteins was phase separation. This occurs when the drop containing the solubilized protein and precipitant separates into two distinct phases. The precipitant effectively drives out the detergent from the aqueous phase. Then, depending on the density of the detergent, the detergent-rich phase either floats on the surface of the aqueous phase or sinks below it. Often, the protein that is now in the oily, detergent-rich phase denatures, and then crystallization is not possible. Only in very few cases have membrane proteins been crystallized from phase-separated droplets.[22]

The problem of phase-separation was solved independently by Michel[23] and Garavito.[24] They realized that the addition of a small, amphiphilic compound to the crystallization drop might induce the formation of mixed micelles with the detergent. In effect, the addition of these compounds shifts the phase diagram of the detergent so that the phase separation point moves above the precipitation point thereby making crystal formation possible. Amphiphiles also change the size and hardness of the detergent micelles. Often, the mixed micelles are smaller than pure detergent micelles and are more easily accommodated within the crystal lattice and therefore promote crystal formation. This means that the choice of the small amphiphile can often be important not only in stopping phase separation but also in controlling the type of crystals that form. For example, the use of benzamidine hydrochloride was the critical factor (Figure 6.2) that allowed the production of well-ordered crystals of the LH2 complex from *Rhodopseudomonas acidophila*.[25]

Over time, many hundreds of different small molecules have been screened for their suitability to be used as additives during membrane protein crystallization trials. Heptane-1,2,3-triol and benzamidine have proven to be the most successful

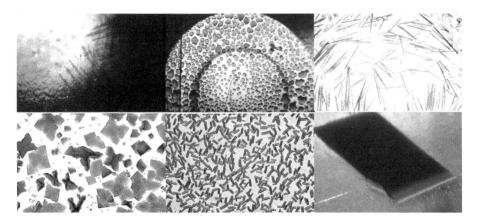

FIGURE 6.2 Some examples of crystallization trials with the light-harvesting complex II (LH2) from *Rhodopseudomonas acidophila* 10050. These photographs illustrate how methodical optimization of conditions can lead to progressively better crystals.
Top left: LH2 crystals grown with (detergent/precipitant/amphiphile) LDAO–phosphate–heptantriol. The heptantriol crystallized into plate-like structures. **Top center:** LH2 protein grown with LDAO/phosphate and no added amphiphile. The importance of an amphiphile is apparent as phase separation has occurred. **Top right:** Needle crystals grown from LDAO–ammonium sulfate–heptantriol. **Bottom left:** LH2 crystals grown with LDAO–phosphate–pipieridine-2-carboxylic acid. **Bottom center:** LH2 crystals grown with LDAO–phosphate–pipieridine-2-carboxylic acid and heptantriol. **Bottom right:** Beautiful, large LH2 crystals could be grown with BOG–phosphate–benzamidine.

compounds in terms of structures so far published. Therefore, prior to screening to determine the necessary conditions for obtaining membrane protein crystals, it is advisable to consider these two compounds at the outset.

6.2.5 CHOICE OF TEMPERATURE AND pH

Temperature and pH are also important factors to consider when setting up crystallization trials. These variables affect the stability of the membrane protein and the actual crystallization process. Often, crystallization can take days or weeks; therefore, long-term stability of the membrane protein under the crystallization conditions is an important consideration. The speed at which the membrane protein is taken through the precipitation point will greatly affect both the size and the quality of the crystals. Temperature is an important parameter to control the rate of this process. However, again, there are no rules, and we have found that it is best to set up replicate crystallization trials at a range of temperatures from 4 to 20°C. It is also more complicated than this, since temperature will also affect other parameters such as the detergent's CMC. Our initial screening of crystallization conditions usually involves using as wide a range of pH values as possible, for example, from pH 4.5 to 9.5. While using the LH2 complex from *Rhodospseudomonas acidophila,* pH was an important factor because the complex was only stable above pH 8.5 for the period of 7 to 21 days during which crystals developed. It is also worth noting that some buffers such as Tris have a marked temperature-dependent pK, and therefore the actual pH in the crystallization trial can vary significantly when the trials are conducted at different temperatures. This means that it is important to keep accurate and extensive records of the precise crystallization conditions of all the trials.

6.3 CRYSTALLIZATION STRATEGIES

If an active membrane protein of high purity has been obtained, how would one go about crystallizing it? First, if possible, it is desirable to attempt to screen as many homologues isolated from different species as possible. To this end, a list of organisms that have had their genomes sequenced, in combination with an easy-to-use search facility to help and identify suitable homologous proteins, is available at http://www.ncbi.nlm.nih.gov/sutils/genom_table.cgi. In our experience, it has been found that some proteins just never crystallize, paradoxically for reasons that might only be elucidated by determining their structure. However, usually with a minimum of four or five homologues one will be successful. It often requires much effort to optimize the initial conditions to produce crystals of sufficient size and order to yield a high-resolution structure. Routinely, the trial begins with the membrane protein at a concentration of between 2 and 10 mg/ml. Sparse matrix screens are then employed to determine the precipitant, pH, type of detergent, and type of amphiphile that give crystals. In our experience, we often find that the concentration of precipitants in these screens is too high, and so we routinely dilute these by 30% prior to screening.[26]

6.3.1 Vapor Diffusion

The most frequently used crystallization method is vapor diffusion. This technique has the advantage that it requires only small amounts of material, is ideal for screening a large number of conditions, and can be automated. The most common form of vapor diffusion experiment used to crystallize membrane proteins is the "sitting drop," where the protein–precipitant solution (the drop) is placed in an elevated well that is above a reservoir solution in a sealed chamber (Figure 6.3). Supersaturation is induced by vapor phase equilibration of water from the drop to the reservoir. The alternative method of "hanging drop," where the drop is placed on an inverted coverslip suspended above the reservoir and sealed in the chamber, is less suitable for membrane proteins because of decreased surface tension caused by the presence of the detergent. Vapor diffusion methods are also very convenient for monitoring the crystallization process, since the standard crystallization vessels can easily be viewed directly under a microscope.

6.3.2 Microdialysis

Slow concentration of the precipitant can also be induced by microdialysis. This method has been successfully used to crystallize porin but has been less popular

FIGURE 6.3 Part of a 24-well Cryschem plate used for sitting drop vapor diffusion crystal trials. The well in the center of the reservoir can accommodate drop sizes of up to 50 μl, enabling growth of large protein crystals of up to a few millimeters in length. A similar 96-well Greiner plate would be used for screening small volume droplets of down to a few nanoliters.

than vapor diffusion. There are probably two reasons for this. First, the microdialysis setup must be dismantled to look for the presence of crystals. Second, it is difficult to automate this method.

Micodialysis has, however, been used several times as a method of introducing crystals of membrane proteins into solutions of cryoprotectants prior to freezing the crystals and placing them into the x-ray beam at a synchrotron. Often, just dipping a crystal into a solution of cryoprotectant can cause a significant osmotic shock that will induce disorder. The microdialysis approach allows the cryoprotectant to diffuse slowly into the crystal and thus avoid any sudden shocks that might reduce crystal order.[27]

6.3.3 CO-CRYSTALLIZATION

As can be seen in Figure 6.1, a type II crystal is formed primarily by hydrophilic contacts between the polar regions of adjacent protein molecules. Hydrophobic interactions also occur between the alkyl chains across adjacent detergent micelles; however, although these mutual interactions may lend a degree of stabilization, they are not sufficient to provide the strength necessary to hold the crystal together. The hydrophilic portion of the protein that extends out from the detergent micelle forms these obligatory strong contacts. Therefore, any protein that lacks protruding hydrophilic domains, for example, certain transporters with very small loops between the transmembrane α-helices, may be difficult to crystallize, since these strong contacts are unlikely to form. Any strategy that promotes or facilitates polar contacts could therefore promote the formation of a well-ordered crystal lattice. A procedure called co-crystallization is able to do this. By attaching an antibody fragment raised specifically to the polar region of the membrane protein in its native conformation, it is possible to dramatically increase the hydrophilic surface and thus facilitate crystallization.[28,29]

Polyclonal antibodies are unsuitable for co-crystallization due to their inherent variability. Native monoclonal antibodies are also unsuitable for two reasons: the linker region between the Fc and Fab domains introduces unwelcome flexibility and the undesirability of divalent binding through each Fab fragment. While taking great care to generate homogeneous preparations, a simple and inexpensive method to produce Fab fragments ($\sim$ 56 kDa) is to subject the native antibody to proteolytic cleavage. Co-crystallization using proteolytically cleaved Fab fragments has now also been successfully applied to the structural determination of the KcsA K$^+$-channel. In this case, the KcsA K$^+$-channel Fab fragment complex produced crystals that diffracted x-rays to a resolution of 2.0 Å.[30]

Although much more expensive, recombinant antibody fragments obtained by cloning the encoding genes from hybridoma cell lines are, from the outset, homogeneous and have the added advantage that an immunoaffinity tag can be attached. Either Fab ($\sim$ 56 kDa) or Fv ($\sim$ 28 kDa) fragments can then be produced that yield a highly purified co-complex that is ready for crystallization after a single-affinity chromatography step. In order to select suitable antibody fragments for use in co-crystallization trials, it has proven important to choose only those fragments that give positive enzyme-linked immunosorbent assay (ELISA) reactions against native, discontinuous epitopes and show negative results in Western blots to denatured protein. This strategy ensures that the Fab or Fv fragments only recognize the hydrophilic domains of the membrane protein of choice.

Co-crystallization was used successfully for the first time to determine the structure to 2.8 Å of cytochrome *c* oxidase (COX) from *Paracoccus denitrificans*.[31,32] The co-crystals contained all four subunits of COX complexed with an Fv fragment that was involved in all crystal contacts. Subsequent well-ordered crystals that diffracted to better than 2.5 Å could be obtained with a COX preparation containing only the two catalytically essential subunits.[33] Different crystal packing resulted in direct protein–protein interactions between the cytoplasmic and periplasmic surfaces of adjacent COX molecules, with all further contacts being mediated by the Fv fragments.

6.3.4 Lipidic Cubic Phase

The Rosenbusch group in Basel[34] was among the first to explore the role of different detergents, and phase science in general, on membrane protein crystallization. This work led to a realization of the importance of phase diagrams when trying to understand membrane biology and the connection between the biophysical chemistry of lipids comprising the membrane and the crystallization of proteins held within it. It was apparent to the researchers in this group that the correct conditions for some detergent-solubilized proteins to crystallize may never be successfully elucidated. Therefore, they developed an alternative concept whereby the protein is provided with an appropriate lipid environment that maintains its native conformation and permits free diffusion in three dimensions. Crystallization can then be induced within this lipid matrix. This method has since become known as lipidic cubic phase crystallization.[5,35]

A common misconception is that lipids can only form a bilayer in aqueous solution. In addition to bilayers, it has been known for many years that lipids are able to form lamellar, hexagonal, and cubic three-dimensional structures. The phase diagrams for the formation of these structures are dependent on the particular architecture of the specific lipid involved. In a lipidic cubic phase system, water-soluble solutes are able to freely diffuse throughout the aqueous compartments and channels. In contrast, lipophilic or amphiphilic compounds such as membrane proteins are able to partition into the lipid phase and diffuse laterally along the bilayer. Upon formation of a nucleation event (the precise mechanism by which this occurs is uncertain), a constant supply of protein is channeled onto the growing surface of the crystal through this free diffusion along the bilayer.

The verification of this method came with the crystallization of bacteriorhodopsin (bR) in bicontinuous lipidic cubic phases,[5] followed by the elucidation of its structure at high resolution from crystals grown in a monoolein-based cubic phase.[36] Monoolein is not, however, a common membrane lipid, and therefore Caffrey and co-workers have evaluated a number of the more common lipids for compatibility with the cubic phase bilayer of hydrated monoolein with the aim of making the methodology more applicable to a wider range of membrane proteins.[37]

In principle, the procedure for setting up a lipidic cubic phase crystallization attempt is relatively straightforward. The membrane protein solution (this method is equally applicable to soluble proteins), small organic molecules, and inorganic salts are mixed vigorously at room temperature in a ratio of 2:3 with lipid (usually monoolein), and consequently, the cubic phase forms spontaneously. The cubic phase is considered formed when the solution is viscous, translucent, and nonbirefringent. Crystallization is induced by the addition of a precipitant, usually a solid salt, and depending on the precise conditions, crystals form within days or weeks. Two excellent mini-reviews of the methods for setting up and exploiting a lipidic cubic phase crystallization experiment and of the equipment and materials required are available.[38,39]

The applicability of this method across a range of membrane proteins has been demonstrated.[40] Therefore, it is hoped that lipidic cubic phase crystallization can be extended to crystallize proteins that have proven difficult to crystallize with other methods. The notable success of the lipidic cubic phase method in obtaining crystals of the seven transmembrane α-helical bacteriorhodopsins (and homologous proteins: halorhodopsin, sensory rhodopsin II, and sensory rhodopsin II with its transducer domain) has led to the intriguing possibility that the extremely important class of nonbacterial hepta-helical proteins, the G protein-coupled receptors (GPCRs), may also be amenable to crystallization by this method.

6.3.5 TYPE III CRYSTALLIZATION METHOD

At the time of writing, the only example of successful structural determination of a membrane protein crystallized by this method was the major light-harvesting protein from spinach chloroplasts.[6] The protein was incubated in the presence of one of the major chloroplast lipids, digalactosyl diacyl glycerol plus N,N-bis- (3-D-glucona-mido propyl) deoxycholamide (DGDG). The crystallization conditions caused the light-harvesting complex to insert into very small liposomes. Within these membrane vesicles, the light-harvesting complexes formed very regular crystalline arrays. Remarkably, these vesicles crystallized together, forming a structure that is reminiscent of crystals formed when icosohedral viruses crystallize. Since the complexes were regularly ordered both within a single vesicle and between vesicles, the crystals diffracted x-rays to high resolution. It remains to be seen whether this approach will be generally applicable to successful crystallization of membrane proteins.

6.4 FUTURE PROSPECTS

As more structures of membrane proteins appear and more research groups lose their reluctance to tackle this challenge, the major bottleneck would be moving back to tackle the challenge of obtaining sufficient, active membrane proteins for structural studies. Some success is now being achieved with the overexpression of prokaryotic membrane proteins, but it is altogether more challenging to overexpress eukaryotic membrane proteins. Hopefully, the major effort that is now ongoing will produce a breakthrough. It is clear that even though producing crystals of membrane proteins that are suitable for x-ray structural analysis is difficult, the methods currently available for crystallizing membrane proteins are steadily improving. If researchers are willing to put in the extra effort that is required to crystallize membrane proteins rather than water-soluble proteins, then this becomes highly possible. Moreover, as robot technology incorporating the capability to produce reproducible nanoliter droplets is being increasingly applied to the initial screening of crystallization conditions and more of the possible crystallization space is accessed, the crystallization of membrane proteins will become a more routine procedure.

ACKNOWLEDGMENTS

The authors wish to acknowledge the support of BBSRC and E-MeP.

REFERENCES

1. Michel, H., Ed. *Crystallisation of Membrane Proteins*. CRC Press, Boca Raton, Florida, 1991.
2. Deisenhofer, J., Epp, O., Miki, K., Huber, R., and Michel, H. Structure of the protein subunits in the photosynthetic reaction centre of *Rhodopseudomonas viridis* at 3 Å resolution. *Nature*, 318, 618–624, 1985.
3. Henderson, R., Baldwin, J.M., Ceska, T.A., Zemlin, F, Beckman, E., and Downing, K.H. Model for the structure of bacteriorhodopsin based on high-resolution electron cryomicroscopy. *J. Mol. Biol.*, 213, 899–929, 1990.
4. Doyle, D.A., Cabral, J.M., Pfuetzner, R.A., Kuo, A., Gulbis, J.M., Cohen, S.L., Chait, B.T., and MacKinnon, R. The structure of the potassium channel: molecular basis of K+ conduction and selectivity. *Structure*, 280, 69–77., 1998.
5. Landau, E.M. and Rosenbusch, J.P. Lipidic cubic phases: a novel concept for the crystallisation of membrane proteins. *Proc. Natl. Acad. Sci. USA*, 93, 14532–14535, 1996.
6. Liu, Z.F., Yan, H.C., Wang, K.B., Kuang, T.Y., Zhang, J.P., Gui, L.L., An, X.M., and Chang, W.R. Crystal structure of spinach major light-harvesting complex at 2.72 Å resolution. *Nature*, 428, 287–292, 2004.
7. De Grip, W.J. and Bovee-Geurts, P.H.M. Synthesis and properties of alkylglucosides with mild detergent action: improved synthesis and purification of β-1-octyl-, β-nonyl- and β-decyl-glucose. Synthesis of β-1-undecylglucose and β-1-dodecylmaltose. *Chem. Phys. Lipids*, 23, 321–335, 1979.
8. Chattopadhyay, A. and London, E. Fluorimetric determination of critical micelle concentration avoiding interference from detergent charge. *Anal. Biochem.*, 139, 408–412, 1984.
9. Gould, R., Ginsberg, B., and Spector A. Effects of octyl-β-glucoside on insulin binding to solubilized membrane receptors. *Biochemistry*, 20, 6776—6781, 1981.
10. Lorber, B., Bishop, J., and DeLucas, L. Purification of octyl-β-D-glucopyranoside and re-estimation of its micellar size. *Biochim Biophys. Acta.*, 1023, 254–365, 1990.
11. Casey, J. and Reithmeier, R. Detergent interaction with band 3, a model polytopic membrane protein. *Biochemistry*,132, 1172–1179, 1993.
12. Shinamoto, T., Saito, S. and Tsuchiya, T. Value of hetyl-β-D-thioglucoside, a new nonionic detergent, in studies on membrane proteins. *J. Biochem.*, 97, 1807–1810, 1985.

13. De Vendittis, G., Palumbo, G., Parlato, V. and Bocchini, V. A fluorimetric method for the estimation of the critical micelle concentration of surfactants. *Anal. Biochem.*, 115, 278–286, 1981.

14. Plusquellec, D., Chevalier, G., Taibart, R. and Wroblewski, H. Synthesis and characterisation of 6-0-(N-heptylcarbamoyl)-methyl-α-D-glucopyranoside, a new surfactant for membrane studies. *Anal. Biochem.*, 179, 145–153, 1989.

15. Yu, F. and McCarty, R.E. Detergent activation of the ATPase activity of chloroplast coupling factor 1. *Arch. Biochem. Biophys.*, 238, 61–68, 1985.

16. Hermann, K.W. Non-ionic cationic micellar properties of dimethyldodecyl-amine oxide. *J. Phys. Chem.*, 66, 295–300, 1962.

17. Sardet, C., Tardieu, A. and Luzzati, V. Size and shape of bovine rhodopsin: A small-angle X-ray scattering study of a rhodopsin-detergent complex. *J. Mol. Biol.*, 105, 383–407, 1976.

18. Kaimoto, H., Shoho, K., Sasaki, S. and Maeda, H. Aggregation numbers of dodecyldimethylamine oxide micelles in salt solutions. *J. Phys. Chem.*, 98, 10243–10248, 1994.

19. Koronakis, V., Scharff, A., Koronakis, E., Luisi, B., and Hughes, C. Crystal structure of the bacterial membrane protein TolC central to multidrug efflux and protein export. *Nature*, 405, 914–919, 2000.

20. Jordan, P., Fromme, P., Klukas, O., Witt, H.T., Senger, W., and Krauss, N. Three dimensional structure of cyanobacterial photosystem I at 2.5 angstroms resolution. *Nature*, 411, 909–917, 2001.

21. Iwata, S., Ed. *Methods and Results in Crystallisation of Membrane Proteins.* International University Line, La Jolla, California, 2003.

22. Garavito, R.M. Crystallising membrane proteins: experiments on different systems, in Michel, H., Ed. *Crystallisation of Membrane Proteins.* CRC Press, Boca Raton, Florida, 1991.

23. Michel, H. Crystallisation of membrane proteins. Trends Biol. Sci., 8, 56–59, 1983.

24. Garavito, R.M., Markovich-Housley, Z. and Jenkins, J. The growth and characterisation of membrane protein crystals. J. Crystal Growth, 76, 701–706, 1986.

25. McDermott, G., Prince, S.M., Freer, A.A., Hawthornthwaite, A.M., Papiz, M.Z., Cogdell, R.J., and Isaacs, N.W. Crystal structure of an integral membrane light-harvesting complex from photosynthetic bacteria. Nature, 374, 517–521, 1995.

26. Howard, T.D., McAuley-Hecht, K.E., and Cogdell, R.J. Crystallisation of membrane proteins, in Baldwin, S.A., Ed. Membrane Transport: A Practical Approach. Practical Approach Series, Oxford University Press, 2000, pp. 269–307.

27. Prince, S.M., McDermott, G., Freer, A.A., Papiz, M.Z., Lawless, A.M., Cogdell, R.J., and Isaacs, N.W. Derivative manipulation in the structure solution of the integral membrane LH2 complex. *Acta. Crys.*, D55, 1428–1431, 1999.

28. Hunte, C. and Michel, H. Crystallisation of membrane proteins mediated by antibody fragments. Curr Opin Struc Biol 2002; 12:503-508.

29. Hunte, C. and Kannt, A. Antibody fragment mediated crystallisation of membrane proteins, in Hunte, C., von Jagow, G., and Schagger, H., Eds. Membrane Protein Purification and Crystallisation: a Practical Guide. Academic Press, San Diego, 2003, pp. 205–218.

30. Zhou, Y., Morais-Cabral, J.H., Kaufman, A., and MacKinnon, R. Chemistry of ion coordination and hydration revealed by a K+ channel-Fab complex at 2.0Å resolution. *Nature*, 414:43-48, 2001.

31. Ostermeier, C., Iwata, S., Ludwig, B., and Michel, H. Fv fragment mediated crystallisation of the membrane protein bacterial cytochrome oxidase. *Nat. Struct. Biol.*, 2, 842–846, 1995.

32. Iwata, S., Ostermeier, C., Ludwig, B., and Michel, H. Structure at 2.8 Å resolution of cytochrome c oxidase from *Parracoccus denitrificans*. *Nature*, 376, 660–669, 1995.

33. Ostermeier, C., Harrenga, A., Ermler, U., and Michel, H. Structure at 2.7 Å resolution of the *Parracoccus denitrificans* cytochrome c oxidase complexed with an antibody Fv fragment. *Proc. Natl. Acad. Sci.U.S.A.*, 94, 10547–10553, 1997.

34. Rosenbusch, J.P. The critical role of detergents in the crystallisation of membrane proteins. *J. Struc. Biol.*, 21, 429–477, 1990.

35. Caffrey, M. Membrane protein crystallization. *J. Struc. Biol.*, 142, 108–132, 2003.

36. Pebay-Peroula, E., Rummel, G., Rosenbusch, J.P., and Landau, E.M. X-ray structure of bacteriorhodopsin at 2.5 angstroms from microcrystals grown in lipidic cubic phases. *Science*, 277, 1676–1681, 1997.

37. Cherezov, V., Clogston, J., Misquitta, Y., Abdel-Gawad, W., and Caffey, M. Membrane protein cyrstallization in meso: lipid type-tailoring of the cubic phase. *Biophys. J.*, 83, 3393–3407, 2002.

38. Landau, E.M. In cubo crystallization of membrane proteins, in Hunte, C., von Jagow, H., and Schagger, H., Eds. *Membrane Protein Purification and Crystallisation: a Practical Guide*. Academic Press, San Diego, 2003, pp. 285–302.

39. Nollert, P. Lipidic cubic phases as matrices for membrane protein crystallisation. *Methods*, 34, 348–353, 2004.

40. Chiu, M.L., Nollert, P., Loewen, M.C., Belrhali, H., Pebay-Peyroula, E., Rosenbusch, J.P., and Landau, E.M. Crystallisation in cubo: general applicability to membrane proteins. *Acta. Crys.*, D56, 781–784, 2000.

7 Signaling through Membrane Proteins

J. Robert Bostwick and Deborah S. Hartman

CONTENTS

7.1   INTRODUCTION

The human genome contains approximately 25,000 genes,[1] including an estimated 1,543 signaling receptors.[2] Stimulation of these receptors can ultimately result in the activation or inhibition of transcription factors, of which there appear to be over 1,800 in humans,[3] which direct the changes in cellular regulatory processes. Cellular signaling networks operate over diverse spatio-temporal scales, with various degrees of interconnectivity and a wide range of structural and dynamic dimensions that are just beginning to be mapped.[4] Extensive efforts are being made to collect, collate, and extract knowledge from the abundant data being generated in this area, for example, the AfCS-Nature Signaling Gateway http://www.signaling-gateway.org, which includes both cell signaling databases and primary research results from around the world.

Receptor signaling occurs via induction of conformational changes in membrane proteins, which can also be associated with covalent modifications of the receptor. Activation of signaling pathways can be mediated via ligand interaction or through alteration of membrane potential in excitable cells, for example, by activating ion channels. Ligands are derived from a wide variety of biological molecules, including proteins (such as thrombin), ions (such as calcium), small organic compounds (such as the neurotransmitters dopamine and serotonin), lipids (such as arachidonic acid), as well as light and odors. Ligands are characterized as agonists (activating), partial agonists, antagonists (inhibitory), inverse agonists (reducing basal activity levels), and allosteric modulators.

Models of receptor activation and modulation continue to be refined in this rapidly expanding field. The molecular mechanisms by which transmembrane signaling occurs are inferred from studies of ligand-induced effects on cells, structural characterization of cell surface receptors, site-directed mutagenesis, and most recently bioinformatics initiatives that provide insight into the relatedness of receptor molecules within and across species. Studies in simpler organisms such as yeast and bacteria, together with heterologous expression technologies for recombinant proteins in mammalian systems, have formed the basis for our understanding of signaling today.

The signaling capabilities of a cell are strongly influenced by the tissue environment, which can represent a high degree of cell differentiation and specialization. Expression levels of signaling proteins, availability of key accessory proteins, and resting membrane potential in heterologous systems are just some of the factors that can dramatically affect the nature and intensity of cell signaling. A major challenge to studying cell surface receptor structure and function is in reproducing the cellular state that exists in the natural environment. This chapter focuses on the molecular mechanisms of membrane protein signaling, indicating only selected examples to illustrate key findings and highlighting concepts that can provide insight into the therapeutic potential of signaling pathway modulation.

7.2   MOLECULAR MECHANISMS OF SIGNALING

There are five basic components involved in cellular signaling: a stimulus, a receptor, a transducer, an effector, and a responder (Figure 7.1). The stimulus is the element to which the cell responds, and it derives from a change in the external environment of the cell. The receptor is a membrane protein that exists in equilibrium among multiple conformational states, found primarily in resting states in the absence of a stimulus. The stimulus stabilizes "active" receptor conformations that are favorable for interactions with a transducing element. The transducer is a molecular switch that converts ligand interactions into an intracellular signal, which may involve production of second messenger molecules. This activity can be intrinsic to the receptor itself, directly coupled to the receptor or contained in a separate molecule recruited to form a complex with the receptor via adapter proteins. The effector, typically an enzyme, initiates a cascade of subsequent reactions that may significantly amplify the signal, possibly altering phosphorylation states of proteins or changing intracellular levels of ions. The responder is the final element in the signal pathway, and it produces a physiologically relevant cellular response such as a change in excitability, metabolism, cell morphology or gene expression.

There are five major types of membrane signaling proteins (Figure 7.2), and the molecular basis of signaling in all cases involves a change in the conformation of the receptor protein.

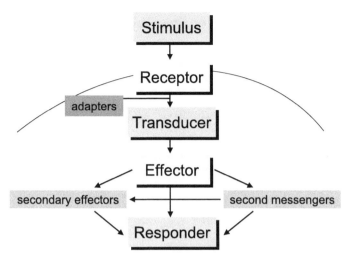

FIGURE 7.1 Schematic representation of receptor signaling (see text).

1. G protein-coupled receptors (GPCRs), which regulate a wide variety of cell functions from metabolism to neurotransmission, signal through allosteric activation of heterotrimeric guanine nucleotide binding–proteins (G proteins).[5–7] Activated G proteins generate intracellular second messengers with high levels of signal amplification and can also interact directly with downstream elements such as potassium channels.

2. Ion channel activation allows ion flux across the cell membrane. Changes in intracellular ion concentrations activate effector molecules through changes in membrane potential or through binding of ions to target proteins. Ion channels mediate rapid signaling such as action potential propagation in neurons.

3. Membrane signaling proteins with a single transmembrane domain include the receptor tyrosine kinases (RTKs), which play a major role in cell differentiation and proliferation. Activation occurs via ligand-induced homodimerization, which promotes autophosphorylation, enabling interaction with accessory adapter proteins that activate effector systems.

4. Cytokine receptors and integrins lack intrinsic enzyme activity but also signal through ligand-induced aggregation of receptor subunits to recruit and activate effectors. These receptors mediate cell–cell interactions responsible for immune and inflammatory responses.

5. Notch-like proteins signal through proteolytic cleavage and release of an intracellular domain that translocates to the cell nucleus, where they are best known to influence the fate of the cell during embryonic development. This subfamily also includes the p75 neurotrophin receptor, which is cleaved by gamma-secretase to release an intracellular domain that migrates to the nucleus to transduce signaling.[8]

Membrane transporters also play an important, although indirect, role in cell signaling through their effects on membrane potential and regulation of extracellular

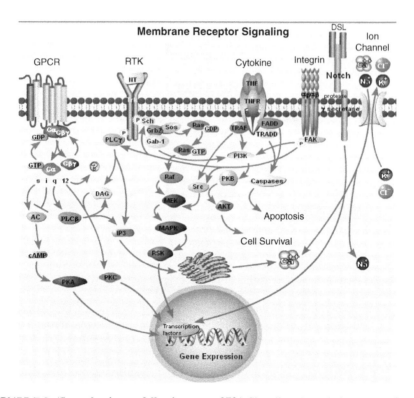

FIGURE 7.2 (See color insert following page 272.) Signaling through five types of membrane receptors. Activation of a GPCR induces exchange of GDP for GTP on the Gα subunit and dissociation of the G protein complex. GTP-loaded Gα subunits of different families (s, i, q, and 12) act as transducers to activate intracellular effector enzymes leading to the production of second messengers and propagation of the intracellular signal to alter gene transcription or cellular physiology. Ligand binding to single membrane-spanning receptors (receptor tyrosine kinases, cytokine and integrin receptors) induces oligomerization of subunits leading to the creation of binding sites for adapter proteins, which bring effector molecules into association with the receptor for subsequent activation and signal propagation, ultimately affecting cell fate. Activation of notch receptors induces proteolytic cleavage of an intracellular domain that translocates to the nucleus for direct activation of transcription factors. Ion channels, activated by ligands or voltage changes, allow for passage of ions across the membrane to alter intracellular levels of ions and membrane excitability.

neurotransmitter levels.[9] Conformational changes in biogenic amine transporters, which are the primary targets of psychostimulants,[10] have been investigated by the substituted cysteine accessibility method (SCAM), the use of engineered Zn^{2+}-binding sites, and, most recently, the use of fluorescence spectroscopy studies.[11] These studies have consistently suggested that transporters are oligomeric structures, and that large-scale conformational changes are involved in transport function. In contrast, recent fluorescence resonance energy transfer (FRET) analysis of the glutamate transporter supports a trimeric structure, but a surprising lack of large-scale conformational changes associated with membrane transport.[12] Instead, relatively small protein movement occurs around the glutamate-binding sites, perhaps contributing to the fast kinetics of glutamate transport. Even more unexpected is the finding that despite the trimeric nature of the glutamate transporter complex, each of the subunits appears to function independently in glutamate transport.

Many transporters, including the high-affinity choline uptake transporter,[13] are regulated by phosphorylation and share clathrin-mediated endocytic pathways with GPCRs. The cystic fibrosis transmembrane regulator (CFTR) chloride channel, which contains two nucleotide-binding sites, is functionally distinct from other transporters. When the protein is phosphorylated by protein kinase A, channel gating is initiated by cytoplasmic adenosine triphosphate (ATP) binding. Intermolecular interactions with both N- and C-terminal ends of the CFTR protein affect both channel function and cellular localization.[14] Well-known disease-associated mutations interfere with domain folding and association, and occur both co- and post-translationally.

Ion channels and ion transporters function as effectors by altering intracellular concentrations of ions and membrane potential. Most effectors involved in signal transduction, however, are enzymes that catalyze the phosphorylation or hydrolysis of proteins or lipids or the formation of second messengers. Membrane proteins are now known to be quite pleitropic in their use of downstream effector systems, and selected examples are detailed in the following sections.

7.2.1 G PROTEIN-COUPLED RECEPTORS

All GPCRs are comprised of seven transmembrane (TM) helical domains and are classified as Group 1, Group 2, and Group 3 receptors according to their sequence homology.[15] The TM domains are in a constant state of motion within the membrane and change position relative to one another, producing different conformational states of the receptor.[16] The N-terminal domain is located on the extracellular surface, and for many GPCRs, it forms a part of the ligand-binding domain. Agonists for Group 3 receptors bind exclusively to this region in a cleft between two lobes of a Venus flytrap module that closes and interacts as a tethered ligand with the extracellular loops. Protein ligands are also known to interact with the extracellular loops between TM2 and TM3 and between TM5 and TM6, while naturally occurring small molecules such as neurotransmitters bind directly within the TM regions. When a GPCR is activated, conformational states are stabilized such that the TM domains (typically involving TM3 and TM6) are repositioned, inducing movement of the cytosolic (C) loops to which they are attached.[17] Movement of these C loops, in turn, induces a conformational change in the G protein complex when bound to the activated receptor.[18]

The G protein complex is composed of the three distinct subunits Gα, Gβ and Gγ. Receptor-induced conformational change promotes exchange of guanosine diphosphate (GDP) for guanosine triphosphate (GTP), which results in dissociation of the complex and separation from the receptor. Both Gα and the dissociated Gβγ complexes act as transducers to activate effector molecules in a cell-specific manner.

Rapid, acute termination is mediated via the intrinsic GTPase activity of the Gα subunit, which hydrolyzes GTP to GDP[19] and allows the reformation of the Gαβγ heterotrimeric complex for reassociation with the receptor.[20] A more sustained inactivation is achieved via receptor downregulation when the receptor is phosphorylated by G protein-coupled receptor kinases (GRKs), allowing the protein arrestin to bind and targeting the receptor for internalization.[21]

GPCRs can associate with and signal through multiple types of G proteins.[22] There are 17 different genes encoding over 20 different Gα subunits currently known, which can be grouped into four isoform families; $G\alpha_s$, $G\alpha_i$, $G\alpha_q$, and $G\alpha_{12}$.[23] $G\alpha_s$ and $G\alpha_i$ activate or inhibit, respectively, adenylyl cyclase (AC), which catalyzes the formation of cyclic adenosine monophosphate (cAMP). cAMP functions as a second messenger to activate protein kinase-A (PKA), which propagates the signal through subsequent phosphorylation of target proteins to alter cellular metabolism or gene expression. cAMP can also activate effector ion channels known as cyclic nucleotide–gated (CNG) ion channels to allow for ion flux across the cell membrane (Figure 7.3).

$G\alpha_q$ directly activates phospholipase Cβ (PLCβ), an effector enzyme that generates two second messengers, inositol 1,4,5 trisphosphate (IP3) and diacyl glycerol (DAG), via hydrolysis of the polar head group from the lipid phosphatidyl inositol 4,5 bisphosphate. IP3 stimulates the release of calcium from intracellular stores, whereas DAG promotes the activation of various isoforms of protein kinase C (PKC). On the other hand, the target effectors of $G\alpha_{12}$ are not well understood. Both $G\alpha_{12}$ and $G\alpha_{13}$ activate Rho, another type of GTPase switch protein, but apparently through different mechanisms. Although both bind to guanine nucleotide exchange factors (GEFs) for Rho, only $G\alpha_{13}$ stimulates GEF activity.[24] The Gβγ subunits can facilitate activation of the effectors AC, PLC, and PI3K, as well as potassium channels.[23]

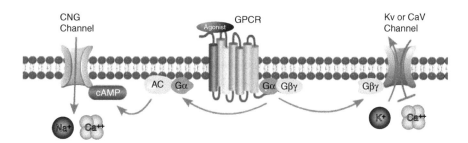

FIGURE 7.3 (**See color insert following page 272.**) GPCR regulation of ion channels. Ion channels can function as downstream effectors of GPCR activation through signal transduction mediated by both Gα and Gβγ subunits. Cyclic nucleotide–gated channels (CNG) can be activated by cAMP produced through $G\alpha_{s/olf}$ activation of adenylate cyclase (AC) thereby increasing intracellular concentrations of sodium and calcium. Gβγ subunits of $G\alpha_{i/o}$-coupled GPCRs can activate inwardly rectifying voltage-gated potassium channels (Kv) to alter membrane potential or inhibit voltage-gated calcium channels (CaV) to prevent entry of calcium into the cell. From Reinscheid, R.K., Kim, J., Zeng, J., and Civelli, O. *Eur. J. Pharmacol.* 478(1), 27–34, 2003; Yamada, M., Inanobe, A., and Kurachi, Y. **G** *Pharmacol. Rev.* 50, 723–757, 1998; Dolphin, A.C. *Pharmacol. Rev.* 55(4), 607–627, 2003. With permission.

7.2.2 Ion Channels

Ion channels detect and induce changes in the cellular environment, including membrane potential, by controlling the flux of monovalent and divalent ions across cell membranes. Activation, inactivation, and desensitization of ion channels are modulated by phosphorylation, which can occur in response to GPCR activation and other signaling events in the cell. There are three major classes of ion channels:

1. *Ligand-gated* ion channels (LGICs) convert extracellular chemical signals into specific intracellular ionic or transmembrane electrical signals. Examples include the nicotinic acetylcholine receptors and ionotropic glutamate receptors, which are permeable to sodium and calcium, as well as receptors activated by the glycine and gamma-aminobutyric acid (GABA) receptors that mediate chloride flux. LGICs are a diverse set of multimeric protein complexes, where ligand binding often occurs at interfaces between subunits on the extracellular side of the membrane.
2. *Voltage-gated* channels are found in excitable cells. They transition between open, closed, and inactivated conformational states in response to changes in membrane potential. Voltage-gated channels flux potassium, sodium, chloride, and calcium, the last of which directly activates additional intracellular calcium-mediated signaling events. This is a key point of intersection between GPCR and ion channel signaling pathways. Sequential activation of voltage-gated ion channels enables propagation of membrane potential changes in excitable cells.
3. *Signal-gated* channels are unique in that they are activated by intracellular signals such as calcium and cyclic nucleotides. In addition, these channels constitute key secondary effector elements in the downstream signaling pathways for GPCRs, since $G\beta\gamma$ subunits released from activated G proteins can also activate these channels.

Ion channels exploit a variety of diverse structures to form ion permeable pores in the cell membrane. In each case, TM domains are assembled such that nonhelical regions form a pore through which ions pass, and ion specificity is conferred by intermolecular coordination between the ion and strategically located residues lining the pore. The pore-forming subunits of voltage-gated Na^+ and Ca^{++} channels, for example, are large monomeric proteins containing four tandem homologous domains, each with six transmembrane α-helices and a nonhelical P region. The fourth (S4) α-helix in each pore-forming domain is a highly positively charged region, containing lysine or arginine at every third or fourth residue, and serves as the voltage-sensing element.[25] When the cell membrane depolarizes, the cytoplasmic side becomes less negatively charged, causing outward movement of the S4 helices that allows the channel to open. The channel is inactivated by translocation of a cytoplasmic receptor subunit domain directly into the inner vestibule of the pore, blocking further movement of ions. When the membrane is repolarized, the channel conformation transitions back to the closed, resting state. Interestingly, both the cyclic nucleotide–gated (CNG) and voltage-gated K^+ channels are similar in overall structure to sodium channels, but these channels are composed of four subunits, each containing six transmembrane α-helices, an S4 voltage sensing element, and a nonhelical P region, which assemble to form the receptor protein complex in the membrane.[26]

7.2.3 SINGLE-MEMBRANE-SPANNING DOMAIN RECEPTORS

The single TM receptors constitute another large group of membrane proteins with an extracellular ligand-binding domain and an intracellular signaling domain. The intracellular domain may have intrinsic enzymatic activity activated upon ligand binding, or it may recruit other intracellular enzymes in a ligand-dependent manner. Similar to GPCRs and ion channels, signaling across the cell membrane is mediated via conformational change. Examples include growth factor receptors with serine–threonine kinase activity, receptor tyrosine kinases (RTKs), and the atrial naturitic factor receptor, which has guanylate cyclase activity.

In the case of RTKs, activation is accompanied by the formation of dimeric or heteromeric receptor complexes, which induce autophosphorylation of specific tyrosine residues. These phosphorylated sites create binding domains for other cytosolic proteins, which are, in turn, phosphorylated to activate the signaling pathway. Several signaling pathways can be activated through a single RTK, as illustrated by the TrkA neurotrophic factor receptor.[27] Dimerization of TrkA is induced by nerve growth factor, and subsequent autophosphorylation of tyrosine residues then occurs in two different regions of the intracellular domain of the receptor.

TrkA autophosphorylation at tyrosine Y785, which is located near the carboxy terminal end of the receptor, enables binding of phospholipase C gamma (PLCγ), and activation of the IP3/DAG pathway. Phosphorylation at tyrosine residue Y490, located near the cytosolic juxtamembrane region, instead enables receptor binding to the adapter protein Shc via an Src homology domain (SH2). Shc is phosphorylated by Trk, creating a binding site for a second adapter protein, Grb2. This leads to the recruitment of either Sos, a guanine nucleotide exchange factor (GEF), or Grb-1, a downstream adapter protein. Sos recruitment will activate Ras, a GTPase tranducing element similar to the G proteins described above, by promoting the exchange of GDP for GTP. This activates the effector enzyme Raf to initiate sequential activation of the kinases MEK and mitogen-activated protein kinase (MAPK) (also called extracellular signal-regulated kinase [ERK]). MAPK activates transcription factors either directly through translocation into the nucleus or indirectly through further sequential kinase activation of ribosomal S6 kinase (Rsk). If Grb-1 binds to Grb-2, it recruits phosphoinositide 3-kinase (PI3K) to the cell surface, where phosphorylation of inositol lipids creates binding sites for pleckstrin homology (PH) domain-containing enzymes such as phosphoinoside-dependent protein kinase (PDK) and Akt (also known as protein kinase B [PKB]), which play a key role in cell survival.

Integrins and cytokine receptors are similar to RTKs in that they consist of single-membrane-spanning subunits that mediate signaling through aggregation. However, these receptors do not possess intrinsic enzyme activity. Rather, it is the conformational change induced by receptor subunit clustering that recruits cytosolic adapter proteins and kinases to activate effector enzymes. Integrins, which connect the intracellular cytoskeleton with the extracellular matrix (ECM) of a cell, mediate signals induced by cell adhesion. Aggregation of integrin receptor subunits induces activation of focal adhesion kinase (FAK), which undergoes autophosphorylation and recruits the SH2 domain proteins PI3K, PLC-γ, Src family kinases, and Shc/GRB-2 for downstream signaling.[28,29]

Cytokine receptors such as those for interleukins and tumor necrosis factor (TNF) also mediate signal transduction through ligand-induced association of multimeric receptor complexes involving both receptor subunits and adapter proteins. The type of adaptor proteins associating with the receptor complex determines the cellular response. For instance, adaptor proteins called TNF receptor-associated factors (TRAFs) interact with TRAF interacting motifs (TIM) in the cytoplasmic tails of TNF receptors to activate kinase cascades (e.g., PI3K and MAPK), which leads to gene expression and cell survival. Alternatively, TNF receptors containing a death domain in their cytoplasmic tails activate caspase cascades leading to apoptosis through recruitment of TRADD (TNFR-associated death domain) and FADD (Fas-associated death domain) adaptor proteins.[30]

The Notch receptors mediate contact-dependent communication between adjacent cells controlling cell fate during development. Activation of Notch receptors occurs through proteolytic cleavage, which releases an intracellular domain that subsequently translocates to the nucleus to activate gene expression. Thus, Notch receptors can be seen as membrane-bound transcription factors. Proteolytic effector enzymes cleave two separate sites on the Notch receptor in a sequence initiated by specific ligand interactions. Delta and serrate/jagged ligands (DSL) expressed on adjacent cells interact with Notch receptors to promote cleavage of the extracellular ectodomain by a metalloprotease, probably through conformational change of the receptor. After shedding the ectodomain, which is subsequently endocytosed into the signaling cell, the remaining C-terminal fragment, NEXT (Notch extracellular truncation), is further processed by the intramembrane proteolytic action of γ-secretase to release the Notch intracellular domain (NICD). NICD translocates to the cell nucleus, where it converts specific DNA–binding proteins from transcriptional repressors to transcriptional activators. Signaling is terminated by degradation of NICD.[31]

7.3 PHARMACOLOGY OF SIGNALING

Membrane receptors exist in different conformational states generally corresponding to resting, active, inactive, or transitional states. The ensemble theory utilizes a probability model to describe receptors as existing in equilibrium among a distribution of multiple conformational states.[32] The pharmacological action of a ligand is therefore defined by the ensemble of conformational states that it stabilizes upon binding.[33–34] Agonists stabilize receptor conformations known as "active states" that are capable of producing a pharmacological effect. Full agonism is defined as the maximum response generated in a cell or tissue system, usually with reference to the natural receptor agonist, if known. Partial agonists produce a submaximal response and often exhibit receptor antagonism at high concentrations.

Some cell systems show constitutive GPCR activity — that is, agonist-independent signal transduction. Inverse agonists have the ability to reduce constitutive activity by stabilizing inactive receptor states, which is referred to as negative intrinsic efficacy. In the absence of constitutive activity, inverse agonists profile as competitive antagonists. Interestingly, "protean" agonists have been identified with both positive and negative efficacies,[35–36] behaving as inverse agonists in constitutively active systems and as conventional agonists in systems without basal activity. Protean agonism occurs when a ligand shifts the distribution of receptor states to those promoting G protein activation but with lower efficacy compared with constitutively active states.[37]

Antagonists prevent agonist-induced redistribution of receptor states. In the earlier two-state model of receptor function, antagonists were said to stabilize an inactive receptor state but lack the ability to induce conformational change. It is now recognized that this is an oversimplification and that many compounds classified as antagonists possess negative intrinsic efficacy. With new assay systems better able to detect constitutive activity, a number of well-known antagonists, including several antipsychotic, cardiovascular, and gastrointestinal drugs, have been found to demonstrate inverse agonism.[38] Neutral antagonists possessing exactly zero efficacy are now thought to be very rare[39] and are particularly attractive as tools in native cell systems because they can block both agonist and inverse agonist effects. Many compounds are known today that possess high affinity binding but no detectable physiological effect in standard assay systems. Kenakin and Onaran[40] have suggested that additional receptor activities, as well as new therapeutic applications, may be revealed by the study of such high affinity compounds in new functional assay systems.

Competitive antagonists interact at the orthosteric, or natural, agonist binding site on the receptor. Maintaining competitive inhibition can be very challenging in cases where receptor antagonists must compete with high concentrations of transiently released ligands,[41] for example, acetylcholine for muscarinic receptors, or histamine for H1-H4 receptors. Noncompetitive antagonists stabilize inactive conformational states in an insurmountable fashion, either through irreversible binding at the orthosteric site or by interacting at an allosteric site that is physically distinct from the orthosteric site. The term "modulator" refers broadly to molecules that bind to allosteric receptor sites and shift the distribution of receptor states toward those promoting (positive modulation) or inhibiting (negative modulation) agonist interaction or signaling efficiency. Negative modulators profile as noncompetitive antagonists. Positive modulators lack intrinsic efficacy but can alter the potency and efficacy of agonists and will be described in greater detail in the following sections.

The pharmacological effect of an agonist is measured in terms of efficacy and potency. Efficacy is an intrinsic property of an agonist that causes the receptor to signal and is operationally measured as the magnitude of a maximal response to which it is proportional. Potency is a measure of pharmacological action in terms of the concentration of agonist needed to produce a given level of response and is a function of both the efficacy and the affinity of the agonist for its receptor. The ensemble theory predicts that each agonist will stabilize a unique set of receptor conformational states and that not all agonists produce the same receptor active state.[42] There is also evidence that an agonist binding to a receptor occurs through a series of conformational intermediates.[43] Agonists manifest different degrees of efficacy and potency that vary depending on the cellular environments in which they are measured.[33]

Measures of agonist efficacy and potency are influenced by many factors, including binding affinity, level of receptor expression, receptor subunit composition, cell state (e.g., resting membrane potential), and the presence of key accessory proteins. Each cell system has its own saturable response limit dictated by the availability and efficiency of the receptor and signaling components. In cells expressing high levels of GPCRs, for example, activation of only a small percentage of total receptors will fully activate the system, and so compounds with low intrinsic efficacy act as full agonists. Agonist profiles are also dependent on where in the signal cascade the response is measured. Detection of responses proximal to the receptor, for example, GTPγS binding assays, will more reliably differentiate partial agonists as compared with downstream measures such as cAMP accumulation, which is associated with substantial signal amplification.

7.3.1 GPCR Agonist-Induced Trafficking

The pleiotropic ability of receptors to signal through multiple pathways by recruitment of different transducers and adaptors, combined with the ability of ligands to stabilize different populations of receptor states, can result in agonist selection of pathway stimulation called agonist-directed trafficking.[44] Agonist-directed trafficking has been demonstrated for several GPCR families, including monoamine, cannabinoid, and peptide receptors.[22] Differential signaling by agonists has also been demonstrated for TrkB receptors, which respond to both brain-derived neurotrophic factor (BDNF) and neurotrophin 4 (NT4). NT4-induced effects at TrkB appear to be more dependent on activation of the Shc-mediated pathways compared with BDNF.[45-47]

Because of differences in the relevant transducer and adaptor protein affinities for ligand-selective receptor conformations, agonist efficacy and potency are unique for each signaling pathway. Thus, an agonist can profile as a full agonist in one pathway and a partial agonist in a second pathway, as demonstrated for adenosine A1[48] and cannabinoid receptors.[49] Likewise, the rank order of potencies for a collection of agonists can also differ depending on the specific signaling response measured, as shown for adrenergic β2[50] and muscarinic acetylcholine receptors.[51] Different conformational states of the neurokinin A receptor have been demonstrated to mediate activation of calcium versus cAMP responses.[52]

Studies using plasmon-waveguide resonance (PWR)[53] confirm that GPCRs exist in multiple conformational states that differ depending on the ligand or G protein bound to the receptor. This technique detects a spectral shift resulting from a change in the molecular orientation of the receptor in a lipid bilayer when a ligand binds. For purified human delta opioid receptors (hDOR), the binding of agonist, antagonist, and inverse agonist each produced characteristic spectral shifts corresponding to different receptor conformations.[54] Additional experiments with hDOR showed that different ligand-bound receptor states interact differently with G proteins. Agonist-occupied receptor had a higher affinity for G protein compared with antagonist-occupied receptor, whereas unliganded receptor bound G protein with intermediate affinity. Receptor occupied by inverse agonists did not bind G protein at all.[55] Furthermore, hDOR exhibited distinct and selective preferences for different G protein subtypes depending on which agonist was bound.[56]

PWR has also been used to characterize the β-adrenergic receptor. Binding of the full agonists (-) isoproterenol and epinephrine and the partial agonist dobutamine each induced unique spectral shifts in PWR, demonstrating that these agonists stabilize different conformational states of the receptor.[57] This observation is consistent with findings in which multiple agonist-induced conformational states of the β_2-adrenergic receptor were detected with fluorescence lifetime.[16] Receptor internalization studies of the β_2-adrenergic receptor have shown that several compounds acting as inverse agonists with respect to adenylyl cyclase activity are instead partial agonists with respect to activation of the MAPK pathway.[58] These observations further support the concept of agonist-directed trafficking, in which different agonists for the same receptor can activate different signaling pathways, possibly by promoting selective interactions with G proteins.

7.3.2 ALLOSTERIC MODULATION

Allosteric modulators exert pharmacological effects through interactions at domains distinct from the orthosteric sites, altering the ensemble distribution of receptor conformations. Allosteric modulators can exert positive or negative effects on receptor activation and affect receptor signaling through reciprocal cooperative effects on agonist binding and efficacy.[59] They are particularly effective as therapeutic agents because they can dampen or enhance phasic activation of receptors produced by the endogenous release of agonists. Furthermore, by interacting at a site away from the agonist, their saturating effects can be limited to the level of exerted cooperativity. They also offer the potential for enhanced receptor subtype selectivity by interacting with less conserved receptor domains.

Therapeutically relevant allosteric modulators acting at both ion channels and GPCRs have been identified. For example, benzodiazepines and barbiturates are well-known molecules that exert clinically useful pharmacological effects by enhancing or inhibiting GABA binding to the GABA-A receptor, respectively.[60] Glycine is an endogenous positive modulator of NMDA receptor channels, and memantine exerts negative modulation at the NMDA receptor. Highly selective subtype-specific allosteric regulators of Group 3 GPCRs have also been described, including negative modulators at mGlu1/5,[61] positive modulators at mGlu2/3,[62] and a positive allosteric modulator of the calcium-sensing GPCR, Cinacalcet, which has been approved for treatment of hyperthyroidism.[63] Amiloride analogues are allosteric modulators of several GPCRs, including adenosine, α-adrenergic receptors, and dopamine receptors,[64] and moduline is an endogenous negative modulator of 5-$HT_{1B/1D}$ receptors.[65] Sodium ions have also been shown to negatively modulate the activation of GPCRs by stabilizing receptor conformations with reduced affinity for G proteins.

A multistate model (cubic ternary complex model) has been developed to describe allosteric modulation of agonist affinity and efficacy for Group 1 GPCRs and LGICs,[66–67] shown in Figure 7.4. In this model, the receptor is depicted as existing in inactive R states and active R* states governed by an isomerization

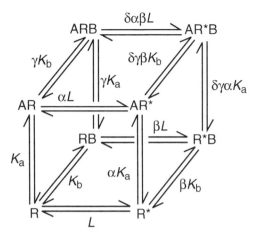

FIGURE 7.4 The cubic ternary model is used to describe the association of orthosteric (A) and allosteric (B) ligands with active (R*) and inactive (R) receptor states. The equilibrium association constants for A and B are Ka and Kb, respectively. L is the isomerization constant governing the transition between the R and R* states. The efficacy and modulatory effects of ligands are determined by the ligand-receptor cooperativity factors $\alpha\gamma$, $\beta\gamma$, δ, and γ (From Christopoulos, A. *Nat. Rev. Drug Disc.* 1, 198–210, 2002; Christopoulos, A. and Kenakin, T. *Pharmacol. Rev.* 54(2), 323–374, 2002. With permission.)

constant L. The binding affinity of a ligand (A) to the orthosteric site on R and R* is described by the affinity constants Ka and αKa, respectively. Similarly, binding of a modulator (B) to a topographically distinct allosteric site is governed by affinity constants Kb and βKb. The efficacy (intrinsic activity) of A and B are governed by α and β, respectively, and will vary depending on the conformational state of R*. The value of α is greater than 1 for conventional agonists with high affinity for R* states. For inverse agonists with higher affinity for R states, α is less than 1. For neutral antagonists having the same affinity for R and R* states, $\alpha = 1$.

Allosteric modulators are generally defined as having no intrinsic activity ($\beta = 1$) although the model allows for ectopic ($\beta > 1$) and inverse ($\beta < 1$) agonism to occur through interaction at a non-orthosteric site in the absence of an orthosteric ligand. Binding of A or B to inactive receptor states will alter the isomerization rate to active receptor states by α and β, respectively. Simultaneous binding of A and B are depicted by ARB and AR*B. Both ligands affect the binding and efficacy of the other in a reciprocal manner. The binding affinity of each ligand for R is altered by the binding cooperativity factor γ, whereas the affinity of each ligand for R* is modulated by γ and the activation cooperativity factor δ. The degree and type of cooperativity between different pairs of allosteric and othosteric ligands varies, since both γ and δ are unique attributes of each ligand pair for a given R* state. In a specific example to illustrate this, an allosteric modulator of agonist binding to muscarinic receptors displays both negative and positive effects depending on which agonist and receptor subtype is used.[68] The degree of cooperativity between ortho-steric and allosteric ligands is independent of their affinities for the receptor and do not correlate. Therefore, a modulator can bind with high affinity to the receptor but exhibit a small cooperative effect on the agonist or conversely, bind with low affinity yet exhibit a large cooperative effect.

The effects of allosteric modulators on receptor signaling differ depending on the values of β, γ, and δ.[59] For modulators with no intrinsic efficacy ($\beta = 1$), the effects on agonist activation of receptors is determined by γ and δ. Modulators with $\gamma < 1$ and $\delta = 1$ for a given agonist stabilize both R and R* receptor conformations with lower affinity for the agonist without altering agonist efficacy, thus decreasing the potency of the agonist without reducing its maximal effect. Likewise, modulators that stabilize receptor conformations with higher affinity for an agonist without altering agonist efficacy ($\gamma > 1$, $\delta = 1$) increase the potency of the agonist but do not change the maximal response. Modulators with $\gamma = 1$ and $\delta < 1$ for a given agonist stabilize R* receptor conformations for which the agonist has both lower affinity and intrinsic efficacy, resulting in a decrease in both the potency and maximal effect of the agonist. Similarly, modulators with $\gamma = 1$ and $\delta > 1$ for a given agonist stabilize R* receptor conformations for which the agonist has both higher affinity and intrinsic efficacy, resulting in an increase in both the potency and maximal effect of the agonist (provided the system maximum is not saturated).

Signaling through voltage-gated ion channels can also be modulated by ligands acting at multiple topographic sites in a state-dependent manner. Binding of ligands to these sites stabilizes different populations of channel conformational states and thus influences activation. The 1,4-dihydropyridine antagonists of the L-type calcium channel are effective antihypertensive agents because they stabilize inactivated channel states that are more prevalent in smooth muscle of depolarized and contracting blood vessels. Phenylalkylamines, another class of L-type calcium antagonists, are effective antiarrythmic agents because they selectively interact with open channel states by binding to the pore region.[69] Such antagonists are called "use-dependent," since the channel must be open in order for the antagonist to gain access to its binding site. Batrachotoxin, a potent neurotoxin, activates sodium channels by binding to a region away from the pore and stabilizing the open state.[70]

7.4   DETECTION METHODS FOR RECEPTOR SIGNALING

A great deal of our understanding of membrane protein signaling comes from assay technologies that detect downstream effects in either biochemical or cell-based assay formats. Some results come from studies in isolated tissues or dissociated cells; however, most of the data available today come from recombinant targets expressed in heterologous systems that facilitate the study of individual membrane proteins. A number of technologies developed for heterologous systems are now amenable to fully automated robotic platforms, which has enabled them to be applied in high-throughput screening of large compound libraries.[71] In a given cellular background, ligand responses may vary significantly depending on the level of signaling that is analyzed and on the assay method used to detect signaling events.[72]

The types of assay technologies used to monitor signal transduction include affinity measure binding assays, most commonly with radioactive or fluorescently labeled ligands[73]; functional signaling assays, which quantify second messengers, including intracellular calcium release, inositol phosphate turnover, and cAMP production[74]; direct activation of heterotrimeric G proteins through GTPγS accumulation[75]; reporter gene assays[76]; ionic flux, measured by electrophysiological techniques, membrane potential dyes, cation-sensitive fluorescent dyes, or atomic absorption spectrometry[77]; and measurements of metabolic/phenotype changes, including kinase-induced phosphorylation of intracellular substrates.[4]

When using signaling assays to study membrane proteins, it is essential to first characterize the pharmacodynamic range of the assay, to consider whether the assay provides steady state or transient measurements, and to understand the level of receptor reserve in the test cell line. Each of these will have important consequences on appropriate interpretation of the data. Functional assays utilizing the primary (and proximal) signaling step, such as GTPγS assays for GPCRs, offer the best approach to understanding the mechanistic aspects of receptor signaling and yet may give a much lower signal intensity versus second messenger assays, which provide signal amplification. Receptor density and stochiometric ratio with the G protein will affect agonists and antagonists in different ways. For example, high levels of receptor reserve can exaggerate agonist potency and compromise antagonist potency in a functional assay, and the limited availability of G proteins may reduce the maximum signal obtained with agonist stimulation. Confirmation of ligand behavior in a recombinant system should be confirmed as early as possible in assays using native receptors and, where possible, intact tissues.

Several recent developments in assay technologies provide new approaches to signal detection for both receptor research and drug discovery. The central tool to study ion channel function is still the patch clamp, a traditionally slow and labor-intensive technology in which an operator can produce less than 10 data points per day. The development of planar patch clamp techniques using chip-based formats has increased screening throughputs over 100-fold and promises to make ion channels accessible as a target class for drug discovery.[78]

There are now new high-throughput cell-based imaging approaches that monitor GPCR association with β-arrestins (Figure 7.5), which bind activated receptors and mediate their desensitization and internalization.[79] Studies have shown that the cholecystokinin (CCK)-β receptor, for example, binds arrestin with a pharmacology mirroring receptor signaling through inositol phosphate and that the rate of arrestin dissociation from internalized receptor mirrors receptor recycling to the plasma membrane.[80] The ability to monitor receptor internalization separately from second messenger activation has revealed that some GPCR ligands, for example, for mu opioid receptors, act as agonists in signaling without inducing receptor endocytosis via a β-arrestin pathway.[81] Another case of unexpected behavior is found with serotonin 5HT2A receptors, in which antagonist compounds can promote receptor endocytosis.[82]

Most GPCRs enter the endocytic pathway in clathrin-coated vesicles via a β-arrestin dependent process. Others, such as vasoactive intestinal peptide type 1 and the endothelin type B receptors, can be internalized by a β-arrestin independent pathway in cholesterol-rich membrane microdomains, known as lipid rafts and

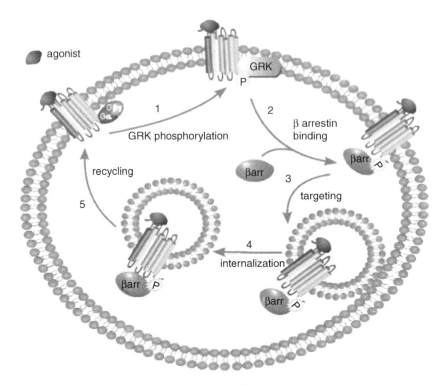

FIGURE 7.5 (See color insert following page 272.) Internalization of G protein-coupled receptors. (1) The binding of agonist to a GPCR induces signaling through dissociation of the G protein and phosphorylation of the receptor by G protein-coupled receptor kinases (GRKs). (2) Cytosolic β-arrestin interacts with the phosphorylated receptor and inactivates the receptor by preventing reassociation of the G protein complex. (3) The receptor is then targeted to clathrin-coated pits and (4) internalized into endosomes, where β-arrestin is released and the GPCR is dephosphorylated and (5) recycled back to the plasma membrane for another round of stimulation. (From Ferguson, S.S.G. *Pharmacol. Rev.* 53, 1–24, 2001.)

caveolae. Both of these internalization pathways involve the action of dynamin. A third pathway independent of β-arrestin and dynamin appears to be involved in the internalization of the secretin, M2 muscarinic, bradykinin type 2, and N-formyl peptide, and 5HT2A receptors. Once internalized, receptors can be either recycled back to the membrane (resensitization) or degraded in lysosomes (desensitization). Chimeric constructs of various receptors exhibiting different patterns of pre- and post-endocytic trafficking have been used to show that molecular motifs in the carboxy terminal tail play a major role in determining the route of internalization and receptor fate.[83,84] Receptor conformation may also determine receptor trafficking subsequent to ligand interaction. The corticotrophin releasing factor type 1 receptor has recently been shown to internalize via the β-arrestin dependent pathway when bound by an agonist and via a β-arrestin independent pathway when bound by an antagonist.[85]

Importantly, the classification of a compound as agonist, antagonist, inverse agonist, or partial agonist using a heterologous expression system can be quite different from the *in vivo* situation.[86] For example, the H3 histamine receptor ligand proxyfan is a partial agonist for the inhibition of cAMP in heterologous cells, but a full agonist for the sleep–wake cycle in the cat, and an inverse agonist for the sleep–wake cycle in the mouse.[35] In addition, a single compound may have differential effects between tissue types or host cell environments. For example, the antipsychotic aripiprazole shows the full range of agonist, antagonist, and partial agonist effects at cloned dopamine D2 receptors depending on the host cell type and receptor signal detected.[87] This highlights the challenges in interpreting data from recombinant systems and the importance of establishing a correlation between recombinant and native receptor systems, for the appropriate selection of candidate drug compounds.

7.5 THE ROLE OF RECEPTOR COMPLEXES IN CELL SIGNALING

Receptor–receptor interactions can be classified as "direct," referring to intermolecular association, or "indirect," where the steric conformation of one receptor is modified in a signaling-dependent manner due to the activation of another receptor. Although cell surface receptors are still widely treated as monomeric entities in pharmacological assays and in models of receptor activation, evidence is strong for the existence of direct interactions between membrane receptors that can affect signaling selectivity and efficacy. These complexes are referred to as "signalosomes" (M. Bouvier, Biochemistry, Université de Montréal, Montréal, Québec, Canada), "receptor mosaics" (K. Fuxe, Department of Neuroscience, Karolinska Institutet, Stockholm, Sweden), and "transductosomes" (J.P. Pin, CNRS, Centre CNRS-INSERM de Pharmacologie-Endocrinologie, Montpellier, France).[88,89] Homo- and heterodimerization of GPCRs, for example, provide new mechanisms for receptor cross-talk and allosteric modulation.[90]

Investigation of the GABA-B receptor led to the first direct evidence for heterodimerization of GPCRs, with the demonstration that the R2 subunit acted as a chaperone for R1, enabling cell surface expression as well as high affinity binding at the R1 subunit.[91] Heterodimerization of alpha-1D and alpha-1B adrenergic receptors has also been shown to be required for the cell-surface expression of the alpha-1D subtype.[92] A completely different approach to pharmacological treatment of disease has been suggested by studies with mutant forms of the vasopressin V2 receptor, responsible for a condition called nephrogenic diabetes insipidus. It was discovered that several V2 receptor ligands were able to act as molecular chaperones, assisting the folding and trafficking of the mutant receptors and rescuing cell surface expression of these receptors.[93]

Quantitative studies continue to provide evidence for additional heterodimers, such as adrenergic $\beta1$-$\beta2$,[94] vasopressin V1a-V2,[95] and mu-delta opioid receptor complexes.[96] Pharmacological consequences of receptor associations are also clear in some cases. For example, coexpression of recombinant mu and delta opioid receptors results in a loss of binding for selective mu and delta ligands, except when both types of ligands are added simultaneously.[96] Epitope–epitope electrostatic interactions have been reported as the mechanism underlying adenosine A2a and dopamine D2 receptor heteromerization.[97] An extensive review of GPCR interactions through heteromerization has been published by Agnati and co-workers.[89]

Signaling through RTKs is also known to involve agonist-induced dimerization, but in this case, multiplexing between target classes can also be required for physiologically relevant cell signaling. For example, TrkB can associate with the sodium channel NaV1.9 to enable BDNF activation of NaV1.9. BDNF can evoke rapid membrane depolarization in hippocampal neurons,[98] an effect normally mediated by gating of ion channels rather than through enzymatic activity of a tyrosine kinase. When NaV1.9 expression is blocked by antisense oligonucleotides, BDNF is no longer able to evoke sodium currents in hippocampal neurons. Also, NaV1.9 coexpressed with TrkB in HEK293 cells is activated by BDNF but not by voltage changes. These findings suggest that TrkB and NaV1.9 associate in such a way that a conformational change in TrkB caused by binding of BDNF induces a conformational change in NaV1.9, allowing it to become permeable to Na^+ ions. Thus, in this reconstitution experiment, NaV1.9 acts as a ligand-gated channel rather than a voltage-gated channel.[99]

Accessory proteins may also alter specificity of receptor signaling.[100] For example, PDZ-binding motifs and homer-binding proteins are involved in trafficking, targeting, and fine-tuning signaling pathways.[101] β-adrenergic receptors can activate both Gi and Go, but the binding of the Na^+–H^+ exchanger regulatory factor (NHERF) via a PDZ domain suppresses coupling to Gi.[102] Another interesting example is the calcitonin receptor-like receptor (CRLR), a heterodimeric receptor consisting of two membrane-bound protein subunits, which acts as a calcitonin-gene-related peptide (CGRP) receptor in the presence of receptor activity modifying protein (RAMP)-1, and as an adrenomedullin receptor in the presence of RAMP-2.[103]

7.6 CONCLUSION

Recent breakthroughs in receptor expression, structural biology, assay technologies, and informatics have dramatically expanded our knowledge of signaling through membrane proteins. At the same time, however, they have revealed a new horizon of unprecedented complexity and signal integration that will challenge scientists for years to come. Whether it be delineating the molecular mechanisms involved in neuronal signal transduction in the brain or uncovering cellular communication events that keep our hearts beating, scientists are faced with the daunting task of modeling the native tissue environment, and translating *in vitro* cell-based results to the *in vivo* situation in the living organism.

Cell surface receptors have a long, successful history as drug targets, and advances in cell signaling are now revealing new pharmacological mechanisms of well-known drugs. For example, compounds thought to act as silent antagonists, including the atypical antipsychotic drug clozapine and the cardiovascular drug metoprolol, have been shown to behave as inverse agonists at several GPCR targets.[38,39] The development of drugs targeting ion channels is particularly demanding due to neuronal and cardiac side effects, and there is hope that alternative approaches based on downstream signaling or allosteric modulation may provide opportunities to influence channel function indirectly to obtain the efficacy, selectivity, and safety required in a therapeutic agent.

In addition to signaling mechanisms, drug discovery approaches must take into consideration receptor localization, receptor reserve in the target tissue, and the possibility of functionally distinguishable receptor polymorphisms in the human population. Increased understanding of membrane receptors and their signaling potential holds the promise of new therapeutic agents with increased potency and specificity, and better safety profiles. Over the next decade, potent and selective small molecule ligands at membrane receptors will provide critical tools for establishing links between specific conformational changes and efficacy profiles, and for unveiling the molecular mechanisms of membrane protein signaling and the control of physiological processes.

REFERENCES

1. Pennisi, E. Human genome. A low number wins the GeneSweep Pool. *Science*, 300, 1484, 2003.
2. Venter J.C., et al. The sequence of the human genome. *Science*, 291, 1304–1351, 2001.
3. Lander E.S., et al. Initial sequencing and analysis of the human genome. *Nature*, 409, 860–921, 2001.
4. Papin, J.A., Hunter, T., Palsson, B.O., and Subramaniam, S. Reconstruction of cellular signaling networks and analysis of their properties. *Nat. Rev. Mol. Cell. Biol.*, 6, 99–111, 2005.
5. Bourne, H.R. How receptors talk to trimeric G proteins. *Curr. Opin. Cell Biol.*, 9, 134–142, 1997.
6. Gether, U. Uncovering molecular mechanisms involved in activation of G protein-coupled receptors. *Endocrin. Rev.*, 21, 90–113, 2000.
7. Hamm, H.E. How activated receptors couple to G proteins. *Proc. Natl. Acad. Sci. USA*, 98, 4819–4821, 2001.
8. Frade, J.M. Nuclear translocation of the p75 neurotrophin receptor cytoplasmic domain in response to neurotrophin binding. *J. Neurosci.*, 25, 1407–1411, 2005.
9. Gainetdinov, R.R. and Caron, M.G. Monoamine transporters: from genes to behavior. *Annu. Rev. Pharmacol. Toxicol.*, 43, 261–284, 2003.
10. Rothman, R.B. and Baumann, M.H. Monoamine transporters and psychostimulant drugs. *Eur. J. Pharmacol.*, 479, 23–40, 2003.
11. Rasmussen, S.G., Adkins, E.M., Carroll, F.I., Maresch, M.J., and Gether, U. Structural and functional probing of the biogenic amine transporters by fluorescence spectroscopy. *Eur. J. Pharmacol.*, 479, 13–22, 2003.
12. Koch, H.P. and Larsson, H.P. Small-scale molecular motions accomplish glutamate uptake in human glutamate transporters. *J. Neurosci.*, 25, 1730–1736, 2005.
13. Sarter, M. and Parikh, V. Choline transporters, cholinergic transmission and cognition. *Nat. Rev. Neurosci.*, 6, 48–56, 2005.
14. Riordan, J.R. Assembly of functional cftr chloride channels. *Annu. Rev. Physiol.*, 67, 701–718, 2005.

15. Muller, G. Towards 3D structures of G protein-coupled receptors: a multidisciplinary approach. *Curr. Med. Chem.*, 7, 861–888, 2000.
16. Ghanouni, P., Gryczynski. Z., Steenhuis, J.J., Lee, T.W., Farrens, D.L., Lakowicz, J.R., and Kobilka, B.K. Functionally different agonists induce distinct conformations in the G protein coupling domain of the beta 2 adrenergic receptor. *J. Biol. Chem.*, 276, 24433–24436, 2001.
17. Swaminath, G., Xiang, Y., Lee, T.W., Steenhuis, J., Parnot, C., and Kobilka, B.K. Sequential binding of agonists to the beta 2 adrenoceptor. Kinetic evidence for intermediate conformational states. *J. Biol. Chem.*, 279, 686–691, 2004.
18. Yeagle, P.L. and Albert, A.D. A conformational trigger for activation of a G Protein by a G protein-coupled receptor. *Biochemistry*, 42, 1365–1368, 2002.
19. Gillman, A.G. G proteins: transducers of receptor generated signals. *Annu. Rev. Biochem.*, 56, 615–649, 1987.
20. Bourne, H.R., Landis, C.A., and Masters, S.B. Hydrolysis of GTP by the α-chain of Gs and other GTP binding proteins. *Proteins*, 6, 222–230, 1989.
21. Ferguson, S.S., Downey, W.E., Colapietro, A.M., Barak, L.S., Menard, L., and Caron, M.G. Role of b-arrestin in mediating agonist-promoted G protein-coupled receptor internalization. *Science*, 271, 363–366, 1996.
22. Hermans, E. Biochemical and pharmacological control of the multiplicity of coupling at G protein-coupled receptors. *Pharmacol. Therapeutic.*, 99, 25–44, 2003.
23. Morris, A.J. and Malbon, C.C. Physiological regulation of G protein linked signaling. *Physiol. Rev.*, 79, 1373–1430, 1999.
24. Sah, V.P., Seasholtz, T.M., Sagi, S.A., and Brown, J,H. The role of Rho in G protein-coupled receptor signal transduction. *Annu. Rev. Pharmacol. Toxicol.*, 40, 459–489, 2000.
25. Catterall, W.A. Molecular mechanisms of gating and drug block of sodium channels. *Novartis Found. Symp.*, 241, 206–218, 2002.
26. Elinder, F., Mannikko, R., and Larsson, H.P. S4 charges move close to residues in the pore domain during activation in a K channel. *J. Gen. Physiol.*, 118(1), 1–10, 2001.
27. Segal, R. Selectivity in neurotrophin signaling: theme and variations. *Annu. Rev. Neuorosci.*, 26, 299–330, 2003.
28. Kumar, C.C. Signaling by integrin receptors. *Oncogene*, 17, 1365–1373, 1998.
29. Parsons, J.T. Focal adhesion kinase: the first ten years. *J. Cell Sci.*, 15(116), 1409–1416, 2003.
30. Dempsey, P.W., Doyle, S.E., He, J.O., and Chang, G. The signaling adaptors and pathways activated by TNF superfamily. *Cytokine Growth Factor Rev.*, 14, 193–209, 2003.
31. Selkoe, D and, Kopan, R. Notch and presenilin: regulated transmembrane proteolysis links development and degeneration. *Annu. Rev. Neurosci.*, 26, 565–597, 2003.
32. Hilser, V.J., Dowdy, D., Oas, T.G., and Freire, E. The structural distribution of cooperative interactions in proteins: analysis of the native state ensemble. *Proc. Natl. Acad. Sci. USA*, 95, 9903–9908, 1998.
33. Kenakin, T. Drug efficacy at G protein-coupled receptors. *Annu. Rev. Pharmacol.*, 42, 349–379, 2002.
34. Kenakin, T. Principles: receptor theory in pharmacology. *Trends Pharmacol. Sci.*, 25, 186–192, 2004.
35. Gbahou, F., Rouleau, A., Morisset, S., Parmentier, R., Crochet, S., Lin, J.S., Ligneau, X., Tardivel-Lacombe, J., Stark, H., Schunack, W., Ganellin, C.R., Schwartz, J.C., and Arrang, J.M. Protean agonism at histamine H3 receptors in vitro and in vivo. *Proc. Natl. Acad. Sci. USA*, 100, 11086–11091, 2003.

36. Kenakin, T. Inverse, protean and ligand-selective agonism: matters of receptor conformation. *FASEB J.*, 15, 598–611, 2001.

37. Leff, P., Scaramellini, C., Law, C., and McKechnie, K. A three-state receptor model of agonist action. *Trends Pharmacol. Sci.*, 18, 355–362, 1997.

38. Milligan, G. Constitutive activity and inverse agonists of G protein-coupled receptors: a current perspective. *Mol. Pharmacol.*, 64, 1271–1276, 2003.

39. Seifert, R. and Wenzel-Seifert, K. Constitutive activity of G protein-coupled receptors: cause of disease and common property of wild-type receptors. *Arch. Pharmacol.*, 366, 381–416, 2002.

40. Kenakin, T. and Onaran, O. The ligand paradox between affinity and efficacy: can you be there and not make a difference? *Trends Pharmacol. Sci.*, 23, 275–280, 2002.

41. Swinney, D.C. Biochemical mechanisms of drug action: what does it take for success? *Nat. Rev. Drug Disc.*, 3, 801–808, 2004.

42. Kenakin, T. Ligand-selective receptor conformations revisited: the promise and the problem. *Trends Pharmacol. Sci.*, 24, 346–354, 2003.

43. Kobilka, B. Agonist binding: a multistep process. *Mol. Pharmacol.*, 65, 1060–1062, 2004.

44. Kenakin, T.P. Agonist-receptor efficacy II: agonist-trafficking of receptor signals. *Trends Pharmacol. Sci.*, 16, 232–238, 1995.

45. Minichiello, L., Casagranda, F., Tatche, R.S., Stucky, C.L., Postigo, A., Lewin, G.R., Davies, A.M., and Klein, R. Point mutation in trkB causes loss of NT4-dependent neurons without major effects on diverse BDNF responses. *Neuron*, 21, 335–345, 1998.

46. Minichiello, L., Calella, A.M., Medina, D.L., Bonhoeffer, T., Klein, R., and Korte, M. Mechanism of Trk-B mediated hippocampal long-term potentiation. *Neuron*, 36, 121–137, 2002.

47. Fan, G., Egles, C., Sun, Y., Minichiello, L., Renger, J.J., Klein, R., Liu, G., and Jaenisch, R. Knocking the NT4 gene into the BDNF locus rescues BDNF deficient mice and reveals distinct NT4 and BDNF activities. *Nat. Neurosci.*, 3, 350–357, 2000.

48. Cordeaux, Y., Briddon, S.J., Megson, A.E., McDonnell, J., Dickenson, J.M., and Hill, S.J. Influence of receptor number on functional responses elicited by agonists acting at the human adenosine A1 receptor: evidence for signaling pathway-dependent changes in agonist potency and relative intrinsic activity. *Mol. Pharmacol.*, 58, 1075–1084, 2000.

49. Bonhaus, D.W., Chang, L.K., Kwan, J., and Martin, G.R. Dual activation and inhibition of adenylyl cyclase by cannabinoid receptor agonists: Evidence for agonist-specific trafficking of intracellular responses. *J. Pharmacol. Exp. Ther.*, 287, 884–888, 1998.

50. Wenzel-Seifert, K. and Seifert, R. Molecular analysis of β2-adrenoceptor coupling to Gs-, Gi- and Gq-proteins. *Mol. Pharmacol.*, 58, 954–966, 2000.

51. Akam, E.C., Challiss, R.A.J. and Nahorski, S.R. G(q/11) and G(i/o) activation profiles in CHO cells expressing human muscarinic acetylcholine receptors: Dependence on agonist as well as receptor-subtype. *Br. J. Pharmacol.*, 132, 950–958, 2001.

52. Palanche, T., Ilien, B., Zoffmann, S., Reck, M.P., Bucher, B., Edelstein, S.J., and Galzi, J.L. The neurokinin A receptor activates calcium and cAMP responses through distinct conformational states. *J. Biol. Chem.*, 276, 34853–34861, 2001.

53. Salamon, Z., Brown, M.F., and Tollin, G. Plasmon resonance spectroscopy: probing interactions within membranes. *Trends Biochem. Sci.*, 24, 213–219, 1999.

54. Salamon, Z., Hruby, V.J., Tollin, G., and Cowell, S. Binding of agonists, antagonists and inverse agonists to the human δ-opioid receptor produces distinctly different conformational states distinguishable by plasmon-waveguide resonance spectroscopy. *J. Peptide Res.*, 60, 322–328, 2002.

55. Alves, I.D., Salamon, Z., Varga, E., Yamamura, H.I., Tollin, G., and Hruby, V.J. Direct observation of G protein binding to the human δ-opioid receptor using plasmon-waveguide resonance spectroscopy. *J. Biol. Chem.*, 278, 48890–48897, 2003.

56. Alves, I.D., Ciano, K.A., Boguslavski, V., Varga, E., Salamon, Z., Yamamura, H.I., Hruby, V.J., and Tollin, G. 2004. Selectivity, cooperativity, and reciprocity in the interactions between the δ-opioid receptor, its ligands, and G proteins. *J. Biol. Chem.*, 279, 44673–44682, 2004.

57. Devanathan, S., Yao, Z., Salamon, Z., Kobilka, B., and Tollin, G. Plasmon-waveguide resonance studies of ligand binding to the human beta 2-adrenergic receptor. *Biochemistry*, 43, 3280–3288, 2004.

58. Azzi, M., Charest, P.G., Angers, S., Rousseau, G., Kohout, T., Bouvier, M., and Pineyro, G. Beta-arrestin mediated activation of MAPK by inverse agonists reveals distinct active conformations for G protein-coupled receptors. *Proc. Natl. Acad. Sci. USA*, 100, 11406–11411, 2003.

59. Jensen, A.A. and Spalding, T.A. Allosteric modulation of G protein coupled receptors. *Eur. J. Pharmacol. Sci.*, 21, 407–420, 2004.

60. Roudolph, U., Crestiani, F., and Mohel, H. GABA_A receptor subtypes: dissecting their pharmacological functions. *Trends Pharmacol. Sci.*, 22, 188–194, 2001.

61. Gasparini, F., Kuhn, R., and Pin, J.P. Allosteric modulators of group I metabotropic glutamate receptors: novel subtype-selective ligands and therapeutic perspectives. *Curr. Opin. Pharmacol.*, 2, 43–49, 2002.

62. Schaffhauser, H., Rowe, B.A., Morales, S., Chavez-Noriega, L.E., Yin, R., Jachec, C., Rao, S.P., Bain, G., Pinkerton, A.B., Vernier, J.M., Bristow, L.J., Varney, M.A., and Daggett, L.P. Pharmacological characterization and identification of amino acids involved in the positive modulation of metabotropic glutamate receptor subtype 2. *Mol. Pharmacol.*, 64, 798–810, 2003.

63. Peacock, M., Bilezikian, J.P., Klassen, P.S., Guo, M.D., Turner, S.A., and Shoback, D. Cinacalcet hydrochloride maintains long-term normocalcemia in patients with primary hyperparathyroidism. *J. Clin. Endocrinol. Metab.*, 90, 135–141, 2005.

64. Christopoulos, A. Allosteric binding sites on cell-surface receptors: Novel targets for drug discovery. *Nat. Rev. Drug Disc.*, 1, 198–210, 2002.

65. Massot, O., Rousselle, J.C., Fillion, M.P., Grimaldi, B., Cloez-Tayarani, I., Fugelli, A., Prudhomme, N., Seguin, L., Rousseay, B., Plantefol, M., Hen, R., and Fillion, G. 5-Hydroxytryptamine-moduline, a new endogenous cerebral peptide, controls the serotinergic activity via its specific interaction with 5-hydroxytryptamine_{1B/1D} receptors. *Mol. Pharmacol.*, 50, 752–762, 1996.

66. Christopoulos, A. and Kenakin, T. G protein-coupled receptor allosterism and complexing. *Pharmacol. Rev.*, 54, 323–374, 2002.

67. May, L.T. and Christopoulos, A. Allosteric modulators of G protein-coupled receptors. *Curr. Opin. Pharmacol.*, 3, 551–556, 2003.

68. Jakubik, J., Bacakova, L., El-Fakahany, E.E., and Tucek, S. Positive cooperativity of acetylcholine and other agonists with allosteric ligands on muscarinic acetylcholine receptors. *Mol. Pharmacol.*, 52, 172–179, 1997.

69. Hering, S., Aczel, S., Kraus, R.L., Berjukow, S., Striessnig, J., and Timin, E.N. Molecular mechanism of use-dependent calcium channel block by phenylalkylamines: role of inactivation. *Proc. Natl. Acad. Sci. USA*, 94, 13323–13328, 1997.

70. Trainer, V.L., Brown, G.B., and Catterall, W.A. Site of covalent labeling by a photo-reactive batrachotoxin derivative near transmembrane segment IS6 of the sodium channel alpha subunit. *J. Biol. Chem.*, 271, 11261–11267, 1996.

71. Walters, W.P. and Namchuck, M. Designing screens: how to make your hits a hit. *Nat. Rev. Drug Disc.*, 2, 259–266, 2003.

72. Niedernberg, A., Tunaru, S., Blaukat, A., Harris, B., and Kostenis, E. Comparative analysis of functional assays for characterization of agonist ligands at G protein-coupled receptors. *J. Biomol. Screen.*, 5, 500–510, 2003.

73. Scheel, A.A., Funsch, B., Busch, M., Gradl, G., Pschorr, J., and Lohse, M.J. Receptor-ligand interactions studied with homogeneous fluorescence-based assays suitable for miniaturized screening. *J. Biomol. Screen.*, 6, 11–18, 2001.

74. Raddatz, R., Wilson, A.E., Artymyshyn, R., Bonini, J.A., Borowsky, B., Boteju, L.W., Zhou, S., Kouranova, E.V., Nagorny, R., Guevarra, M.S., Dai, M., Lerman, G.S., Vaysse, P.J., Branchek, T.A., Gerald, C., Forray, C., and Adham, N. Identification and characterization of two neuromedin U receptors differentially expressed in peripheral tissues and the central nervous system. *J. Biol. Chem.*, 275, 32452–32459, 2000.

75. Lanau, F., Zenner, M.T., Civelli, O., and Hartman, D.S. Epinephrine and norepinephrine act as potent agonists at the recombinant human dopamine D4 receptor. *J. Neurochem.*, 68, 804–812, 1997.

76. Rees, S., Martin, D.P., Scott, S.V., Brown, S.H., Fraser, N., O'Shaughnessy, C., and Beresford, I.J. Development of a homogeneous MAP kinase reporter gene screen for the identification of agonists and antagonists at the CXCR1 chemokine receptor. *J. Biomol. Screen.*, 6, 19–27, 2001.

77. Scott, C.W., Wilkins, D., Trivedi, S., and Crankshaw, D.J. A medium-throughput functional assay of KCNQ2 potassium channels using rubidium efflux and atomic absorption spectrometry. *Anal. Biochem.*, 319, 251–257, 2003.

78. Wood, C., Williams, C., and Waldron, G. Patch clamping by numbers. *Drug Disc. Today*, 9, 434–441, 2004.

79. Oakley, R.H., Laporte, S.A., Holt, J.A., Barak, L.S., and Caron, M.G. Molecular determinants underlying the formation of stable intracellular G protein-coupled receptor-beta-arrestin complexes after receptor endocytosis. *J. Biol. Chem.*, 276, 19452–19460, 2001.

80. Barak, L.S., Oakley, R.H., and Shetzline, M.A. G protein-coupled receptor desensitization as a measure of signaling: modeling of arrestin recruitment to activated CCK-B receptors. *Assay Drug Dev. Technol.*, 1, 409–424, 2003.

81. Whistler, J.L., Chuang, H.H., Chu, P., Jan, L., and von Zastrow, M. Functional dissociation of mu-opioid receptor signaling and endocytosis: implications for the biology of opiate tolerance and addiction. *Neuron*, 23, 737–746, 1999.

82. Willins, D.L., Alsayegh, L., Berry, S.A., Backstrom, J.R., Sanders-Bush, E., Friedman, L., Khan, N., and Roth, B.L. Serotonergic antagonist effects on trafficking of serotonin 5HT2A receptors *in vivo* and *in vitro*. *Ann. N.Y. Acad. Sci.*, 861, 121–127, 1998.

83. Kristiansen, K. Molecular mechanisms of ligand binding, signaling, and regulation within the superfamily of G protein-coupled receptors: molecular modeling and mutagenesis approaches to receptor structure and function. *Pharmacol. Ther.*, 103, 21–80, 2004.

84. Claing, A., Laporte, S.A., Caron, M.G., and Lefkowitz, R.J. Endocytosis of G protein coupled receptors: roles of G protein-coupled receptor kinases and β-arrestin proteins. *Prog. Neurobiol.*, 66, 61–79, 2002.

85. Perry, S.J., Jungar, S., Kohout, T.A., Hoare, S.M.J., Struthers, R.S., Grigoriadis, D.E., and Maki, R.A. Distinct conformations of the corticotrophin releasing factor type 1 receptor adopted following agonist and antagonist binding are differentially regulated. *Biol. Chem.*, 280, 11560–11568, 2005.

86. Kenakin, T. Predicting therapeutic value in the lead optimization phase of drug discovery. *Nat. Rev. Drug Disc.*, 2, 429–438, 2003.

87. Shapiro, D., Renock, S., Arrington, E., Chiodo, L.A., Liu, L.-X., Sibley, D.R., Roth, B.L., and Mailman, R. Aripiprazole, a novel atypical antipsychotic drug with a unique and robust pharmacology. *Neuropsychopharmacology*, 28, 1400–1411, 2003.

88. Nature Reviews Drug Discovery GPCR Questionnaire Participants. The state of GPCR research in 2004. *Nat. Rev. Drug Disc.*, 3(7), 575, 577–626, 2004.

89. Agnati, L.F., Ferre, S., Lluis, C., Franco, R., and Fuxe, K. Molecular mechanisms and therapeutical implications of intramembrane receptor/receptor interactions among heptahelical receptors with examples from the striatopallidal GABA neurons. *Pharmacol. Rev.*, 55, 509–550, 2003.

90. Bai, M. Dimerization of G protein-coupled receptors: roles in signal transduction. *Cell Signal.*, 16, 175–186, 2004.

91. Bettler, B., Kaupmann, K., Mosbacher, J., and Gassmann, M. Molecular structure and physiological functions of GABA(B) receptors. *Physiol. Rev.*, 84, 835–867, 2004.

92. Hague, C., Uberti, M.A., Chen, Z., Hall, R.A., and Minneman, K.P. Cell surface expression of alpha-1D adrenergic receptors is controlled by heterdimerization with alpha-1B receptors. *J. Biol. Chem.*, 279, 15541–15549, 2004.

93. Morello, J.P. Pharmacological chaperones rescue cell surface expression and function of misfolded V2 vasopressin receptor mutants. *J. Clin. Invest.*, 105, 887–895, 2000.

94. Mercier, J.F., Salahpour, A., Angers, S., Breit, A., and Bouvier, M. Quantitative assessment of beta 1- and beta 2-adrenergic receptor homo- and heterodimerization by bioluminescence resonance energy transfer. *J. Biol. Chem.*, 277, 44925–44931, 2002.

95. Terrillon, S., Durroux, T., Mouillac, B., Breit, A., Ayoub, M.A., Taulan, M., Jockers, R., Barberis, C., and Bouvier, M. Oxytocin and vasopressin V1a and V2 receptors form constitutive homo- and heterodimers during biosynthesis. *Mol. Endocrinol.*, 17, 677–691, 2003.

96. Jordan, B.A. and Devi, L.A. G protein-coupled receptor heterodimerization modulates receptor function. *Nature*, 399, 697–700, 1999.

97. Ferre, S., Ciruela, F., Canals, M., Marcellino, D., Burgueno, J., Casado, V., Hillion, J., Torvinen, M., Fanelli, F., de Benedetti, P., Goldberg, S.R., Bouvier, M., Fuxe, K., Agnati, L.F., Lluis, C., Franco, R., and Woods, A. Adenosine A2a-dopamine D2 receptor-receptor heteromers. Targets for neuro-psychiatric disorders. *Parkinsonism and Related Disorders*, 10, 265–271, 2004.

98. Kafitz, K.W., Rose, C.R., Thoenen, H., and Konnerth, A. Neurotrophin-evoked rapid excitation through TrkB receptors. *Nature*, 401, 918–921, 1999.

99. Blum, R., Kafitz, K.W., and Konnerth, A. Neurotrophin-evoked depolarization requires the sodium channel Na(V)1.9. *Nature*, 419, 687–693, 2002.

100. Neubig, R.P. and Siderovski, D.P. Regulators of G protein signaling as new central nervous system drug targets. *Nat. Rev. Drug Disc.*, 1, 187–197, 2002.

101. Bockaert, J., Marin, P., Dumuis, A., and Fagni, L. The "magic tail" of G protein-coupled receptors: an anchorage for functional protein networks. *FEBS Lett.*, 546, 65–72, 2003.

102. Xiang, Y. and Kobilka, B. The PDZ-binding motif of the beta2-adrenergic receptor is essential for physiologic signaling and trafficking in cardiac myocytes. *Proc. Natl. Acad. Sci. USA*, 100, 10776–10781, 2003.

103. Sexton, P.M., Albiston, A., Morfis, M., and Tilakaratne, N. Receptor activity modifying proteins. *Cell Signal*, 13, 73–83, 2001.

8 Expression of Membrane Proteins in Yeasts

Christoph Reinhart and Christoph Krettler

CONTENTS

8.1 INTRODUCTION

During the last decades it has become clear that membrane proteins play a dominant role in many cellular functions and are involved in cardiovascular, metabolic, neurological, psychiatric and oncological diseases. The sequencing of the human genome revealed that approximately 30% of human genes code for membrane proteins.[1,2] From the viewpoint of pharmacology, membrane proteins are the most important protein family. Many pharmaceuticals mediate intracellular effects by targeting membrane proteins and, as a result, the study of membrane proteins has become one of the most active fields in biochemical and pharmacological research.

Based on topology, there are a plethora of structurally different membrane proteins. Monotopic proteins such as prostaglandin H2 synthase-1 (cyclo-oxigenase

1)[3,4] inserted into, but not crossing the membrane; pores such as aquaporin which have six membrane-spanning a-helices;[5,6] to ATP-binding cassette (ABC) transporters with 17 membrane-spanning a-helices[7] are just a few examples. Due to their pharmacological relevance, G protein-coupled receptors (GPCRs) belong to the most intensively studied family of membrane proteins. Almost 50% of pharmaceuticals act on this family whose members share the common topology of seven transmembrane spanning a-helices. Transport proteins such as the super-family of ABC transporters are also the focus of active research, as they cause multidrug resistance when overproduced in tumor cells, thus reducing the efficacy of chemotherapy. The importance of many membrane proteins has led to increasing interest in their three-dimensional structure, aimed at revealing the molecular mechanisms involved in their function. However, with only few solved structures thus far, this important class of proteins is not adequately represented in the protein structure data bases (reviewed in 8).

One of the major challenges in solving structures of membrane proteins is the production of sufficient quantities of pure, homogenous protein. Except for rare examples such as bovine rhodopsin[9] and the nicotinic acetylcholine receptor (nAChR) from *Torpedo marmorata*[10,11] most eukaryotic membrane proteins are present in native tissues in only minor quantities. In the case of the neuropeptide Y_2 receptor, for example, 1000 pig brains were needed to obtain 190 μg of purified receptor for functional studies.[12] This example illustrates how time-consuming and labor-intensive the purification of a membrane protein from native tissue can be. Characterization of membrane proteins from native tissue can be confounded by the presence of multiple protein subtypes.

Therefore, the overproduction of a membrane protein in homologous or heterologous expression systems is better suited for structural and functional studies. When prokaryotic membrane proteins are produced in prokaryotic systems (in most cases *Escherichia coli*), the success rates regarding production level, as well as functionality, are quite reasonable. Many bacterial membrane proteins have been produced in large amounts.[13,14] However, attempts to produce eukaryotic membrane proteins in prokaryotic systems have yielded dramatically poorer results. Among the few examples of successful expression of mammalian membrane proteins in *E. coli* are GPCRs such as the rat neurotensin receptor[15,16] and the human adenosine A_{2a} receptor[17]. In many cases, eukaryotic membrane proteins produced in *E. coli* end up in a denatured form in inclusion bodies. There are some examples of successful refolding from these inclusion bodies;[18] however, this technique is not yet well established.[19] Therefore, the use of eukaryotic systems seems to be often the only choice for successful overproduction of eukaryotic membrane proteins.[20]

The predominant eukaryotic expression systems include yeast, insect cells and mammalian cells. Even though insect and mammalian cells have been used successfully to produce a large variety of eukaryotic membrane proteins, yeast systems are particularly attractive because they combine ease of growth and handling relative to mammalian and insect cells. Genetic manipulations in yeast are, furthermore, relatively straightforward and similar to those for bacteria. A further advantage is that almost all appropriate posttranslational modifications can be achieved in yeast. In particular, the use of *Saccharomyces cerevisiae* for functional assays as well as protein-protein interaction studies of membrane proteins has been well established.

8.2 YEAST EXPRESSION SYSTEMS USED FOR MEMBRANE PROTEIN PRODUCTION

Although yeasts such as *Hansenula polymorpha, Klyveromyces lactis, Pichia methanolica, Candida maltosa,* or *Yarrowia lipolytica* have been successfully used for recombinant production of many soluble proteins, only three yeast strains have been regularly used for heterologous expression of membrane proteins. These are *Saccharomyces cerevisiae, Schizosaccharomyces pombe,* and *Pichia pastoris*, summarized in the following sections.

8.2.1 *SACCHAROMYCES CEREVISIAE*

S. cerevisiae is probably the most intensively studied eukaryotic model organism. Consequently, it was also the first eukaryote genome to be sequenced,[21] which identified some 5400 genes distributed on 12 chromosomes.[22]

Even though *S. cerevisiae* is more complex than bacteria with a genome 3.5-fold larger than that of *E. coli*, it possesses many biotechnological advantages such as rapid growth (doubling time 90 min) relative to other eukaryotic species, simple handling, rapid isolation and characterization of mutants, a well-characterised genetic system, and the possibility of efficient DNA transformation. Furthermore, *S. cerevisiae* is non-pathogenic, which makes handling it in the laboratory unproblematic.

Many different expression vectors are available for use in *S. cerevisiae*. Recombinant DNA can be introduced either as self-replicating episomal plasmids or stably integrated into the yeast genome *via* homologous recombination.

In addition to the haploid strain S288C (*MATα SUC2 mal mel gal2 CUP1 flo1 flo8-1 hap1*), which was sequenced in 1996, many other strains with various genetic markers are available.[23]

8.2.2 *SCHIZOSACCHAROMYCES POMBE*

S. pombe, sequenced in 2002,[24] replicates both by asexual sporulation and by medial fission. No budding is observed for *S. pombe* as was previously described for *S. cerevisiae* and *P. pastoris*. *S. pombe* grows as a haploid organism with two mating types (*h+* and *h-*), whereas *S. cerevisiae* grows as a diploid. For both yeast species, secreted pheromones are needed to initiate the mating process (a- and α-factors for *S. cerevisiae* / M- and P-factors for *S. pombe*). Similar to *S. cerevisiae*, *S. pombe* is a fast replicating organism with properties such as simple handling and opportunity for rapid genetic manipulation. Specifically, *S. pombe* tolerates foreign DNA well, which makes it an attractive host for expression of higher eukaryotic genes. An advantage of *S. pombe* over *S. cerevisiae* is that initiation of transcription is similar to that in higher eukaryotes.[25] Interestingly, in addition to specific yeast promoters, some mammalian promoters such as SV40 and human cytomegalovirus (hCMV) are active in this yeast.[26] However, the *S. pombe* expression system is much less developed than those of *E. coli* or *S. cerevisiae*. Nevertheless, several *S. pombe* expression vectors have become available for recombinant protein production.[27]

8.2.3 PICHIA PASTORIS

In 1969, Ogata and co-workers discovered a yeast strain that was able to grow on methanol as a sole carbon source.[28] Only a few years later, during the 1970s, the biotechnological understanding of this organism increased as a result of Phillips Petroleum Company adapting it to large-scale fermentation in an attempt to produce yeast biomass as a high protein animal feed.[29]

Unlike *S. cerevisiae*, which grows as a diploid organism, *P. pastoris* cells remain haploid unless forced to mate. In contrast to *S. cerevisiae* system, only vectors for genomic integration are available in *P. pastoris*, and these chiefly utilize the methanol-inducible alcohol oxidase promoter (*AOX1*), one of the strongest known promoters. Recently, vectors containing the constitutive promoter of the glyceraldehyde-3-phosphate dehydrogenase (*GAP*) gene have also become commercially available.[30–33]

8.3 PRODUCTION OVERVIEW

The following section summarizes some examples, hints and references about using the different yeast systems for production of integral membrane proteins. Yeast systems have been widely used for membrane protein production with the aim of performing functional assays and *in vivo* characterization. However, only a few examples of large-scale production and purification have been reported. By definition, intergral membrane proteins have one or more membrane spanning domains, typically α-helical except for the β-barrel proteins in bacteria and mitochondria. Integral membrane proteins include enzymes that perform catalytic functions localized to the lipid bilayer in which they reside, transport proteins (*i.e.*, ion channels and transporters) that facilitate mass (ions and molecules) movement across cellular membranes, while receptors communicate signals across membranes. Whereas most membrane proteins are specialized for one of these functions, some can serve multiple functions such as is the case for members of the major facilitator superfamily (MFS).[35] Of these categories of membrane proteins, this review will only cover expression of transport proteins and receptors. Receptors are found both at the cell surface (plasma membrane) and in intracellular membranes (subcellular organelles). They can be roughly grouped into four categories based on their function which include ion channel receptors, tyrosine kinase-coupled receptors, receptors with intrinsic enzyme activity (*i.e.*, receptor tyrosine kinases, RTKs), and G protein-coupled receptors (GPCRs). We begin with a fairly comprehensive review of GPCR production in yeast systems.

8.3.1 RECEPTORS

Yeast expression systems have been used mainly for production of receptors of the GPCR superfamily.[36] GPCRs form the largest family of integral membrane proteins found in eukaryotic organisms. These receptors are involved in the recognition of light, extracellular ions and signaling molecules such as odorants, nucleotides, lipids, peptides, amino acids, and biogenic amines. GPCRs mediate their functions by interaction with heterotrimeric GTP-binding proteins (G proteins), which then trigger intracellular responses. Many of the receptors are also capable of direct interaction with alternative signaling proteins (a more detailed description of these is found in Chapter 7). GPCRs are both directly and indirectly involved in many diseases and are, therefore, attractive targets for drug development. A more detailed knowledge of the three-dimensional structure of GPCRs will provide major insight into the mechanisms of receptor activation and signal transduction and, thereby, facilitate and improve drug design.

The yeasts *S. cerevisiae*, *S. pombe,* and *P. pastoris* have all been used as expression systems for recombinant proteins for several years and a number of membrane proteins have already been successfully produced in these yeasts. In the following section, details of GPCR production in each of these yeast hosts are given (a more complete list can be found in Table 8.1).

8.3.1.1 Production of GPCRs in *S. cerevisiae*

The first GPCR produced in yeast was the human muscarinic M1 receptor.[37] The receptor gene was under the control of the alcohol dehydrogenase promoter *(ADH)* in the vector pVT102-U, which resulted in the production of a single class of binding sites with a receptor density of 20 fmol/mg corresponding to approximately 40 receptors/yeast cell. This is in agreement with experiments performed by Erlenbach and coworkers,[38] revealing production yields of approximately 5 fmol/mg for the wild-type M1, M3, and M5 receptors. The deletion of the central portion of the third intracellular loop (i3) in all three receptors resulted in a ten- to forty-fold increase in yield (5 fmol/mg for the wild-type and 53-214 fmol/mg for the Δi3 receptors). In contrast, the human β_2 adrenergic receptor was produced in this yeast at much higher levels, reaching up to 115 pmol/mg membrane protein.[39] Here, the receptor gene was under the transcriptional control of the galactose inducible *GAL*1 promoter using the multicopy vector YEp24. To reach such a high expression level, several rounds of optimization were necessary. First, the gene product of the transcriptional transactivator gene *(LAC9)* was co-expressed. Second, the N-terminal region of the β_2 adrenergic receptor was replaced by the first 14 amino acids of the yeast Ste2 receptor. Furthermore, the cells were induced with galactose in the late exponential growth phase, and the medium was supplemented with a β_2 adrenergic receptor antagonist. The conditions for fermentation of the β_2 adrenergic receptor producing yeast cells had to be subsequently established and optimized.[40] Conditions found to be of importance for production of the receptor in 1–15 liter fermentors were the maintenance of the pH at 7.2 –7.5 during growth, the use of glucose-free medium and the presence of a receptor antagonist in the growth medium. Using a 15-liter fermentor, a production level of approximately 40 pmol/mg was reached, corresponding to approximately 20–30 mg of functional receptor. This optimized approach was also used for large-scale production of the α_2-C2-adrenergic receptor.[40] In contrast to the β_2 adrenergic receptor, the α_2-C2-adrenergic receptor was not N-terminally fused to the first 14 amino acids of the Ste2 receptor. In this case, a

TABLE 8.1
Production of G Protein-Coupled Receptors in Yeast

Receptor	Species	Host	Promoter	Size [kDa]	Fusion /Tag	Production Level /Functional Assay	Ref.
Adenosine A_1	Human	*S. cerevisiae*	*n. g.*	36	÷	ST	62
Adenosine A_{2A}	Human	*S. cerevisiae*	*n. g.*	44	÷	ST	62
Adenosine A_{2A}	Human	*S. cerevisiae*	*Gal 1/10*	45	Pre-pro (synthetic) C-myc	0.6 pmol/mg (crude cell lysate) LB	63
Adenosine A_{2A}	Rat	*S. cerevisiae*	*ADH1*	44	÷	0.45 pmol/mg LB, ST	64
Adenosine A_{2B}	Human	*S. cerevisiae*	*PGK1*	36	÷	ST	65
Adrenergic a_2-C2	Human	*S. cerevisiae*	*Gal 1*	47	N-H6	30–60 pmol/mg LB	40
Adrenergic β_2	Human	*P. pastoris*	*AOX1*	46	α-MF N-FLAG C-Bio	25 pmol/mg LB	53
Adrenergic β_2	Human	*S. cerevisiae*	*Gal 1*	46	Ste2	115 pmol/mg LB, ST	39
Adrenergic β_2	Human	*S. pombe*	*ADH1*	46	Ste2	7.5 pmol/mg LB	47
Adrenergic β_2	Human	*S. cerevisiae*	*CYC-GAL*	46	÷	0.16 pmol/mg LB	66
Adrenergic β_2	Human	*S. cerevisiae*	*CYC-GAL*	43	÷	0.18 pmol/mg LB	66
Cannabinoid CB2	Human	*P. pastoris*	*AOX1*	47	α-MF C-myc C-H6	1.7 pmol/mg LB	55
Chemoattr. C5a	Human	*S. cerevisiae*	*n. g.*	39	C-CFP C-YFP	ST	67
Dopamine D_{1A}	Human	*S. cerevisiae*	*GPD*	49	Ste2 C-H6 C-FLAG	0.12 pmol/mg LB	68
Dopamine D_{2L}	Rat	*S. pombe*	*nmt1*	50	÷	1 pmol/mg LB	45
Dopamine D_{2S}	Human	*P. pastoris*	*AOX1*	51	α-MF N-FLAG N-H6 C-H10 C-Bio	100 pmol/mg LB	51
Dopamine D_{2S}	Human	*S. cerevisiae*	*PMA1*	40	STE2	1–2 pmol/mg LB	44
Dopamine D_{2S}	Human	*S. cerevisiae*	*Gal 10*	40	÷	2.8 pmol/mg LB	69

(continued)

TABLE 8.1 (CONTINUED)
Production of G Protein-Coupled Receptors in Yeast

Receptor	Species	Host	Promoter	Size [kDa]	Fusion /Tag	Production Level /Functional Assay	Ref.
Dopamine D_{2S}	Human	*S. pombe*	*nmt1*	40	÷	14.6 pmol/mg LB	44
Edg2 (Vzg-1)	Human	*S. cerevisiae*	UAS_{GAL}	41	÷	ST	70
Endothelin ET_A	Human	*P. pastoris*	*AOX1*	49	α-MF N-FLAG N-myc	5.3 pmol/mg LB	71
Endothelin ET_B	Human	*P. pastoris*	*AOX1*	49	α-MF N-FLAG C-Bio	60 pmol/mg LB	54
Endothelin ET_B	Human	*P. pastoris*	*AOX1*	49	α-MF N-FLAG C-Bio C-H10 C-GFP N-H10	60 pmol/mg LB	56
GHRH	Human	*S. cerevisiae*	*GAL 1/10*	47	÷	ST	72
Melatonin Mel_{1a}	Human	*S. cerevisiae*	*PGK*	39	÷	ST	73
Melatonin Mel_{1B}	Human	*S. cerevisiae*	*PGK1*	40	÷	ST	65
Metabotropic glutamate mGluR1a	Rat	*S. cerevisiae*	*UPR-ICL*	133	Ste2 C-H6	÷	74
Muscarinic acetylcholine M1	Human	*S. cerevisiae*	*ADH*	51	÷	0.02 pmol/mg LB, ST	37
Muscarinic acetylcholine M3	Human	*S. cerevisiae*	*GPD*	66	N-HA	0.002–0.2 pmol/mg LB, ST	38
Muscarinic Acetylcholine M5	Rat	*S. cerevisiae*	*α-factor*	60	α-MF	0.13 pmol/mg LB	75
Neurokinin NK2	Human	*S. pombe*	*nmt1*	44	aP	1.16 pmol/mg LB	46
Opsin	Bovine	*S. cerevisiae*	*GAL1*	40	÷	2 ± 0.5mg/10^{10} cells	76
Opsin	Bovine	*P. pastoris*	*AOX1*	40	÷	0.3 mg/L	77
Prostanoid FP2a	Human	*S. cerevisiae*	*GAL1*	40	÷	ST	78
Purinergic $P2Y_{12}$	Human	*S. cerevisiae*	*GPD1*	39	÷	ST	79
Purinergic $P2Y_{12}$	Human	*S. cerevisiae*	*GPD1*	39	÷	ST	79
Purinergic $P2Y_1$	Human	*S. cerevisiae*	*PGK1*	42	÷	ST	65
Purinergic $P2Y_2$	Human	*S. cerevisiae*	*PGK1*	42	÷	ST	65

(continued)

TABLE 8.1 (CONTINUED)
Production of G Protein-Coupled Receptors in Yeast

Receptor	Species	Host	Promoter	Size [kDa]	Fusion /Tag	Production Level /Functional Assay	Ref.
Serotonin 5HT$_{5A}$	Mouse	*P. pastoris*	*AOX1*	41	a-MF aP C-myc	22 pmol/mg LB	50
Serotonin 5HT$_{5A}$	Mouse	*S. cerevisiae*	*PRB1*	41	Bm	16 pmol/mg LB	42
Serotonin 5-HT$_{1A}$	Human	*S. cerevisiae*	*PGK1*	46	÷	ST	65
Serotonin 5-HT$_{1Da}$	Human	*S. cerevisiae*	*PGK1*	41	÷	ST	65
Serotonin 5HT$_{5A}$	Mouse	*P. pastoris*	*AOX1*	41	α-MF N-FLAG C-Bio C-myc	40 pmol/mg LB	53
Somatostatin SST$_2$	Human	*S. cerevisiae*	*PGK1*	41	÷	ST	65
Somatostatin SST$_5$	Human	*S. cerevisiae*	*PGK1*	39	÷	ST	65
Somatostatin SST$_2$	Rat	*S. cerevisiae*	*GAL1/10*	41	÷	0.2 pmol/mg ST	80
Vasopressin V2	Human	*S. cerevisiae*	*GAP*	40	N-HA	ST	81
VPAC1	Rat	*S. cerevisiae*	*CYCGAL*	49	α-MF C-HSV	70 pmol/mg LB	41
α-Factor Ste2	S. cerevisiae	*S. cerevisiae*	*GDP*	60	C-FLAG C-H6	350 pmol/mg LB	43
μ Opioid	Human	*P. pastoris*	*AOX1*	44	α-MF Ste2	0.4 pmol/mg LB	82
μ Opioid	Human	*P. pastoris*	*AOX1*	44	α-MF N-GFP C-myc C-H6	1 pmol/mg LB	57

C-, C-terminal; N, N-terminal; n. g., not given. **Promoters:** *ADH*, constitutive promoter for alcohol dehydrogenase; *AOX1*, alcohol oxidase 1 promoter; *CYC-GAL*, hybrid promoter containing the *GAL10* upstream activating sequence fused to the 59-nontranslated leader of the cytochrome-1 gene; *GAL1, GAL1* promoter; *GAL10, GAL10* promoter; *GAP*, glyceraldehyd-3-phosphate dehydrogenase promoter; GHRH, growth hormone–releasing hormone receptor; *GPD1*, glycerol-3-phosphate dehydrogenase promoter; *nmt1, nmt1* promoter (no message in thiamine); *PGK1*, phosphoglycerate kinase promoter; *PMA1, PMA1* promoter; *PRB1, PRB1* endopeptideas B promoter; *UAS$_{GAL}$*, hybrid promoter (upstream activation sequence of GAL1); UPR-ICL, isocitrate lyase promoter. **Fusions/Tags:** a-MF, *Sacchraromyces cerevisiae* mating type factor a; aP, acidic phosphatase; Bm, *Bacillus macerans* signalsequence; Ste2, α-factor receptor fusion; Bio, Biotinylation domain of the transcarboxylase from *P. shermanii*; CFP, cyan fluorescent protein; GFP, green fluorescent protein; YFP, yellow fluorescent protein; FLAG, FLAG-tag; H6, hexa-histidine-tag; H10, deca-histidine-tag; HA, hemagglutinin-tag; HSV, HSV-tag; myc, c-*myc*-tag. **Assays:** LB, ligand binding assay; ST, signal transduction/G-protein coupling.

receptor antagonist (phentolamine) was added to the medium during cultivation, resulting in receptor yields of up to 70 pmol/mg, which corresponds to 10–15 mg receptor from a 15-liter fermentor. The recombinant α_2-C2-adrenergic receptor appeared pharmacologically indistinguishable from the receptor expressed in mammalian cells.

Another receptor which was produced in large quantities in *S. cerevisiae* was the VPAC1 receptor.[41] The receptor gene was fused to the α-factor signal sequence, and the galactose-inducible hybrid CYCGAL promoter was used to drive expression. By using two selection markers (URA3/leud-d), clones containing up to 200 copies of the vector were obtained. Further optimization including use of amino acid-supplemented minimal medium, a temperature shift from 30 to 15°C and the use of the protease deficient yeast strain PAP1500 resulted in a production yield of approximately 66 pmol/mg membrane protein. As reported for other GPCRs produced in yeast cells, the pharmacological profile (relative affinity for various ligands) was similar to the wild-type receptor, although the K_d for ligand binding was elevated.

For production of the mouse 5-HT$_5$ receptor[42] a construct containing an N-terminal fusion to the (1–3, 1–4)-β–glucanase signal sequence from *Bacillus macerans* and a C-terminal myc-tag for immunological detection was engineered. The receptor gene was cloned under the control of the vacuolar endopeptidase B (*PRB*1) promoter, induced upon glucose depletion in the medium. Using a protease-deficient yeast strain as host, ligand binding studies using [^{3}H]-lysergic acid diethylamine (LSD) revealed production of approximately 16 pmol/mg. Furthermore, the ligand binding characteristics of the receptor produced in *S. cerevisiae* were similar to those of the receptor produced in mammalian cells. Interestingly, the production yield of functional receptor doubled after heat shock treatment.

One of the highest production levels for a GPCR ever reported was for the *S. cerevisiae* Ste2 receptor.[43] The FLAG- and His6-tagged fusion protein was homologously produced at up to 350 pmol/mg membrane protein in the protease-deficient strain BJ2168. In this case, solubilization and purification from approximately 60 g recombinant yeast cells yielded approximately 1 mg of receptor. Reconstitution into phospholipid vesicles (3:2 of POPC:POPG) was necessary to perform radioligand-binding assays, however, only 6% of the expected ligand binding was observed. Ligand binding was significantly enhanced upon addition of solubilized yeast membranes (strain lacking Ste2), suggesting that lipids other than POPC and POPG are important for activity.

8.3.1.2 Production of GPCRs in *S. pombe*

Although many GPCRs have been expressed in *S. cerevisiae* both for functional assays and large-scale production, only a few types of receptors have been expressed in the fission yeast *Schizosaccharomyces pombe*. The first GPCR produced in this yeast was the dopamine D$_{2S}$ receptor.[44] The receptor was placed under the control of the strong thiamine-repressible *nmt1* promoter. As was revealed by ligand binding, approximately 14.6 pmol receptor/mg membrane protein was produced, a higher production level than that reached in *S. cerevisiae* (~ five-fold increase). Furthermore, the affinities of tested ligands were two-fold higher in *S. pombe* than in *S. cerevisiae*, indicating that the former is better suited for recombinant production of D_{2S} receptor. In contrast, heterologous production in *S. pombe* of the rat D_{2L} receptor and the human neurokinin NK2 receptor yielded only 1 pmol/mg and 1.17 pmol/mg, respectively.[45,46]

The human β_2 adrenergic receptor was also produced in this yeast.[47] One construct replaced the first 14 amino acids by the corresponding sequence of the *STE2* gene (α-factor receptor) as previously described by King and co-workers,[39] while another construct had no fusion partner. Both constructs were placed under the constitutive promoter of the *S. pombe* alcohol dehydrogenase (*ADH*) gene. Addition of 40 µM of the receptor antagonist alprenolol to the medium resulted in an increased yield (approximately 7.5 pmol/mg membrane protein) over the production level achieved in the absence of antagonist. The pharmacological profile was similar to that of the receptor in its native tissue.

8.3.1.3 Production of GPCRs in *P. pastoris*

During the last decade, the methylotrophic yeast *P. pastoris* has become an extremely useful organism for the overproduction of recombinant proteins. Transformation into this yeast results in stable integration of the expression cassette into the host genome. Additionally *P. pastoris* has been shown to perform typical eukaryotic post-translational modifications. A large number of soluble recombinant proteins have already been produced using the strong and tightly regulated alcohol oxidase I (*AOXI*) promoter.[30] Production levels of up to several grams of recombinant soluble protein per liter culture have been obtained.[33,48] Furthermore, the possibility of high cell density fermentation[49] makes this yeast an attractive host also for heterologous production of membrane proteins. Thus far, only a few GPCRs have been produced in *P. pastoris*, however, in relatively large quantities.

The mouse 5-HT$_5$ receptor, the first GPCR expressed in *P. pastoris*, revealed a production level of approximately 22 pmol receptor/mg membrane protein.[50] In this study, three different plasmids were constructed. The receptor gene was cloned into the plasmid PHIL-D2 without modifications, whereas the two other vectors contained constructs bearing a c-myc-tag along with the *P. pastoris* acid phosphatase (*PHO*)*1* signal sequence or the *S. cerevisiae* α-factor signal sequence. It was found that fusion to the α-factor signal sequence increased the receptor yield two-fold, whereas the use of the *PHO1* signal sequence resulted in a decrease in receptor yield. These results agree with those obtained for the human D_2 receptor;[51] a construct containing no signal sequence yielded approximately 1170 receptors per cell, whereas fusion of the α-factor signal sequence increased the level by more than 20-fold. The *S. cerevisiae* α-factor signal sequence is a pre-propeptide, which is processed in two steps. The pre-peptide is cleaved from the fusion protein by a signal peptidase during insertion into the endoplasmatic recticulum membrane. The remaining a-factor prosequence is then cleaved by the Kex2 protease located in the late Golgi.[33,52] Nevertheless, it was reported for the mouse 5-HT$_5$ receptor,[50] as well as for other receptors such as the human D_2 receptor,[51] the human β2 adrenergic receptor[53], the human ET$_B$ receptor[54] and the cannabinoid CB2[55] receptor, that correct Kex processing failed. For some receptors it was shown that deletion of putative *N*-glycosylation sites at the N-terminus of the receptor resulted in improved or proper processing of the receptor. Likely, carbohydrates close to the Kex2 protease site resulted in steric hindrance and lack of Kex2 processing. N-terminal addition of tags such as the His10-tag, the FLAG-tag, the myc-tag, or C-terminal fusions of the receptor to other proteins such as the enhanced green fluorescent protein (eGFP) from the jellyfish *Aequorea victoria* did not result in significant changes in the expression levels of the GPCRs in *P. pastoris*,[53] Reinhart et al. in preparation.[51,56,57] On the other hand, for the human ET_B receptor[56] and the human β_2 adrenergic receptor (Weiß *et al.* unpublished results), C-terminal fusion of the His10-tag decreased the production level two- to three-fold when compared to a non-tagged receptor. For the human D_2 receptor, no significant changes in production yields were observed when the His-tag was placed either N- or C-terminally.[51] Interestingly, a significantly increased production level was observed when the biotinylation domain (bio-tag) of the trans-carboxylase from *Propionibacterium shermanii*[58] was fused to the C-terminus of various receptors. For instance, for the mouse 5-HT_{5A} receptor,[53] the D_{2S} receptor,[51] the β_2 adrenergic receptor (Reinhart *et al.* in preparation) and the ETB receptor[56] the number of receptors per cell was more than doubled. The same effect was observed for recombinant GPCRs produced in baculovirus-infected insect cells. Addition of the bio-tag likely stabilized the recombinant receptor either by protecting the receptor from direct degradation or by maintaining folding fidelity to avoid the unfolded protein response.

As in *S. cerevisiae*, many recombinant GPCRs expressed in *P. pastoris* exhibit a pharmacological profile similar to that of the receptor in native tissue, whereas the affinity for specific ligands was often found to be lower in yeast than in mammalian cells.[51,53,55] This might be due to the different complement of lipids in yeast relative to mammalian cell membranes (*e.g.*, ergosterol in yeast substitutes for cholesterol in mammals).[59–61] In mammalian cells, GPCRs are mainly located in the plasma membrane. In contrast, the recombinant GPCRs produced in *P. pastoris* have been localized by immunogold electron microscopy techniques, to chiefly endoplasmatic reticulum and Golgi membranes with only a minor fraction of the receptor population in the plasma membrane. Additionally, it was found that the human D_{2S}[51] and the human ET_B receptors[54] were localized to atypically induced stacked membranes located in the *P. pastoris* cytoplasm. A further prerequisite for successful overproduction of GPCRs is the choice of an appropriate host strain. In almost all cases, production yields in a protease-deficient *P. pastoris* strain were superior those obtained in the wild-type strain. There was, however, no difference in μ-opioid receptor production between the protease-deficient strain SMD1163 and the wild-type strain GS115.[57]

Further optimization of production levels can be achieved by using multicopy vectors. It has been shown for *P. pastoris* that multiple insertions of the expression cassette into the genome have a positive effect on production levels. For the β2 adrenergic receptor the highest production level was reached with four copies of the expression cassette. For the 5-HT_{5A} receptor, as well as for the D_2 receptor, the maximum production level was achieved with a copy number of two.[51] In contrast, for the μ-opioid receptor, the copy number had no effect on the production level.[57] As already reported by King and co-workers[39] and others,[40,51,53] supplementation of specific receptor agonists or antagonists to the medium during methanol induction resulted, in many cases, in significant increase in production level.

To date, high-density fermentation has been achieved for the β2 adrenergic receptor (Reinhart *et al.* in preparation), the mouse 5-HT$_5$ receptor (Weiß *et al.* personal communication) and the ET$_B$ receptor.[56] For the ETB receptor, a production level of 45 pmol/mg (corresponding to 9 mg/liter) was achieved by high density fermentation resulting in ~3 kg cells harvested from an 8 L fermentor. Subsequent solubilization and purification in the presence of the agonist peptide ET-1 resulted in a highly pure preparation. Nevertheless, as revealed by analytical gel filtration, the purified receptors aggregated in some of the preparations. In addition to the ETB receptor and the β$_2$ adrenergic receptor (Reinhart et al. in preparation), the cannabinoid CB2[55] receptor was solubilized and purified after production in *P. pastoris*. However, no ligand binding was observed for purified CB2.

8.3.2 TRANSPORT PROTEINS

Genome sequencing projects have revealed that the number of transport proteins (*i.e.* ion channels and transporters) found in various organisms is roughly proportional to the size of their genome.[83,84] Combined information from sequencing projects and functional studies reveals that, for organisms on all evolutionary levels, similar mechanisms for energy-coupling and similar designs of transport proteins are found.[85] This underlines the importance of this class of membrane proteins.

Why produce transport proteins in yeast? Ease of genetic manipulation, rapid growth on inexpensive media, as well as rapid and efficient screening for phenotypes make yeast ideal experimental systems for performing advanced techniques that elucidate protein function, interactions and structure. One example of this, functional complementation, is a method to determine the specific function of an uncharacterized protein by using host deletion stains with discernable non-native phenotypes.[86] Well characterized, easy to use yeast genetics enabled the creation of strains with deficiencies in almost every discrete transport function. Expression of any gene of interest, either homologous or heterologous, may complement for the missing function. The screening process is then greatly simplified by requiring only screening for growth. Furthermore, this screening can be carried out in parallel as a high throughput approach. To this end, Zhang and colleagues applied functional complementation in a high throughput (HT) manner to screen cDNA libraries.[87] Specifically, they constructed a *S. cerevisiae* system with two regulated promoters, allowing tight control of a variety of distinct and essential yeast genes. By screening three partial human cDNA libraries, they were able to isolate six human cDNAs capable of complementing essential yeast genes. As a result, screening appropriate cDNA-libraries for complementation of yeast strains with defined transport deficiencies now enables one to identify and characterize unknown transport proteins. Boles and coworkers[88] further developed this approach to identify hexose transport proteins previously uncharacterized in *Pichia stipitis*. They created the *hxt* null mutant, an *S. cerevisiae* strain completely devoid for its endogenous hexose transporters. To create this mutant phenotype, seventeen hexose transporters (Hxt1-11, Hxt13-17, Gal2) and, additionally, three maltose transporter homologues (Agt1, Mph2, Mph3) were deleted. By screening for restored growth of the complemented *hxt* null mutant in the presence of glucose as the sole carbon source, they identified and characterized *Pichia stipitis* genes SUT1, SUT2 and SUT3, encoding glucose transporters. Substrate specificity, as well as kinetic parameters of all three heterologously produced Sut proteins, was subsequently determined. Furthermore, the authors used functional complementation in order to study mammalian glucose transporters. Unexpectedly, initial trials expressing mammalian GLUT1 or GLUT4 in the *S. cerevisiae hxt* null mutant did not restore uptake of glucose. Introduction of additional mutations either in the yeast genome or in the exogenous GLUT1 or GLUT4 genes were required to restore growth on glucose.[89] Moreover, they determined the kinetics of glucose transport for intact yeast cells by influx measurements. Finally, the yeast glycerol efflux channel Fps1p is another example of the use of functional complementation to study transport proteins. Bill and colleagues combined functional complementation (using a yeast strain deleted for endogenous *FPS1*) with mutational analysis to obtain insight into the structure and function of Fps1p.[90]

In addition to proteins facilitating transport of molecules as for the glucose transporters mentioned above, several ion-channel proteins were expressed using yeast systems in combination with functional complementation. To characterize the putative K^+-channel KAT1 from *Arabidopsis thaliana*, Anderson and colleagues[91] expressed the channel in a *S. cerevisiae* strain lacking TRK1 and TRK2, two proteins of the K^+-transport system. To enable growth, this trk1Δ/trk2Δ double deletion mutant requires approximately 100-fold higher K^+ concentration in the growth medium compared to the wild-type. Anderson and colleagues showed that expression of KAT1 in the double deletion strain re-established the ability to grow in low K^+ media. Growth inhibition of the KAT1-complemented trk1Δ/trk2Δ strain was evident in the presence of tetraethylammonium (TEA$^+$) or Ba$^+$ ions which block K^+ ion flux. This was the first indication that KAT1 belongs to the shaker-related family of inward rectifying K^+-channels. Direct proof that KAT1 is an inward rectifying K^+-channel was obtained in 1995 by Bertl and colleagues.[92] They performed patch-clamp measurements on spheroplasted yeast cells expressing KAT1 and found channel properties typical for this class. In another study, a different *A. thaliana* K^+ channel, AKT1, was characterized by Sentenac and co-workers in 1992.[93] They showed that expression of AKT1 complemented the trk1Δ deletion in *S. cerevisiae*. Functional expression of the putative K^+-channel was confirmed by K^+ and Rb$^+$ uptake assays.

Beside plant channels, mammalian K^+ channels have been expressed in trk-deletion strains. In 1995, Tang and colleagues complemented the trk1Δ/trk2Δ strain by expression of the Guinea pig K^+ channel gpIRK1.[94] Two-electrode voltage clamp recording of gpIRK1 expressed in *Xenopus* oocytes demonstrated channel properties characteristic of inwardly rectifying K^+ channels.

In addition to functional complementation, other advanced techniques addressing functional and structural analysis that are well suited to yeast expression systems include functional assays (requiring correct subcellular localization), scanning cysteine mutagenesis and random mutagenesis. Specific examples of these techniques, as well as the use of yeast systems for large scale production of protein for structural analysis follow.

One complication that can arise in heterologous expresion of membrane proteins is targeting to incorrect subcellular compartment membranes potentially compromising their function. Another feature making yeast a very attractive host system for the study of transport proteins is the fact that targeted localization of membrane protein can be achieved. Correct localization of a heterologously produced transport protein is often a precondition in order to perform functional assays. Studying the insect aquaporin AQPcic, Lagrée and co-workers[95] found that one mutant failed to be expressed in active form in *Xenopus* oocytes, possibly due to mistargeting. Using *S. cerevisiae* as the host system, the same mutant was functionally expressed and could be characterized.

To investigate the structure-function relationship of yet uncharacterized membrane proteins, mutational analysis is a key technique suited to yeast systems. Cysteine-scanning mutagenesis is specially applied to localize structural features of proteins, whereas, random mutagenesis, as a more general approach, points to specific structure-function characteristics of proteins. To map membrane protein topology and surface accessibility, cysteine-scanning mutagenesis is an established approach. Stephens and co-workers[96] used *S. cerevisiae*-based cysteine-scanning mutagenesis to investigate the yeast mitochondrial F_1F_0-ATP synthase. They constructed 46 single cysteine substitution mutants distributed throughout subunit 8. Each of the mutants was expressed in a subunit 8 deficient *S. cerevisiae* strain, which led to restoration of a functional phenotype. Using a fluorescent thiol-modifying probe, they tested the aqueous environment accessibility of each mutated cysteine. This study, in combinantion with results from cross-linking experiments, concluded that subunit 8 is an integral component of the stator stalk of F_1F_0-ATP synthase.

Random mutagenesis approaches require efficient mutant screening capabilities. Yeasts are ideal for this purpose. Soteropoulos and Perlin[97] used random mutagenesis to study an *S. cerevisiae* plasma membrane H^+-ATPase. Mutations of residue Ile 183 located in the stalk region of the enzyme lead to partial uncoupling of ATP hydrolysis and proton transport. The focus was now to explore the role of the side group at this position and it was found that substitutions were tolerated, although three of these required secondary site mutations for stabilization at the C-terminus. These suppressor mutants provided valuable topological information, suggesting the close proximity of both mutation sites.

In vivo selection of suppressor mutations as a tool for study of protein tertiary structure was also used by Proff and Kolling[98] to investigate the yeast a-factor transporter Ste6p, a protein closely related to the human multidrug resistance protein 1 (MDR1). Using a random mutagenesis approach, a set of 26 Ste6 mutants was constructed, many of which were correctly targeted to the plasma membrane. To identify individual residues in Ste6 that interact with the ABC transporter signature motif (LSGGQ), Proff and Kolling screened for intragenic revertants of the LSGGQ mutant S507N. They identified suppressor mutations localized in TM12, as well as upstream of TM6. Surprisingly these mutants also suppressed the effect of mutation G397D in the Walker A site that should impair ATP-binding and hydrolysis at the nucleotide binding domain 1. Further analysis of suppressor mutants demonstrated that the two nucleotide binding domains of this ABC transporter are not equivalent and that they interact.

In the examples given thus far, the aim of transport protein production was structure-function analysis in native membranes. Following, we discuss some examples of overproduction in yeast for isolation and purification of recombinant transport proteins for *in vitro* function analysis and structure determination. Although only a few studies have been published to date, yeast systems have enabled the production of milligram amounts of functional, homogenous membrane protein.

Lerner-Marmarosh and colleagues[99] used high-density fermentation of *P. pastoris* for large-scale production of the P-glycoprotein (Pgp), an ABC transporter encoded by the mouse *mdr3* gene. From a single fermentation run, a yield of more than 4 kg of cells (425 g cells/liter) was obtained. After microsomal fraction preparation, Pgp was solubilized with n-dodecyl-β-D-maltoside and purified by Ni-NTA and subsequent DEAE-cellulose chromatographies. Starting from a 120 g fermentation batch approximately 6 mg of recombinant Mdr3 Pgp was purified to homogeneity. Nucleotide binding was analyzed using 2'-(3')-*O*-(2,4,6-trinitrophenyl)adenosine tri- and diphosphates. The addition of asolectin resulted in an approximately five-fold increase in activity.

The water channel aquaporin was heterologously produced in *P. pastoris* and purified by Karlsson and co-workers.[100] Aquaporin PM28A, which is one of the most prominent integral proteins in the plasma membrane of spinach leaf cells, was produced as a C-terminally His- and myc-tagged protein. After French press cell lysis and subsequent membrane preparation, peripheral membrane proteins were removed by a urea/alkali treatment. PM28A was solubilized with octyl-β-D-thioglucopyranoside and purification achieved by single step affinity chromatography using Ni-NTA resin. Twenty milligrams of integral membrane protein yielded ~8 mg of PM28A. Furthermore, it was possible to characterize PM28A in reconstituted proteoliposomes by stopped-flow light spectroscopy.

Finally, the rat neuronal delayed rectifying voltage-sensitive K^+ channel was heterologously produced in *P. pastoris*, purified and subsequently used for cryo-EM structural analysis of both single-particles and reconstituted 2D-crystals. Parcej and Eckhardt-Strelau[101] reconstructed a 3D volume from single particle analysis of this channel at 2.1 nm resolution. Voltage-dependent (Kv) K^+ channels have been shown to function as octamers comprised of four α and four β subunits of various isoform composition. The a subunits are encoded by genes of the Kv1 subfamily, with Kv1.2 being the most abundant subtype, whereas the β subunits are encoded by the Kvb gene family, the major isoform being Kvβ2. High-affinity binding of ^{125}I-labeled α-dendrotoxin[101] demonstrated proper folding of the Kv1.2 subunit. Co-expression of both subunits resulted in Kv1.2 and Kvβ2 heterooligomers, resembeling the organization of the native channel. Expression of affinity tagged constructs enabled scale-up production of homo- and hetero-oligomeric channels to milligram quantities. Production of mutant channels, deleting putative N-glycosylation and phosphorylation sites, yielded a homogeneous preparation. Crude membranes were prepared using a Microfluidiser and subsequently solubilised. In order to selectively purify heterooligomers, an N-terminal polyhistidine tag on the Kv1.2 subunit and a StrepII-tag on the Kv1.2 subunit were used to facilitate a two step purification using Ni-NTA and Streptactin resins. Additional examples for transport proteins produced in yeast are cited in Table 8.2.

TABLE 8.2
Production of Transport Proteins in Yeasts

Protein	Species	Host	TMs	Size [kDa]	Promoter	Fusion/Tag	Production Level / Assay	Ref.
ABC transporter MRP6	Rat	*P. pastoris*	17	165	*AOX1*	C-HA-H6	EAA	102
Antimalarial drug resistance protein Pfcrt	Plasmodium falciparum	*P. pastoris*	10	57.7	*AOX1*	Bad	EAA	103
Antimalarial drug resistance protein Pfcrt	Plasmodium falciparum	*S. cerevisiae*	10	57.7	*Ste6*	Bad	EAA	103
Aquaporin AQP1 Tandem dimers	Human	*S. cerevisiae*	24	112	*GALi*	÷	UA	104
Aquaporin PM28A	Spinach leaf	*P. pastoris*	6	32.4	*AOX1*	myc H6	25 mg/l UA	100
Bacteriorhodopsin	*Halobacterium halobium*	*S. pombe*	7	26	*ADH*	÷	EAA	105
Brest cancer resistance protein BCRP	Human	*P. pastoris*	6	62	*AOX1*	C-H10	EAA	106
Ca²⁺-ATPase SERCA1a	Rabbit	*S. cerevisiae*	10	110	*GAL10*	C-H6	300 µg purifed/l culture EAA	107
Ca²⁺-channel OsTPC1	Oryza sativa	*S. cerevisiae*	12	n. g.	*GAP*	÷	Ca²⁺ accumulation	108
cGMP-gated cation channel subunit 1	Bovine	*S. cerevisiae*	n. g.	78	*PMA1*	÷	÷	109
Drug efflux pump CDr1p	Candida albicans	*S. cerevisiae*	12	170	*PDR5*	÷	Drug resistance assay EAA, UA	110

Glucose transporter GLUT1	Human	S. cerevisiae	12	36	HXT2	+	UA Zero-trans influx meassurements	89
Glucose transporter GLUT4	Human	S. cerevisiae	12	50	HXT2	+	UA Zero-trans influx meassurements	89
Glucose transporter SUT1	Pichia stipitis	S. cerevisiae	12	61	n. g.	+	UA	88
Glucose transporter SUT2	Pichia stipitis	S. cerevisiae	12	61	n. g.	+	UA	88
Glucose transporter SUT3	Pichia stipitis	S. cerevisiae	12	61	n. g.	+	UA	88
Intestinal peptide transporter hPEPT1	Human	P. pastoris	n. g.	71	GAP	C-myc-His	64 pmol/mg membrane protein transport assay	111
K+ channel αβENaC	Mammalian	S. cerevisiae	n. g.	n. g.	HSE	+	GA	112
K+ channel AKT1	A. thaliana	S. cerevisiae	6	95.4	ADH1, MET25	N-NubG C-CubPLV	mbSUS	113
K+ channel AKT1	A. thaliana	S. cerevisiae	6	95.4	PGK	+	functional complementation, UA	93
K+ channel AKT2	A. thaliana	S. cerevisiae	6	n. g.	ADH1, MET25	C-CubPLV	mbSUS	113
K+ channel GIRK2	Mammalian	S. cerevisiae	2	47.5	MET25	+	functional complementation	114
K+ channel gpIRK1	Guinea pig	S. cerevisiae	6	47	PGK GAL1	+	functional complementation	94
K+ channel KAT1	A. thaliana	S. cerevisiae	6	78	ADH1, MET25	C-CubPLV C-NubG	mbSUS	113

(continued)

TABLE 8.2 (CONTINUED)
Production of Transport Proteins in Yeasts

Protein	Species	Host	TMs	Size [kDa]	Promoter	Fusion /Tag	Production Level / Assay	Ref.
K+ channel KAT1	A. thaliana	S. cerevisiae	6	78	GAL	÷	functional complementation	91
K+ channel KAT1	A. thaliana	S. cerevisiae	6	78	GAL	÷	whole cell patch-clamp	92
K+ channel Kir 6.1	Mammalian	S. cerevisiae	n. g.	n. g.	PGK	÷	functional complementation, drug sensitivity assay	115
K+ channel Kir 6.2	Mouse	S. cerevisiae	n. g.	n. g.	PGK	GFP	functional complementation, drug sensitivity assay	115
K+ transporter HKT1	T. aestivum	S. cerevisiae	10-12	58.9	GAL	÷	UA, functional complementation	116
Mitochondrial F_1F_0-ATPase Subunit 8	S. cerevisiae	S. cerevisiae	n. g.	5.5	n. g.	N-HA C-FLAG	functional complementation	96
Monamine oxidase A	Human	P. pastoris	u. c.	60.5	AOX1	÷	330 mg/L (Fermenter) 115 mg/L purified EAA	117
Monamine oxidase A	Human	S. cerevisiae	u. c.	60.5	n. g.	n. g.	7.5 mg/L (Fermenter)	117*
Monamine oxidase B	Human	P. pastoris	u. c.	60.5	AOX1	÷	100 mg/L (purified) EAA	118

Protein	Source			Promoter	Tag	Yield/Detection	Ref
Multidrug resistance protein 1 MPR1	Human	17	165	*PMA1*	N-FLAG N-H9 C-GFP C-FLAG C-H10 C-H9 C-H6	0.9 mg/l UA	119
Multidrug resistance protein 1 MPR1	Human	17	165	*AOX1*	C-HA-H6	n.d. UA	120
Na$^+$, K$^+$ -ATPase (α1, β1)	Pig	α 10 β 1	α1:112 β1: 47	*CYC-GAL*	÷	54 μg/g calls EAA	121
Na$^+$, K$^+$ -ATPase (α1, β1)	Pig	α 10 β 1	α1:112 β1: 47	*AOX1*	÷	50 pmol/mg EAA	122
Nicotinic acetyl-choline receptor	*Torpedo californica*	20	280	*ADC1*	÷	n.d. Immunoblot	123
Nucleotide sugar transporter SLC35D2	Human	10	37	*CUP1*	C-HA	UA	124
Oxidosqualene cyclase (hOSC)	Human	MT	85.6	*AOX1*	C-Myc C-His6	120 mg/L (purified) EAA	125
P-Glycoprotein MDR1	Human	12	170	*PMA1*	C-H10	2.5 mg Pgp purified from 1 g membrane EAA	126
P-Glycoprotein P-gp MDR3	Mouse	12	140	*AOX1*	C-H6	5 mg Pgp / 100 g cells EAA	99
P-Glycoprotein P-gp MDR3	Mouse	12	140	*AOX1*	C-H6-Bad	0.8 mg from 100 mg membranes EAA	127
Receptor tyrosine kinase ErbB3	Mammalian	1	148	*CYC1*	÷	MbYTH	128

S. cerevisiae ... *P. pastoris* ... *S. cerevisiae* ... *P. pastoris* ... *S. cerevisiae* ... *S. cerevisiae* ... *P. pastoris* ... *S. cerevisiae* ... *P. pastoris* ... *P. pastoris* ... *S. cerevisiae*

(continued)

TABLE 8.2 (CONTINUED)
Production of Transport Proteins in Yeasts

Protein	Species	Host	Size [kDa]	TMs	Promoter	Fusion /Tag	Production Level / Assay	Ref.
Serotonin transporter rSERT	Rat	P. pastoris	75	12	AOX1	÷	2-3 mg/l (non-functional) UA	129
Transferrin receptor	Human	S. cerevisiae	95 x2	n. g.	ADHi	÷	1.1 pmol/10^8 cells LB	130
Vanilloid receptor	Human	S. cerevisiae	95	6	PMA1	C-H12 C-eGFP	1mg purified from 16 l fermentation Ca^{2+} UA	131
Voltage- sensitive K+ Channel Kv1.2/Kvb2	Rat	P. pastoris	56.7	6	AOXi	N-His (Kv1.2) N-StrepII (Kvb2)	72 pmol/mg LB	101
Yeast a-factor transporter Ste6	S. cerevisiae	S. cerevisiae	145	12	n. g.	÷	Quantitative mating assay	98

* Nandigama unpublished results.

C-, C-terminal; N-, N-terminal; TM, transmembrane spanning regions; n.d., not determined; n. g., not given. **Promoters:** *ADC1*, yeast alcohol dehydrogenase I promoter; *ADH*, constitutive promoter for alcohol dehydrogenase; *AOX1*, alcohol oxidase 1 promoter; *CYC1*, cytochrome-C oxidase promoter; *CYC-GAL*, hybrid promoter containing the GAL10 upstream activating sequence fused to the 59-nontranslated leader of the cytochrome-1 gene; *GAL1, GAL1* promoter; *GAL10, GAL10* promoter; *GAP*, glyceraldehydes-3-phosphate dehydrogenase promoter; *HSE*, heat shock element; *HXT2*, yeast hexose transporter *HXT2* promoter; *MET25*, Met-repressible *MET25* promoter; *PDR5*, multidrug resistance transporter promoter (*S. cerevisiae*); *PGK*, phosphoglycerate kinase promoter; *PMA1, PMA1* promoter; *Ste6*, Yeast a-factor transporter promoter. **Fusions/Tags:** α-MF; *S. cerevisiae* mating type factor α; aP, acidic phosphatase; Bad, Biotin acceptor domain; CubPLV, C-terminal fragment of ubiquitin fused to protein A-LexA-VP16; eGFP, enhanced green fluorescent protein; FLAG, FLAG-tag; HA, haemagglutinin-tag; H6, hexa-histidine-tag; H9, nona-histidine-tag; H10, deca-histidine-tag; myc, c-*myc*-tag; NubG, N-terminal fragment of ubiquitin (wild-type isoleucine residue at position 13 replaced with glycine); StrepII, Strep-tag II. **Assays:** EAA, enzymatic activity assay; GA, growth assay; LB, ligand binding assay; u. c., uncertain classification; mbSUS, mating-based split ubiquitin system; MbYTH, membrane based yeast two hybrid system; UA, uptake assay.
Source: From Mathai, J.C. and Agre, P. *Biochemistry,* 38, 923–928, 1999. With permission.

8.4 FUNCTIONAL ASSAYS IN YEAST SYSTEMS

8.4.1 PHARMACOLOGICAL LIGAND SCREENING ASSAYS FOR GPCRs

In the past, pharmacological ligand screening of GPCRs was performed in animal tissues. Currently, the introduction of highly sensitive cell-based assays suitable for high-throughput screening (HTS) has replaced animal tissue screens. HTS approaches enable fast and selective identification of lead compounds for GPCRs. Cell-based assays use functional coupling of reporter genes with heterologously produced GPCRs. Usually, modified mammalian cell-based systems are used for this purpose.[132] However, *S. cerevisiae*–based assays have also been developed as sensitive and robust alternatives to mammalian cell culture.

Several yeasts, including *S. cerevisiae*, have endogenous signaling pathways that are based on GPCRs and G proteins.[133] Therefore, non-native GPCRs, such as those from mammalian cells, can be studied in yeast. In fact, it is now possible to couple an exogenous GPCR to an endogenous yeast signaling pathway.

First we explain the endogenous signaling pathways in *S. cerevisiae* and then describe the development of modifications necessary to study heterologously expressed GPCRs.

S. cerevisiae grows as a diploid organism, but starvation-induced meiosis generates two haploid mating types (MATa/MATα). Both types secrete type-specific tridecapeptide pheromone (a- or α-factor), which can bind to the corresponding receptor (a-factor receptor/Ste3p or α-factor receptor/Ste2p) on the complementary cell. Upon ligand binding, the activated receptor interacts with the α-subunit of a heterotrimeric G protein. The heterotrimeric G protein is composed of the α-subunit Gpa1p (also named Scg1), the β-subunit Ste4 and the γ-subunit Ste18p. Activation of the α-subunit Gpa1p by the receptor (Ste2p or Ste3p) leads to exchange of GTP for GDP on the α-subunit and concomitant dissociation of βγ-heterodimer from the α-subunit. The βγ-heterodimer activates a MAP kinase module consisting of Ste11p (MEK kinase), Ste7p (MAP-kinase kinase or MEK), Fus3p/Kss1 and Ste5. Additionally, the βγ-dimer activates protein kinase Ste20. Fus3p, as part of the kinase module, activates the cyclin-dependent kinase inhibitor Far1p, as well as the transcription factor complex Ste12/Dig1/Dig2. This leads to cell-cycle arrest in G1 and the expression of genes such as *FUS1, FUS2, FIG1* required for the mating process (for detailed review, see[134]).

Ditzel and co-workers[135] showed that the rat G protein subunit $G_{\alpha s}$, expressed under the control of the yeast *CUP1* promoter, was able to couple to downstream effectors in a *scg1* null mutant and complemented growth as well as morphological defects in this deletion strain. Three years later, this study was extended by the work of Kang and co-workers,[136] who introduced other G protein subunits such as $G_{\alpha i2}$ and $G_{\alpha o}$ as well as hybrids consisting of yeast Scg1 and mammalian G_α subunits. All three hybrids Scg-αs, Scg-αi und Scg-αo were able to complement the *scg1* growth defect, whereas the resulting cells were sterile. Surprisingly, the Scg-αs hybrid, consisting of mammalian $G_{\alpha s}$ (which is able to bind yeast βγ) and Scg1 failed to produce a clearly defined phenotype.

Several ways have been developed for coupling heterologously expressed (non-native) GPCRs with the native or modified yeast transduction pathways. In the simplest case a heterologously expresses GPCR will couple directly to *S. cerevisiae* G protein pathways and, hence, ligand binding can be detected. Co-transformation of a mammalian GPCR along with its complementary mammalian G protein can also activate an endogenous yeast signaling pathway. Alternatively, when ligand-bound GPCR can not activate an endogenous signaling pathway, it has been shown that a chimera consisting of mammalian and yeast α-subunit sequences can induce signal transduction. The following examples illustrate both approaches.

King and co-workers were able to demonstrate functional coupling of a human GPCR to the pheromone-response pathway of yeast.[39] The β2 adrenergic receptor was co-transformed with the rat G protein Gαs into an *S. cerevisiae* GPA1 deletion strain under the control of the galactose-inducible *Gal1* promoter. The strain was modified by the integration of a *FUS1-lacZ* gene fusion as a marker into the yeast genome. Radioligand binding assays on yeast membranes showed high levels of recombinant receptor (115 pmol/mg membrane protein). Addition of the agonist isoproterolol resulted in a change in cell morphology as well as in cell-cycle arrest. Furthermore, increased galactosidase-activity was measured confirming coupling. The induction of the *FUS1-lacZ* reporter gene was blocked by addition of the antagonist Alprenolol. As the full-length Gα-protein exhibits only a low affinity to the yeast G$_{\beta\gamma}$ heterodimer, high expression levels were necessary to quench G$_{\beta\gamma}$ signaling.

Another receptor, the rat somatostatin receptor type 2 (SSTR2), could also be coupled to the yeast pheromone pathway. It was shown that this receptor couples to the G proteins Gα_{i2} and Gα_{i3}.[137,138] For functional assays several modifications were introduced: the FAR1 gene was deleted in order to allow continued cell growth, and the transcriptional control element of the FUS1 was fused with the HIS3 gene. With these modifications, the expression of HIS3 was now under the control of the pheromone response pathway. Therefore, receptor activation by agonist binding led to expression of the HIS3 gene through induction of the *FUS1* promoter, enabling the cells to grow on medium lacking histidine. Furthermore GPA1 was deleted and replaced by a rat G$_\alpha$ gene or a chimera of Gpa1-Gα_{i2}.[139] Growth was induced in transformed yeast cells bearing the SSTR2 expression plasmid by adding the agonistic peptide S-14. This agonist-promoted response could be increased by inactivating the SST2 gene. Interestingly, yeast cells bearing only Gpa1p and not the chimera Gpa1-Gαi2 also showed growth response to the agonist S-14, implying that SSTR2 functionally couples to the yeast G protein, therefore stimulating the downstream signaling pathway. Yeast cells containing the pheromone receptor Ste2p and SSTR2 as well as the chimera Gpa1-Gαi2 showed growth response to the a-pheromone peptide, suggesting coupling between chimeric G protein and the yeast GPCR Ste2p.

Brown and co-workers[65] showed that chimeras consisting of yeast Gpa1p and five C-terminal amino acids of a mammalian G-α subunit (so called Gpa1/G$_\alpha$ transplants) were able to couple to the yeast pheromone pathway with higher efficiency than chimeras with a larger portion of the mammalian G$_\alpha$ protein. In their experiments, 8 human GPCRs (somatostatin SST$_2$ and SST$_5$, serotonin 5-HT$_{1A}$ and

5-HT$_{1D}$, melatonin ML$_{1B}$, purinergic P2Y$_1$, and P2Y$_2$ and adenosine A2B receptors) were co-expressed with a large number of Gpa1/G$_\alpha$ chimeras and Gpa1/G$_\alpha$ transplants. Using chimeras derived from G$_{\alpha i/o}$, G$_{\alpha s}$, G$_{\alpha q}$ and G$_{\alpha 12}$ subtypes representing the major G protein classes, all GPCRs except P2Y$_1$ and P2Y$_2$ where able to couple to the yeast pheromone response pathway upon agonist activation. Nevertheless, the Gpa1/G$_\alpha$ transplants showed much greater agonist sensitivity than the chimeras and were able to couple all GPCRs. Furthermore, the Gpa1/G$_{\alpha 16}$ transplant was able to couple to six out of eight human receptors due to its promiscuous nature,[140,141] potentially a valuable tool for deorphanizing receptors in a high throughput manner.

In addition to ligand screening, this yeast system was also used by Noble and coworkers[142] for interaction studies between GPCRs and G protein-coupled receptor kinases (GRKs). GRKs specifically bind and phosphorylate agonist-activated GPCRs. Subsequent arrestin binding leads to desensitization of the receptor.[143] As little was known about GRK/receptor interaction, a yeast based assay was developed to examine these processes in more detail. The mammalian somatostatin receptor subtype 2 was coupled to the yeast pheromone response pathway as described by Price and co-workers.[139] The G protein-coupled receptor kinases GRK2, GRK5 and their catalytically inactive counterparts GRK2-K220R and GRK5-K215R were expressed under the control of a galactose-inducible, glucose-repressible promoter. In the presence of galactose, GRK2 and GRK5 were able to block SSTR2 promoted growth, whereas the inactive forms did not, confirming correct binding and function of the kinase. Based on these experiments a model was proposed whereby the N-terminal part of GRK5 mediates phosphorylation and is therefore indispensable for receptor phosphorylation.

In the past, mutant analysis of GPCRs was carried out by site-directed mutagenesis. Many valuable results, especially information about amino acid residues involved in ligand binding or G protein coupling, were obtained by this method. With random mutagenesis followed by a yeast genetic screen, it was possible to rapidly screen thousands of mutant receptors for their ability to couple to certain G proteins. This approach was used for the vasopressin V2 receptor, which normally couples to the G protein G$_{\alpha s}$ but not to G$_{\alpha q}$.[81] The second intracellular loop (i2 loop), which was thought to be involved in G protein coupling, was subjected to random mutagenesis. Subsequently, a yeast library was created in which individual mutant receptors were co-expressed with the G protein chimera Gpa1p/G$_{\alpha q}$ which failed to couple to the wild-type V2 receptor. The screening of the mutant library, totalling around 30,000 mutants, yielded 4 mutant receptors able to functionally couple to the Gpa1p/G$_{\alpha q}$ protein. Interestingly, it was found that all 4 mutants showed modification at position Met145, which is in the center of the i2 loop, supporting the model in which this loop is responsible for selective coupling to the G protein.

8.4.2 Detection of Interactions of Membrane Proteins: The Split-Ubiquitin Membrane Yeast Two-Hybrid System

Genome-sequencing projects have provided the primary sequences of an overwhelming number of genes from a variety of organisms. The major task now has shifted from identifying new genes to assigning functions to the thousands of proteins encoded by as yet uncharacterized genes[144]. As protein activity can be modulated, in part, by protein-protein interactions, this is now beginning to be investigated.

The major techniques available to study protein-protein interactions can be divided into biochemical and genetic. Biochemical methods such as co-fractionation/co-purification, co-immunoprecipitation and cross-linking, directly assess identity of the interacting proteins. As this generally, and for membrane proteins in particular, requires extensive optimization for each protein complex to be studied, until recently, biochemical methods were considered to be too difficult and time-consuming for large numbers of targets. However, at least for soluble proteins, this has improved significantly with the introduction of mass spectrometry-based interactive proteomics (for a review see 145). In this approach, affinity-tagged proteins are expressed in yeast as baits to capture and identify interacting proteins. Isolated complexes are analyzed by gel electrophoresis and mass spectrometry. Even though this method has proven to be powerful and sufficiently robust for high throughput approaches,[146,147] there are several drawbacks: firstly, distinguishing between specific and nonspecific interactions remains difficult and, secondly, this system is not yet applicable to membrane proteins. Other biochemical techniques such as stable isotope labeling of amino acids in cell culture (SILAC), SH2 profiling and target-assisted iterative screening (TAIS) are reviewed elsewhere.[148]

Genetic methods are based on indirect in vivo detection of protein interactions by outputs generated through the manipulation of already existing or engineered genetic networks (thus abolishing all *in vitro* handling of proteins). These systems have become popular tools in analyzing protein-protein interactions because they require little individual optimization and are well suited for HT approaches.[149] The yeast two-hybrid system (YTH) developed by Fields and Song in 1989,[150] opened a wide field. Their system is based on the modular nature of transcription factors, which can be separated into a DNA-binding Domain (DBD) and a transcriptional activation domain (TAD). To test for interaction of the two proteins of interest (termed bait and prey), one of them is fused to the TAD while the other is fused to the DBD. If the proteins interact, TAD and DBD are brought in close proximity, reconstituting the transcription factor activity. Monitoring the activity of an exogenous reporter gene reveals the protein interaction.

To enable study of more complex interactions involving three partners, different forms of three-hybrid systems have been developed (reviewed in 151). The different forms can be used to investigate a variety of multiple interactions of which proteins are capable. All these formats have in common that they employ two hybrid proteins, one of them including the TAD while the other one the DBD. A third (macro)molecule interacting with both hybrid proteins brings them into close proximity and reconstitutes the transcription factor activity. The nature of this third partner which is probed for interaction can vary considerably; thus far, proteins, peptides, small organic molecules and hybrid RNAs have been used.

Although the yeast two- and three-hybrid systems are useful and quite powerful techniques, there are inherent limitations:[152] as transcription factors are used as sensors for the protein interaction, all baits and preys have to be targeted to the nucleus, excluding all proteins from the assay that can not be expressed or correctly folded in the nuclear milieu. It is obvious that this restriction especially affects integral membrane proteins as they tend to form inactive aggregates when expressed in the nucleus. To make membrane proteins accessible to this method, they have to be portioned into fragments as they are re-localized to the nucleus. Examples for using YTH to investigate membrane proteins in this fractionated fashion are the detection of biochemical interactions between the dopamine D_3 receptor and elongation factor-1B$\beta\gamma$[153] as well as the exclusion of homo- or hetero-dimerisation of the muscarinic receptor subtypes by direct protein-protein interaction through intracellular and extra cellular regions.[154] In a homolog approach, the peptide ligand three-hybrid system has been used to study binding of the soluble extracellular domains of the growth hormone receptor as well as the vascular endothelial growth factor receptor to their native peptide ligands.[155]

To overcome these limitations regarding membrane proteins, Stagljar and co-workers developed the split-ubiquitin membrane yeast two-hybrid (MbYTH) system in 1998.[156] This system is specifically designed for detection of interactions between two membrane proteins as well as between a membrane protein and a soluble protein. It can be used to confirm the dimerisation of known membrane proteins, identifying critical amino acids involved in interaction, or for identifying novel interactions by screening cDNA libraries against the protein of interest. Instead of using artificial transcription factors as yeast two- and three-hybrid system do, the MbYTH assay is based on complementation between separable domains of ubiquitin. The small, highly conserved protein ubiquitin (Ub) plays a central role in the protein degradation system: covalent attachment of Ub to other proteins marks them for degradation by the 26S proteasome. While the modified proteins undergo degradation, the Ub itself is preserved by cleavage from the target protein by ubiquitin-specific proteases (UBPs). Recognition and subsequent cleavage of the Ub-target protein complex by the UBPs is dependent on the properly folded Ub structure. In 1994, Johnson and Varshavsky[157] demonstrated that Ub can be split into N-terminal (amino acids 1-34, Nub) and C-terminal (amino acids 35-76, Cub) halfs. Co-expressing these two Ub moieties in yeast results in their spontaneous re-association to form quasi-native Ub that is subsequently recognized by UBPs. Changing the wild-type isoleucine residue at position 13 of Nub to glycine (NubG) abolishes the intrinsic affinity of the two halves for each other, hindering re-association after expression. Fusing Cub and NubG with two interacting proteins, a bait and a prey, however, forces them into close proximity, resulting in a partial re-association of quasi-native Ub and subsequent recognition by UBPs. If a reporter protein is fused to the C-terminus of Cub, the re-association of Cub and NubG upon interaction of bait and prey will lead to the cleavage and release of this reporter by UBPs. The reporter used in the MbYTH system by Stagljar and co-workers is a hybrid transcription factor (TF) that is composed of the bacterial LexA protein and the Herpes simplex VP16 transactivator domain. As this artificial TF is fused to Cub (Cup-TF), it is required that the bait fused to Cub-TF is a membrane protein or at least membrane associated, to prevent the fusion construct from entering the nucleus. On the other hand, the prey, which is fused to Nub, can either be a soluble protein or a membrane protein. Upon re-association of quasi-native Ub through prey-bait interaction, the TF is cleaved off from Cub by UBPs, releasing it from the membrane, allowing it to enter the nucleus and activating the reporter gene expression. Using the yeast reporter strain L40, the artificial TF activates two reporter genes, *lacZ* and *His*3. This allows direct detection of protein interaction by either growing the yeast on a selective medium lacking histidine or by a simple X-Gal test. For detailed protocols on using the MbYTH system including information on bait- and prey-vectors refer to publications by the groups of Fetchko and Thaminy.[158,159]

Stagljar and co-workers had to prove the functionality and specificity of their novel system.[156] Therefore, they chose Wbp1p and Ost1p, two well characterized components of the oligosaccharyl transferase membrane protein complex. As a negative control, they included Alg5 in their assay, a protein that also localizes to the membrane of the endoplasmatic reticulum (ER), but does not interact with the oligosaccharyl transferase. Using the MbYTH system, they were able to detect specific interactions between Wbp1p and Ost1p, but no interactions between Wbp1p and Alg5p were found.

Massaad and co-workers applied the system to two more proteins localized in the ER membrane, Msn1p and Rer1p.[160] The α-1,2-mannosidase Msn1p is a type II membrane protein of the ER involved in the N-glycosidic pathway of *S. cerevisiae*. Its ER-localization depends on retrieval from the Golgi, a process involving Rer1p, a 22 kDa protein with four putative transmembrane domains, that is structurally and functionally conserved in yeast, humans and plants. Rer1p was already shown to be involved in recycling of Sec12p, Sec63p and Sec71p, representing membrane proteins with different topologies. Using MbYTH, Massaad and colleagues could prove a direct interaction between Msn1p and Rer1p, further indicating that Rer1p is likely to be part of a common recycling mechanism for retrieving membrane proteins from the Golgi to the ER.

Reinders and co-workers[161] used the system to demonstrate intra- and intermolecular interactions in sucrose transporters at the plasma membrane. First, they expressed the sucrose transporter StSUT1 in two parts and were able to detect a specific signal from the MbYTH system as well as sucrose transport activity similar to the intact protein, demonstrating spontaneous re-association of the two parts under native conditions. Furthermore, they co-expressed the N-terminal half of the low-affinity sucrose transporter LeSUT2 and the C-terminal part of SUT1. Again, they were able to detect specific interactions between the two portions *via* MbYTH and measured re-established sucrose transport activity. However, since the N-terminus of SUT2 determines affinity for sucrose, the transport rate of the reconstituted chimera was lower than that of wild-type SUT1.

Cervantes and co-workers[162] used the MbYTH assay to investigate the dimerisation properties of presenilins, highly conserved polytropic membrane proteins primarily located in the endoplasmic reticulum and early Golgi. Mutations in human presenilins 1 and 2 are known to cause dominant early-onset familial Alzheimer's disease (FAD). Presenilins are endoproteolytically processed *in vivo* to N- and C-terminal fragments (NTFs and CTFs), which subsequently build up high molecular weight complexes as the functional units. The domains responsible for presenilin complex formation and other intramolecular interactions were not known. By the use of MbYTH Cervantes and colleagues could detect NTF:CTF heterodimerisation as well as NTF:NTF and CTF:CTF homodimerisation. Moreover, they were able to address the effect of FAD-linked mutations on NTF interactions. Using a quantitative X-Gal assay in combination with MbYTH, they could demonstrate that the mutation E120K in the human PS1 gene impaired NTF:NTF *in situ* homodimerisation, indicating the corresponding domain to be involved in dimerisation.

Finally, Thaminy and co-workers demonstrated, that the MbYTH system can be successfully used to identify novel interacting partners for a mammalian membrane protein in a generally applicable screening approach.[128] By screening human cDNA libraries against the human ErbB3 receptor, they identified three previously unknown ErbB3-interacting proteins. It was possible to verify one of the newly found interactions (between ErbB3 and the membrane-associated RGS4 protein) by co-immunoprecipitation of the two proteins from human cells.

8.5 CONCLUSION

Yeast-based expression systems can be valuable tools for studies of integral membrane proteins, allowing recombinant protein production for functional characterization as well as for large-scale production for subsequent purification and structural studies. The inherent advantages of yeast systems are reflected in the rapid growth in the number of publications in this field. This is due to fast and easy handling, comparable with that of *E. coli*, combined with advantages found in eukaryotic systems including: appropriate folding, processing, targeting, and post-translational modifications such as phosphorylation and palmitoylation. Glycosylation, however, differs from that in mammalian cells. While hyperglycosylation with more than 50 sugar residues has been reported for recombinant proteins in *S. cerevisiae*, this problem has not been found in *P. pastoris* where, in the majority of cases, not more than 14 mannose residues were detected.[163] A major problem with yeast systems derives from proteases present in the vacuole. However, using protease-deficient yeast strains lacking one or more proteases, improvements in heterologous protein production have been achieved in many cases.

When using yeast for heterologous expression it has to be taken into consideration that differences in codon usage can result in reduced production levels[164–167] and this may require optimization (preferred codon usage for various organisms can be found on the www[168]).

Also it has been shown that some higher eukaryotic proteins are not compatible with the yeast secretory pathway,[32] potentially reducing the yield of functional membrane protein. To detect incorrectly folded membrane proteins, Griffith and co-workers introduced a system based on the unfolded protein response (UPR) pathway.[169] They coupled this protein quality control mechanism, operating in the endoplasmatic reticulum (ER), to a suitable reporter system, thus creating a tool to simplify optimization of expression conditions.

The different lipid and sterol composition of yeast cells compared with that of higher eukaryotes[170] has led to altered properties of some recombinant proteins compared to the properties of native systems.[60,61,77] The importance of this host system property is also highlighted by the fact that all crystallized membrane proteins have defined lipid–protein interactions.

Large-scale production of membrane proteins in yeast is facilitated by rapid growth in inexpensive media as well as the option of fermentor culture. *P. pastoris* is especially suited for high cell-density fermentation, yielding up to 500 OD_{600} units/liter[33] and thus can serve as the basis for obtaining large quantities of a homogenous protein preparation. Examples of successful structure determination of proteins derived from yeast expression systems include the voltage-sensitive K-channel which was used for single particle analysis[101] and the monoamine oxidase B (MAO B), for which the crystal structure was determined to a 1.7 Å resolution.[171] This raises the expectation that, in the near future, high-resolution structures of membrane proteins will be more frequently determined using yeast expression systems.

ACKNOWLEDGMENTS

We are grateful to H. Reiländer, D. Parcej, and CG. Bevans for critical reading of the manuscript.

REFERENCES

1. Wallin, E. and von Heijne, G. Genome-wide analysis of integral membrane proteins from eubacterial, archaean, and eukaryotic organisms. *Prot. Sci.*, 7, 1029–1038, 1998.
2. Liu, Y., Donald, M., Engelman, D.M., and Gerstein, M. Genomic analysis of membrane protein families: abundance and conserved motifs. *Genome Biol.*, 3, 1–12, 2002.
3. Picot, D., Loll, P.J., and Garavito, R.M. The x-ray crystal structure of the membrane protein prostaglandin H2 synthase-1. *Nature*, 367, 243–249, 1994.
4. Malkowski, M.G., Ginell, S.L., Smith, W.L., and Garavito, R.M. The productive conformation of arachidonic acid bound to prostaglandin synthase. *Science*, 289, 1933–1937, 2000.
5. Verkman, A.S. and Mitra, A.K. Structure and function of aquaporin water channels. *Am. J. Physiol. Renal. Physiol.*, 278, 13–28, 2000.
6. Schrier, R.W., Chen, Y.C., and Cadnapaphornchai, M.A. From finch to fish to man: role of aquaporins in body fluid and brain water regulation. *Neuroscience*, 129, 897–904, 2004.
7. Hipfner, D.R., Deeley, R.G., and Cole, S.P.C. Structural, mechanistic and clinical aspects of MRP1. *Biochim. Biophys. Acta.*, 1461, 359–376, 1999.
8. http://www.mpibp-frankfurt.mpg.de/michel/public/memprotstruct.html (accessed 2005).
9. Palczewski, K., Kumasaka, T., Hori, T., Behnke, C.A., Motoshima, H., Fox, B.A., Le Trong, I., Teller, D.C., Okada, T., Stenkamp, R.E., Yamamoto, M., and Miyano, M. Crystal structure of rhodopsin: a G protein-coupled receptor. *Science*, 289, 739—745, 2000.
10. Stroud, R.M. and Finer-Moore, J. Acethylcholine receptor structure, function, and evolution. *Ann. Rev. Cell Biol.*, 1, 317–351, 1985.
11. Miyazawa, A., Fujiyoshi, Y., and Unwin, N. Structure and gating mechanism of the acetylcholine receptor pore. *Nature*, 423, 949–955, 2003.
12. Wimalawansa, S.J. Purification and biochemical characterization of neuropeptide Y_2 receptor. *J.Biol. Chem.*, 270, 18523–18530, 1995.

13. Loll, P.J. Membrane protein structural biology: the high throughput challenge. *J.Struc. Biol.*, 142, 144–153, 2003.
14. Bannwarth, M. and Schulz, G.E. The expression of outer membrane proteins for crystallization. *Biochim. Biophys. Acta.*, 1610, 37–45, 2003.
15. Grisshammer, R., Duckworth, R., and Henderson, R. Expression of a rat neurotensin receptor in *Escherichia coli*. *Biochem. J.*, 295, 571–576, 1993.
16. Tucker, J. and Grisshammer, R. Purification of a rat neurotensin receptor expressed in *Escherichia coli*. *Biochem. J.*, 317, 891–899, 1996.
17. Weiß, M. and Grisshammer, R. Purification and characterization of the human adenosine A2a receptor functionally expressed in *Escherichia coli*. *Eur. J. Biochem.*, 269, 82–92, 2002.
18. Kiefer, H., Krieger, J., Olszewski, J.D., von Heijne, G., Prestwich, G.D., and Breer, H. Expression of an olfactory receptor in *Escherichia coli*: purification, reconstitution, and ligand binding. *Biochemistry*, 35, 16077–16084, 1996.
19. Booth, P.J., Templer, R.H., Meijberg, W., Allen, S.J., Curran, A.R., and Lorch, M. *In vitro* studies of membrane protein folding. *Crit. Rev. Biochem. Mol. Biol.*, 36, 501–603, 2001.
20. Grisshammer, R. and Tate, C.G. Overexpression of integral membrane proteins for structural studies. *Quart. Rev. Biophys.*, 28, 315–422, 1995.
21. Goffeau, A. et al. The Yeast Genome Directory. *Nature*, 387, 5, 1997.
22. Mackiewicz, P., Kowalczuk, M., Mackiewicz, D., Nowicka, A., Dudkiewicz, M., Laszkiewicz, A., Dudek, M.R., and Cebrat, S. How many protein-coding genes are there in the *Saccharomyces cerevisiae* genome? *Yeast*, 19, 619–629, 2002.
23. http://www.lgcpromochem.com/atcc/ (accessed March 2005).
24. Giga-Hama, Y. and Kumagai, H. Foreign gene expression in fission yeast *Schizosaccharomyces pombe*. Springer-Verlag, TX, U.S.A., Berlin, Heidelberg, New York, 1997.
25. Bharathi, A., Ghosh, A., Whalen, W.A., Yoon, J.H., Pu, R., Dasso, M., and Dhar, R. The human *RAE1* gene is a functional homologue of *Schizosaccharomyces pombe rae1* gene involved in nuclear export of Poly(A)+ RNA. *Gene*, 198, 251–258, 1997.
26. Wood V. et al. The genome sequence of *Schizosaccharomyces pombe*. *Nature*, 415, 871–880, 2002.
27. Siam, R., Dolan, W.P., and Forsburg, S.L. Choosing and using *Schizosaccharomyces pombe* plasmids. *Methods*, 33, 189–198, 2004.
28. Ogata, K., Nishikawa, H., and Ohsugi, M. A yeast capable of utilizing methanol. *Agri. Biol. Chem.*, 33, 1519–1520, 1969.
29. Wegner, G. Emerging applications of methylotrophic yeast. *FEMS Microbiol. Rev.*, 87, 279–284, 1990.
30. Cregg, J.M., Vedvick, T.S., and Raschke, W.C. Recent advances in the expression of foreign genes in *Pichia pastoris*. *Bio/Technology*, 11, 905–910, 1993.
31. Romanos, M. Advances in the use of *Pichia pastoris* for high-level gene expression. *Curr. Opin. Biotechnol.*, 6, 527–533, 1995.
32. Cregg, J.M. Expression in the methylotrophic yeast *Pichia pastoris,* in Hoeffler J, Fernandez J, eds. *Nature: The Palette for the Art of Expression.* Academic Press, San Diego, CA, 1998, pp. 157–191.
33. Lin-Cereghino, J. and Cregg, J.M. Heterologous protein expression in the methylotrophic yeast *Pichia pastoris*. *FEMS Microbiol. Rev.*, 24, 45–66, 2000.
34. Bill, R.M. Yeast — a panacea for the structure-function analysis of membrane proteins. *Curr. Genet.*, 40, 157–171, 2001.

35. Lalonde, S., Boles, E., Hellmann, H., Barker, L., Patrick, J.W., Frommer, W.B., and Ward, JM. The dual function of sugar carriers. Transport and sugar sensing. *Plant Cell*, 11, 707–726, 1999.

36. Reiländer, H., Reinhart, C., and Szmolenszky, A large-scale expression of receptors in yeasts, in Haga, T. and Bernstein, G., eds. *CRC Methods in Signal Transduction: G Protein-Coupled Receptors*. Boca Raton, Florida, CRC Press Inc., 1999, pp. 281–322.

37. Payette, P., Gossard, F., Whiteway, M., and Dennis, M. Expression and pharmacological characterization of the human M1 muscarinic receptor in *Saccharomyces cerevisiae*. *FEBS Lett.*, 266, 21–25, 1990.

38. Erlenbach, I., Kostenis, E., Schmidt, C., Hamdan, F.F., Pausch, M.H., and Wess, J. Functional expression of M1, M3 and M5 muscarinic acetylcholine receptors in yeast. *J. Neurochem.*, 77, 1327–1337, 2001.

39. King, K., Dohlman, H.G., Thorner, J., Caron, M.G., and Lefkowitz, R.J. Control of yeast mating signal transduction by a mammalian β_2-adrenergic receptor and G_{Sa} subunit. *Science*, 250, 121–123, 1990.

40. Sizmann, D., Kuusinen, H., Keränen, S., Lomasney, J., Caron, M.G., Lefkowitz, R.J., and Keinänen, K. Production of adrenergic receptors in yeast. *Receptors & Channels*, 4, 197–203, 1996.

41. Hansen, M.K., Tams, J.W., Fahrenkrug, J., and Pedersen, P.A. Functional expression of rat VPAC1 receptor in *Saccharomyces cerevisiae*. *Receptors & Channels*, 6, 271–281, 1999.

42. Bach, M., Sander, P., Haase, W., and Reiländer, H. Pharmacological and biochemical characterization of the mouse 5HT$_{5A}$ serotonin receptor heterologously expressed in the yeast *Saccharomyces cerevisiae* strains. *Receptors & Channels*, 4, 129–139, 1996.

43. David, N.E., Gee, M., Andersen, B., Naider, F., Thorner, J., and Stevens, R.C. Expression and purification of the *Saccharomyces cerevisiae* a-factor receptor (Ste2p), a 7-transmembrane-segment G protein-coupled receptor. *J. Biol. Chem.*, 272, 15553–15561, 1997.

44. Sander, P., Grünewald, S., Maul, G., Reiländer, H., and Michel, H. Constitutive expression of the human D$_{2S}$ dopamine receptor in the unicellular yeast *Saccharomyces cerevisiae*. *Biochim. Biophys. Acta.*, 1193, 255–262, 1994.

45. Presland, J. and Strange, P.G. Pharmacological characterisation of the D2 dopamine receptor expressed in the yeast *Schizosaccharomyces pombe*. *Biochem. Pharmacol.*, 56, 577–582, 1998.

46. Arkinstall, S., Edgerton, M., Payton, M., and Maundrell, K. Co-expression of the neurokinin NK2 receptor and G protein components in the fission yeast *Schizosaccharomyces pombe*. *FEBS Lett.*, 375, 183–187, 1995.

47. Ficca, A.G., Testa, L., and Tocchini-Valentini, G.P. The human β_2-adrenergic receptor expressed in *Schizosaccharomyces pombe* retains its pharmacological properties. *FEBS Lett.*, 377, 140–144, 1995.

48. Sreekrishna, K., Brankamp, R.G., Kropp, K.E., Blankenship, D.T., Tsay, J.T., Smith, P.L., Wierschke, J.D., Subramaniam, A., and Birkenberger, L.A. Strategies for optimal synthesis and secretion of heterologous proteins in the methylotrophic yeast *Pichia pastoris*. *Gene*, 190, 55–62, 1997.

49. Cereghino, G.P.L., Cereghino, J.L., Ilgen, C., and Cregg, J.M. Production of recombinant proteins in fermentor cultures of the yeast *Pichia pastoris*. *Curr. Opin. Biotechnol.*, 13, 1–4, 2002.

50. Weiß, H.M., Haase, W., Michel, H., and Reiländer, H. Expression of functional mouse 5-HT$_{5A}$ serotonin receptor in the methylotrophic yeast *Pichia pastoris*: pharmacological characterization and localization. *FEBS Lett.*, 377, 451–456, 1995.

51. Grünewald, S., Haase, W., Molsberger, E., Michel, H., and Reiländer, H. Production of the human D2S receptor in the methylotrophic yeast *P. pastoris*. *Receptors & Channels*, 10, 37–50, 2004.

52. Julius, D., Schekman, R., and Thorner, J.. Glycosylation and processing of prepro-alpha-factor through the yeast secretory pathway. *Cell*, 36, 309–318, 1984.

53. Weiß, H.M., Haase, W., Michel, H., and Reiländer, H. Comparative biochemical and pharmacological characterization of the mouse 5HT$_{5A}$ 5-hydroxytryptamine receptor and the human β$_2$-adrenergic receptor produced in the methylotrophic yeast *Pichia pastoris*. *Biochem. J.*, 330, 1137–1147, 1998.

54. Schiller, H., Haase, W., Molsberger, E., Janssen, P., Michel, H., and Reiländer, H. The human ET$_B$ endothelin receptor heterologously produced in the methylotrophic yeast *Pichia pastoris* shows high-affinity binding and induction of stacked membranes. *Receptors & Channels*, 7, 93–107, 2000.

55. Feng, W., Cai, J., Pierce, W.M. Jr, and Song, Z. Expression of CB2 cannabinoid receptor in *Pichia pastoris*. *Prot. Exp. Purif.*, 26, 496–505, 2002.

56. Schiller, H., Molsberger, E., Janssen, P., Michel, H., and Reilander, H. Solubilization and purification of the human ETB endothelin receptor produced by high-level fermentation in *Pichia pastoris*. *Receptors & Channels*, 7, 453–469, 2001.

57. Sarramegna, V., Demange, P., Milon, A., and Talmont, F. Optimizing functional versus total expression of the human μ-opioid receptor in *Pichia pastoris*. *Prot. Express. Purif.*, 24, 212–220, 2002.

58. Cronan, J.E. Jr. Biotination of proteins *in vivo*. A post-translational modification to label, purify, and study proteins. *J. Biol. Chem.*, 265, 10327–10333, 1990.

59. Gimpl, G., Burger, K., and Fahrenholz, F. Cholesterol as modulator of receptor function. *Biochemistry*, 36, 10959–10974, 1997.

60. Gimpl, G., Burger, K., Politowska, E., Ciarkowski, J., and Fahrenholz, F. Oxytocin receptors and cholesterol: interaction and regulation. *Exp. Physiol.*, 85, 41–49, 2000.

61. Pucadyil, T. and Chattopadhyay, A. Cholesterol modulates ligand binding and G protein coupling to serotonin1A receptors from bovine hippocampus. *Biochim. Biophys. Acta*, 1663, 188–200, 2004.

62. Campbell, R.M., Cartwright, C., Chen, W., Chen, Y., Duzic, E., Fu, J., Loveland, M., Manning, R., McKibben, B., Pleiman, C.M., Silverman, L., Trueheart, J., Webb, D.R., Wilkinson, V., Witter, D.J., Xie, X., and Castelhano, L. Selective A1-Adenosine receptor antagonists identified using yeast *Saccharomyces cerevisiae* functional assays. *Bioorg. Med. Lett.*, 9, 2413–2418, 1999.

63. Butz, J.A., Niebauer, R.T., and Robinson, A.S. Co-expression of molecular chaperones does not improve the heterologous expression of mammalian G protein coupled receptor expression in yeast. *Biotech. Bioeng.*, 84, 292–304, 2003.

64. Price, L.A., Strnad, J., Pausch, M.H., and Hadcock, J.R. Pharmacological characterization of the rat A$_{2a}$ adenosine receptor functionally coupled to the yeast pheromone response pathway. *Mol. Pharmacol.*, 50, 829–837, 1996.

65. Brown, A.J., Dyos, S.L., Whiteway, M.S., White, J.H., Watson, M.A., Marzioch, M., Clare, J.J., Cousens, D.J., Paddon, C., Plumpton, C., Romanos, M.A., and Dowell, S.J. Functional coupling of mammalian receptors to the yeast mating pathway using novel yeast/mammalian G protein alpha-subunit chimeras. *Yeast*, 16, 11–22, 2000.

66. Duport, C., Loeper, J., and Strosberg, A.D. Comparative expression of the human b2 and b3 adrenergic receptors in *Saccharomyces cerevisiae. Biochim. Biophys. Acta.*, 1629, 34–43, 2003.

67. Floyd, D.H., Geva, A., Bruinsma, S.P., Overton, M.C., Blumer, K.J., and Baranski, T.J. C5a receptor oligomerization. II. Fluorescence resonance energy transfer studies of a human G protein-coupled receptor expressed in yeast. *J. Biol. Chem.*, 278, 35354–35361, 2003.

68. Andersen, B. and Stevens, R.C. The human D_{1A} dopamine receptor: heterologous expression in *Saccharomyces cerevisiae* and purification of the functional receptor. *Prot. Exp. Purif.*, 13, 111–119, 1998.

69. Sander, P., Grünewald, S., Reiländer, H., and Michel, H. Expression of the human D_{2S} dopamine receptor in the yeasts *Saccharomyces cerevisiae* and *Schizosaccharomyces pombe*: a comparative study. *FEBS Lett.*, 344, 41–46, 1994.

70. Erickson, J.R., Wu, J.J., Goddard, G., Tigyi, G., Kawanishi, K., Tomei, L.D., and Kiefer, M. Edg-2/Vzg-1 couples to the yeast pheromone response pathway selectively in response to lysophosphatidic acid. *J. Biol. Chem.*, 273, 1506–1510, 1998.

71. Cid, G.M., Nugent, P.G., Davenport, A.P., Kuc, R.E., and Wallace, B.A. Expression and characterization of the human endothelin-A-receptor in *Pichia pastoris*: influence of N-terminal epitope tags. *J. Cardiovasc. Pharmacol.*, 36, 55–57, 2000.

72. Kajkowski, E.M., Price, L.A., Pausch, M.H., Young, K.H., and Ozenberger, B.A. Investigation of growth hormone releasing hormone receptor structure and activity using yeast expression technologies. *J. Recept. Signal Transduct. Res.*, 17, 293–303, 1997.

73. Kokkola, T., Watson, M.A., White, J., Dowell, S., Foord, S.M., and Laitinen, J.T. Mutagenesis of human Mel1a melatonin receptor expressed in yeast reveals domains important for receptor function. *Biochem. Biophys. Res. Comm.*, 249, 531–536, 1998.

74. Sugiyama, K., Niki, T.P., Inokuchi, K., Teranishi, Y., Ueda, M., and Tanaka, A. Heterologous expression of metabotropic glutamate receptor subtype 1 in *Saccharomyces cerevisiae. Appl. Microbiol. Biotechnol.*, 64, 531–536, 2004.

75. Huang, H.J., Liao, C.F., Yang, B.C., and Kuo, T.T. Functional expression of rat M5 muscarinic acetylcholine receptor in yeast. *Biochem. Biophys. Res. Comm.*, 182, 1180–1186, 1992.

76. Mollaaghababa, R., Davidson, F.F., Kaiser, C., and Khorana, H.G. Structure and function in rhodopsin: expression of functional opsin in *Saccharomyces cerevisiae. Proc. Natl. Acad. Sci. USA*, 93, 11482–11486, 1996.

77. Abdulaev, N.G., Popp, M.P., Smith, W.C., and Ridge, K.D. Functional expression of bovine opsin in the methylotrophic yeast *Pichia pastoris. Prot. Exp. Purif.*, 10, 61–69, 1997.

78. Kong, J.L., Panetta, R., Song, W., Somerville, W., and Greenwood, M.T. Inhibition of somatostatin receptor 5-signaling by mammalian regulators of G protein signaling (RGS) in yeast. *Biochim. Biophys. Acta.*, 1542, 95–105, 2002.

79. Pausch, M.H., Lai, M., Tseng, E., Paulsen, J., Bates, B., and Kwak, S. Functional expression of human and mouse $P2Y_{12}$ receptors in *Saccharomyces cerevisiae. Biochem. Biophys. Res. Comm.*, 324, 171-177, 2004.

80. Price, L.A., Kajkowski, E.M., Hadcock, J.R., Ozenberger, B.A., and Pausch, M.H. Functional coupling of a mammalian somatostatin receptor to the yeast pheromone response pathway. *Mol. Cell Biol.*, 15, 6188–6195, 1995.

81. Erlenbach, I., Kostenis, E., Schmidt, C., Serradeil-Le Gal, C., Raufaste, D., Dumont, M.E., Pausch, M.H., and Wess, J. Single amino acid substitutions and deletions that alter the G protein coupling properties of the V2 vasopressin receptor identified in yeast by receptor random mutagenesis. *J. Biol. Chem.*, 276, 29382–29392, 2001.

82. Talmont, F., Sidobre, S., Demange, P., Milon, A., and Emorine, L.J. Expression and pharmacological characterization of the human m-opioid receptor in the methy-lotrophic yeast *Pichia pastoris*. *FEBS Lett.*, 394, 268–272, 1996.

83. Paulsen, I.T., Nguyen, L., Sliwinski, M.K., Rabus, R., and Saier, M.H. Microbial genome analyses: comparative transport capabilities in eighteen prokaryotes. *J. Mol. Biol.*, 301, 75–100, 2000.

84. Van Belle, D. and Andre, B. A genomic view of yeast membrane transporters. *Curr. Opin. Cell Biol.*, 13, 389–398, 2001.

85. http://www.membranetransport.org (accessed March 2005).

86. Agnan, J., Korch, C., and Selitrennikoff, C. Cloning heterologous genes: problems and approaches. *Fungal Genet. Biol.*, 21, 292–301, 1997.

87. Zhanga, N., Osbornb, M., Gitshama, P., Yena, K., Millerb, J.R., and Olivera, S.G. Using yeast to place human genes in functional categories. *Gene*, 303, 121–129, 2003.

88. Weierstall, T., Hollenberg, C.P., and Boles, E. Cloning and characterization of three genes (SUT1-3) encoding glucose transporters of the yeast *Pichia stipitis*. *Mol. Microbiol.*, 31, 871–883, 1999.

89. Wieczorke, R., Dlugai, S., Krampe, S., and Boles, E. Characterization of mammalian GLUT glucose transporters in a heterologous yeast expression system. *Cell Physiol. Biochem.*, 13, 123–134, 2003.

90. Bill, R.M., Hedfalk, K., Karlgren, S., Mullins, J.G.L., Rydström, J., and Hohmann, S. Analysis of the pore of the unusual MIP channel, yeast Fps1p. *J. Biol. Chem.*, 276, 36543–36549, 2001.

91. Anderson, J.A., Huprikar, S.S., Kochian, L.V., Lucas, W.J., and Gaber, R.F. Functional expression of a probable *Arabidopsis thaliana* potassium channel in *Saccharomyces cerevisiae*. *Proc. Natl. Acad. Sci. USA*, 89, 3736–3740, 1992.

92. Bertl, A., Anderson, J.A., Slayman, C.L., and Gaber, R.F. Use of *Saccharomyces cerevisiae* for patch-clamp analysis of heterologous membrane proteins: character-ization of Kat1, an inward-rectifying K^+ channel from *Arabidopsis thaliana*, and comparison with endogeneous yeast channels and carriers. *Proc. Natl. Acad. Sci. USA*, 92, 2701–2705, 1995.

93. Sentenac, H., Bonneaud, N., Minet, M., Lacroute, F., Salmon, J.M., Gaymard, F., and Grignon, C. Cloning and expression in yeast of a plant potassium ion transport system. *Science*, 256, 663–665, 1992.

94. Tang, W., Ruknudin, A., Yang, W.P., Shaw, S.Y., Knickerbocker, A., and Kurtz, S. Functional expression of a vertebrate inwardly rectifying K+ channel in yeast. *Mol. Biol. Cell*, 6, 1231–1240, 1995.

95. Lagrée, V., Pellerin, I., Hubert, J.F., Tacnet, F., Cahérec, F., Roudier, N., Thomas, D., Gouranton, J., and Deschamps, S. A yeast recombinant aquaporin mutant that is not expressed or mistargeted in *Xenopus* oocyte can be functionally analyzed in recon-stituted proteoliposomes. *J. Biol. Chem.*, 273, 12422–12426, 1998.

96. Stephens, A.N., Khan, M.A., Roucou, X., Nagley, P., and Devenish, R.J. The molec-ular neighborhood of subunit 8 of yeast mitochondrial F_1F_0–ATP synthase probed by cysteine scanning mutagenesis and chemical modification. *J. Biol. Chem.*, 278, 17867–17875, 2003.

 97. Soteropoulos, P. and Perlin, D.S. Genetic probing of the stalk segments associated with M2 and M3 of the plasma membrane H⁺-ATPase from *Saccharomyces cerevisiae*. *J. Biol. Chem.*, 273, 26426–26431, 1998.

 98. Proff, C. and Kolling, R. Functional asymmetry of the two nucleotide binding domains in the ABC transporter Ste6. *Mol. Gen Genet.*, 264, 883–893, 2001.

 99. Lerner-Marmarosh, N., Gimi, K., Urbatsch, I.L., Gros, P., and Senior, A.E. Large scale purification of detergent-soluble P-glycoprotein from *Pichia pastoris* cells and characterization of nucleotide binding properties of wild-type, Walker A, and Walker B mutant proteins. *J. Biol. Chem.*, 274, 34711–34718, 1999.

100. Karlsson, M., Fotiadis, D., Sjövall, S., Johansson, I., Hedfalk, K., Engel, A., and Kjellbom, P. Reconstitution of water channel function of an aquaporin overexpressed and purified from *Pichia pastoris*. *FEBS Lett.*, 537, 68-72, 2003.

101. Parcej, D.N. and Eckhardt-Strelau, L. Structural characterization of neuronal voltage-sensitive K⁺ channels heterologously expressed in *Pichia pastoris*. *J. Mol. Biol.*, 333, 103–116, 2003.

102. Cai, J., Daoud, R., Alqawi, O., Georges, E., Pelletier, J., and Gros, P. Nucleotide binding and nucleotide hydrolysis properties of the ABC transporter MRP6 (ABCC6). *Biochemistry*, 41, 8058–8067, 2002.

103. Zhang, H., Howard, E.M., and Roepe, P.D. Analysis of the antimalarial drug resistance protein Pfcrt expressed in yeast. *J. Biol. Chem.*, 277, 49767–49775, 2002.

104. Mathai, J.C. and Agre, P. Hourglass pore-forming domains restrict aquaporin-1 tetramer assembly. *Biochemistry*, 38, 923–928, 1999.

105. Hildebrandt, V., Polakowski, F., and Büldt, G. Purple fission yeast: overexpression and processing of the pigment bacteriorhodopsin in *Schizosaccharomyces pombe*. *Photochem. Photobiol.*, 54, 1009–1016, 1991.

106. Mao, Q, Conseil, G., Gupta, A., Cole, S.P., and Unadkat, J.D. Functional expression of the human breast cancer resistance protein in *Pichia pastoris*. *Biochem. Biophys. Res. Comm.*, 320, 730–737, 2004.

107. Lenoir, G., Menguy, T., Corre, F., Montigny, C., Pedersen, P.A., Thines, D., le Maire, M., and Falson, P. Overproduction in yeast and rapid and efficient purification of the rabbit SERCA1a Ca(2+)-ATPase. *Biochim. Biophys. Acta.*, 1560, 67–83, 2002.

108. Hashimoto, K., Saito, M., Matsuoka, H., Iida, K, and Iida, H. Functional analysis of a rice putative voltage-dependent Ca2+ channel, OsTPC1, expressed in yeast cells lacking its homologous gene CCH1. *Plant Cell Physiol.*, 45, 496–500, 2004.

109. Marheineke, K., Bach, M., Haase, W. and Reiländer, H. High level production and localization of bovine rod cGMP-gated cation channel subunit 1 in baculovirus-infected insect cells and *Saccharomyces cerevisiae*. *Biochem. Biophys. Res. Comm.*, 215, 961–967, 1995.

110. Nakamura, K., Niimi, M., Niimi, K, Holmes, A.R., Yates, J.E., Decottignies, A., Monk, B.C., Goffeau, A., and Cannon, R.D. Functional expression of *Candida albicans* drug efflux pump Cdr1p in a *Saccharomyces cerevisiae* strain deficient in membrane transporters. *Antimicrob. Agents Chemother.*, 45, 3366–3374, 2001.

111. Theis, S., Döring, F., and Daniel, H. Expression of the myc/His-tagged human peptide transporter hPEPT1 in yeast for protein purification and functional analysis. *Prot. Exp. Purif.*, 22, 436–442, 2001.

112. Gupta, S.S. and Canessa, C.M. Heterologous expression of a mammalian epithelial sodium channel in yeast. *FEBS Lett.*, 481, 77–80, 2000.

113. Obrdlik, P., El-Bakkoury, M., Hamacher, T., Cappellaro, C., Vilarino, C., Fleischer, C., Ellerbrok, H., Kamuzinzi, R., Ledent, V., Blaudez, D., Sanders, D., Revuelta, J.L., Boles, E., Andre, B., and Frommer, W.B. K+ channel interactions detected by a genetic system optimized for systematic studies of membrane protein interactions. *Proc. Natl. Acad. Sci. USA*, 101, 12242–12247, 2004.

114. Yi, B.A., Lin, Y-F., Jan, Y.N., and Jan, L.Y. Yeast screen for constitutively active mutant G protein-activated potassium channel. *Neuron*, 29, 657–667, 2001.

115. Graves, F.M. and Tinker, A. Functional expression of the pore forming subunit of the ATP-sensitive potassium channel in *Saccharomyces cerevisiae. Biochem. Biophys. Res. Comm.*, 272, 403–409, 2000.

116. Schachtman, D.P. and Schroeder, J.I. Structure and transport mechanism of a high-affinity potassium uptake transporter from higher plants. *Nature*, 370, 655–658, 1994.

117. Li, M., Hubálek, F., Newton-Vinson, P., and Edmondson, D.E. High-level expression of human liver monoamine oxidase A in *Pichia pastoris*: comparison with the enzyme expressed in *Saccharomyces cerevisiae. Prot. Exp. Purif.*, 24, 152–162, 2002.

118. Newton-Vinson, P., Hubalek, F., and Edmondson, D.E. High-level expression of human liver monoamine oxidase B in *Pichia pastoris. Prot. Exp. Purif.*, 20, 334–345, 2000.

119. Lee, S.H. and Altenberg, G.A. Expression of functional multidrug-resistance protein 1 in *Saccharomyces cerevisiae*: effects of N- and C-terminal affinity tags. *Biochem. Biophys. Res. Comm.*, 306, 644–649, 2003.

120. Cai, J., Daoud, R., Georges, E., and Gros, P. Functional expression of multidrug resistance protein 1 in *Pichia Pastoris. Biochemistry*, 40, 8307–8316, 2001.

121. Pedersen, P.A., Rasmussen, J.H., and Jorgensen, P.L. Expression in high yield of pig a1b1 Na,K-ATPase and inactive mutants D369N and D807N in *Saccharomyces cerevisiae. J. Biol. Chem.*, 271, 2514–2522, 1996.

122. Strugatsky, D., Gottschalk, K.E., Goldshleger, R., Bibi, E., and Karlish, S.J.D. Expression of Na+,K+-ATPase in *Pichia Pastoris. J. Biol. Chem.*, 278, 46064–46073, 2003.

123. Jansen, K.U., Conroy, W.G., Claudio, T., Fox, T.D., Fujita, N., Hamill, O., Lindstrom, J.M., Luther, M., Nelson, N., Ryan, K.A., Sweet, M.T., and Hess, G.P. Expression of the four subunits of the *Torpedo californica* nicotinic acetylcholine receptor in *Saccharomyces cerevisiae. J. Biol. Chem.*, 264, 15022–15027, 1989.

124. Ishida, N., Kuba, T., Aoki, K., Miyatake, S., Kawakita, M., and Sanai, Y. Identification and characterization of human Golgi nucleotide sugar transporter SLC35D2, a novel member of the SLC35 nucleotide sugar transporter family. *Genomics*, 85, 106–116, 2005.

125. Ruf, A., Müller, F., D'Arcy, B., Stihle, M., Kusznir, E., Handschin, C., Morand, O.H., and Thoma, R. The monotopic membrane protein human oxidosqualene cyclase is active as monomer. *Biochem. Biophys. Res. Comm.*, 315, 247–254, 2004.

126. Mao, Q. and Scarborough, G.A. Purification of functional human P-glycoprotein expressed in *Saccharomyces cerevisiae. Biochim. Biophys. Acta.*, 1327, 107–118, 1997.

127. Julien, M., Kajiji, S., Kaback, R.H., and Gros, P. Simple purification of highly active biotinylated P-glycoprotein: enantiomer-specific modulation of drug-stimulated ATPase activity. *Biochemistry*, 39, 75-85, 2000.

128. Thaminy, S., Auerbach, D., Arnoldo, A., and Stagljar, I. Identification of novel ErbB3-interacting factors using the split-ubiquitin membrane yeast two-hybrid system. *Genome Res.*, 13, 1744–1753, 2003.

129. Tate, C.G., Haase, J., Baker, C., Boorsma, M., Magnani, F., Vallis, Y., and Williams, D.C. Comparison of seven different heterologous protein expression systems for the production of the serotonin transporter. *Biochim. Biophys. Acta.*, 1610, 141–153, 2003.

130. Terng, H.J., Geßner, R., Fuchs, H., Stahl, U., and Lang, C. Human transferrin receptor is active and plasma membrane-targeted in yeast. *FEMS Microbiol. Lett.*, 160, 61–67, 1998.

131. Moiseenkova, V.Y., Hellmich, H.L., and Christensen, B.N. Overexpression and purification of the vanilloid receptor in yeast (*Saccharomyces cerevisiae*). *Biochem. Biophys. Res. Comm.*, 310, 196–201, 2003.

132. Stratowa, C., Himmler, A., and Czernilofsky, A.P. Use of a luciferase reporter system for characterizing G protein-linked receptors. *Curr. Opin. Biotechnol.*, 6, 574–581, 1995.

133. Dohlman, H.G. and Thorner, J.W. Regulation of G protein-initiated signal transduction in yeast: paradigms and principles. *Annu. Rev. Biochem.*, 70, 703–754, 2001.

134. Bardwell, L. A walk-through of the yeast mating pheromone response pathway. *Peptides*, 25, 1465–1476, 2004.

135. Dietzel, C. and Kurjan, J. The yeast *SCG1* gene: a Gα-like protein implicated in the a- and α-factor response pathway. *Cell*, 50, 1001–1010, 1987.

136. Kang, Y.S., Kane, J., Kurjan, J., Stadel, J.M., and Tipper, D.J. Effects of expression of mammalian G alpha and hybrid mammalian-yeast G alpha proteins on the yeast pheromone response signal transduction pathway. *Mol. Cell Biol.*, 10, 2582–2590, 1990.

137. Eppler, C.M., Zysk, J.R., Corbett, M., and Shieh, H.M. Purification of a pituitary receptor for somatostatin. *J. Biol. Chem.*, 267, 15603–15612, 1992.

138. Luthin, D.R., Eppler, C.M., and Linden, J. Identification and quantification of Gi-type GTP-binding proteins that copurify with pituitary somatostatin receptor. *J. Biol. Chem.*, 268, 5990–5996, 1993.

139. Price, L.A., Kajkowski, E.M., Hadcock, J.R., Ozenberger, B.A., and Pausch, M.H. Functional coupling of a mammalian somatostatin receptor to the yeast pheromone response pathway. *Mol. Cell Biol.*, 15, 6188–6195, 1995.

140. Milligan, G., Marshall, F., and Rees, S. G_{16} as a universal G protein adapter: implications for agonist screening strategies. *Trends Pharmacol. Sci.*, 17, 235–237, 1996.

141. Offermanns, S. and Simon, M.I. G_{a15} and G_{a16} couple a wide variety of receptors to phospholipase C. *J. Biol. Chem.*, 270, 15175–15180, 1995.

142. Noble, B., Kallal, L.A., Pausch, M.H., and Benovic, J.L. Development of a yeast bioassay to characterize G protein-coupled receptor kinases. Identification of an NH2-terminal region essential for receptor phosphorylation. *J. Biol. Chem.*, 278, 47466–47476, 2003.

143. Luttrell, L.M. and Lefkowitz, R.J. The role of β-arrestins in the termination and transduction of G protein-coupled receptor signals. *J. Cell Sci.*, 115, 455–465, 2002.

144. Auerbach, D., Thaminy, S., Hottiger, M.O., and Stagljar, I. The post-genomic era of interactive proteomics: facts and perspectives. *Proteomics*, 2, 611–623, 2002.

145. Aebersold, R. and Mann, M. Mass spectrometry-based proteomics. *Nature*, 422, 198–207, 2003.

146. Gavin, A.C., Bösche, M., Krause, R., Grandi, P., Marzioch, M., Bauer, A., Schultz, J., Rick, J.M., Michon, A.M., Cruciat, C.M., Remor, M., Höfert, C., Schelder, M., Brajenovic, M., Ruffner, H., Merino, A., Klein, K., Hudak, M., Dickson, D., Rudi, T., Gnau, V., Bauch, A., Bastuck, S., Huhse, B., Leutwein, C., Heurtier, M.A., Copley, R.R., Edelmann, A., Querfurth, E., Rybin, V., Drewes, G., Raida, M., Bouwmeester, T., Bork, P., Seraphin, B., Kuster, B., Neubauer, G., and Superti-Furga, G. Functional organization of the yeast proteome by systematic analysis of protein complexes. *Nature*, 415, 141–147, 2002.

147. Ho, Y., Gruhler, A., Heilbut, A., Bader, G.D., Moore, L., Adams, S.L., Millar, A., Taylor, P., Bennett, K., Boutilier, K., Yang, L., Wolting, C., Donaldson, I., Schandorff, S., Shewnarane, J., Vo, M., Taggart, J., Goudreault, M., Muskat, B., Alfarano, C., Dewar, D., Lin, Z., Michalickova, K., Willems, A.R., Sassi, H., Nielsen, P.A., Rasmussen, K.J., Andersen, J.R., Johansen, L.E., Hansen, L.H., Jespersen, H., Podtelejnikov, A., Nielsen, E., Crawford, J., Poulsen, V., Sùrensen, B.D., Matthiesen, J., Hendrickson, R.C., Gleeson, F., Pawson, T., Moran, M.F., Durocher, D., Mann, M., Hogue, C.W.V., Figeys, D., and Tyers, M. Systematic identification of protein complexes in *Saccharomyces cerevisiae* by mass spectrometry. *Nature*, 415, 180–183, 2002.

148. Stagljar, I. Finding partners: emerging protein interaction technologies applied to signaling networks. *Sci. STKE*, 213 1–5, 2003.

149. Vidal, M. and Legrain, P. Yeast forward and reverse 'n'-hybrid systems. *Nucleic Acids Res.*, 27, 919–929, 1999.

150. Fields, S. and Song, O. A novel genetic system to detect protein-protein interactions. *Nature*, 340, 245–246, 1989.

151. Brachmann, R.K. and Boeke, J.D. Tag games in yeast: the two-hybrid system and beyond. *Curr. Opin. Biotechnol.*, 8, 561–568, 1997.

152. Stagljar, I. and Fields, S. Analysis of membrane protein interactions using yeast-based technologies. *Trends Biochem. Sci.*, 27, 559–563, 2002.

153. Cho, D.I., Oak, M.H., Yang, H.J., Choi, H.K., Janssen, G.M.C., and Kim, K.M. Direct and biochemical interaction between dopamine D_3 receptor and elongation factor-1Bβγ. *Life Sci.*, 73, 2991–3004, 2003.

154. Kang, Y.K., Yoon, T., Lee, K., and Kim, H.J. Homo- or hetero-dimerization of muscarinic receptor subtypes is not mediated by direct protein-protein interaction through intracellular and extracellular regions. *Arch. Pharm. Res.*, 26, 846–854, 2003.

155. Ozenberger, B.A. and Young, K.H. Functional interaction of ligands and receptors of the hematopoietic superfamily in yeast. *Mol. Endocrinol.*, 9, 1321–1329, 1995.

156. Stagljar, I., Korostensky, C., Johnsson, N., and Heesen, S. A genetic system based on split-ubiquitin for the analysis of interactions between membrane proteins *in vivo*. *Proc. Natl. Acad. Sci. USA*, 95, 5187–5192, 1998.

157. Johnsson, N. and Varshavsky, A. Split ubiquitin as a sensor of protein interactions *in vivo*. *Proc. Natl. Acad. Sci. USA*, 91, 10340–10344, 1994.

158. Fetchko, M. and Stagljar, I. Application of the split-ubiquitin membrane yeast two-hybrid system to investigate membrane protein interactions. *Methods*, 32, 349–362, 2004.

159. Thaminy, S., Miller, J., and Stagljar, I. The split-ubiquitin membrane-based yeast two-hybrid system. *Meth. Mol. Biol.*, 261, 297–312, 2004.

160. Massaad, M.J. and Herscovics, A. Interaction of the endoplasmatic reticulum alpha 1,2-mannosidase Mns1p with Rer1p using the spit-ubiquitin system. *J. Cell Sci.*, 114, 4629–4635, 2001.

161. Reinders, A., Schulze, W., Thaminy, S., Stagljar, I., Frommer, W.B., and Ward, J.M. Intra- and intermolecular interactions in sucrose transporters at the plasma membrane detected by the split-ubiquitin system and functional assays. *Structure*, 10, 763–772, 2002.

162. Cervantes, S., Gonzalez-Duarte, R., and Marfany, G. Homodimerization of presenilin N-terminal fragments is affected by mutations linked to Alzheimer's disease. *FEBS Lett.*, 505, 81–86, 2001.

163. Grinna, L.S. and Tschopp, J.F. Size distribution and general structural features of N-linked oligosaccharides from the methylotrophic yeast, *Pichia pastoris. Yeast*, 5, 107–115, 1989.

164. Sinclair, G. and Choy, F.Y.M. Synonymous codon usage bias and the expression of human glucocerebrosidase in the methylotrophic yeast, *Pichia pastoris. Prot. Exp. Purif.*, 26, 96–105, 2002.

165. Woo, J.H., Liu, Y.Y., Mathias, A., Stavrou, S., Wang, Z., Thompson, J., and Neville, D.M. Jr. Gene optimization is necessary to express a bivalent anti-human anti-T cell immunotoxin in *Pichia pastoris. Prot. Exp. Purif.*, 25, 270–282, 2002.

166. Outchkourov, N.S., Stiekema, W.J., and Jongsma, M.A. Optimization of the expression of equistatin in *Pichia pastoris. Prot. Exp. Purif.*, 24, 18–24, 2002.

167. Gustafsson, C., Govindarajan, S., and Minshull, J. Codon bias and heterologous protein expression. *Trends Biotechnol.*, 22, 346–353, 2004.

168. http://gcua.schoedl.de/ (accessed March 2005).

169. Griffith, D.A., Delipala, C., Leadsham, J., Jarvis, S.M., and Oesterhelt, D. A novel yeast expression system for the overproduction of quality-controlled membrane proteins. *FEBS Lett.*, 553, 45–50, 2003.

170. Opekarova, M. and Tanner, W. Specific lipid requirements of membrane proteins — a putative bottleneck in heterologous expression. *Biochim. Biophys. Acta.*, 1610, 11–22, 2003.

171. Binda, C., Li, M., Hubalek, F., Restelli, N., Edmondson, D.E., and Mattevi, A. Insights into the mode of inhibition of human mitochondrial monoamine oxidase B from high-resolution crystal structures. *Proc. Natl. Acad. Sci. USA*, 100, 970–975, 2003.

9 Expression of Functional Membrane Proteins in the Baculovirus–Insect Cell System: Challenges and Developments

Giel J.C.G.M. Bosman and Willem J. de Grip

CONTENTS

9.1 INTRODUCTION

The baculovirus–insect cell system has become a widely used tool for protein expression. Because of its ease to use in combination with the processing potential of eukaryotic cells (Table 9.1 and Table 9.2), this system is especially useful for expression of integral membrane proteins in a structurally relevant, functional conformation. In addition, the baculovirus genome tolerates large insertions (30 kb) enabling coexpression of multiple domains or constituents of multimeric complexes. Finally, baculovirus allows selection of medium to strong promoters that are active during various periods after infection, can be obtained in high titers, and — last but not least — has a very low pathogenicity. Nevertheless, a PubMed search indicates that after a strong increase between 1987 and 1995, the number of publications in which the baculovirus–insect cell system was used for expression of membrane

TABLE 9.1
A Comparison of Expression Systems with Regard to Their Potential in Functional Expression of Membrane Proteins

Processing	E. coli	Yeast	Insect Cells	Mammalian Cells
Folding	+/–	+/–	++	+++
Glycosylation	–	++	++	+++
Cleavage	+/–	+/-	++	+++
Phosphorylation	–	++	+++	+++
Acylation	–	+	+++	+++
Amidation	–	–	++	+++
Maximal yield (% of total protein)	5	1	2	1

TABLE 9.2
A Comparison of Heterologous Expression Systems with Regard to Their Technical Potential

Process	E. coli	Yeast	Insect Cells	Mammalian Cells
Suspension culture	+++	+++	++	+/–
Serum/protein-free culture media	+++	+++	++	+/–
Stable isotope labeling	++	+	+/-	+/–
Scale up	++	++	++	+/–
Ease of control	++	++	++	+/–

proteins has remained constant or may have started to decline since 2001 (Figure 9.1). We interpret this trend as correlated with the difficulties to obtain high-resolution data on the structure of membrane proteins. The availability of functional membrane proteins is the bottleneck for obtaining two-dimensional (2D) and three-dimensional (3D) crystals. The increase in knowledge on the processes that influence folding and trafficking of membrane proteins in the baculovirus system has lagged far behind the development of molecular biology methodology. Also, standardized methods for solubilization, purification, and reconstitution of membrane proteins with preservation of function are only now beginning to emerge. Against this background, we present here our view of the state of the baculovirus–insect cell expression system, with emphasis on the new developments required to meet the need for detailed knowledge of structure and function of membrane proteins in the current genomic and proteomic landslide.

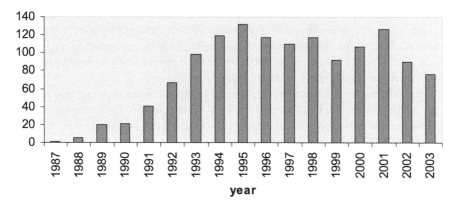

FIGURE 9.1 The number of publications during 1987–2003 in which the baculovirus–insect cell expression system was used for expression of membrane proteins (receptors, channels, and transporters).

9.2 THE BACULOVIRUS SYSTEM

9.2.1 RECOMBINANT VIRUS

The baculovirus expression system is based on heterologous protein production in insect cells with the baculovirus *Autographa californica* multicapsid nucleo-polyhedrovirus (AcMNPV) as the expression vector. During the life cycle of wild-type baculovirus, two forms of progeny are produced: (1) a budded virus for rapid secondary infection of neighboring cells, and (2) an occluded virus for survival outside and horizontal transmission between hosts. The occluded forms are encap-sulated in polyhedra, with the structural protein polyhedrin as their major com-ponent. Since the polyhedra are not active in viral replication in cultured cells, interference restricted to this late phase of infection allows expression of up to four foreign proteins within the same recombinant virus, under the control of late-phase promoters such as the polyhedrin or p10 promoter.[1] Most commercial systems use the very strong, late-phase polyhedrin promoter to drive expression. Comparable expression levels may be obtained with the late p10 promoter, but the use of the early basic protein promoter may be less successful (Vissers and De Grip, unpublished observations). In most applications, recombinant virus is generated by either (1) homologous recombination after co-transfection of insect cells with linearized viral DNA and a transfer vector containing a plasmid for bacterial replication, a viral promoter, the cDNA insert, and flanking sequences; or by (2) site-specific transposition in order to insert cDNA into bacmids, that is, baculovirus shuttle vectors that replicate in *Escherichia coli* as a low-copy plas-mid.[2] The latter method allows rapid selection of recombinant viruses in bacteria. Most commercially available kits are variations on these themes, including the possibility to add various N- and C-terminal purification and detection tags (His,

GFP, etc.). Lately, we have encountered the loss of productive, but not necessarily of infectious, recombinant virus, generated with the bacmid system, after only a few rounds of virus amplification in Sf9 cells. This may be due to an inherent instability of recombinant virus obtained by transposition,[3] or an increased sensitivity of bacmid-generated viruses to the "passage effect," that is, accumulation of viral particles with a defective genome and a replication advantage that interfere with normal viral replication and protein production.[4,5] Since high-frequency recombination between various viruses may be involved in this phenomenon,[6] we have taken the precaution of plaque purification in combination with a check on recombinant protein production as early as possible during the various rounds of virus amplification. Even though relatively laborious plaque purification and plaque assays remain the recommended method for isolating homogeneous virus and measuring its titer, other methods, such as end-point dilutions,[7] immunoassays, or cell viability assays,[8] have, in our hands, a high interassay and interuser variability.

9.2.2 CELL LINES

Although the *Spodoptera frugiperda* cell lines Sf9 and Sf21 are used in most cases, for some proteins the functional yield may be higher in other cell lines. This may be related to differences in glycosylation or chaperoning potential, and in the case of G protein-coupled receptors (GPCRs), in endogenous production of G proteins.[9] In addition to protein yield, growth in suspension and serum- and protein-free culture may be important parameters.[10] In our hands, for expression of the integral membrane protein rhodopsin, a prototypical GPCR, Sf cell lines give the best results (Table 9.3).

TABLE 9.3
Rhodopsin Production in Various Insect Cell Lines

| | | Culture Parameters | | |
Cell Line	Rhodopsin* Production	Cell Growth	Suspension Culture	Protein-Free Culture
Mamestra brassica	+	+	–	+/–
Spodoptera frugiperda (Sf9/Sf21)	+/++	++	++	++
Trichoplusia ni ("High Five")	+/++	++	+	+

* +, 5 pmoles/10^6 cells; ++, 20 pmoles/10^6 cells; 10^6 Sf9 cells yield 0.20–0.25 mg membrane protein.

9.2.3 OPTIMIZATION

In our opinion, the optimal conditions for obtaining the highest yields of functional membrane proteins are those that are focused on keeping the cells as healthy as possible, as long as possible, within the limits of the transient infection system. This is based mostly on experience with rhodopsin and related visual pigments, where the highest yields of functional receptor are obtained when (1) cells are cultured in controlled conditions in suspension culture in a bioreactor in which ambient temperature, stirring speed, oxygen supply and removal of CO_2 by headspace purging are optimized; (2) cells are infected when they are actively dividing; (3) cells are infected at a low multiplicity of infection (MOI), varying between 0.01 and 0.05. Low MOIs also minimize the interference of defective virus particles[5]; and (4) cells are harvested just before they start to disintegrate, which is under the preceding conditions at four or five days post infection. These results confirm the validity of the "cell yield concept."[11] Comparable results were obtained for GPCRs.[12] Online determination of the cumulative oxygen use is a sensitive indicator of the maximum protein yield, with identical kinetics as those of the adenosine triphosphate (ATP) content (Olejnik et al.[13] and Bosman et al., unpublished observations). For some proteins, the presence of a protease inhibitor may be highly beneficial (Figure 9.2). Protease inhibitors probably act through inhibition of virus-encoded proteases and inhibition of proteases that are activated or liberated when the cells start to desintegrate[14] and thus may prevent protein degradation directly but also by retardation of cell death.[15] For proteins that are extremely vulnerable to proteolytic degradation, nonlytic baculoviruses or stable transfection may be alternative approaches.[16,17] In this respect, RNA interference may be considered to minimize the expression of nondesired viral proteins.[18]

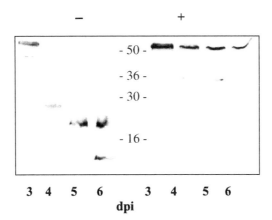

FIGURE 9.2 Expression of the histamine H1 receptor in the baculovirus–insect cell expression system. Sf9 cells were infected at an MOI of 0.1, and samples were taken at 3–6 days post infection in the presence (+) or absence (–) of 5 µM leupeptin. H1 was detected using an anti-His antibody. Molecular weight markers are indicated.

9.3 GLYCOSYLATION

One of the advantages of the baculovirus-based protein expression system, in comparison with those based on bacteria and yeasts, is the capacity of insect cells for N-linked protein glycosylation and, to a lesser extent, O-linked glycosylation. Although the function of carbohydrates on mature glycoproteins may be very diverse, the early steps of the glycosylation process are intimately linked with the quality control of correct folding and trafficking of most membrane proteins through the endoplasmic reticulum (ER) and Golgi apparatus on their way to the plasma membrane.[19] As correct folding is a bottleneck in recombinant protein expression, there is a growing attention for the glycosylation processes in insect cells. In most eukaryotic cells, N-glycosylation begins with the transfer of a glycan precursor $(Glc_3Man_9GlcNAc_2)$ to an asparagine residue in a Asn-X-Ser/Thr sequence on the nascent polypeptide. This precursor is trimmed and elongated in the ER and Golgi apparatus by the consecutive actions of α-glucosidases, α-mannosidases that can remove up to six mannose residues, and glycosyltransferases that can add N-acetylglucosamine, fucose, galactose, and sialic acid residues. The final outcome of this processing depends not only on the primary structure of the protein but also on the tissue and species of origin.[19] In mammalian cells, this outcome is mostly in the form of multi-antennary, sialylated complex glycans. Insect cells have an analogous N-glycan processing pathway with a unique form of mannosidase,[20] but have low, if any, levels of galactosyl- and sialyltransferases.[21] Therefore, most glycan structures produced by insect cells are of the paucimannosidic $(Man_3[Fuc]GlcNAc_2)$ type, although there may be cell line-, medium-, and baculovirus-related variations.[22] This may be due to insect-specific, Golgi-associated GlcNAcase-mediated removal of GlcNAc from the precursor N-glycan, thereby preventing its galactosylation and terminal sialylation.[23] Indeed, inhibition of GlcNAcase has been shown to allow secretion of sialylated glycoproteins.[23] Also, addition of N-acetylmannosamine to and the presence of sialoglycoproteins in the culture medium have been reported to result in the production of mature (sialylated) glycoproteins.[24,25] In another approach, insect cell lines that express up to seven different mammalian galactosyl- and sialyltransferases have been engineered.[26] Such transgenic cell lines, sometimes in combination with a baculovirus in which expression of different transferases is driven by different promoters, do, indeed, produce recombinant glycoproteins with more authentic biantennary, complex, terminally sialylated N-linked glycans.[27] Thus, although insect cell lines do seem to have the potential for glycoprotein sialylation, metabolic and genetic engineering are necessary for the realization of this potential. However, the studies underlying this conclusion are almost all performed with secreted, soluble glycoproteins. Although most membrane proteins do require the positioning of N-glycans in critical regions for optimal folding during synthesis, expression of a functional membrane protein does not seem to require the presence of mature, sialylated N-glycans, unless these are critical for partner recognition.[28] In addition, the presence of complex, often strongly charged and mobile carbohydrate chains is likely to hamper crystal formation. In this respect, the paucomannosidic carbohydrate chains of insect cell-produced membrane proteins may be beneficial factors. Thus, large-scale expression of functional membrane proteins is likely to benefit more from optimization of steps in the early processes in the ER (translocation, folding, trafficking) than from efforts to produce fully mature, sialylated complex N-linked glycans.

9.4 IMPROVING THE YIELD OF FUNCTIONAL PROTEINS

It is evident that correct folding is the major bottleneck in overexpression of membrane proteins for the purpose of determining molecular structures as well as for high-throughput expression as a tool for structural and functional proteomics. It is also evident that insect cells do possess the capacity for the production of correctly folded, functional integral membrane proteins and targeting to their membrane. The production capacity of insect cells for recombinant membrane proteins easily exceeds 50 pmoles/10^6 cells (ca 3×10^7 copies per cell). However, the functional amount can vary strongly. Under optimal conditions, insect cells produce the prototypical GPCR rhodopsin at a functional level of 2×10^7 copies per cell with a functional ratio of 80%, giving a yield of 4 to 6 mg protein per liter of culture medium.[29] The yield varies for GPCRs substantially from 0.3 to 20 pmol/10^6 cells, where only 6 out of more than 30 GPCRs have expression levels of higher than 5 pmol/10^6 cells.[30] However, in our hands, expression levels over 5 pmoles/10^6 cells (> 0.5 mg/L) have been achieved for at least 5 out of the 7 GPCRs tested so far in large-scale production conditions. The functionality of GPCRs has mainly been determined by the capacity to bind ligands, whereas data on coupling with intracellular signalling pathways are much scarcer. However, where described, correct ligand binding in wild-type recombinant receptors always was associated with the ability to activate the corresponding G protein.

GPCRs constitute more than 75% of the membrane proteins expressed with baculovirus. Successful expression in the baculovirus-insect cell expression system of other membrane proteins, such as ion channels and transporters, has been described mainly in qualitative rather than in quantitative terms. Where quantitative data are available, the yield of functional channels and transporters is very low. One example is a voltage-gated potassium channel, determined by specific toxin binding to be expressed at 0.02 pmol/10^6 cells.[31] Similar data were obtained with a related voltage-gated potassium channel,[32] with two sodium-dependent anion transport proteins (1 µg transporter per 10 mg membrane protein[33]), with a membrane-spanning domain of CFTR,[34] and with various ABC transporters.[35] In all these cases the baculovirus–insect cell expression system seems to offer considerable advantages for determining functional characteristics in comparison with other expression systems, such as the absence of endogenous channels,[36,37] and the ease with which gigaohm seals can be obtained.[38] In these cases, the yield of functional protein seems to be directly related to the fraction that is present in the plasma membrane and thus inversely correlated with the fraction that is apparently stuck in the endoplasmic reticulum. Notable exceptions are the Na$^+$K$^+$–ATPase, which is already fully functional in the ER if expressed in its heterodimeric form,[39] and the vanilloid receptor, for which the ER is the natural habitat.[37]

Thus, improvement of the folding and trafficking processes, with expression of a fully functional protein at the plasma membrane as its measurable outcome, seems to be the first and foremost challenge. In this respect, it is well documented that oligomerization occurs during expression of GPCRs in the baculovirus system,[40,41] and although it cannot be excluded that the observed associations are caused by the overexpression conditions, oligomers may actually represent physiological, functional units.[42]

In general, glycosylation does not seem to be the weakest link in these processes. As discussed above, insect cells do have the capacity for complex N-linked glycosylation, and in most cases, underglycosylation seems to be rather a consequence than a cause of inadequate folding. We find, for example, upon expression of mutants of rhodopsin, that the yield of total protein is mostly comparable with that of the wild-type rhodopsin but that relatively large quantities of unglycosylated protein are generated, which do not bind ligand. In addition, the yield of functional protein is related to the amount of protein that is present in the plasma membrane rather than with the amount of protein that is glycosylated.[43] Hence, complete use of glycosylation sites does not guarantee completely correct folding, and immature species probably are retained in the endoplasmic reticulum as well.[44] On the other hand, the presence of mature, complex glycans does not seem to be required for correct folding, and immature glycosylation, resulting in the paucimannosidic glycans that form the majority in baculovirus-infected cells, does not hamper targeting, folding, and trafficking. In general, the machinery for correct targeting, folding, and trafficking is present in the baculovirus–insect cell system but, for most membrane proteins, at too low a capacity. Thus, constructs in fusion with a signal sequence, for example, of the baculovirus gp64 protein or of influenza hemaglutinin, have been described to enhance translocation of membrane proteins across the ER membrane and enhance production.[45,46]

Under certain conditions, the use of weaker promoters such as the basic protein promoter rather than the commonly used polyhedrin promoter may result in a higher yield of functional protein.[31] We could not confirm this for certain GPCRs. Hence, it is not clear whether this effect is due to lower expression levels with a less overloaded ER machinery or due to activation of the promoter earlier after infection, when the ER quality control is still in better shape.

Coexpression of molecular chaperone proteins, an obvious consequence of the conclusion that too low a capacity for correct folding may be a bottleneck in the baculovirus expression system, has been described to be beneficial so far only in the combination of a voltage-gated potassium channel and calnexin[31] and the serotonin transporter with calnexin or BiP.[47] Another approach for increasing the yield of functional proteins is suggested by the observation that some receptor mutants for gonadotropin-releasing hormone and vasopressin can be rescued by specific ligands.[48,49] Apparently, the presence of these "pharmacopherones" corrected misfolding and correctly routed the mutant proteins to the plasma membrane, resulting in receptors with the functional characteristics (specificity and affinity for ligands, and coupling to effector protein) similar to those of the wild type.[48,49] The conclusion that this approach might be generally applicable for correcting a broad array of "protein-folding diseases" including Alzheimer's disease, cataract, retinitis pigmentosa, and so on,[50] may be overly enthusiastic, but various examples warrant the putative feasibility of this stratagem for improving the yield of correctly folded and routed membrane proteins in overexpression situations. Our own experience with this approach has been disappointing so far, but possibly the finding of a suitable pharmacopherone is a matter of serendipity for every single receptor.

Thus, in order to overcome the bottleneck for expression of functional integral membrane proteins, the primary goal seems to be the identification of chaperones and related proteins that provide correct translocation, folding, and trafficking. Together with a molecular description of the limiting step in overexpressing conditions (disulfide bridge formation, glycosylation, folding kinetics, stability of folding intermediates, and so on), this goal will most likely be obtained using a structural strategy, consisting of coexpression of a proper chaperone protein or enzyme in the proper time window, in combination with a supply of chemical or pharmacological stabilizers. The observed large range in yields of functional membrane proteins, together with the few data that are presently available, strongly suggest that a fully generic approach will not be feasible and that specific strategies may be restricted to individual protein families.

9.5 NEW DEVELOPMENTS

One major advantage of the baculovirus system is that it enables one to achieve fine control of recombinant protein expression at various levels: variation of virus particles per cell, the expression time, and the use of early and late promoters. Furthermore, the possibility for multiple infections, as well as — more efficiently — the combination of two or more cDNAs in one viral vector, allows relatively easy expression of two or more proteins at a desired ratio and time of expression. Thus, using fluorescent tags, the combination of baculovirus and fluorescence resonance energy transfer (FRET) can be developed into a new assay system for detecting protein–protein interaction in living cells.[51] In addition, coexpression of GPCRs with various G proteins resulting in functional reconstitution has been described to strongly increase the fraction of high-affinity binding sites[52] and may constitute a valuable tool for ligand screening.[53] The Bac-to-Bac system has recently been combined with Cre-mediated recombination at loxP sites for the sequential integration of multiple dual-expression cassettes into a transfer vector, followed by recombination into a bacmid containing the baculovirus genome.[54] This has been noted as a timely technology, given the current limitations in coexpression of multiple recombinant proteins in eukaryotic cells.[55]

Such analyses are not necessarily restricted to the insect cell environment, since baculoviruses are capable of efficiently transducing mammalian cells and cell lines, and to express genes under the control of mammalian promoters.[56–60] In the past, their use in large-scale production seemed to be restricted by the need for relatively high MOIs to obtain good infection rates. Recently, however, the range and extent of the infectivity of baculovirus for mammalian cells have been strongly improved by surface expression of specific tags (VSV, avidin). Thanks to their very low pathogenicity, baculovirus vectors may be highly suitable alternatives for other viral expression systems.

In general, the baculovirus system is less suited for high-throughput expression screening, if this includes generation and titration of individual recombinant viruses, since these are technically demanding and time-consuming procedures. However, for certain forms of expression screening, the system could be quite useful. For instance, baculoviruses have recently been used in the production of cDNA libraries for expression cloning of genes encoding membrane proteins.[61] This example, together with the efforts devoted to optimization of insect cell growth and development of a high-throughput insect cell expression screen,[62] illustrates the unique potential of the baculovirus–insect cell system and emphasizes the putative importance of the baculovirus system in the area of high-throughput screening (for a recent review see Hunt[63]).

A major goal that remains is the large-scale production of membrane proteins in a quantity and quality that make structural analyses not only possible but also worthwhile and affordable. Detailed and structurally relevant information can be obtained from proteins labelled with stable isotopes (^{15}N, ^{13}C, ^{1}H) in amounts of 5 to 10 mg with the newest solid-state nuclear magnetic resonance (NMR) techniques.[64] Insect cells can be adapted relatively easily for the growth and production of recombinant proteins in completely defined media. Such media are presently being developed, in combination with methods to produce specifically labelled amino acids based on biomass from microorganisms, using simple — and thus affordable — starting materials and culture techniques.[65]

Another very promising development is to utilize expression of integral membrane proteins on the membrane of budded viruses. This has been used for functional coupling of receptor with effector proteins, resulting in a strong increase in ligand affinity.[66] The same approach was also successfully used for reconstitution of a functional γ-secretase complex (involved in Alzheimer-related amyloid formation) consisting of four integral membrane proteins.[67] In addition, vaccination with antigens that were exposed on recombinant baculoviruses or infected insect cells (probably also containing budding viruses) has recently been shown to be much more protective, that is, more effective in eliciting neutralizing antibodies, than vaccination with the purified antigens.[68] Exposure of recombinant proteins at the surface of budded viruses can be strongly increased by fusion to an extra copy of the major coat protein gp64, or to its membrane anchor.[69] In a similar way, occlusion bodies containing recombinant proteins may be powerful vaccines.[70] These examples indicate that budded baculovirus particles may be selectively enriched in properly folded forms of membrane proteins, possibly because they contain intracellular membranes.[71] This hitherto largely unexplored aspect of the baculovirus-insect cell system adds to its value as a tool for structural and functional analyses of membrane protein complexes.

9.6 CONCLUSION

Taken together, these observations warrant the conclusion that the horizon for the baculovirus expression system now considerably extends beyond the scope of large-scale protein expression and includes the development of functional high-throughput screening assays and clinical applications. We emphasize that for such developments, the need for rational and generic procedures of purification and reconstitution is as high as for a high-yield production process. However, these procedures are beyond the scope of this review and have recently been discussed elsewhere.[72]

REFERENCES

1. Summers, M.D. and Smith, G.E. A manual of methods for baculovirus vectors and insect culture procedures. *Texas Exp. Station Bull.*, Volume 1555, 1987.
2. www.invitrogen.com (accessed 2005).
3. Pijlman, G.P., Van Schijndel, J.E., and Vlak, J.M. Spontaneous excision of BAC vector sequences from bacmid-derived baculovirus expression vectors upon passage in insect cells. *J. Gen. Virol.*, 84, 2669–2678, 2003.
4. Pijlman, G.P., Van den Born, E., Martens, D.E., and Vlak, J.M. *Autographa californica* baculoviruses with large genomic deletions are rapidly generated in infected insect cells. *Virology*, 283, 132–138, 2001.
5. Van Lier, F.L.J., Van de End, E.J., De Gooijer, C.D., Vlak, J.M., and Tramper, J. Continuous production of baculovirus in a cascade of insect cell reactors. *Appl. Microbiol. Biotechnol.*, 33, 43-47, 1990.
6. Kamita, S.G., Maeda, S., and Hammock, B.D. High-frequency homologous recombination between baculoviruses involves DNA replication. *J. Virol.*, 77, 13053–13061, 2003.
7. O'Reilly, D.R., Miller, L.K., and Luckow, V.A. *Baculovirus Expression Vectors: A Laboratory Manual.* WH Freeman, New York, 1992.
8. Mena, J.A., Ramirez, O.T., and Palomares, L.A. Titration of non-occluded baculovirus using a cell viability assay. *BioTechniques*, 34, 260–264, 2003.
9. Knight, P.J.K. and Grigliatti, T.A. Diversity of G proteins in lepidopteran cell lines: partial sequences of six G protein alpha subunits. *Arch. Insect Biochem. Physiol.*, 57, 142–150, 2004.
10. Rhiel, M., Mitchell-Logean, C.M., and Murhammer, D.W. Comparison of *Trichoplusia ni* BTI-Tn-5B1-4 (High Five) and *Spodoptera frugiperda* Sf-9 insect cell line metabolism in suspension cultures. *Biotechnol. Bioeng.*, 55, 909–920, 1997.
11. Wong, K.T.K., Peter, C.H., Greenfield, P.F., Reid, S., and Nielsen, L.K. Low multiplicity infection of insect cells with a recombinant baculovirus: the cell yield concept. *Biotechnol. Bioeng.*, 49, 659–666, 1996.
12. Carpentier, E., Lebesgue, D., Kamen, A.A., Hogue, M., Bouvier, M., and Durocher, Y. Increased production of active human beta(2)-adrenergic/G(alphaS) fusion receptor in Sf-9 cells using nutrient limiting conditions. *Prot. Exp. Purif.*, 23, 66–74, 2001.
13. Olejnik, A.M., Czaczyk, K., Marecik, R., Grajek, K., and Jankowski, T. Monitoring the progress of infection and recombinant production in insect cell cultures using intracellular ATP measurement. *Appl. Microbiol. Biotechnol.*, 65, 18–24, 2004.
14. Ratnala, V.R., Swarts, H.G., VanOostrum, J., Leurs, R., DeGroot, H.J., Bakker, R.A., and DeGrip, W.J. Large-scale overproduction, functional purification and ligand affinities of the His-tagged human histamine H1 receptor. *Eur. J. Biochem.*, 271, 2636-2646, 2004.
15. Hu, Y.C. and Bentley, W.E. Enhancing yield of infectious bursal disease virus structural proteins in baculovirus expression systems: focus on media, protease inhibitors, and dissolved oxygen. *Biotechnol. Progr.*, 15, 1065–1071, 1999.

16. Ho, Y., Lo, H.R., Lee, T.C., Wu, C.P.Y., and Chao, Y.C. Enhancement of correct protein folding *in vivo* by a non-lytic baculovirus. *Biochem. J.*, 382, 695–702, 2004.

17. McCarroll, L. and King, L.A. Stable insect cell cultures for recombinant protein production. *Curr. Opin. Biotechnol.*, 8, 590–594, 1997.

18. Agrawal, N., Malhotra, P., and Bhatnagar, R.K. siRNA-directed silencing of transgene expressed in cultured insect cells. *Biochem. Biophys. Res. Commn.*, 320, 428–434, 2004.

19. Roth, J., Zuber, C., Guhl, B., Fan, J.Y., and Ziak, M. The importance of trimming reactions on asparagine-linked oligosaccharides for protein quality control. *Histochem. Cell Biol.*, 117, 159–169, 2002.

20. Kawar, Z., Karaveg, K., Moremen, K.W., and Jarvis, D.L. Insect cells encode a class II α-mannosidase with unique properties. *J. Biol. Chem.*, 276, 16335–16340, 2001.

21. Marchal, I., Jarvis, D.L., Cacan, R., and Verbert, A. Glycoproteins from insect cells: sialylated or not? *Biol. Chem.*, 382, 151–159, 2001.

22. Joshi, L., Davis, T.R., Mattu, T.S., Rudd, P.M., Dwek, R.A., Shuler, M.L., and Wood, H.A. Influence of baculovirus-host cell interactions on complex N-linked glycosylation of a recombinant human protein. *Biotechnol. Progr.*, 16, 650–656, 2000.

23. Watanabe, S., Kokuho, T., Takahashi, H., Takahashi, M., Kubota, T., and Inumaru, S. Sialylation of N-glycans on the recombinant proteins expressed by a baculovirus-insect cell system under beta-N-acetylglucosaminidase inhibition. *J. Biol. Chem.*, 277, 5090–5093, 2002.

24. Joshi, L., Shuler, M.L., and Wood, H.A. Production of sialylated N-linked glycoproteins in insect cells. *Biotechnol. Progr.*, 17, 822–827, 2001.

25. Hollister, J., Conradt, H., and Jarvis, D.L. Evidence for a sialic acid salvaging pathway in lepidopteran insect cells. *Glycobiology*, 13, 487–495, 2003.

26. Aumiller, J.J., Hollister, J.R., and Jarvis, D.L. A transgenic insect cell line engineered to produce CMP-sialic acid and sialylated glycoproteins. *Glycobiology*, 13, 497–507, 2003.

27. Tomiya, N., Howe, D., Aumiller, J.J., Pathak, M., Park, J., Palter, K.B., Jarvis, D.L., Betenbaugh, M.J., and Lee, Y.C. Complex-type biantennary N-glycans of recombinant human transferrin from *Trichoplusia ni* insect cells expressing mammalian beta-1,4-galactosyltransferase and beta-1,2-N-acetylglucosaminyltransferase II. *Glycobiology*, 13, 23–34, 2003.

28. Tate, C.G., Haase, J., Baker, C., Boorsma, M., Magnani, F., Vallis, Y., and Williams, D.C. Comparison of seven different heterologous protein expression systems for the production of the serotonin transporter. *Biochim. Biophys. Acta.*, 1610, 141–153, 2003.

29. Klaassen, C.H.W. and De Grip, W.J. Baculovirus expression system for expression and characterization of functional recombinant visual pigments. *Meth. Enzymol.*, 315, 12–29, 2000.

30. Massotte, D. G protein-coupled receptor overexpression with the baculovirus-insect cell system: a tool for structural and functional studies. *Biochim. Biophys. Acta.*, 1610, 77–89, 2003.

31. Higgins, M.K., Demir, M., and Tate, C.G. Calnexin co-expression and the use of weaker promoters increase the expression of correctly assembled Shaker potassium channel in insect cells. *Biochim. Biophys. Acta.*, 1610, 124–132, 2003.

32. Gaymare, F., Cerutti, M., Horeau, C., Lemaillet, G., Urbach, S., Ravallec, M., Devauchelle, G., Sentenac, H., and Thibaud, J.B. The baculovirus/insect cell system as an alternative to Xenopus oocytes. *J. Biol. Chem.*, 271, 22863–22870, 1996.

33. Fucentese, M., Winterhalter, K.H., Murer, H., and Biber, J. Functional expression and purification of histidine-tagged rat renal Na/phosphate (NaPi-2) and Na/sulfate (NaSi-1) cotransporters. *J. Membr. Biol.*, 160, 111–117, 1997.

34. Ramjeesingh, M., Ugwu, F., Li, C., Dhani, S., Huan, L.J., Wang, Y., and Bear, C.E. Stable dimeric assembly of the second membrane-spanning domain of CFTR (cystic fibrosis transmembrane conductance regulator) reconstitutes a chloride-selective pore. *Biochem. J.*, 375, 633–641, 2003.

35. Cai, J. and Gros, P. Overexpression, purification, and functional characterization of ATP-binding cassette transporters in the yeast, *Pichia pastoris. Biochim. Biophys. Acta.*, 1610, 63–76, 2003.

36. Salvador, C., Mora, S.I., Ordaz, B., Antaramian, A., Vaca, L., and Escobar, L.I. Basal activity of GIRK5 isoforms. *Life Sci.*, 72, 1509–1518, 2003.

37. Wisnoskey, B.J., Sinkins, W.G., and Schilling, W.P. Activation of vanilloid receptor type I in the endoplasmic reticulum fails to activate store-operated Ca^{2+} entry. *Biochem. J.* 372, 517–528, 2003.

38. Astill, D.St.J., Rychkov, G., Clarke, J.D., Hughes, B.P., Roberts, M.L., and Bretag, A.H. Characteristics of skeletal muscle chloride channel ClC-1 and point mutant R304E epxressed in Sf-9 insect cells. *Biochim. Biophys. Acta.*, 1280, 178–186, 1996.

39. Gatto, C., McLoud, S.M., and Kaplan, J.H. Heterologous expression of Na+-K+-ATPase in insect cells: intracellular distribution of pump subunits. *Am. J. Physiol. Cell Physiol.*, 281, C982–C992, 2001.

40. Fukushima, Y., Asano, T., Saitoh, T., et al. Oligomer formation of histamine H2 receptors expressed in Sf9 and COS7 cells. *FEBS Lett.*, 402 :283–286, 1997.

41. Cheung, T.C. and Hearn, J.P. Development of a baculovirus-based fluorescence resonance energy transfer assay for measuring protein–protein interaction. *Eur. J. Biochem.*, 270, 4973–4981, 2003.

42. Park, P.S., Filipek, S., Wells, J.W., and Palczewski, K. Oligomerization of G protein-coupled receptors: past, present and future. *Biochemistry*, 43, 15643–15656, 2004.

43. De Caluwé, G. Structure-function relationships in bovine rhodopsin: a study using recombinant baculovirus mediated expression. Ph.D. dissertation, Radboud University Nijmegen, 1995.

44. Ellgaard, L. and Helenius, A. Quality control in the endoplasmic reticulum. *Nat. Rev. Mol. Cell Biol.*, 4, 181–191, 2003.

45. Massotte, D., Pereira, C.A., Pouliquen, Y., and Pattus, F. Parameters influencing human mu opioid receptor over-expression in baculovirus-infected insect cells. *J. Biotechnol.*, 69, 39–45, 1999.

46. Guan, X.M., Kobilka, T.S., and Kobilka, B.K. Enhancement of membrane insertion and function in a type IIIb membrane protein following introduction of a cleavable signal peptide. *J. Biol. Chem.*, 267, 21995–21998, 1992.

47. Tate, C.G., Whiteley, E.M., and Betenbaugh, M.J. Molecular chaperones stimulate the functional expression of the cocaine-sensitive serotonin transporter. *J. Biol. Chem.*, 274, 17551–17558, 1999.

48. Janovick, J.A., Maya-Nunez, G., and Cohn, P.M. Rescue of hypogonadotropic hypogonadism-causing and manufactured GnRH receptor mutants by a specific protein-folding template: misrouted proteins as a novel diseased etiology and therapeutic target. *J. Clin. Endocrinol. Metab.*, 87, 4825–4828, 2002.

49. Morello, J.P., Salahpour, A., Laperriere, A., Bernier, V., Arthus, M.F., Lonergan, M., Petaja-Repo, V., Angers, S., Morin, D., Bichet, D.G., and Bouvier, M. Pharmacological chaperones rescue cell surface expression and function of misfolded V2 vasopressin receptor mutants. *J. Clin. Invest.*, 105, 887–895, 2000.

50. Cohn, P.M., Leanos-Miranda, A., and Janovick, J. Protein origami: therapeutic rescue of misfolded gene products. *Mol. Interven.*, 2, 308–316, 2002.

51. Cheung, T.C. and Hearn, J.P. Development of a baculovirus-based fluorescence resonance energy transfer assay for measuring protein–protein interaction. *Eur. J. Biochem.*, 270, 4973–4981, 2003.

52. Maeda, Y., Kuroki, R., Haase, W., Michel, H., and Reilander, H. Comparative analysis of high-affinity ligand binding and G protein coupling of the human CXCR1 chemokine receptor and of a CXCR1-G12α fusion protein after heterologous production in baculovirus-infected insect cells. *Eur. J. Biochem.*, 271, 1677–1689, 2004.

53. Uustare, A., Nasman, J., Akerman, K.E.O., and Rinken, A. Characterization of M2 muscarinic receptor activation of different G protein subtypes. *Neurochem. Int.*, 44, 119–124, 2004.

54. Berger, I., Fitzgerald, D.J., and Richmond, T.J. Baculovirus expression system for heterologous multiprotein complexes. *Nat. Biotechnol.*, 22, 1583–1587, 2004.

55. Roy, P. Baculovirus solves a complex problem. *Nat. Biotechnol.*, 22, 1527–1528, 2004.

56. Abe, T., Takahashi, H., Hamazaki, H., Myano-Kurosaki, N., Matsuura, Y., and Takaku, H. Baculovirus induces an innate immune respons and confers protection from lethal influenza virus infection in mice. *J. Immunol.*, 171, 1133–1139, 2003.

57. Bilello, J.P., Cable, E.E., Myers, R.L., and Isom, H.C. Role of paracellular junction complexes in baculovirus-mediated gene transfer to non-dividing rat hepatocytes. *Gene Ther.*, 10, 733–749, 2003.

58. Viswanathan, P., Venkayah, B., Kumar, M.S., Rasheedi, S., Vrati, S., Bashyam, M.D., and Hasnain, S.E. The homologous region sequence (hr1) of *Autographa californica* multinucleocapsid polyhedrosis virus can enhance transcription from non-baculoviral promoters in mammalian cells. *J. Biol. Chem.*, 278, 52564–52571, 2003.

59. Liang, C.Y., Wang, H.Z., Li, T.X., Hu, Z.H., and Chen, X.W. High efficiency gene transfer into mammalian kidney cells using baculovirus vectors. *Arch. Virol.*, 149, 51–60, 2004.

60. Lehtolainen, P., Tyynela, K., Kannasto, J., Airenne, K.J., and Yla-Herttuala, S. Baculoviruses exhibit restricted cell type specificity in rat brain: a comparison of baculovirus- and adenovirus-mediated intracerebral gene transfer *in vivo*. *Gene Ther.*, 9, 1693–1699, 2002.

61. Urano, Y., Yamaguchi, M., Fukuda, R., Masuda, K., Takahashi, K., Uchiyama, Y., Iwanari, H., Jiang, S.Y., Naito, M., Kodama, T., and Hamakubo, T. A novel method for viral display of ER membrane proteins on budded baculovirus. *Biochem. Biophys. Res. Commn.*, 308, 191–196, 2003.

62. Bahia, D., Cheung, R., Buchs, M., Geisse, S., and Hunt, I. Optimisation of insect cell growth in deep-well blocks: development of a high-throughput insect cell expression screen. *Prot. Exp. Purif.*, 39, 61–70, 2005.

63. Hunt, I. From gene to protein: a review of new and enabling technologies for multi-parallel protein expression. *Prot. Exp. Purif.*, 40, 1–22, 2005.

64. Luca, S., Heise, H., and Baldus, M. High-resolution solid-state NMR applied to polypeptides and membrane proteins. *Account. Chem. Res.*, 36, 858–865, 2003.

65. www.proteinlabelling.nl (accessed 2005).

66. Masuda, K., Itoh, H., Sakihama, T., Akiyama, C., Takahashi, K., Fukuda, R., Yokomizo, T., Shimizu, T., Kodama, T., and Hamakubo, T. A combinatorial G protein-coupled receptor reconstitution system on budded baculovirus. *J. Biol. Chem.*, 278, 24552–24562, 2003.

67. Hayashi, I., Urano, Y., Fukuda, R., Isoo, N., Kodama, T., Hamakubo, T., Tomita, T., and Iwatsubo, T. Selective reconstitution and recovery of functional γ-secretase complex on budded baculovirus particles. *J. Biol. Chem.*, 279, 38040–48046, 2004.
68. Tami, C., Peralta, A., Barbieri, R., Berinstein, A., Carrillo, E., and Taboga, O. Immunological properties of FMDV-gP64 fusion proteins expressed on SF9 cell and baculovirus surfaces. *Vaccine*, 23, 840–845, 2004.
69. Grabherr, R. and Ernst, W. The baculovirus expression system as a tool for generating diversity by viral surface display. *Comb. Chem High Throughput Screen.*, 4, 185–192, 2001.
70. Je, Y.H., Jin, B.R., Park, H.W., Roh, J.Y., Chang, J.H., Seo, S.J., Olszewski, J.A., O'Reilly, D.R., and Kang, S.K. Baculovirus expression vectors that incorporate the foreign protein into viral occlusion bodies. *BioTechniques*, 34, 81–87, 2003.
71. Urano, Y., Yamaguchi, M., Fukuda, R., Masuda, K., Takahashi, K., Uchiyama, Y., Iwanari, H., Jiang, S.Y., Naito, M., Kodama, T., and Hamakubo, T. A novel method for viral display of ER membrane proteins on budded baculovirus. *Biochem. Biophys. Res. Commn.*, 308, 191–196, 2003.
72. Selinsky, B.S. Membrane protein protocols. *Meth. Mol. Biol.*, 228, 2003.

10 Expression of Membrane Proteins in Mammalian Cells

Kenneth H. Lundstrom

CONTENTS

10.1 INTRODUCTION

The completion of the sequencing of human and other genomes has provided a vast amount of material for gene expression and gene function studies. Recombinant protein expression plays an important role in these activities. During the past few years, alternative expression systems based on both prokaryotic and eukaryotic vectors have been engineered. Two excellent, albeit now to some extent outdated, reviews have previously been published on heterologous expression of integral transmembrane proteins[1] and more specifically G protein-coupled receptors (GPCRs).[2] In a more recent review, the focus is on large-scale production of GPCRs applying various expression systems.[3] Bacterial vectors have often been favored due to their simplicity, low cost, and scalability. In this context, many soluble proteins have been efficiently expressed in bacteria, especially in *Escherichia coli*, independent of whether their origin was prokaryotic or eukaryotic. Integral membrane proteins have, however, presented a different challenge. The higher complexity of membrane proteins, as compared with soluble proteins, has made their structural characterization significantly more difficult. Despite that, relatively good success has been obtained for the expression of various bacterial membrane proteins,[4] which has formed the basis for the generation of several high-resolution structures for prokaryotic targets.[5] In contrast, expression of eukaryotic and especially mammalian membrane proteins in bacteria has been less successful, although recently several GPCRs have given reasonable yields when expressed in either *E. coli* inclusion bodies[6] or membranes.[7] A more detailed description of bacterial expression can be found in Chapter 3 through Chapter 5.

Eukaryotic expression has relied, to a large extent, on yeast-based systems due to their ease to work with and the possibility to produce large biomasses. Both *Saccharomyces cerevisiae*[8] and *Pichia pastoris*[9] have been widely used as hosts for recombinant protein production, which is described in detail in Chapter 8. Although yeast cells possess many similar post-translational modification properties as do mammalian cells, there are obviously some differences, which might, under certain conditions, be less favorable. For this reason, much effort has been dedicated to the development of expression systems for higher eukaryotes such as insect cells. The most popular approach has been to use baculovirus vectors for the expression of recombinant membrane proteins in Sf9 cells and other insect cells. Very high yields of recombinant GPCRs have been obtained applying the baculovirus expression system.[10] Application of baculovirus vectors for membrane protein expression is described in detail in Chapter 9.

Ideally, mammalian cell lines should be used for the expression of mammalian proteins, since that provides the closest resemblance to the native situation. Not surprisingly, major efforts have been invested in the development of suitable mammalian expression systems. Obviously, the more complex composition and culturing requirements of mammalian cells, as compared with prokaryotes, has set limitations and required compromises in costs and time spent. The various approaches to establish expression of recombinant membrane proteins in mammalian cells are described in this chapter (Table 10.1). The applications of different procedures for transient expression are also discussed. Generation of stable cell lines, for which gene expression can be conditionally turned on or off, is presented with several examples. Moreover, the use of viral vectors such as adenoviruses, alphaviruses, lentiviruses, and vaccinia viruses is also described.

10.2 TRANSIENT TRANSFECTION

Transient expression has generally been characterized by the rapid production of small protein quantities to verify the functionality of recombinant proteins and to optimize expression conditions.[11] Many cell lines such as BHK-21, CHO-K1, COS-7, and HEK293 cells have been used for this purpose. Recently, large-scale transient transfection methods have also been established.[12]

Many membrane proteins have been verified for expression from transient mammalian expression vectors. Generally, the approach has been to apply strong promoters and develop transfection reagents that provide good delivery of plasmid DNA to the host cells. In this context, several GPCRs have provided relatively high specific binding activity with Bmax values in the range of 1 to 20 pmol receptor per milligram protein. Among GPCRs, the cholecystokinin (CCK) A receptor was expressed in COS cells[13] and in HEK293 cells.[14] The major concern of transient transfection in mammalian cells has been the complications and expenses in scaling up the process.[15]

TABLE 10.1

Expression of Integral Membrane Proteins in Various Mammalian Systems

Vector	Membrane protein	Quantity	Reference
Non-viral			
Transient	CCK-A receptor	binding, functional activity	13, 14
	GABA transporter	functional activity	17
	muI Na$^+$ channel	functional activity	18
	GLUT 1, 4 transporters	functional activity	19
Stable	opioid receptors	0.27 pmol/mg	21
	SERT	250,000 / cell, 293-EBNA	23
		80,000 / cell, MEL cells	23
		250,000 / cell, pCytTS	23
		400,000 / cell, T-Rex	23
	opsin receptor	6 mg/L	22
Viral			
Adenovirus	opioid receptor	1-3 pmol/mg	26
	β2-adrenergic receptor	4 pmol/mg	27
	large number of GPCRs	50 pmol/mg	28
Lentivirus	RGR opsin	functional activity	31
	GPCRs	binding activity	32
Semliki Forest virus	serotonin 5HT3	60 pmol/mg; 2 mg/L	35, 42
	purinoreceptor p2X	150 pmol/mg	36, 37
	neurokinin-1 receptor	100 pmol/mg; 10 mg/L	45
	adrenergic α2B receptor	176 pmol/mg; 5 mg/L	39
	+ 100 other GPCRs	10-170 pmol/mg	40, 41
	transporters	functional activity	34
Vaccinia virus	neuropeptide Y receptor	5-10 million / cell	29
	Dopamine D2, D4 receptors	binding activity	30

CCK-A, cholecystokinin-A; GABA, gamma-aminobutyric acid; GLUT, glucose transporter; SERT, serotonin transporter; RGR, retinal G protein-coupled receptor.

However, a modified calcium–phosphate co-precipitation method has been developed for suspension cultures of HEK293 EBNA cells at a 100-L scale.[16] Although more than 0.5 g of monoclonal antibodies could be produced this way, the yields for receptor expression have been relatively low. In addition to GPCRs, other mammalian membrane proteins such as ion channels and transporters have been evaluated. For instance, the cDNA for GAT-1, which codes for the Na$^+$- and Cl$^-$-coupled γ-aminobutyric acid (GABA) transporter, was transiently expressed in Ltk$^-$ cells.[17] A band of approximately 70 kD was visualized in Western blots on GAT-1 transfected cells using an anti-transporter antibody. The functional expression observed for the GABA transporter was abolished in the presence of tunicamycin. Other transiently expressed targets are the muI Na⁺ channel expressed in HEK293 cells[18] and the glucose transporters GLUT1 and GLUT4 in COS-7 cells.[19]

10.3 STABLE CELL LINES

Although the procedure to generate stable cell lines is relatively time-consuming, this approach has been widely used. The popularity of stable cell lines relies, to a large extent, on the ease of continued production of the recombinant proteins once the stable construct has been established. However, experience has revealed that in the stable clones, there is a tendency of insert removal even when associated with antibiotic selection pressure. One way to prevent this phenomenon has been to engineer inducible vector systems, where the transgene expression is only turned on after induction.[20]

   The expression levels obtained from stable cell lines are generally lower than those seen for transient expression. For instance, the κ-opioid receptor stably expressed from CHO cells showed binding activity of only 266 fmol/mg.[21] Recently, the levels for stably expressed GPCRs has improved to 1 to 20 pmol/mg, and in large scale production in suspension cultures, modest quantities of 0.1 mg/L culture are common. Exceptionally high levels of wild-type and mutant opsin were demonstrated in a mutant HEK293S cell line, which was inducible by tetracycline.[22] The N-glycan defect HEK293S cells produced 6 mg receptor in a 1.1-L suspension culture. The expression of other membrane proteins has also been evaluated in stable mammalian cells. A good example is the human serotonin transporter (SERT), which was tested in parallel in four stable mammalian cell lines.[23] The expression of SERT from a constitutively active promoter required the addition of SERT inhibitors to the cell culture medium to obtain stable clones. Addition of cocaine and imipramine resulted in the 293-EBNA-Coca270 and the 293-EBNA-Imi270 cell lines, respectively, which contained approximately 250,000 copies of SERT per cell. The inducible mammalian expression system based on mouse erythroleukemic (MEL) cells, which uses the ability of MEL cells to differentiate in the presence of DMSO,[24] was also evaluated for SERT expression. In this case, the number of SERT copies per cell was only 80,000. Two other inducible mammalian systems were tested for SERT. The cold-inducible pCytTS system, which is based on a temperature-sensitive Sindbis virus replicase,[25] generated some 250,000 SERT molecules per cell. The highest expression levels for SERT were obtained from the T-Rex expression system, which is based on a tetracycline-inducible system. T-Rex-based SERT expression generated 400,000 copies per cell. In all cases of stable expression, the recombinant SERT protein was functionally active and located at the plasma membrane.

10.4 ADENOVIRUS

Viral vectors have been attractive for transgene expression both *in vitro* and *in vivo* due to their strong promoters and wide host range. Adenoviruses have also been frequently used for the expression of membrane proteins. For instance, the mu (MOR) and kappa (KOR) opioid receptors have been expressed from replication-deficient adenovirus vectors.[26] The opioid vectors were expressed alone or as fusion proteins with C-terminal green fluorescent protein (GFP). CHO cells expressing the truncated version of the coxsackie–adenovirus receptor (CAR), were efficiently transduced by Ad5-MOR and Ad5-KOR recombinant viruses. The Kd values were similar to published values, and the Bmax values were 1 to 3 pmol/mg, two to threefold higher than observed for stable cell lines.[21] Previously, the β_2 adrenergic receptor has been expressed from adenovirus vectors in rabbit myocytes and generated up to 4 pmol receptor/mg.[27] The Joint Center for Structural Genomics (JCSG) based in California (see details in Chapter 16) has included in its program a study on more than 100 GPCRs, which will be expressed among other vectors also from adenoviruses.[28] Preliminary results indicate that many GPCRs were well expressed from adenovirus vectors and are suitable for structural biology applications.

10.5 VACCINIA VIRUS

Vaccinia virus vectors have also been used for the recombinant expression of membrane proteins, specifically GPCRs. For instance, the neuropeptide Y (NPY) receptor was expressed in vaccinia virus-infected mammalian cell lines.[29] Saturation and competition binding experiments resulted in 5 to 10 million binding sites per cell. Localization studies demonstrated the presence of NPY receptors on the plasma membrane. In another study, the human dopamine D2 and D4 receptors were expressed in rat-1 cells from vaccinia virus vectors showing high specific binding activity, with affinity constants consistent with previously reported values.[30] Functional coupling measured by adenylyl cyclase activity could be prevented in pertussis toxin-treated cells. The GABA transporter expressed after transient transfection of plasmid DNA (see section on transient transfection, above) has also been introduced into HeLa cells by vaccinia virus infection.[17] The properties of the vaccinia-based GABA transporter expression were similar to those obtained from transient transfections.

10.6 LENTIVIRUS

Lentiviruses have been applied for the expression of mammalian membrane proteins. RGR opsin, the human retinal pigment epithelium (RPE) retinal G protein-coupled receptor, was introduced into a replication-deficient lentivirus vector, and COS-7 cells and a retinal pigment epithelium, ARPE-19 cell line, were transduced.[31] The evaluation of expression levels was carried out by immunodetection (Western blots) and [^{3}H] all-*trans*-retinal binding assays. The expression levels in ARPE-19 cells were 100-fold higher than in COS-7 cells. Additionally, the expression levels increased steadily until day 10 post infection in ARPE-19 cells, and stable expression was observed for more than six months. Within Tranzyme Pharma, a novel lentivirus based expression system named TranzExpression Technology (TExT™) has been developed.[32] This technology has been applied to set up a proprietary functional biology platform that is based on gene delivery and controlled expression *in vitro*, *ex vivo,* and *in vivo.* GPCRs and transporters have been successfully expressed using the TExT™ system.

10.7 SEMLIKI FOREST VIRUS

Alphaviruses, and particularly Semliki Forest virus (SFV), have frequently been employed for overexpression of recombinant proteins in mammalian cell lines, primary cell cultures, and *in vivo.*[33] The features that have made SFV attractive are the rapid high-titer (up to 10^{10} infectious particles/ml) virus production, broad host range, and extreme levels of transgene expression. A large number of membrane proteins have been expressed using SFV vectors.[34] Among ligand-gated ion channels, the mouse serotonin 5-HT$_3$ receptor showed a specific binding activity of 60 pmol/mg in BHK and CHO cells, and electrophysiological recordings were established on whole cells.[35] Moreover, several members of the purinoreceptor p2X family have demonstrated very high binding activity (more than 150 pmol/mg)[36] and subtype-specific eletrophysiological responses after whole-cell patch-clamp recordings.[37] Additionally, the neuronal Kv1 channel subunits Kv1.1 and 1.2 have been expressed in CHO cells from SFV vectors demonstrating functional assembly.[38] Additionally, a large number of GPCRs have also been expressed from SFV vectors in a variety of mammalian cell lines generating very high binding activity, the highest Bmax values reaching 287 pmol receptor per milligram[34,39] Moreover, in a structural genomics initiative called MePNet (Membrane Protein Network), 100 GPCRs have been overexpressed from SFV vectors. In this network, 95% of the targets showed positive signals in immunoblots, and about 60 targets could be considered as structural biology compatible.[40,41] The aim of MePNet is to move forward with a number of selected targets in an attempt to purify and crystallize GPCRs (described in more detail in Chapter 16).

The first integral membrane protein expressed in large scale with SFV was the mouse serotonin 5-HT$_3$ receptor. Suspension cultures of BHK cells were infected with SFV vectors carrying the C-terminally hexa-histidine-tagged 5-HT$_3$ receptor gene in a 11.5-L fermentor volume; this resulted in similar binding activity as that observed in adherent small-scale cultures and yielded up to 20 mg receptor.[42] The recombinant 5-HT$_3$ receptor was solubilized and purified to high homogeneity demonstrating a molecular weight of 62 kD, which was in good agreement with a glycosylated subunit of the channel.[43] Preliminary structural characterization by gel filtration chromatography indicated a size of 280 kD, which is the expected size of the proposed pentameric channel. Moreover, single particle imaging also provided confirmation of a channel structure with five subunits.

A couple of GPCRs have also been expressed in large scale. The adrenergic α_2B receptor (α_2B-AR) with a C-terminal GFP fusion was expressed from SFV in CHO cells and analyzed for cellular localization by fluorescence microsocopy. The main signal for α_2B-AR was located in internal structures, and only a minority of the receptor was on the plasma membrane.[39] Despite that, binding activity of 176 pmol/mg was obtained. In a similar way, the His-FLAG tagged human neurokinin-1 receptor (hNK1R) was demonstrated by immuno electron microscopy and cell fractionation studies to be located mainly in internal membranes. Although the transport to the plasma membrane was incomplete, the NK1 receptors showed very strong binding activity. In this context, the hNK1R was fused to the SFV capsid gene, which contains a translation enhancement signal.[44] This resulted in a 5- to 10-fold increase in expression levels. Moreover, due to the autocatalytic cleavage properties of the capsid protein, the hNK1R was efficiently released from its fusion partner.[45] The scale-up of the SFV-based hNK1R expression was carried out in suspension cultures of CHO cells in spinner flasks with a maximum volume of 6 L. The expression levels obtained were similar to those earlier described in adherent cells. The hNK1R was purified to high homogeneity, and crystallization attempts resulted in generation of two-dimensional crystals, although of insufficient quality for obtaining high-resolution diffraction.

10.8 CONCLUSION

The expression of integral membrane proteins has been evaluated in many different mammalian expression systems. These include nonviral transient and stable systems, as well as a number of viral vectors such as adenovirus, alphavirus, lentivirus, and vaccinia virus vectors. The nonviral vectors are generally time-consuming and expensive to use, and the yields of recombinant membrane proteins are relatively low. However, in contrast to viral vectors, the biosafety risks are low. Several of the viral vector systems applied have resulted in high yields of recombinant membrane proteins. Especially, SFV vectors have been applied for the expression of several ion channels and more than 100 GPCRs. A few targets have also been subjected to scale-up production and purification for structural studies. For instance, preliminary characterization by chromatography and cryo-EM further confirmed the suggested pentameric structure postulated for the serotonin 5-HT$_3$ receptor. A number of GPCRs have also been purified, and two-dimensional crystals have been obtained initially, albeit of too low a quality for diffraction studies.

Although relatively high levels of expression have been obtained for several membrane proteins in various expression systems, no novel structures have been reported so far. However, as more material can be provided for structural studies and methods for automation and miniaturization are developed, the possibilities to obtain new structures will increase. Another important point is that when mammalian membrane proteins are expressed in mammalian expression systems, the most authentic expression conditions resembling those of native membrane proteins will be achieved. Despite this, recombinantly expressed targets are not necessary transported, folded, and modified exactly as in the native state. The importance of these requirements should not be underestimated. For this reason, much effort should be dedicated to these areas of research, which eventually might lead to a higher success rate in obtaining novel high-resolution structures of membrane proteins.

REFERENCES

1. Grisshammer, R., and Tate, C.G. Overexpression of integral membrane proteins for structural studies. *Q. Rev. Biophys.*, 28, 315–422, 1995.
2. Tate, C.G. and Grisshammer, R. Heterologous expression of G protein-coupled receptors. *Trends Biotechnol.*, 14, 426–430, 1996.

3. Sarramegna, V., Talmont, F., Demange, P., and Milon, A. Heterologous expression of G protein-coupled receptors: comparison of expression systems from the standpoint of large-scale production and purification. *Cell Mol. Life Sci.*, 60, 1529–1546, 2003.

4. Drew, D., Fröderberg, L., Baars, L., and De Gier, J.W.L. Assembly and overexpression of membrane proteins in *Escherichia coli. Biochim. Biophys. Acta.*, 1610, 3–10, 2003.

5. Abramson, J., Smirnova, I., Kasho, V., Verner, G., Kaback, H.R., and Iwata, S. Structure and mechanism of the lactose permease of *Escherichia coli. Science*, 301, 610–615, 2003.

6. Baneres, J.L., Martin, A., Hullot, P., Girard, J.P., Rossi, J.C., and Parello, J. Structure-based analysis of GPCR function: conformational adaptation of both agonist and receptor upon leukotriene B4 binding to recombinant BLT1. *J. Mol. Biol.*, 329, 801–814, 2003.

7. Weiss, H.M. and Grisshammer, R. Purification and characterization of the human adenosine A2a receptor functionally expressed in *Escherichia coli. Eur. J. Biochem.*, 269, 82–92, 2002.

8. Andersen, B. and Stevens, R.C. The human D1A dopamine receptor: heterologous expression in *Saccharomyces cerevisiae* and purification of the functional receptor. *Prot. Exp. Purif.*, 13, 111–119, 1998.

9. Weiss, H.M., Haase, W., Michel, H., and Reilander, H. Comparative biochemical and pharmacological characterization of the mouse 5HT5A 5-hydroxytryptamine receptor and the human beta2-adrenergic receptor produced in the methylotrophic yeast *Pichia pastoris. Biochem. J.*, 330, 1137–1147, 1998.

10. Possee, R.D. Baculoviruses as expression vectors. *Curr. Opin. Biotechnol.*, 8, 569–572, 1997.

11. Makrides, S.C. Vectors for gene expression in mammalian cells, in *Gene Transfer and Expression in Mammalian Cells*, Makrides, S.C., ed., Elsevier Science B.V., pp. 9–26, 2003.

12. Meissner, P., Pick, H., Kulangara, A., Chatellard, P., Friedrich, K., and Wurm, F.M. Transient gene expression: recombinant protein production with suspension-adapted HEK293-EBNA cells. *Biotechnol. Bioeng.*, 75, 197–203, 2001.

13. Ulrich, C.D., Ferber, I., Holicky, E., Hadac, E., Buell, G., and Miller, L.J. Molecular cloning and functional expression of the human gallbladder cholecystokinin A receptor. *Biochem. Biophys. Res. Commn.*, 193, 204–211, 1993.

14. Reuben, M., Rising, L., Prinz, C., Hersey, S., and Sachs, G. Cloning and expression of the rabbit gastric CCK-A receptor. *Biochim. Biophys. Acta.*, 1219, 321–327, 1994.

15. Gu, H., Wall, S.C., and Rudnick, G. Stable expression of biogenic amine transporters reveals differences in inhibitor sensitivity, kinetics, and ion dependence. *J. Biol. Chem.*, 269, 7124–7130, 1994.

16. Girard, P., Derouazi, M., Baumgartner, G., Bourgeois, M., Jordan, M., Jacko, B., and Wurm, F. 100-Liter transient transfection. *Cytotechnology*, 38, 15–21, 2001.

17. Keynan, S., Suh, Y.J., Kanner, B.I., and Rudnick, G. Expression of a cloned gamma-aminobutyric acid transporter in mammalian cells. *Biochemistry*, 31, 1974–1979, 1992.

18. Ukomadu, C., Zhou, J., Sigworth, F.J., and Agnew, W.S. muI Na+ channels expressed transiently in human embryonic kidney cells: biochemical and biophysical properties. *Neuron*, 8, 663–676, 1992.

19. Schurmann, A., Monden, I., Joost, H.G., and Keller, K. Subcellular distribution and activity of glucose transporter isoforms GLUT1 and GLUT4 transiently expressed in COS-7 cells. *Biochim. Biophys. Acta.*, 1131, 245–252, 1992.

20. Walter, C.A., Humphrey, R.M., Adair, G.M., and Nairn, R.S. Characterization of Chinese hamster ovary cells stably transformed by a plasmid with an inducible APRT gene. *Plasmid*, 25, 208–216, 1991.

21. Prather, P.L., McGinn, T.M., Claude, P.A., Liu-Chen, L.Y., Loh, H.H., and Law, P.Y. Properties of a κ-opioid receptor expressed in CHO cells: interaction with multiple G proteins is not specific for any individual Gα subunit and s similar to that of other opioid receptors. *Mol. Brain Res.*, 29, 336–346, 1995.

22. Reeves, P.J., Callewaert, N., Contreras, R., and Khorana, H.G. Structure and function in rhodopsin: high-level expression of rhodopsin with restricted and homogeneous N-glycosylation by a tetracycline-inducible N-acetylglucosaminyltransferase I-negative HEK293S stable mammalian cell line. *Proc. Natl. Acd. Sci. USA*, 99, 13419–13424, 2002.

23. Tate, C.G., Haase, J., Baker, C., Boorsma, M., Magnani, F., Vallis, Y., and Williams, D.C. Comparison of seven different heterologous protein expression systems for the production of the serotonin transporter. *Biochim. Biophys. Acta.*, 1610, 141–153, 2003.

24. Needham, M., Egerton, M., Millest, A., Evans, S., Popplewell, M., Cerillo, G., McPheat, J., Monk, A., Jack, A., and Johnstone, D. Further development of the locus control region/murine erythroleukemia expression system: high level expression and characterization of recombinant human calcitonin receptor. *Prot. Exp. Purif.*, 6, 124–131, 1995.

25. Boorsma, M., Nieba, L., Koller, D., Bachmann, M.F., Bailey, J.E., and Renner, W.A. A temperature-regulated replicon-based DNA expression system. *Nat. Biotechnol.*, 18, 429–432, 2000.

26. Zhen, Z., Bradel-Thretheway, B.G., Dewhurst, S., and Bidlack, J.M. Transient overexpression of kappa and mu opioid receptors from recombinant adenovirus vectors. *J. Neurosci. Meth.*, 136, 133–139, 2004.

27. Drazner, M.H., Peppel, K.C., Dyer, S., Grant, A.O., Koch, W.J., and Lefkowitz, R.J. Potentiation of beta-adrenergic signaling by adenoviral-mediated gene transfer in adult rabbit ventricular myocytes. *J. Clin. Invest.*, 99, 288–296, 1997.

28. www.jcsg.org (accessed 2005).

29. Walker, P., Munoz, M., Combe, M.C., Grouzmann, E., Herzog, H., Selbie, L., Shine, J., Brunner, H.R., Waeber, B., and Wittek, R. High level expression of human neuropeptide Y receptors in mammalian cells infected with a recombinant vaccinia virus. *Mol. Cell Endocrinol.*, 91, 107–112, 1993.

30. Bouvier, C., Bunzow, J.R., Guan, H.C., Unteutsch, A., Civelli, O., Grandy, D.K., and Van Tol, H.H. Functional characterization of the human dopamine D4.2 receptor using vaccinia virus as an expression system. *Eur. J. Pharmacol.*, 290, 11–17, 1995.

31. Yang, M., Wang, X.G., Stout, J.T., Chen, P., Hjelmeland, L.M., Appukuttan, B., and Fong, H.K. Expression of a recombinant human RGR opsin in lentivirus-transduced cultured cells. *Mol. Vis.*, 6, 237–242, 2000.

32. www.tranzyme.com (accessed 2005).

33. Lundstrom, K., Schweitzer, C., Rotmann, D., Hermann, D., Schneider, E.M., and Ehrengruber, M.U. Semliki Forest virus vectors: efficient vehicles for in vitro and in vivo gene delivery. *FEBS Lett.*, 504, 99–103, 2001.

34. Lundstrom, K. Semliki Forest virus vectors for rapid and high-level expression of integral membrane proteins. *Biochim. Biophys. Acta.*, 1610, 90–96, 2003.

35. Werner, P., Kawashima, E., Reid, J., Hussy, N., Lundstrom, K., Buell, G., Humbert, Y., and Jones, K.A. Organization of the mouse 5-HT3 receptor gene and functional expression of two splice variants. *Brain Res. Mol. Brain Res.*, 26, 233–241, 1994.

36. Michel, A.D., Miller, K.J., Lundstrom, K., Buell, G.N., and Humphrey, P.P. Radio-labeling of the rat P2X4 purinoceptor: evidence for allosteric interactions of purinoceptor antagonists and monovalent cations with P2X purinoceptors. *Mol. Pharmacol.*, 51, 524–532, 1997.

37. Michel, A.D., Lundstrom, K., Buell, G.N., Surprenant, A., Valera, S., and Humphrey, P.P. A comparison of the binding characteristics of recombinant P2X1 and P2X2 purinoceptors. *Br. J. Pharmacol.*, 118, 1806–1812, 1996.

38. Shamotienko, O., Akhtar, S., Sidera, C., Meunier, F.A., Ink, B., Weir, M., and Dolly, J.O. Recreation of neuronal Kv1 channel oligomers by expression in mammalian cells using Semliki Forest virus. *Biochemistry*, 38, 16766–16776, 1999.

39. Sen, S., Jaakola, V.P., Heimo, H., Engstrom, M., Larjomaa, P., Scheinin, M., Lundstrom, K., and Goldman, A. Functional expression and direct visualization of the human alpha 2B-adrenergic receptor and alpha 2B-AR-green fluorescent fusion protein in mammalian cell using Semliki Forest virus vectors. *Prot. Exp. Purif.*, 32, 265–275, 2003.

40. Lundstrom, K. Structural genomics on membrane: Mini review. *Comb. Chem. High Throughput Screen.*, 7, 431–439, 2004.

41. Hassaine, G., Wagner, R., Kempf, J., Cherouati, N., Hassaine, N., Prual, C., André, N., Reinhart, C., Pattus, F., and Lundstrom, K. Semliki Forest virus vectors for over-expression of 101 G protein-coupled receptors in mammalian host cells. *Prot. Exp. Purif.*, 2005.

42. Lundstrom, K., Michel, A., Blasey, H., Bernard, A.R., Hovius, R., Vogel, H., and Surprenant, A. Expression of ligand-gated ion channels with the Semliki Forest virus expression system. *J. Recept. Signal Transduct. Res.*, 17, 115–126, 1997.

43. Hovius, R., Tairi, A.P., Blasey, H., Bernard, A., Lundstrom, K., and Vogel, H. Characterization of a mouse serotonin 5-HT3 receptor purified from mammalian cells. *J. Neurochem.*, 70, 824–834, 1998.

44. Sjoberg, E.M., Suomalainen, M., and Garoff, H. A significantly improved Semliki Forest virus expression system based on translation enhancer segments from the viral capsid gene. *Biotechnology* (N Y), 12, 1127–1131, 1994.

45. Lundstrom, K. Semliki Forest virus vectors: versatile tools for efficient large-scale expression of membrane receptors, in *Perspectives on Solid State NMR in Biology*; Kühne, S.R., de Groot, H.J.M, eds., Kluwer Academic Publishers, Netherlands, pp. 131–139, 2001.

11 Solubilization and Purification of Membrane Proteins

Bernadette Byrne and Mika Jormakka

CONTENTS

11.1 INTRODUCTION

The purification of membrane proteins to near homogeneity is prerequisite to a number of analytical, spectroscopic, and structural studies. Prior to purification, the protein must be in a soluble state, that is, no longer associated with the lipid bilayer. Hence, it is necessary to disrupt the interactions between the protein and the lipid while maintaining the structural and functional integrity of the protein. This can, however, be complicated due to the requirement for particular lipid molecules for folding and due to the function of many membrane proteins.[1-3] So, how do we solubilize a membrane protein while keeping all features intact? Once we have a soluble membrane protein, how can we proceed with purification and preliminary analysis of the protein? Are there unique issues associated with the purification of membrane proteins not encountered with soluble proteins? We shall cover these and other issues in this chapter.

11.2 MEMBRANE PARTITIONING

There are many thousands of proteins in each cell. One simple and relatively straightforward way to exclude some of them from a given sample of cells is to isolate the cell membrane. This acts as a first purification step by separating the soluble proteins from the membrane-bound proteins and provides a suitable starting point for solubilization procedures. Figure 11.1 gives an example of a simple method used for preparing cell membranes from *Escherichia coli* in our laboratory.

1. The cells are treated with a 1% lysozyme solution to strip away the outer membrane.
2. Cells are then harvested and resuspended in a hypotonic solution of 5 mM EDTA (ethylenediaminetetraacetic acid), treated with DNAase and protease inhibitors and then sonicated. This treatment should result in effective lysis of almost all the cells.
3. Unbroken cells are removed by low-speed centrifugation.
4. Membranes are harvested by centrifugation at $100,000 \times g$ and can be stored at $-80°C$ until further use.

Processing of the cells by sonication is relatively rapid; however, other methods such as the use of the French Press can sometimes be more effective at cell breakage.

11.3 SOLUBILIZATION

The next step is to solubilize the target protein. Solubilization involves the removal of a membrane protein from the lipid bilayer. However, it is not as straightforward as simply removing the lipid molecules that surround the protein. Because the integral membrane portions of the proteins are extremely hydrophobic, once they are no longer protected by the lipid environment, the protein denatures or aggregates. Therefore, it is necessary to replace the lipids that surround the protein with other molecules that also protect the hydrophobic regions from solvent while maintaining the native conformation. The molecules that are used to fulfill both these roles are called *detergents*. There are many different detergents that are used to solubilize membrane proteins, but they all act in much the same way.[4]

11.3.1 DETERGENTS

Determining the detergent that is best suited for solubilization of the target protein can be a complicated and tedious process, mainly due to the intricate and complex relationship between the protein and the lipid bilayer. Detergents have the same dual properties as lipid molecules and membrane proteins in that they have both hydrophobic tails and hydrophilic head groups. Detergents vary in both the nature of the

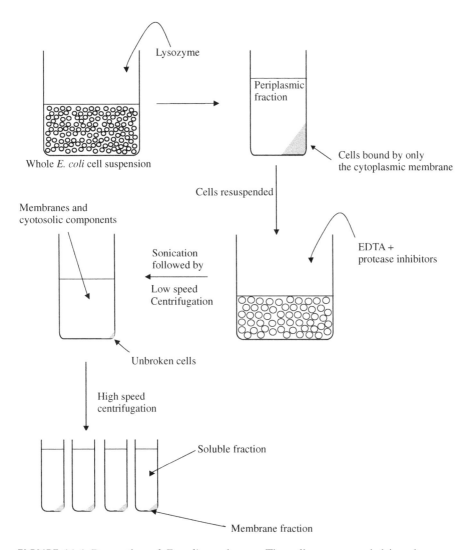

FIGURE 11.1 Preparation of *E. coli* membranes. The cells are suspended in a lysozyme treatment buffer, which strips away the outer membrane. The periplasmic fraction is removed by centrifugation and the cells resuspended in a hypotonic solution of EDTA. The cells are sonicated to induce lysis and any unbroken cells removed by low-speed centrifugation. High-speed centrifugation performs the final separation into membrane and soluble fractions. The membranes can be used immediately or, depending on the stability of the target protein, frozen and stored at –80°C.

hydrophilic head group (sugar-based, phospholipid-like) and the length and composition of the hydrophobic alkyl tail (See Figure 11.2 and Table 11.1). At low concentrations, the detergents exist as monomers in solution; however, at a specific concentration called the critical micelle concentration (CMC), the detergent molecules form micelles due to the hydrophobic effect (Figure 11.3). The CMC is

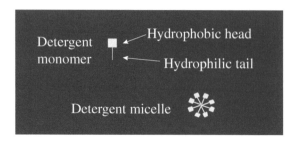

FIGURE 11.2 Diagrammatic representation of a detergent molecule and a detergent micelle. At low concentrations, the detergent molecules exist as monomers. At a higher concentration, specific to each individual detergent, the detergent molecules spontaneously form micelles. In this micellar form, the detergent is able to effectively solubilize the membrane.

TABLE 11.1
List of Commonly Used Detergents and Their CMC Values

Detergent	CMC (mM)
3-[(3-cholamidopropyl) dimethylamonio]-1-propane-sulfonate (CHAPS)	8.0
Triton X-100	0.23
n-Octylglucoside	19.0
n-Nonylglucoside	6.5
n-Dodecylmaltoside	0.17
HECAMEG	19.5
FOS-CHOLINE™ 12	1.5

TABLE 11.2
List of Common Alkyl Chain Lengths and the Name Given to the Relevant Detergent

Alkyl Chain Length	Name
7	Heptyl
8	Octyl
9	Nonyl
10	Decyl
11	Undecyl
12	Dodecyl

inversely related to the length of the alkyl chain; thus; the longer the alkyl chain, the smaller is the value. However, the precise value is different for each different detergent and must be experimentally determined (Table 11.2 gives a range of commonly used detergents and corresponding CMCs).

At the CMC, a detergent can effectively disrupt the interactions between a membrane protein and the lipid bilayer. In other words, this is the concentration at

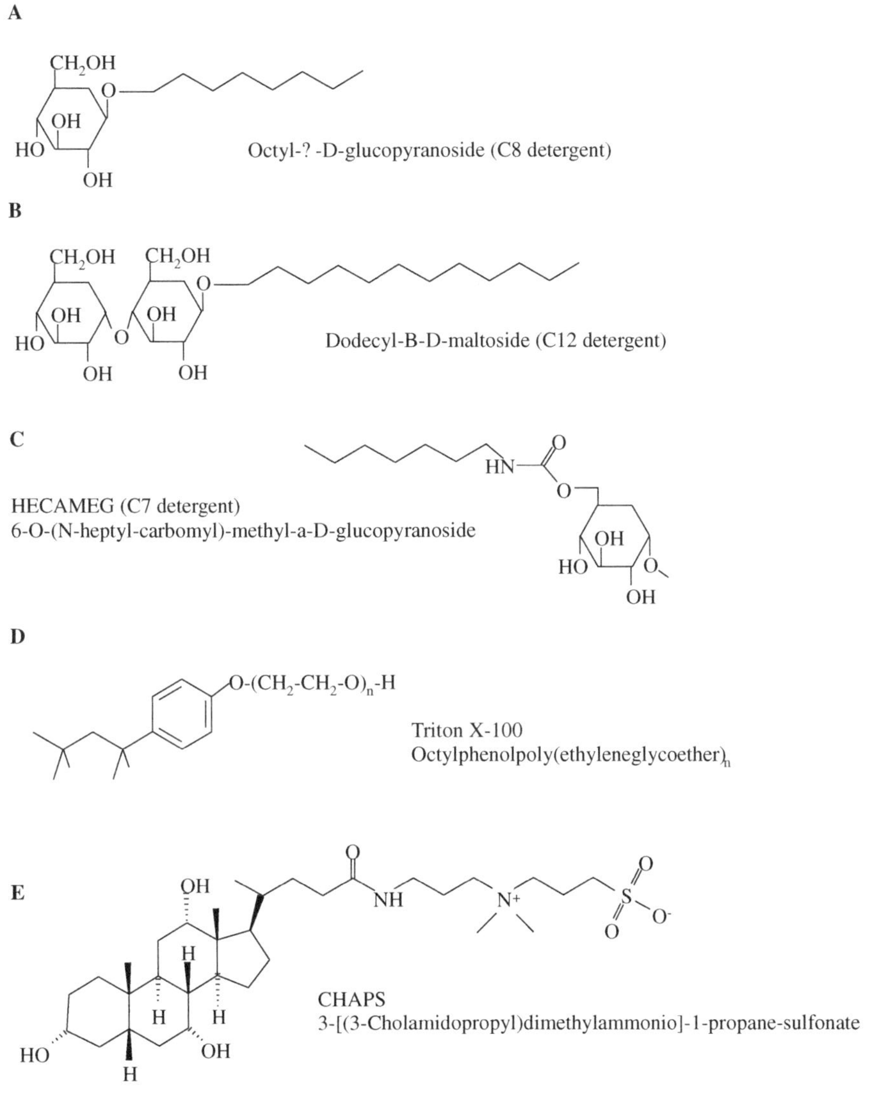
A
CH_2OH
O
OH
HO
OH
Octyl-? -D-glucopyranoside (C8 detergent)
B
CH_2OH
CH_2OH
O
OH
OH
HO
O
OH
OH
Dodecyl-B-D-maltoside (C12 detergent)
C
O
HN
O
HECAMEG (C7 detergent)
6-O-(N-heptyl-carbomyl)-methyl-a-D-glucopyranoside
OH
HO
O
OH
D
$O\text{-}(CH_2\text{-}CH_2\text{-}O)_n\text{-}H$
Triton X-100
Octylphenolpoly(ethyleneglycoether)$_n$
E
OH
O
NH
N+
O
S
O-
O
H
H
H
HO
OH
H
CHAPS
3-[(3-Cholamidopropyl)dimethylammonio]-1-propane-sulfonate

FIGURE 11.3 Structures of five commonly used detergents differing in both head group and length of alkyl chain. Detergents are available with the same head group but alternative lengths of alkyl chain (see Table 11.2).

which a detergent breaks the membrane into detergent–lipid and detergent–protein micelles and hence effectively solubilizes a membrane protein. The detergent not only disrupts the interactions between the protein and the lipid, but because of its dual hydrophobic–hydrophilic nature it also effectively replaces the lipid (Figure 11.3). The hydrophobic tails of the detergent molecules associate with the hydrophobic portions of the protein, leaving the hydrophilic head groups in contact with the solvent.

The CMC is an important feature of detergents, since this value can have major implications on how the solubilized protein is processed after purification (see section on detergent removal/exchange). It is also important to keep in mind that the CMC varies with differing salt concentrations and temperature, two important variables for many functional assays and crystallization trials.

11.3.2 TYPES OF DETERGENTS

As expected, different detergents have different properties depending on the length of the alkyl chain and the differing head groups. Figure 11.3 shows the chemical structure of some of the more commonly used detergents for membrane protein solubilization. Some of the most popular detergent families include the alkyl glucosides and maltosides, polyoxyethylenes, alkyldimethylamine oxides, and cholate derivatives. When selecting a detergent for a given protein, the most important criterion is maintaining structure and function. Unfortunately, it is not possible to predict exactly how a protein is going to behave in a particular detergent. In the last five years or so, there has been a significant increase in the amount of research on membrane proteins, and we can begin to develop a more rational approach to obtaining a soluble membrane protein. For instance, a detergent such as dodecyl-B-D-maltoside (DDM) has a relatively long alkyl chain (12 C; see Figure 11.3), which means that a micelle of DDM is larger than a micelle of a detergent with a shorter alkyl chain such as octylglucoside (OG, C8; see Figure 11.3). Thus, a micelle of DDM covers more of the hydrophobic portions of the protein than does a micelle of OG. As a result of this, membrane proteins tend to be more stable in DDM than in OG, as DDM is a better mimic of the native membrane. Sugar-based detergents such as OG and DDM are available with a wide range of different lengths of alkyl chain (Table 11.2) and in general, the longer the hydrophobic tail and the larger the head group, the more stable the protein tends to be.

Points to keep in mind are that it is possible to vary the size of the detergent micelle by varying the detergent and that it may be necessary to solubilize in one detergent but change to another detergent in order to maintain the membrane protein in a stable state.

11.3.3 SOLUBILIZATION PROCEDURES

Membranes are usually the starting material for solubilization; however, for some procedures such as electrophoresis, it is possible to simply add a sodium dodecyl sulphate (SDS)-based buffer to whole cells to solubilize all proteins (see below for further details on performing SDS-PAGE (polyacrylamide gel electrophoresis) analysis on membrane proteins). SDS is an ionic detergent and is classed as harsh in that it disrupts hydrogen bonds (H-bonds) and the hydrophobic interactions of all proteins. Detergents used for membrane protein work need to be much gentler. There are hundreds of such detergents available from a number of specialized companies. Choosing the most efficient detergent is often a process of trial and error. While uncharged (non-ionic) detergents are most commonly used, there are a few examples of charged (zwitterionic) detergents that have been successfully used for membrane

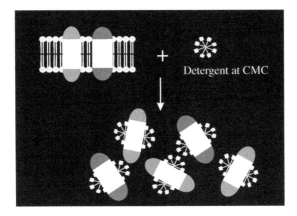

FIGURE 11.4 (See color insert following page 272.) Solubilization of membrane proteins. Detergents are added to a membrane preparation at or above the CMC value. The detergent disrupts the interactions between the protein and the lipid bilayer at the same time, replacing the lipid molecules to form protein-detergent micelles.

solubilization, particularly 3-[(3-Cholamidopropyl) dimethylammoniol-1-prppane-sulfonate] (CHAPS) and 3-[(3-Cholamidopropyl) dimethylammoniol-2-hydroxy-1-propanesulfonate] (CHAPSO).

Practically, solubilization involves the addition of detergent at a concentration usually 10 to 100 times the CMC — to ensure effective solubilization of as many protein molecules as possible — to a sample of membranes in a suitable buffer (Figure 11.4). The protein concentration of the membrane sample can be critical for effective solubilization. Generally 1 to 5 mg/ml total protein in a buffer supplemented with protease inhibitors is a good starting point. The samples are then usually mixed slowly for about 30 minutes to an hour at 4°C. The solubilized material is separated from insoluble material by high-speed centrifugation, typically 100,000 × g for an hour.

As mentioned previously, some detergents are suitable for extraction of certain membrane proteins but either are ineffectual at solubilizing others or result in denaturation. Therefore, it may be necessary to screen a number of detergents to optimize recovery of soluble material. In terms of assessing recovery, such screening is greatly facilitated by the application of a functional assay. Table 11.3 shows the recovery of a G protein-coupled receptor (GPCR) from mammalian cell membranes in the soluble fraction after solubilization trials with a range of detergents. The recovery of functional protein was assessed by a receptor-specific ligand-binding assay and compared with the binding to an equivalent amount of nonsolubilized membranes. As can be seen from these results, in most cases, the solubilization procedure resulted in inactive receptor. This may be due to loss of lipids essential for correct folding or function of the protein, or it is possible that certain detergent micelles leave too much of the protein exposed to the aqueous environment resulting in denaturation. The exceptions to this loss of protein were 1% Tween, which did not effectively solubilize the receptor, and 4 mM CHAPS, which yielded approxi-

TABLE 11.3

Recovery of Functional GPCR in the Soluble Fraction after Solubilization with a Range of Commonly Used Detergents

	Recovery %	
Detergent	Supernatant (soluble fraction)	Pellet (insoluble fraction)
4 mM CHAPS	27.4 ± 6.8	40.0 ± 15.1
1% DDM	2.8	1.4
1% OG	5.8	3.1
1% Triton-X	2.0	0.0
1% Nonidet P40	8.0	0.0
1% Tween	1.2	82.9

mately 25% functional soluble receptors. However, a significant amount of protein is lost in the insoluble fraction, indicating that some of the protein is resistant to solubilization. Changing CHAPS concentration or increasing the solubilization incubation time had no effect on this loss.

Similar results can be observed in other studies. Bannerjee and co-workers[5] investigated solubilization of the serotonin 5-HT1A receptor. They not only assessed the effective extraction of functional protein but also investigated the lipid–protein ratio in the extracted micelles. CHAPS (see Figure 11.3) and the related zwitterionic detergent CHAPSO were most effective at extracting the protein in a functional form. Interestingly, these detergents also resulted in the highest lipid–detergent ratio. Detergents such as Triton X-100 that extracted mostly protein and little lipid resulted in a loss of function. This is a clear example of the importance of lipids in the function of membrane proteins and how optimization of solubilization needs to take into account not only the effective extraction of protein but also that of associated lipids. Recently, a quick method for assessing the lipid:protein and detergent: protein ratios in a solubilized sample using Fourier-transform infrared spectroscopy (FTIR) has been developed,[6] which may facilitate the rational optimization of solubilization procedures. Such approaches may be key to producing soluble, stable membrane protein for downstream biochemical and structural analyses.

However, from the example of the GPCR, it is clear that even after optimization losses of protein can be significant. One explanation may be lipid rafts, patches of detergent-resistant lipids,[7] which are believed to have important roles in protein-sorting and cell-signaling events. It is possible that these lipid rafts account for some of this loss of receptor into the insoluble fraction, since they can be prepared using CHAPS. Perhaps as we learn more about the structure and biochemistry of these lipid rafts, we may be able to better exploit them as a source of expressed membrane protein.

Although, in some cases, it is possible to effectively solubilize a membrane protein with a single detergent, some researchers have had to develop relatively complicated initial solubilization procedures. For example, for the human neurotensin receptor expressed in *E. coli*, it was necessary to use a combination of CHAPS and DDM together with cholesteryl hemisuccinate, a cholesterol derivative.[8] This example illustrates the complexity of solubilization and the trial and error approach that must be applied.

Thus far, membrane protein solubilization seems a rather inexact science: put some detergent in, and get some soluble membrane protein back. However, recent work has focused on understanding exactly what is happening at a molecular level during the solubilization procedure. Techniques such as dynamic light scattering (DLS), nuclear magnetic resonance (NMR), and electron paramagnetic resonance (EPR) have all been applied to the detailed study of detergent-membrane interactions during solubilization. A detailed analysis of the results falls outside the scope of this overview, but interested readers are referred to the reviews by Goni and Alonso and Le Marie et al.[9,10]

11.3.4 ALTERNATIVE MEANS OF SOLUBILIZING MEMBRANE PROTEINS

Detergents have been utilized for the solubilization of membrane proteins for many years; however, there are a number of disadvantages to their use. Detergents generally have to be added in excess in order to maintain a protein in the soluble state; in other words, free detergent is present in any protein–detergent sample. This excess of detergent can interfere with analysis of the solubilized protein and can have detrimental effects on crystallization attempts. In recent years, work has focused on developing alternatives to detergents for membrane protein work. One notable success is the amphipols.[11,12] These are linear amphipathic polymers made up of alternating hydrophilic and hydrophobic side chains, which, like detergents, form a belt around the hydrophobic regions of the protein. The inner surface of the belt forms a hydrophobic contact with the integral membrane regions, while the outer surface is made up of the hydrophilic side chains exposed to the aqueous environment (Figure 11.5). The major advantage is that in contrast to detergent, stoichiometric quantities of amphipols can be used to maintain the protein in the soluble state. Furthermore, studies on the diacylglycerol kinase (DAGK), a membrane-bound enzyme, have demonstrated that amphipol solubilization can maintain membrane protein function. Interestingly, the use of the zwitterionic amphipol, PMAL-B-100 (see Figure 11.5) not only supported the activity of the protein but also maintained function in the absence of the otherwise essential lipid cardiolipin.[13] Other amphipols screened were less effective at maintaining DAGK activity, indicating that it is necessary to screen for the most effective amphipol for each individual membrane protein, as is often the case with detergents.

To date, significant efforts have been made to develop and characterize these amphipols as alternatives to conventional detergents. Although preliminary results show that these amphipathic molecules can maintain the structural and functional integrity of the membrane protein while avoiding some of the disadvantages of detergents, their use as a viable alternative to detergents has yet to be fully evaluated.

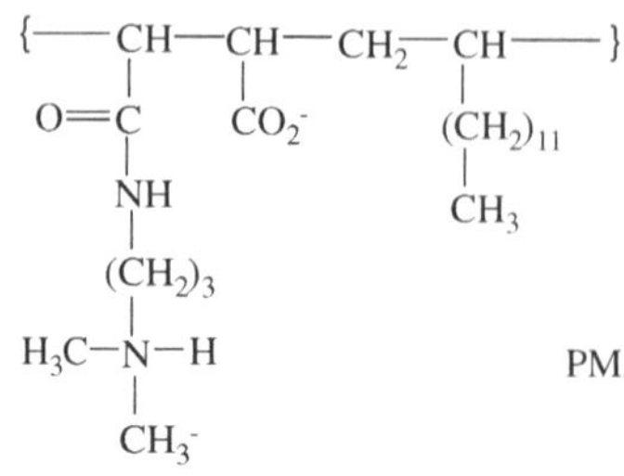

$$\{-\!\!-\!CH\!-\!CH\!-\!CH_2\!-\!CH\!-\!\!-\}n$$

with substituents: $O\!=\!C$, CO_2^-, $(CH_2)_{11}$, NH, CH_3, $(CH_2)_3$, $H_3C\!-\!N\!-\!H$, CH_3^-

PMAL-B-10

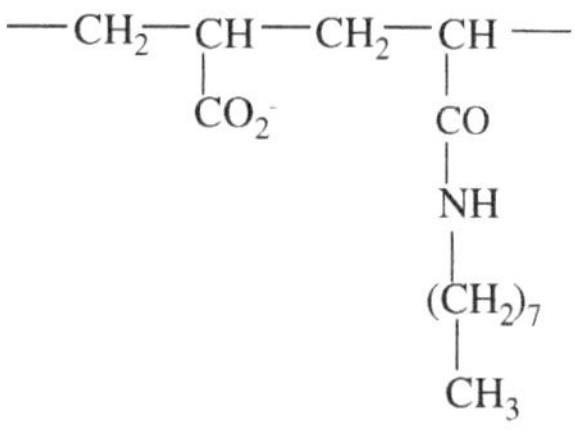

$$-\!CH_2\!-\!CH\!-\!CH_2\!-\!CH\!-$$

with substituents: CO_2^-, CO, NH, $(CH_2)_7$, CH_3

OAPA-20

FIGURE 11.5 Structures of two amphipols. (From Gorzelle, B.M., Hoffman, A.K., Keyes, M.H., Gray, D.N., Ray, D.G., and Sanders, C.R. *J. Am. Chem. Soc.*, 124, 11594–11595, 2002. With permission.)

11.4 PURIFICATION

Once sufficiently high levels of soluble target protein have been established, one can proceed to purification. Purification of membrane proteins is generally carried out by column chromatography, which exploits the interaction between the protein of interest and some other factor immobilized onto a chromatographic media. The intrinsic properties of the protein of interest in terms of size, charge, post-translational modifications, and so on can also be utilized, and until relatively recently, this was the only way to isolate membrane proteins. Indeed, this still remains an important means for obtaining membrane proteins from natural sources and will be discussed later in this chapter.

11.4.1 AFFINITY CHROMATOGRAPHY

Recently, advances in molecular biology have allowed engineering of either the gene coding for the protein of interest or a suitable expression vector to specifically include affinity tags. The most common affinity tag is the histidine tag (His-tag) which exploits the relatively specific interaction between a series of histidine residues and a chromatographic media charged with Ni^{2+}. A gradually increasing concentration of imidazole competes for interaction with the charged resin, thus eluting the bound protein. The technique of using a metal ion for purification is called immobilized metal affinity chromatography (IMAC). This technique has been extensively applied to soluble proteins but is also becoming a routine method for membrane protein isolation. Theoretically, this is a one-step purification procedure isolating only the His-tagged protein; however, proteins also contain intrinsic His-residues and, indeed, other charged residues that can result in nonspecific binding to the column.

Effective purification of the integral membrane protein by IMAC depends on a number of factors, such as:

1. The His-tag must be accessible to the resin; optimal positioning of the His-tag is protein dependent and often must be determined by trial and error.
2. The length of the His-tag can determine the specificity of the interaction and thus ultimately the purity of the eluted protein. Original studies with His-tags utilized six His-residues (a hexa-His-tag), but up to 10 His-residues have been used successfully.[14,15] A higher number of His-residues should ensure a stronger interaction with the resin, allowing washes with buffer containing higher concentrations of imidazole to remove more of the contaminant proteins.

Rational studies on the length and position of the His-tag for optimal expression and recovery of pure material have been carried out for individual proteins. One such study on the human neurotensin receptor showed that position and composition of the His-tag had dramatic effects on expression level.[8] More recently, researchers have focused on the rational design of expression constructs of the *E. coli* water-channel protein aquaporin Z (AqpZ) specifically to optimize recovery from the affinity column.[16] Expression constructs that incorporated 6- or 10His-tags at either the N or C terminus of the protein were generated. All gave similar levels of expression according to Western blot analysis, demonstrating that the length and position of the His-tag had no significant effect on expression in *E.coli*.

Although the presence of the different tags had little discernable effect on expression, localization, and solubility of the AqpZ, there were effects on the purification profiles of the different constructs. The AqpZ modified with an N-terminal 10His-tag bound much more specifically to the chromatographic media, requiring a fivefold increase in the concentration of imidazole to initiate elution compared with the other constructs. This relative increase in interaction strength is not simply a function of the number of His-residues of the tag, since the C-terminal 10His-tag eluted at a much lower concentration of imidazole but is likely due to the exposure of the tag to media. Interestingly, despite the stronger interaction the overall yield of the N-terminal 10His-tagged AqpZ was lower than that of the N-terminal 6His-tagged protein, although the sample from the column did appear a little cleaner. The 6His-tag at either the N or C terminus gave higher yields than did the 10His-tag at either position, but overall, the highest yield came from the N-terminal 6His-tagged construct. It is not clear why the N-terminal 10His-tag construct produces lower yields of AqpZ compared with the N-terminal 6His-tag, but this study does underline the unpredictability of the effects of His-tags not only on expression but also on purification of membrane proteins. It is often a good idea to check at least the N- and C-terminal positioning of the His-tag.

Although, in some cases, a very pure sample of membrane protein can be obtained using only IMAC, it is usually necessary to perform a further purification step in order to obtain protein of sufficient purity to allow spectroscopic analysis or successful crystal growth. This second step can be ion exchange, gel filtration, or a further affinity chromatography step involving a different affinity medium. The major advantage to IMAC is that it cuts down the steps necessary for the effective purification of a membrane protein, thus minimizing losses. This is particularly important as low expression levels and losses at the solubilization stage of preparation mean that obtaining usable quantities of pure membrane protein is a challenge. Affinity purification of His-tagged proteins also has distinct advantages for the isolation of multi-subunit membrane proteins. The assembly of these proteins often involves the formation of noncovalent interactions between the individual subunits. Purification of these proteins through a series of chromatographic columns can result in loss of some of the weakly associated subunits. In addition, the His-tag allows the detection and monitoring of the protein through the complete preparation procedure. Hence, the His-tag–Ni^{2+}–NTA system is highly advantageous due to ease of use, low cost, and the fact that it is very well characterized. For most groups, therefore, it remains the system of choice for expression and purification.

11.4.2 DISADVANTAGES OF HIS-TAGS

The disadvantages of the His-tag system such as effects on expression levels and low affinity can sometimes preclude the use of this tag. In addition, His-tagged membrane proteins can undergo nonspecific cleavage during expression. This is particularly frustrating because high levels of protein can often be expressed but in a potentially unpurifiable form. The removal of the His-tag may be time-dependent. In the case of cytochrome bo_3 ubiquinol oxidase, the longer the culturing time, the higher is the percentage of protein that lost the His-tag. In this case, the untagged protein bound to the Ni^{2+}–NTA column, although with lower affinity than the tagged protein. Two peaks eluted from the column: the first at low imidazole concentrations, the untagged protein; and the second at high imidazole concentrations (Figure 11.6). Following the loss of the His-tag, most proteins of interest will not bind nonspecifically to the resin resulting in a protein that is either very difficult, if not impossible, to purify. However, it may be possible to limit proteolysis of the His-tag by changing the expression conditions, for example, temperature, duration of induction, or media.

The disadvantages of the His-tag system have encouraged the development of alternative tag systems. The eight-amino-acid, streptavidin-binding sequence, or the Strep tag,[17] consists of a "milder" sequence (WSHPQFEK) than that of the His-tag and seems a viable option for large-scale purification. Although, to date, no structure has been determined for a Strep-tagged membrane protein, this may be due to researcher preference for the better characterized His-tag system. Other tags such as the hemaglutinin (HA; YPYDVPDYA) tag rely on antibody affinity chromatography methods, which are too expensive for large-scale production but may be useful for initial characterization or for the production of positive control protein.

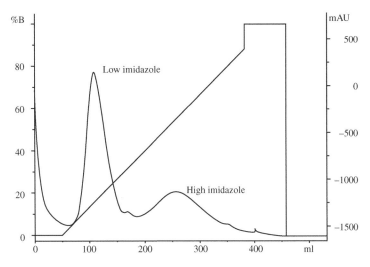

FIGURE 11.6 Purification of cytochrome bo_3 on a Ni^{2+}–NTA column. Two clear peaks can be seen corresponding to the non–His-tagged low imidazole peak (first peak) and the His-tagged high imidazole peak (second peak).

11.4.3 OTHER MEANS OF PURIFICATION

Traditionally, protein complexes were purified using classical biochemical methods such as ammonium sulfate precipitation, sucrose gradients, and other techniques that exploit the intrinsic properties of the protein of interest. Some of these methods remain popular, particularly, ion exchange chromatography, which separates proteins with minute differences in charge, and size exclusion chromatography. Generally, proteins purified by these techniques are naturally highly abundant, for example, respiratory complexes, photosystems, or rhodopsin. To purify membrane proteins using these classical approaches is often quite time-consuming, since a large amount of starting material and several purification steps are needed. Careful consideration of the characteristics of the target protein is important when designing a purification protocol in order to maximize yield and minimize the number of steps involved. In the next few sections, we shall discuss some of the major techniques applied to the isolation of membrane proteins.

11.4.4 ION EXCHANGE CHROMATOGRAPHY

Because a protein can exist either as a cation or as an anion, ion-exchange chromatography provides a custom-designed system for the separation of proteins. The type of ionic sites on the protein will vary with pH, and their number will vary with the protein type and its tertiary structure. Hence, for ion exchange chromatography, one has to consider the pI of the protein and the pH range in which the target protein is stable. The buffer used in the purification should be at least 1 pH unit higher or lower (anion–cation exchange) than the pI of the protein to ensure proper binding to the resin. For example, for the respiratory complex nitrate reductase, with a pI of 6.5, 30 mM Tris-HCl at pH 7.5 was used as the loading buffer, with the protein eluting at 280 mM sodium chloride (NaCl) concentration. Medium to high concentrations of NaCl can cause precipitation of highly hydrophobic proteins, which should be taken into consideration when selecting the pH range of the buffer. Finding the salt concentration that gives the best resolution of the target protein is a matter of trial and error; however, selecting buffer pH according to the pI of the protein and eluting with a gradient of 0 to 500 mM salt is a good starting point. The salt concentration can then be optimized according to the elution profile. In addition, the choice of ion exchanger can affect sample separation. Strong ion exchangers such as Q Sepharose FF media and MonoQ (Amersham Biosciences) allow work over a broad pH range and give different peak separation compared with weak ion exchangers.

11.4.5 SIZE EXCLUSION

Size exclusion (also called gel filtration) chromatography separates proteins based on their molecular weights. There is a large selection of media to choose from, providing an optimal separation range of proteins ranging from 10 to 200 kDa. Stationary phases for size exclusion achieve separation by allowing selective penetration into the pores of the media. The pore size dictates that small molecules can pass freely in and out of the pores, while large molecules that do not penetrate the pores elute from the column first. Parameters to consider for selection of media are the size of the target protein, resolution required, sample volume, and hydrophobicity of the protein. Small sample volume is an important factor for resolution. However, highly concentrated membrane protein samples can be quite viscous, which means that individual proteins run through the size exclusion column as smears rather than tight bands. This is a particular issue if working with high concentrations of detergent. The only answer to this problem is to dilute the sample. In addition, membrane proteins, due to their hydrophobic nature, often interact with the chromatographic media more than does an equivalently sized soluble protein, resulting in aberrant mobility. The effects of the slightly negative charge of many matrices can be compensated for by adding salts to the buffer. Generally, 0.05 to 0.2 M of sodium carbonate ($NaPO_4$) or NaCl in the buffer work well, although care should be taken not to use too high a salt concentration, since this increases hydrophobic interactions.

In addition to these two commonly used methods, hydrophobic interaction and reverse-flow chromatography remain important alternative or complementary techniques to IMAC. The benefits of IMAC have, however, made it the predominant way of purifying membrane proteins. Membrane proteins are usually expressed at low levels, and the loss of protein through a long series of columns can have serious consequences on yield. Another drawback with more "traditional" purification protocols is that a new set of purification steps has to be designed for each individual protein. Nevertheless, the production of proteins via such methods has resulted in a number of high-resolution structures[18–21] (the references given do not describe the high resolution structures but rather the purification methods used in order to obtain protein for the growth of well-diffracting crystals).

11.5 AGGREGATION OF MEMBRANE PROTEINS

For further functional, spectroscopic, or structural analysis of the purified membrane protein, it is necessary to maintain the protein in a monomerically dispersed state. In other words, the protein needs to exist in solution as single copies or particles. Membrane proteins are prone to aggregation in an effort to protect their hydrophobic regions from the aqueous solution. The amount of aggregation is related to the amount of exposed hydrophobic region and thus related to the detergent used to maintain the protein in a soluble state. Although aggregation occurs slowly over time for most membrane proteins, it can be a real problem for some proteins almost immediately after isolation. One way around this is to use glycerol, which reduces the concentration of water and increases the hydrophobicity of the solution. Glycerol also increases the pressure in the system, generating a more native-like surrounding for membrane proteins and has, on many occasions, been shown to be important for maintaining activity. Due to these benefits, glycerol is a popular additive and is included in purification and sample buffers by default in many labs. However, it is important to recognize that glycerol, while reducing aggregation, can also preclude nucleation for crystal formation.

It is possible to assay for the presence of protein aggregates by the use of a technique called dynamic light scattering (DLS). DLS measures the intensity of light scattered by molecules in solution. This intensity is related to the hydrodynamic radius of the molecule by the Stokes-Einstein equation. Aggregation is measured as an increase in the hydrodynamic radius of the molecules. A DLS profile provides the range of molecule sizes in a given protein sample. A monomerically dispersed sample will give a single peak corresponding to one hydrodynamic radius, while a heavily aggregated sample will give a number of peaks corresponding to different hydrodynamic radii. DLS is a widely used diagnostic tool, which can provide information on the potential for a protein sample to yield crystals. However, it has very limited use with membrane proteins, since the free detergent also scatters light, making it difficult to interpret the data.

Another useful technique for checking the aggregation status of a given membrane protein sample is to perform single-particle electron microscopy. This simply involves mounting a small amount of the purified soluble protein onto a carbon-coated grid, staining the protein and examining it under an electron microscope. What we hope to see is a single population of the same-sized particles. However, what is often observed is a mixture of different-sized particles corresponding to the protein in various states of aggregation (Figure 11.7). It is possible to remove these large protein aggregates by high speed centrifugation (10,000 to 150,000 × g centrifugation) for about 30 minutes. The heavier aggregates are removed in the form of a small pellet, leaving a sample of protein that contains largely monomerically dispersed protein. Centrifugation time and speed are important variables that need to be optimized for each protein and sometimes even for each sample of a given protein. Centrifugation at too high a speed or for too long may lead to loss of too much of the protein in question, whereas centrifugation at too low a speed or for too short a time may not effectively remove the protein aggregates. It is possible to further analyze the protein sample by electron microscopy after centrifugation in

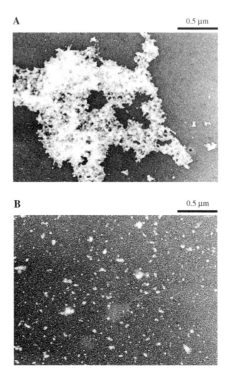

FIGURE 11.7 Sample of pure SQR before (A) and after (B) high-speed centrifugation visualized by negative-stain electron microscopy. (From Horsefield, R., Yankovskaya, V., Tornroth, S., Luna-Chavez, C., Stambouli, E., Barber, J., Byrne, B., Cecchini, G., and Iwata, S. *Acta. Crystallogr. D. Biol. Crystallogr.*, 59, 600–602, 2003. With permission.)

order to assess the status (see Figure 11.7). An elegant way in which this technique was used is described by Horsefield and co-workers.[22] In this case, it had been possible to obtain crystals of the target protein succinate ubiquinone oxidoreductase (SQR) for a number of years. However, these were invariably small and diffracted poorly. Removal of the aggregates facilitated the growth of well-ordered crystals and led to the first reports of a high-resolution structure.

11.6 SDS-PAGE ANALYSIS OF MEMBRANE PROTEINS

SDS-PAGE is a well-established technique for analysis of proteins. We will not cover this method in detail here, since in-depth descriptions are given in most biochemistry textbooks. However, it is pertinent to say that proteins are initially solubilized in an SDS-based buffer. The SDS disrupts H-bonds and hydrophobic interactions important in maintaining the native structure of proteins. SDS-solubilized samples separate on an SDS-PAGE gel according to their molecular weights. The larger the protein, the slower it runs on the gel. It is possible to run a series of molecular-weight standards to obtain an estimate of the size of the protein of interest.

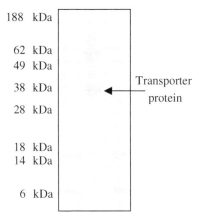

FIGURE 11.8 SDS-PAGE analysis of a 12-transmembrane transport protein. The molecular weight of this protein is 47.5 kDa. However, as can be seen on the gel image, the band corresponding to the monomeric form of the transporter runs at approximately 36 kDa.

The exceptions to this useful rule are highly hydrophobic membrane proteins. Membrane proteins associate with a larger number of SDS molecules, which means that the whole protein SDS micelle migrates faster through the gel than an equivalently sized soluble protein. The result is a smaller-than-expected band on the gel (see, for example, Figure 11.8). In the initial stages of working with a particular membrane protein of interest, this can lead to significant confusion, since it is difficult to know:

1. Whether the band on the gel corresponds to the correct protein; and
2. If the band is the correct protein, it represents the intact protein or is a proteolytically degraded fragment.

There are several ways to resolve the first issue. It is possible to use protein-specific or affinity-tag-specific antibodies to identify the correct band through Western blot analysis. In addition, if a reasonably pure sample is available, then it is possible to perform N-terminal sequencing on the protein band(s). So, how can the problem of ascertaining if the protein is intact be solved? A protein-specific assay can show whether the protein is functionally intact, and a combination of N- and C-terminal sequencing or mass spectroscopy provides a definitive answer.

Many membrane proteins are heavily glycosylated. When these proteins are separated on a gel, they tend to run with a slower mobility and as smears rather than sharp bands of specific molecular weights. This can also lead to confusion and should be taken into account when interpreting the mobility in stained gels.

11.7 REMOVAL/EXCHANGE OF DETERGENT

The way in which the soluble purified membrane protein is treated at this point is clearly dependent on the downstream application. We may wish to reconstitute the protein into liposomes in order to investigate function or obtain two-dimensional

(2D) crystals. It is possible to remove the detergent by dialysis, but this can be difficult depending on the CMC of the detergent. Detergents with low CMC values are dialyzed relatively poorly. A more effective technique involves the use of Bio-beads, hydrophobic polystyrene beads.[23] The nature of these beads would suggest that the specific removal of detergent rather than lipid or protein would be problem-atic. However, studies have shown that it is possible to optimize removal of detergent while minimizing loss of either protein or lipid. Biobeads have a much lower adsorptive capacity for lipid (100 to 200 times less) than does detergent; therefore, direct losses of lipid are relatively low. Lipid loss can be further minimized by pretreating the beads with lipid. These lipid-saturated beads have the same capacity for detergent as nontreated beads but bind much less lipid. No protein appears to bind to the beads, demonstrating a detergent-specific effect as assessed for a number of membrane proteins of varying relative hydrophobicities, including a 12-trans-membrane (TM) sugar transporter, mellibiose permease; P-glycoprotein; and pho-tosystem I and Ca^{2+} ATPase. The use of these beads has been experimentally tested on a number of different detergents, including CHAPS, Triton X-100, Hecameg, dodecyloctaethylene glycol monoether ($C_{12}E_8$), DDM, and OG, making this an appropriate process for a number of membrane proteins.

Alternatively, the solubilizing detergent may need to be exchanged for another detergent more suitable for growth of three-dimensional (3D) crystals. There are a number of ways this can be achieved; it is possible to use dialysis, although, for the same reasons stated for removal of detergent, this can be a relatively inefficient process. It is also possible to perform exchange through a molecular weight cut-off filter involving a series of concentration and wash steps with buffer containing the target detergent. Although, theoretically, this method should remove the initial detergent, it is difficult to achieve complete exchange, since a certain amount (dependent on the micelle size of the detergent and the cut-off limit of the filter) will remain in the sample. The most effective way of ensuring complete detergent exchange is to apply the protein sample to a column and wash through with buffer containing the second detergent. This can be performed at the final purification step or an extra step can be added, for example, a 1-ml Q-sepharose column washed with at least five column volumes of buffer. So, hopefully, we now have a pure, soluble sample of a membrane protein of interest in a suitable sample buffer. We can now proceed to functional and structural analysis, which will be covered in other chapters in this text.

11.8 FUTURE PROSPECTS

The solubilization and isolation of membrane proteins in a functional state remains a challenge. However, the increase in high-resolution structures over the last five years or so has provided a critical mass of information allowing rational screening of detergents both for initial solubilization and downstream structure determination. Advances in molecular biology have facilitated affinity chromatography methods for optimized protein purification, methods that have been readily applied to mem-brane proteins. While it is clear that there is no "one method for all" application, these advances have facilitated membrane protein production processes. In the future, we anticipate that further development and characterization of solubilization agents and purification methods specific to the most challenging membrane proteins will provide significant advances in structure determination.

REFERENCES

1. Lee, A.G. How lipids interact with an intrinsic membrane protein: the case of the calcium pump. *Biochim. Biophys. Acta.*, 1376, 381–390, 1998.
2. Jormakka, M., Tornroth, S., Byrne, B., and Iwata, S. Molecular basis of proton motive force: structure of formate dehydrogenase. *Science*, 295, 1863–1868, 2002.
3. Jones, M.R., Fyfe, P.K., Roszak, A.W., Isaacs, N.W., and Cogdell, R.J. Protein-lipid interactions in the purple bacterial reaction centre. *Biochim. Biophys. Acta.*, 1565, 206–214, 2002.
4. Keyes, M.H., Gray, D.N., Kreh, K.E., and Sanders, C.R. Solubilizing detergents for membrane proteins, in *Methods and Results in Crystallisation of Membrane Proteins*, Iwata, S., ed., IUL, San Diego, pp. 17-38, 2003.
5. Bannerjee, P., Joo, J.B., Buse, J.T., and Dawson, G. Differential solubilisation of lipids along with membrane proteins by different classes of detergents. *Chem. Phys. Lipids*, 77, 65–78, 1995.
6. Da Costa, C.J.B. and Baenziger, J.E. A rapid method for assessing lipid:protein and detergent:protein ratios in membrane-protein crystallization. *Acta. Cryst. D.*, 59, 77–83, 2003.
7. Schuck, S., Honsho, M., Ekroos, K., Shevchenko, A., and Simons, K. Resistance of cell membranes to different detergents. *Proc. Natl. Acad. Sci. USA*, 100, 5795–5800, 2003.
8. Tucker, J. and Grisshammer, R. Purification of a rat neurotensin receptor expressed in Escherichia coli. *Biochem. J.*, 317, 891–899, 1996.
9. Goni, F.M. and Alonso, A. Spectroscopic techniques in the study of membrane solubilization, reconstitution, and permeabilization by detergents. *Biochem. Biophys. Acta.*, 1508, 51–68, 2000.
10. Le Marie, M., Campeil, P., and Moller, J.V. Interaction of membrane proteins and lipids with solubilizing detergents. *Biochim. Biophys. Acta.*, 1508, 86–111, 2000.
11. Tribet, C., Audebert, R., and Popot, J.L. Amphipols: polymers that can keep membrane proteins soluble in aqueous solutions. *Proc. Natl. Acad. Sci USA*, 98, 15047–15050, 1996.
12. Prata, C., Giusti, F., Gohon, Y., Pucci, B., Popot, J.L., and Tribet, C. Nonionic amphiphilic polymers derived from Tris (Hydroxymethyl)-acrylamidomethane keep membrane proteins soluble in the absence of detergent. *Biopolymers*, 56, 77–84, 2001.
13. Gorzelle, B.M., Hoffman, A.K., Keyes, M.H., Gray, D.N., Ray, D.G., and Sanders, C.R. Amphipols can support the activity of a membrane enzyme. *J. Am. Chem. Soc.*, 124, 11594–11595, 2002.
14. Grisshammer, R. and Tucker, J. Quantitative evaluation of neurotensin receptor purification by immobilised metal affinity chromatography. *Prot. Exp. Purif.*, 11, 53–60, 1997.
15. Rumbley, J.N., Furlong Nickels, E., and Gennis, R.B. One-step purification of histidine-tagged cytochrome bo_3 from Escherichia coli and demonstration that associated quinone is not required for the structural integrity of the oxidase. *Biochim. Biophys. Acta.*, 1340, 131–142, 1997.

16. Mohanty, A.K. and Weiner, M.C. Membrane protein expression and production: effects of polyhistidine tag length and position. *Prot. Exp. Purif.*, 33, 311–325, 2004.

17. Schmidt, T.G. and Skerra, A. One-step affinity purification of bacterially produced proteins by means of the "Strep tag" and immobilized recombinant core streptavidin. *J. Chromatogr. A.,* 676, 337–345, 1994.

18. Jormakka, M., Tornroth, S., Abramson, J., Byrne, B., and Iwata, S. Purification and crystallization of the respiratory complex formate dehydrogenase-N from *Escherichia coli. Acta. Cryst. D.*, 58, 160–162, 2002b.

19. Okada, T., Le Trong, I., Fox, B.A., Behnke, C.A., Stenkamp, R.E., and Palczewski, K. X-ray diffraction analysis of three-dimensional crystals of bovine rhodopsin obtained from mixed micelles. *J. Struct. Biol.*, 130, 73–80, 2000.

20. Luna-Chavez, C., Iverson, T.M., Rees, D.C., and Cecchini, G. Overexpression, purification, and crystallization of the membrane-bound fumarate reductase from *Escherichia coli. Prot. Exp. Purif.,* 19, 188–196, 2000.

21. Berry, E.A., Shulmeister, V., Huang, L., and Kim, S. A new crystal form of bovine heart ubiquinol:cytochrome c oxidoreductase-determination of space group and unit cell parameters. *Acta. Crystallogr. D.*, 51, 235–239, 1995.

22. Horsefield, R., Yankovskaya, V., Tornroth, S., Luna-Chavez, C., Stambouli, E., Barber, J., Byrne, B., Cecchini, G., and Iwata, S. Using rational screening and electron microscopy to optimize the crystallization of succinate:ubiquinone oxidoreductase from *Escherichia coli. Acta. Crystallogr. D. Biol. Crystallogr.,* 59, 600–602, 2003.

23. Rigaud, J.L., Mosser, G., Lacapere, J.J., Olofsson, A., Levy, D., and Ranck, J.L. Biobeads: an efficient strategy for two-dimensional crystallisation of membrane proteins. *J. Struct. Biol.,* 118, 226–235, 1997.

12 Fluorescent Labelling of Membrane Proteins in Living Cells

*Ruud Hovius, Bruno H. Meyer,
Emmanuel G. Guignet, and Horst Vogel*

CONTENTS

12.1 INTRODUCTION

"Life in all its diversity became possible after nature had found the trick with the membrane."[1] These self-organizing systems surround the cells, separating the cellular lumen from the (chemically) hostile environment. Moreover, cells are divided by cellular membranes into different organelles, which have specific compositions reflecting their specific functions. All these membranes, however, are more than just barriers; they are involved in many functions such as sensing, recognition, transport, and energy production, storage, and conversion. These processes are performed and regulated by proteins embedded within the membranes.

The importance of membrane proteins is perhaps best illustrated, on the one hand, by their abundance and diversity, since about one-fourth of the genes in the genomes of several species appears to encode for membrane proteins.[2] On the other hand, many disorders are linked to membrane protein malfunction, and more than 70% of the currently used medicines are directed toward this class of proteins.

It is thus of great interest to understand (1) the structure and function of the individual proteins, (2) their interactions with other molecules within the cell, (3) their spatial and temporal distribution and organization within the membrane, and (4) their targeting to the plasma membrane, internalization, and recycling.

Investigations should ideally be made within a living organism; however, for several reasons, it is preferable to work with cellular model systems. To distinguish a certain protein inside a live cell, it is necessary to attach a label in order to observe this protein specifically. The label should be nonperturbing and be detectable by a noninvasive technique, which is both compatible with real-time visualization to follow dynamic cellular processes and highly sensitive to perceive few molecules. Fluorescence methods do meet these requirements and are applied massively within the live sciences to study cellular processes.

Visualizing biological processes in live cells is not easy, and to be successful, three expertises have to be combined:

1. *Molecular and cellular biology* to express to natural abundance functional proteins that can be labelled without affecting function. It is of prime importance that the function of the proteins is evaluated, both before and after labelling, and compared with the wild-type to exclude, or at least to characterize, potential artefacts. Moreover, in the case that protein–protein interactions are investigated using fluorescence methods such as fluorescence resonance energy transfer (FRET) or co-localization, it is important to achieve expression of the different proteins at relevant stoichiometry and location.

2. *Instrumentation and methodology* to accurately acquire fluorescence signals. Especially in the case of further treatment such as emission ratios or co-localization, it is of utmost importance to dispose of both properly aligned and well-characterized microscopy instrumentation. Fluorescence images can be acquired with either a point scanning microscope (confocal) or a wide-field microscope (camera). Recently, many reviews have appeared discussing in detail the different techniques and modalities for data acquisition and evaluation. Instead of treating these issues here, we refer the reader to other reviews for general[3] and high-resolution[4,5] fluorescence imaging. Protein–protein interactions are mainly studied using either FRET[6–8] or fluorescence lifetime imaging (FLIM).[9,10] The mobility of protein molecules is quantified by studying ensembles of molecules by fluorescence recovery after photobleaching (FRAP),[11,12] by imaging of individual single molecules,[13–15] or by monitoring the diffusion of molecules through a confocal observation volume by fluorescence fluctuation methods.[16,17] Processes located at the basolateral membrane or at adhesion sites of cells can be selectively investigated by total internal reflection fluorescence (TIRF) microscopy.[18,19]

3. *Labelling* of proteins, that is, introduction of an optically detectable probe in the molecules, is imperative to distinguish the protein of interest within a complex cellular context. Many methods are available to do so; in general, these are complementary and can be used in parallel such that

often multiple protein species can be monitored simultaneously. Mastering different labelling methods is of utmost importance, especially when more than one protein species is studied simultaneously or when only few or even single molecules are investigated. We will therefore expand on the current and novel methods by discussing their principal characteristics, advantages, and limitations. It has to be stated that some methods are rather novel, which implies that their range of applications is still to be demonstrated. Due to the enormous research in recent years in this area, here we present a concise account of methods that appeared in the last decade for the labelling of proteins during or after biosynthesis.

12.2 BIOSYNTHETICAL LABELLING

The coding sequence for a protein can be modified to introduce into the encoded protein a fluorescent tag. This labelling scheme has the great advantage that, indeed, only the protein of interest is labelled and at a specific site. Disadvantages are that both the choice of labels and their photo-physical properties are limited. Moreover, the addition of the label might hamper protein expression and intracellular targeting.

12.2.1 Fluorescent Proteins

By far the most used modification is the addition of a fluorescent protein (FP); for reviews on FPs and their applications see reports by Tsien and Zimmer.[20,21] The prototypical FP is the green fluorescent protein (GFP), which was identified in the 1960s as the origin of the green fluorescence of the jellyfish *Aequorea victoria*. The structure of this 26-kDa protein is composed of a β-barrel in which the chromophore is located, thus shielded from the environment, resulting in a stable fluorescence. The chromophore is formed by the oxidative rearrangement of three key amino acids: 65Ser–66Tyr–67Gly. With the advent of genetic engineering methods, FPs were fused to proteins of interest and have revolutionized cellular biology because they allowed, for the first time, visualization of protein expression and trafficking in live cells. The addition of GFP has, in several cases, been demonstrated to influence neither the function nor the trafficking of the proteins in the cells, and thus despite its considerable size, it is an "inert" label. A major advantage of FP is that labelling is quantitative, that is, each expressed fusion protein contains one fluorophore.

Through mutagenesis and discovery of FPs in other species, many variants have been obtained, resulting in (1) a complete coverage of the entire visible spectrum, (2) optimized monomeric species with strongly reduced oligomerization tendency as compared with wild-type FPs, and (3) FPs with changing spectroscopic properties. Two examples of the last group are of particular interest. First, the GFP mutant "Timer," the fluorescence of which changes from green to red in a time frame of about 20 hours, allowing one to follow gene expression and localization in time.[22] Second, FP mutants the fluorescence properties of which can be manipulated by photoactivation, enabling pulse-chase as well as double labeling and tracking experiments in cells.[23,24]

There are two GFP labeling approaches based on complementation that are noteworthy. It has been demonstrated that complementary N- and C-terminal fragments of GFP can recombine to form a fluorescent GFP.[25] This method has been used to detect protein–protein interactions both of soluble proteins[25] and of subunits of trimeric membrane-associated G proteins.[26] An advantage of complementation is that fluorescence is only due to the interaction of molecules, and no background of noninteracting proteins is observed.

12.2.2 ARTIFICIAL AMINO ACIDS

Non-natural amino acids can be introduced into a protein using nonsense stop codon suppression and extended genetic code methods or by the use of tRNAs for natural amino acids but loaded with a modified residue.[27,28] Amino acids with a large variety of non-natural side chains or leading to an altered backbone chemistry have been successfully introduced into proteins, either *in vitro*, in *Escherichia coli* or in *Xenopus* oocytes as reviewed by Wang and Schultz and by Dougherty.[27,28] Introduction of fluorescent groups into nascent proteins is mainly limited by the compatibility of the modified amino acids with the ribosome. This method was successfully pioneered by Turcatti,[29] introducing 7-nitrobenz-2-oxa-1,3-diazole (NBD) and 5-dimethylaminonaphthalene1-sulfonyl (dansyl) conjugated amino acids into the neurokinin-2 receptor in oocytes; this was done to probe the orientation of the fluorescently labelled peptide agonist in the binding site of this receptor. Another fluorophore tested is 6-dimethylamino-2-acylnaphthalene to probe the electrostatic potential in the pore of potassium channel proteins expressed in oocytes.[30] Commercially available from Promega is the option to introduce the chromophore Bodipy-FL randomly at lysine residues *in vitro* using ^{Lys}tRNA (FluoroTect™ GreenLys *in vitro* Translation Labelling System). To date, larger and brighter fluorophores could not be introduced into proteins by this suppressor method.

To circumvent these limits imposed by the ribosome, amino acids carrying small recognition elements have been employed. For fluorescence labelling purposes, it is of importance to mention residues carrying either biotin or an acetyl–phenyl group for post-labelling with streptavidin or hydrazine-functionalized fluorophores.[31] However, both these methods have disadvantages. The former is hampered by the large size of streptavidin (65 kDa), which precludes labelling to extracellular domains; and the latter is hampered by the slow reaction between the acetyl and hydrazine groups, which yields an unstable bond, which can, however, be converted to a stable chemical bond upon treatment with the strong reducing agent NaCNBH$_4$. Nevertheless, the outer membrane protein LamB was thus labelled in *E. coli*.[31]

12.2.3 C-TERMINAL LABELLING

A fluorescent group can be introduced *in vitro* at the C-terminus of the nascent polypeptide chain when an mRNA without stop codon is used. Upon completion of peptide synthesis, the new protein will stay on the ribosome and could be covalently labelled with fluorescein[32] or with Cy5[33] conjugated puromycin. Application of this method *in vivo* will result in nonspecific labelling of all proteins being synthesized by the cell at the moment of labelling, that is, it is possible to obtain insight into the protein expression pattern at a given time.[34] Combined with a strong overexpression system using, for example, Semliki Forest virus vectors, this should result in an almost specific labelling of the membrane protein of interest.[35]

12.3 POSTBIOSYNTHETIC LABELLING

In the last decade, a number of methods have been developed for labelling of proteins in two steps: First, a genetically encoded element is incorporated in the protein under investigation, which is, after completion of the biosynthesis, fluorescently labelled in a second step. The fluorophore can be attached either through a covalent bond or via a reversible interaction. These developments were stimulated, on the one hand, by the success of FPs showing the enormous potential of fluorescence labelling for *in vivo* investigations and, on the other hand, by the limitations of the FPs.

The genetically encoded element can be a protein that acts as enzyme, substrate, or binding site or a small oligopeptide that serves as substrate or recognition element for a specific binding protein.

12.3.1 Fusion Proteins with Enzymatic Activity

In this category, an enzyme is fused to the protein of interest; several variants have been reported. In a first approach, the human O^6-alkylguaninine-DNA alkyltransferase (hAGT) was used as an N- or C-terminal fusion.[36] The 207-residue-long hAGT is a DNA-repair protein, which removes nucleotide-modifying groups from O^6 of guanine, transferring this group to a reactive cysteine in its active site. The protein is an enzyme with only a single turnover. This property was elegantly exploited by adding a fluorophore-conjugated benzylguanide to a hAGT-fusion protein, resulting in a labelled hAGT fusion. After labelling, excess substrate is removed by washing. The system has been shown to work *in vitro* and *in vivo,* in both bacteria and mammalian cells. One limitation is that the host cells should be devoid of endogenous AGT activity. However, recently, the same group generated a mutant of hAGT, which is resistant against an inhibitor of the wild-type hAGT; this allowed specific fluorescent labelling of the mutant-hAGT upon specific blockage of endogenous AGT.[37] A range of fluorescent groups has thus been introduced in receptor proteins located in the plasma membrane and nucleus. Moreover, double labelling using cell permeable and impermeable substrates and pulse labelling to follow expression in time are possible with this method. Since the beginning of 2005, the system is commercially available under the trademark "SNAP-tag™" (www.Covalys.com).

Recently, Promega launched a similar approach called "HaloTag,™" where an optimized dehalogenase is used as fusion protein. Upon cytosolic expression, this 33-kDa protein can be irreversibly labelled by chloro-alkane derivatives carrying fluorescein or tetramethylrhodamine, under similar conditions as for the hAGT system.[38] The only data available so far, presented by Promega, are those of a p65-HaloTag fusion.

12.3.2 PROTEIN FUSIONS AS SUBSTRATES

An elegant variation on the above approach is the use of fusions of the acyl carrier protein (ACP), developed by the Johnsson group.[39] Here, the short ACP (only 77 residues) is endowed with a label added as coenzyme A–adduct upon the action of the enzyme 4'–phosphopantotetheine transferase (PPTase). A major advantage of this method is that cell surface proteins can be labelled (1) with great specificity, (2) with any fluorophore, and (3) without intracellular background (Figure 12.1).

12.3.3 FUSION PROTEINS WITH BINDING ACTIVITY

Reversible attachment of labelling moieties has recently been attempted using fusions with proteins that bind with high-affinity, small, fluororescently labelled molecules.

Dihydrofolate reductase (DHFR), a monomeric 18-kDa protein, binds with sub-nanomolar affinity to its inhibitor methotraxate (Mtx). Fluorophores can be linked to the primary amino group of Mtx without its binding to DHFR. Fusion proteins of DHFR can thus be labelled *in cellulo* by incubation with fluorescent Mtx; this is followed by washout of excess ligand, due to the low dissociation rate constant of about 10^{-4} s^{-1}.[40] An advantage of this system is that DHFR and its interaction with Mtx are well characterized and several fluorescent Mtx are commercially available. This method is prohibited by the presence of endogenous DHFR. However, this problem was solved recently using an *E. coli* DHFR as fusion partner and fluorescent trimethoprim conjugates, resulting in specific labelling in mammalian cells.[41]

Similarly, the FK506 binding protein (FKBP) was exploited.[42] The immuno-suppressant FK506 binds with high affinity to the intracellular FKBP. To enable specific cellular labelling in the presence of endogenous FKBP, the researchers designed both an FK506 analogue and a mutant FKBP* that interact with picomolar affinity but do not bind to wild-type FKBP. Fusion proteins with FKBP* could thus be stained selectively *in cellulo* with a fluorescein-labelled FK506 analogue.

12.3.4 RECOGNITION TAGS

The structure, function, and cellular localization of a protein might be compromised by fusion to any of the proteins mentioned above. It is therefore desirable to engineer small genetically encoded tags composed of a maximum of two dozen amino acids, which can be labelled specifically.

Epitopes, which are recognized specifically by antibodies, have been used for a long time. The hemaglutinin HA (YPYDVPDYA), vesicular stomatitis virus (VSV; YTDIEMNRLGK), and FLAG (DYKDDDDK) tags are well-known examples. In general, they can be attached N- or C-terminally to proteins without (too many) negative effects. Specific high-affinity antibodies against a variety of epitopes are commercially available. Fluorescence labelling is done either by using fluorescently labelled primary antibodies, which recognize the tags themselves, or by secondary antibodies directed against the Fc domain of the primary antibodies. A limitation is that only surface-exposed antigens can be labelled in live cells; intracellularly located epitopes can only be accessed upon transduction or cell permeabilization. A major

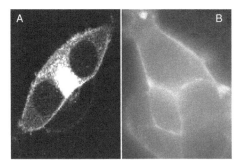

FIGURE 12.1 Fluorescent labelling of the neurokinin-1 receptor through a fusion either (A) of cyan fluoresent protein (CFP) to the C-terminus or (B) of acyl carrier protein to the N-terminus. Both constructs were expressed in HEK293 cells. The ACP-fusion protein was fluorescently labelled with Cy3. Localization of the receptor constructs was performed by confocal (A) and wide-field (B) fluorescence microscopy. Comparison of the images demonstrates that for the CFP-fusion, significant intracellular fluorescence is observed, whereas for the ACP-fusion, an ideal plasma membrane localized receptor labelling is observed.

drawback of this labelling method is the large size of antibodies (approximately 150 kDa). However, several fluorophores can be attached to one antibody, which results in a very bright labelling.

The octa-peptide sequence WSHPQFEK called "*Strep-tag*" binds with micromolar affinity to streptavidin.[43] The sequence can be engineered N- or C-terminally to proteins, and fluorescent streptavidin can subsequently be used to label the protein. This method has comparable advantages and limitations as antibody epitopes.

Recently a *split GFP* approach was presented.[44] When linked to a protein of interest, a 16-residue C-terminal fragment of GFP is recognized by the complementary part of GFP (residues 1–214) yielding a fluorescent GFP-labelled protein. This system was demonstrated in *E. coli*. A major advantage is that no background fluorescence is observed due to free labelling agent, as both GFP fragments are nonfluorescent. Moreover, the method requires only simple genetic engineering approaches. A drawback of the system is that full development of the fluorescence takes several hours.

Biotin ligase was recently used to couple a ketone isostere of biotin to the Lys of the acceptor peptide GLNDIFEAQKIEWHE fused either to the N-terminus or the C-terminus of a target protein.[45] The ketone reacts selectively with hydrazine derivatives of fluorophores, yielding a hydrazone adduct that has to be stabilized by reduction with sodium cyanoborohydride. Drawbacks of this three-step labelling protocol are that the kinetics are slow and millimolar concentrations of hydrazine fluorophores are needed. Therefore, for the labelling of cell surface receptor proteins, the authors decided to rather introduce normal biotin, followed by labelling with fluorescent streptavidin.[45] The sensitivity of this method should allow the detection of cell surface proteins at expression of 10^6 copies per cell, which corresponds to relatively high overexpression levels.[45]

A virtually nonfluorescent *biarsenical derivative* of fluorescein called FlAsH was shown to interact specifically with a tetra-cysteine motif CC-XX-CC, with a concomitant 100-fold increase of the fluorescence intensity.[46] The most favorable conformation of this motif is a hairpin,[47] and it can be introduced to the termini of proteins and also inserted in loops. This approach allows intracellular labelling and can be reversed upon the addition of millimolar concentrations of ethylenedithiol. Differently colored variants have been reported,[47,48] which has allowed pulse labelling study of connexin trafficking in living cells.[49] Despite these interesting observations, the system has several limitations, such as (1) very slow binding and dissociation kinetics[46]; (2) low specificity[50]; (3) the need of high concentrations of thiol reagents during labelling to suppress nonspecific interactions.

Oligo-histidine sequences have been used for protein purification through the specific interaction with immobilized Ni-NTA for more than 15 years.[51] Fluorescent analogues of NTA, originally introduced as indicators for divalent metal ions,[52] were shown to label hexa-histidine-tagged proteins *in vitro*[53,54] with micromolar affinities. The labelling is complete within about 30 seconds and can be fully reversed within 5 minutes by addition of a strong chelator such as EDTA. Despite its moderate affinity, the feasibility to apply this method *in vivo* was demonstrated for hexa-histidine tagged versions of both a ligand-gated ion channel and a G protein-coupled receptor (GPCR).[54] Structural information was obtained on the position of the His-tags within the channel protein. The affinity of the NTA-fluorophore for His-tagged proteins could be significantly enhanced by extension of the tag from 6 to 10 histidines. For the deca-histidine-tagged 5-HT$_3$ receptors an apparent affinity of 166 nM was observed.[54] The advantages of this labelling scheme are its rapidity, full reversibility, and virtually unlimited choice of probes. The versatility of the method was demonstrated for labelling of the 5-HT$_3$ receptor such that only few receptors per cell could be specifically labelled, allowing the detection of the diffusion individual receptor molecules (Figure 12.2).

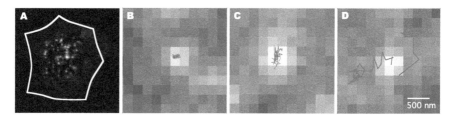

FIGURE 12.2 Deca-histidine-tagged 5-HT$_3$ receptors expressed in HEK293 cells were labeled with subnanomolar concentrations of NTA-Atto647. (A) Sensitive wide-field fluorescence detection revealed several individual receptor proteins on a HEK293 cell (circumference indicated by the white line). Time-series of images were acquired and the trajectories (red lines) of single 5-HT$_3$ receptors could be reconstituted (B–D). It appeared that some receptor molecules are immobile (B), whereas others move within confined regions of about 200 nm (C) or diffuse freely within the membrane (C). Image A is 15 × 15 µm; for images B–D a scale bar of 500 nm is indicated.

12.4 CONCLUSION

In summary, each of the methods presented has its advantages and limitations. Through side-by-side evaluation, the best suited method can be identified for a given experimental system. The different methods are not just complementary to each other but could also be applied in parallel, allowing specific double, or even multiple, labelling. This opens the possibility to investigating, in live cells, intricate biological processes such as molecular interactions and trafficking receptors.

REFERENCES

1. Sackmann, E. Biological membranes architecture and function, in *Handbook of Biological Physics*, *Volume 1A, Structure and Dynamics of Membranes: from Cells to Vesicles*. Lipowsky, R. and Sackmann, E., eds., Elsevier, Amsterdam, pp. 1–63, 1995.
2. Boyd, D., Schierle, C., and Beckwith, J. How many membrane proteins are there? *Prot. Sci.*, 7, 201–205, 1998.
3. Michalet, X., Kapanidis, A.N., Laurence, T., Pinaud, F., Doose, S., Pflughoefft, M., and Weiss, S. The power and prospects of fluorescence microscopies and spectroscopies. *Annu. Rev. Biophys. Biomol. Struct.*, 32, 161–182, 2003.
4. Gustafsson, M.G.L. Extended resolution fluorescence microscopy. *Curr. Opin. Struct. Biol.*, 9, 627–634, 1999.
5. Egner, A. and Hell, S.W. Fluorescence microscopy with super-resolved optical sections. *Trends Cell Biol.*,15, 207–215, 2005.
6. Eidne, K.A., Kroeger, K.M. and Hanyaloglu, A.C. Applications of novel resonance energy transfer techniques to study dynamic hormone receptor interactions in living cells. *Trends Endocrinol. Metab.*,13, 415–421, 2002.
7. Jares-Erijman, E.A. and Jovin, T.M. FRET imaging. *Nat. Biotech.*, 21, 1387–1395, 2003.
8. Yan, Y. and Marriott, G. Analysis of protein interactions using fluorescence technologies. *Curr. Opin. Chem. Biol.*, 7, 635–640, 2003.
9. Bastiaens, P.I. and Squire, A. Fluorescence lifetime imaging microscopy: spatial resolution of biochemical processes in the cell. *Trends Cell Biol.*, 9, 48–52, 1999.
10. Sekar, R.B. and Periasamy, A. Fluorescence resonance energy transfer (FRET) microscopy imaging of live cell protein localizations. *J. Cell Biol.*, 160, 629–633, 2003.
11. Carrero, G., McDonald, D., Crawford, E., De Vries, G., and Hendzel, M.J. Using FRAP and mathematical modeling to determine the *in vivo* kinetics of nuclear proteins. *Methods*, 29, 14–28, 2003.
12. Lippincott-Schwartz, J., Altan-Bonnet, N., and Patterson, G.H. Photobleaching and photoactivation: following protein dynamics in living cells. *Nat. Cell Biol.*, S7–S14, 2003.
13. Weiss, S. Measuring conformational dynamics of biomolecules by single molecule fluorescence spectroscopy. *Nat. Struct. Biol.*, 7, 724–729, 2000.
14. Sako, Y. and Yanagida, T. Single-molecule visualization in cell biology. *Nat. Rev. Mol. Cell Biol.*, SS1–SS5, 2003.
15. Lommerse, P.H., Spaink, H.P. and Schmidt, T. *In vivo* plasma membrane organization: results of biophysical approaches. *Biochim. Biophys. Acta.*, 1664, 119–131, 2004.

16. Chen, Y., Müller, J.D., Tetin, S.Y., Tyner, J.D., and Gratton, E. Probing ligand protein binding equilibria with fluorescence fluctuation spectroscopy. *Biophys. J.*, 79, 1074–1084, 2000.

17. Bacia, K. and Schwille, P. A dynamic view of cellular processes by *in vivo* fluorescence auto- and cross-correlation spectroscopy. *Methods*, 29, 74–85, 2003.

18. Toomre, D., and Manstein, D.J. Lighting up the cell surface with evanescent wave microscopy. *Trends Cell Biol.*, 11, 298–303, 2001.

19. Sako, Y. and Uyemura, T. Total internal reflection fluorescence microscopy for single-molecule imaging in living cells. *Cell Struct. Funct.*, 27, 357–365, 2002.

20. Tsien, R.Y. The green fluorescent protein. *Annu. Rev. Biochem.*, 67, 509–544, 1998.

21. Zimmer, M. Green fluorescent protein (GFP): applications, structure, and related photophysical behavior. *Chem. Rev.*, 102, 759–781, 2002.

22. Terskikh, A., Fradkov, A., Ermakova, G., Zaraisky, A., Tan, P., Kajava, A.V., Zhao, X., Lukyanov, S., Matz, M., Kim, S., Weissman, I., and Siebert, P. "Fluorescent timer": protein that changes color with time. *Science*, 290, 1585–1588, 2000.

23. Patterson, G.H. and Lippincott-Schwartz, J. A photoactivatable GFP for selective photolabeling of proteins and cells. *Science*, 297, 1873–1877, 2002.

24. Chusakov, D.M., Belousov, V.V., Zaraisky, A.G., Novoselov, V.V., Staroverov, D.B., Zorov, D.B., Lukyanov, S., and Lukyanov, K.A. Kindling fluorescent proteins for precise *in vivo* photolabeling. *Nat. Biotech.*, 21, 191–194, 2003.

25. IIu, C.D., Chinenov, Y., and Kerppola, T.K. Visualization of interactions among bZIP and Rel family proteins in living cells using bimolecular fluorescence complementation. *Mol. Cell*, 9, 789–798, 2002.

26. Hynes, T.R., Tang, L., Mervine, S.M., Sabo, J.L., Yost, E.A., Devreotes, P.N., and Berlot, C.H. Visualization of G protein βγ dimers using bimolecular fluorescence complementation demonstrates roles for both beta and gamma in subcellular targeting. *J. Biol. Chem.*, 279, 30279–30286, 2004.

27. Wang, L. and Schultz, P.G. Expanding the genetic code. *Angew Chem. Int. Ed.* 44, 34–66, 2005.

28. Dougherty, D.A. Unnatural amino acids as probes of protein structure and function. *Curr. Opin. Chem. Biol.*, 4, 645–652, 2000.

29. Turcatti, G., Nemeth, K., Edgerton, M., Meseth, U., Talabot, F., Peitsch, M., Knowles, J., Vogel, H., and Chollet, A. Probing the structure and function of the tachykinin NK2 receptor through biosynthetic incorporation of fluorescent amino acids at specific sites. *J. Biol. Chem.*, 271, 19991–19998, 1996.

30. Cohen, B.E., McAnaney, T.B., Park, E.S., Jan, Y.N., Boxer, S.G., and Jan, L.Y. Probing protein electrostatics with a synthetic fluorescent amino acid. *Science*, 296, 1700–1703, 2002.

31. Zhang, Z., Smith, B.A.C., Wang, L., Brock, A., Cho, C., and Schultz, PG. A new strategy for the site-specific modification of proteins *in vivo*. *Biochemistry*, 42, 6735–6746, 2003.

32. Nemoto, N., Miyamoto-Sato, E., and Yanagawa, H. Fluorescence labeling of the C-terminus of proteins with a puromycin analogue in cell-free translation systems. *FEBS Lett.*, 462, 43–46, 1999.

33. Yamaguchia, J., Nemotob, N., Sasakib, T., Tokumasub, A., Mimori-Kiyosuec, Y., Yagid, T., and Funatsua, T. Rapid functional analysis of protein–protein interactions by fluorescent C-terminal labeling and single-molecule imaging. *FEBS Lett.*, 502, 79–83, 2001.

34. Starck, S.R., Green, H.M., Alberola, J., and Roberts, R.W. A general approach to detect protein expression *in vivo* using fluorescent puromycin conjugates. *Chem. Biol.*, 11, 999–1008, 2004.

35. Lundström, K., Blasey, H.D., Hovius, R., Vogel, H., and Bernard, A.R. The Semliki Forest virus system : receptor expression and scale up. *Genet. Engineer Biotechnologist*, 17, 101–106, 1997.

36. Keppler, A, Gendreizig, S., Gronemeyer, T., Pick, H., Vogel, H., and Johnsson, K. A general method for the covalent labeling of fusion proteins with small molecules *in vivo*. *Nat. Biotechnol.*, 23, 86–89, 2003.

37. Juillerat, A., Heinis, C., Sielaff, I., Barnikow, J., Jaccard, H., Kunz, B., Terskikh, A., and Johnsson, K. Engineering substrate specificity of O6-alkylguanine-DNA alkyltransferase for specific protein labeling in living cells. *ChemBioChem*, 6, 1-8, 2005.

38. Los, G. HaloTag Interchangeable labeling technology for cell imaging and protein capture. *Cell Notes*, 11, 2–6, 2005.

39. George, N., Pick, H., Vogel, H., Johnsson, N., and Johnsson, K. Specific labeling of cell surface proteins with chemically diverse compounds. *J. Am. Chem. Soc.*, 126, 8896–8897, 2004.

40. Miller, L.W., Sable, J., Goelet, P., Sheetz, M.P., and Cornish, V.W. Methotrexate conjugates: a molecular *in vivo* protein tag. *Angew Chem. Int. Ed.*, 43, 1672–1675, 2004.

41. Miller, L.W., Cai, Y., Sheetz, M.P., and Cornish, V.W. *In vivo* protein labeling with trimethoprim conjugates: a flexible chemical tag. *Nat. Meth.*, 2, 255–257, 2005.

42. Marks, K.M., Braun, P.D., and Nolan, G.P. A general approach for chemical labeling and rapid, spatially controlled protein inactivation. *Proc. Natl. Acad. Sci. USA*, 101, 9982–9987, 2004.

43. Schmidt, T.G.M., Koepke, J., Frank, R., and Skerra, A. Molecular interaction between the Strep-tag affinity peptide and its cognate target, Streptavidin. *J. Mol. Biol.*, 255, 753–766, 1996.

44. Cabantous, S., Terwilliger, T.C., and Waldo, G.S. Protein tagging and detection with engineered self-assembling fragments of green fluorescent protein. *Nat. Biotechnol.*, 23, 102–107, 2005.

45. Chen, I., Howarth, M., Weying, L., and Ting, A. Site-specific labelling of cell surface proteins with biophysical probes using biotin ligase. *Nat. Meth.*, 2, 99–104, 2005.

46. Griffin, B.A., Adams, S.R., and Tsien, R.Y. Specific covalent labeling of recombinant protein molecules inside live cells. *Science*, 281, 269–272, 1998.

47. Adams, S.R., Campbell, R.E., Gross, L.A., Martin, B.R., Walkup, G.K., Yao, Y., Llopis, J., and Tsien, R.Y. New biarsenical ligands and tetracysteine motifs for protein labeling *in vitro* and *in vivo*: synthesis and biological applications. *J. Am. Chem. Soc.*, 124, 6063–6076, 2002.

48. Nakanishi, J., Nakajima, T., Sato, M., Ozawa, T., Tohda, K., and Umezawa, Y. Imaging of conformational changes of proteins with a new environment-sensitive fluorescent probe designed for site-specific labeling of recombinant proteins in live cells. *Anal. Chem.*, 73, 2920–2928, 2001.

49. Gaietta, G., Deerinck, T.J., Adams, S.R., Bouwer, J., Tour, O., Laird, D.W., Sosinsky, G.E., Tsien, R.Y., and Ellisman, M.H. Multicolor and electron microscopic imaging of connexin trafficking. *Science*, 296, 503–507, 2002.

50. Stroffekova, K., Proenza, C., and Beam, K.G. The protein-labeling reagent FLASH-EDT2 binds not only to CCXXCC motifs but also non-specifically to endogenous cysteine-rich proteins. *Pflugers Arch.*, 442, 859–866, 2001.

51. Hochuli, E., Dobeli, H., and Schacher, A. new metal chelate absorbant selective for proteins and peptides containing neighbouring histidine residues. *J. Chromatogr.*, 411, 177–184, 1987.

52. Stora, T., Hovius, R., Dienes, Z., Pachoud, M., and Vogel, H. Metal ion trace detection by a chelator-modified gold electrode: a comparison of surface to bulk affinity. *Langmuir*, 13, 5211–5214, 1997.

53. Kapanidis, A.N., Ebright, Y.W., and Ebright, R.H. Site-specific incorporation of fluorescent probes into protein: hexahistidine-tag-mediated fluorescent labeling with (Ni^{2+})nitrilotriacetic acid (n)-fluorochrome conjugates. *J. Am. Chem. Soc.*, 123, 12123–12125, 2001.

54. Guignet, E.G., Hovius, R., and Vogel, H. Reversible site-selective labeling of membrane proteins in live cells. *Nat. Biotechnol.*, 22, 440–444, 2004.

13 Membrane Protein NMR

Xiang-Qun (Sean) Xie

CONTENTS

13.1 INTRODUCTION

Among the postgenomic structural studies, membrane proteins remain one of the most challenging frontiers of structural and chemical biology. In their membrane-associated domains, mainly two types of secondary structures have been observed: the rigid pore β-barrels or β-strands types are found in the outer membranes of

Gram-negative bacteria, mitochondria, and chloroplasts, and the single or bundled transmembrane α-helices exist in a broader range of functionalities and complexities of integral membrane proteins (IMPs), helical membrane proteins (HMPs), or trans-membrane proteins (TMPs). Actually, one-third of a typical genome consists of open reading frames corresponding to helical membrane proteins. These membrane pro-teins represent a vast number of important potential therapeutic drug targets as they are involved in almost every biological process in the cells. Among them, G protein-coupled receptors (GPCRs) are one of the most important transmembrane protein families (Figure 13.1).

GPCRs act as the primary detectors of cell signals, and are responsible for the majority of cellular responses to hormones and neurotransmitters as well as light, odor, and taste senses [Spiegel, 2004, 176]. GPCRs are the most common drug targets representing ~ 65% of current drugs worldwide,[2] including such best-selling drugs as Claritin, Zantac, and Cozaar. In particular, the completion of sequencing of the human genome and the initiation of proteome initiatives have identified that the genes encoding these receptors occupy a hefty 3% of the human genome; this has become a driving force for the growing role of GPCRs as drug targets and is expected to generate more blockbuster drugs of tomorrow.

In appreciation of the importance of membrane proteins, which represent one-third of the genes predicted from human genome sequencing, however, only about 1% of the three-dimensional (3D) structures deposited in public protein databases are membrane-associated proteins.[3,4] In fact, high-resolution structures of GPCRs have been rare due to the lack of well-ordered and high-quality crystals suitable for x-ray crystallography. Only one high-resolution x-ray structure of a GPCR is avail-able to date, that is, the bovine rhodopsin with an x-ray resolution of 2.8 Å.[5] This represents the accumulation of over 30 years of work by many groups to find a crystal suitable for x-ray diffraction. Even in the case of rhodopsin, however, parts

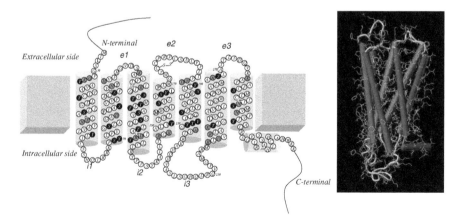

FIGURE 13.1 Two-dimensional sequence model (left) and three-dimensional structure model (right) of cannabinoid receptor CB$_2$ generated by multiple sequence alignment and computer homology. (From Xie, X.Q., Chen, J.Z., and Billings, E.M. *Prot. Struct. Funct. Genet.,* 53(2), 307–319, 2003. With permission.)

of the cytoplasmic regions are either still missing or have a very high B-factor.[6] It is also possible that the membrane protein conformation may be distorted due to crystal packing force and may not be exactly relevant to its *in vivo* counterpart. The lack of these 3D receptor structures has greatly impeded the process of receptor-based rational drug design.

Nuclear magnetic resonance (NMR) has certain advantages over x-ray or other methods. It does not require pure crystals, and protein samples are measured in solution. With the incorporation of perdeuterated micelles or bicelles, the protein can be studied under a more membrane-mimetic environment similar to biological conditions. Additionally, advanced isotope-aided heteronuclear 3D NMR techniques can be carried out on the recombinant membrane proteins in the membrane-mimetic environment to acquire structural information otherwise unobtainable. Although solution and solid-state NMR approaches have been applied to solve many 3D structures of many important water-soluble proteins with a size over 60 kDa.[9–13] However, the reported NMR studies on the full length transmembrane proteins, such as GPCRs, are still limited.[14] The structural studies of transmembrane proteins still face many challenges associated with the intrinsic hydrophobic properties of membrane proteins, including the size of membrane proteins and their micelle-bound state, solubilization, and refolding to native states.

Among many important membrane proteins, the focus of this chapter is on the integral membrane proteins containing multiple transmembrane segments, sometimes with substantial extramembrane domains. Unfortunately, it is for this class of proteins, particularly the GPCR family, that the progress of NMR has been slowest, but most needed from the aspect of basic and applied biomedical sciences.[15] In this chapter, Section 13.2 presents a brief overview of the membrane protein expression and purification. Section 13.3 summarizes NMR sample preparations through membrane protein solubilization and renature by comparing different solvent systems. NMR characterization of the recombinant membrane proteins will be addressed in Section 13.3 and Section 13.4. Finally, Section 13.5 gives examples of NMR studies and development of transmembrane proteins.

13.2 PRODUCTION OF MEMBRANE PROTEINS FOR NMR STUDIES

Overall, large-scale expression and purification of transmembrane proteins on a preparative scale are still considered very challenging, especially in the case of mammalian GPCRs. Overexpression, isotope labeling, and purification are considered the main bottlenecks for NMR structural studies of membrane proteins. Detailed membrane protein expression and purification are described in other chapters in this book; here we briefly review NMR-related expression methods along with isotope labeling methods.

13.2.1 Overview of the Membrane Protein Expression Systems

Several studies have reported functional as well as nonfunctional expression of GPCRs, including *Escherichia coli*,[16] insect cells,[17] and mammalian cells,[18] and each method has its pros and cons.[19,20] Mammalian cells have proven to be the most authentic hosts for functional studies. Recent development of novel mammalian expression vectors encoding affinity tags has facilitated expression and purification in transient or stable cell lines. However, generation and amplification of stable cell lines is a time-consuming procedure; scale-up and isotope labeling are still rather arduous and costly processes. Alternatively, expression of GPCRs in insect cells using recombinant baculovirus has given high expression levels of functional proteins.[21,22] However, virus production and maintenance of cell cultures are still time-consuming and labor-intensive. Isotopic labeling can be very expensive. Moreover, the expressed protein may result in heterogeneity, which might be difficult to adapt to NMR studies. Yeast has advantages over the more expensive and intractable higher eukaryotic systems in its ease of control, growth, and adaptation to large scale. As yeasts possess endogenous GPCRs and G proteins, functional coupling assays can be established. However, compared with soluble proteins, relatively few studies have been described for the production of recombinant membrane proteins in yeast.[23–25]

On the other hand, *E. coli*–based expression is different from eukaryotic systems in many aspects, as it lacks essential post-translational modifications (such as N-linked glycosylation), which are required for correct folding of mammalian proteins. Although many eukaryotic membrane proteins are glycosylated, their function may not be affected by bacterial expression; instead, it might provide a more homogeneous product with superior properties for structural studies.[26,27] Additionally, insertion of recombinantly expressed foreign membrane proteins into bacterial membranes may cause severe toxicity to host cells, which reduces cell growth and subsequently the expression levels. Although tens of milligrams of soluble proteins have routinely been produced in *E. coli,* membrane protein yields have generally been disappointingly low. However, *E. coli* is still considered the primary expression system for production of adequate amounts of isotope-labeled membrane proteins for NMR structural studies for the following reasons. *E. coli* is the most thoroughly studied organism. It has a wide range of expression vectors and host strains. When proteins are synthesized, proteolysis is extensively avoided and high yield may be easier to achieve.[28–30] Unlike insect and mammalian expression systems, *E. coli* is simpler to use, cost-effective, rapid to culture, and easy to scale up.[31] More importantly, the ability to express membrane proteins in bacteria provides the opportunity to incorporate a variety of isotopic labeling schemes into the overall experimental strategy, which can be used for NMR investigations. Although isotope labeling has been demonstrated for yeast and baculovirus systems neither can allow as easy and inexpensive isotope labellings as in *E. coli*.[32–34]

Actually, few GPCRs have been functionally expressed in *E. coli* to study ligand binding and G protein interactions,[35–37] screen in situ for mutations by replica filter assay,[38,39] analyze receptor topology,[40–42] and test the expression in cell-free translation systems[43] and as a source for protein purification.[44,45] Recently, there was a report of a new approach where membrane-integrating sequences for translation of membrane protein constructs for *E. coli* expression of 18 membrane proteins.[14]

In order to achieve high yield expression of TMPs, several factors need to be considered, including the choice of expression vectors and cells, fusion partner, and growth conditions (i.e., media, temperature, induction, and so on).[46,47] Application of synthetic DNA allows optimization of the codon usage for *E. coli*. Opella and co-workers found the use of fusion proteins to be essential for the expression of small membrane proteins in *E. coli*.[48] Recent studies showed that membrane proteins with His-tags have been expressed without any additional fusion proteins.[49] Generally, bacterial expression has been carried out in three stages: growth in (1) Luria-Bertani (LB) medium; (2) M9 medium for optimization of expression conditions; and (3) optimized isotope-enriched M9 media in which ^{15}N-$(NH_4)_2SO_4$ and/or ^{13}C-glucose were used as the sole nitrogen and carbon sources, respectively, for uniformly ^{15}N and ^{13}C-labeled proteins in large scale.[44,50] Furthermore, selective isotope labeling can be applied for certain isotope-labeled specific amino acids as nitrogen or carbon sources[51] for specific NMR investigations.[52,53] Recent studies showed that the membrane protein vasopressin V2 receptor (V2R) was also successfully expressed from a His-tagged vector without any additional fusion proteins, which led to high-level production (ca 5 mg of V2R per liter of culture).[49]

13.2.2 OVERVIEW OF MEMBRANE PROTEIN PURIFICATION

When the fusion protein expression approach is applied, it is recommended that the target protein can be cleaved off the fusion partner, using either chemical or enzymatic cleavage methods. Both have advantages and drawbacks. (1) Site-specific proteases such as Factor Xa, can cleave the protein in a restrictive manner at the site of a particular signal amino acid sequence (IEGR),[44,54–56] and other cleavage enzymes were also used.[55,57–62] On the other hand, the enzyme digests the fusion protein under mild aqueous conditions with a high level of specificity. (2) Chemical digestion will ensure the efficient cleavage of hydrophobic peptides. [63] The typical example is to use cyanogen bromide (CNBr) to cleave the protein at the methionine (Met) residue introduced between the fusion tag and the target protein.[31] This method is very efficient, even for highly hydrophobic peptides, in which the cleavage site might not be fully exposed. However, this method does not allow the presence of any other Met residues in the target protein and requires additional engineering to replace Met residues with Leu.[64,65]

Purification procedures of hydrophobic membrane proteins may vary depending on the different designs of expression vectors. It is common to engineer 6×, 9×, or 10× His-tags into the recombinant membrane protein to facilitate the purification using nickel affinity chromatography,[66,67] also called immobilized metal affinity chromatography (IMAC). His-tags rarely affect the structure of NMR-detected proteins to a significant degree and are therefore usually not removed. The 9× His-residues impart a remarkable affinity for matrices containing nickel or copper. The fact that binding can occur under native as well as under denaturing conditions distinguishes this affinity purification method from the others. The strong affinity of the 9× His-tag tolerates denaturing conditions that facilitate the removal of nonspecific contaminants often associated with recombinant proteins. Elution is accomplished under mild conditions by either reducing the pH or adding imidazole as a competitor. IMAC has been widely used in purification of GPCR membrane proteins.[31,68–70]

After the initial purification using IMAC, further purifications can be performed by using preparative fast protein liquid chromatography (FPLC), high pressure liquid chromatography (HPLC) chromatography methods to separate pure and homogeneous membrane proteins from the protein mixture. Complementary to IMAC, reverse-phase chromatography (RPC) is widely used in the downstream purification method as final steps for separation.[44] Actually, RPC is often used as a tool not only for purification but also for protein analysis and characterization.[71–73] In addition to estimating the retention time of hydrophobic proteins and polypeptides, the role of the hydrophobic effect in protein folding and stabilization can be investigated. The hydrophobic surface layer of the stationary phase of RPC (silica-based matrices with covalently attached hydrophobic ligands) can somewhat mimic the natural environment of membrane proteins. Retention is achieved through discrete interaction between the nonpolar ligands of the station phase and hydrophobic patches that are accessible on the surface of the protein or peptide molecules. The separation is based on differences in the hydrophobic interaction between the hydrophobic regions of the proteins and the hydro-carbonaceous ligands of the stationary phase. Other separation approaches, including size exclusion chromatography (SEC),[74–77] ion-exchange chromatography (IEC),[78,79] hydrophobic interaction chromatography (HIC),[80–84] and liquid chromatography-mass spectometry (LC-MS)[27] used for soluble proteins, are also applicable for membrane protein purifications.[85]

13.2.3 MEMBRANE PROTEIN SOLUBILIZATION AND RENATURATION

Recent developments in solution NMR techniques have made it possible to extend NMR studies on proteins in the 30- to 50-kDa size range to the 50- to 150-kDa range and possibly beyond.[86–88] However, the major challenge when studying transmembrane proteins by solution NMR arises from the fact that they usually need to be incorporated into membrane-mimetic micelles in order to correspond to their native state. Ideally, micelle or bicelle systems for solution NMR should ensure the structural and functional integrity of the transmembrane protein and, at the same time, are sufficiently small to allow rapid Brownian motions.[89,90] However, the micelle-bound protein may typically have an effective particle size beyond the conventional NMR size limit, may tumble slowly in solution, and generate broad NMR lines, all of which effectively destroys the resolution of NMR spectrum and results in poor signal-to-noise ratios as a consequence of rapid transverse spin relaxation.[90,91] These consequences have led NMR biophysicists to devote their efforts to developing other possible strategies: (1) to establish native or native-like solvent systems for membrane protein NMR sample preparation; (2) to study target membrane proteins or truncated protein segments to reduce the overall total molecular mass in micelles or bicelles; (3) to apply special liquid-state NMR methods to reduce the problem of increasing line-broadening and to overcome the size limitation; and (4) to develop solid-state NMR, which does not depend on molecular motion and which has no upper limit in the size of the micelle-/bicelle-bound protein complexes. After all, many special requirements and challenges still remain for transmembrane protein NMR studies due to the hydrophobic properties and the unknown folding process.[15] This section presents a review of membrane protein solubilization and NMR sample preparation methods.

The solubilization procedure is critical to retain the transmembrane protein in active form for NMR experiments. In general, *in vitro* refolding techniques using detergents for solubilization of overexpressed membrane protein aggregates have been reported.[92–94] Usually, *strong* detergents are initially used, and followed by application of *mild and neutral* detergents or lipids to reconstitute the "harsh" detergent and to fine-tune the conditions required for protein renaturation. Alternatively, organic solvents or chaotropic denaturants can be used to solubilize and unfold protein aggregates, the secondary structure induced in solvents such as TFE, and finally the protein reconstituted in a suitable artificial membrane system. More often, these two approaches can be used interactively. It is not the protein folding but rather the *in vitro* refolding technique that is critical for obtaining native structures of recombinant membrane proteins or their fragments. Table 13.1 summarizes the detergents available for membrane protein refolding studies. NMR solvent systems for membrane proteins are described as quasi-native, native, and natural membrane environments, as given below.

TABLE 13.1
Illustration of Strong and Mild Detergents for Membrane Protein Solubilization

Protein	Strong Detergents	Mild Lipid/Detergent Micelles	Ref.
Mitochondrial transporters	NLS	Triton X-100 and phospholipids	95
Rat olfactory receptor OR5	NLS	DM, POPC, & POPG	96
Bacteriorhodopsin	SDS	CHAPS and DMPC	97
Light-harvesting complex LHCP	-	OG, Triton X-100 and DPPG	66
Light-harvesting complex DAGK	DPC	DPC and POPC	98
Mouse serotonin 5-HT$_3$ Receptor	N/A	C$_{12}$E$_9$	99
HIV Receptor CCR5	N/A	Triton X-100 and DM	100
E.coli Glycerol Facilitator	SDS	DM or OG	101
E.coli ammonium transport protein AmtB	DM	LDAO	102
Vasopressin V2 receptor (V2R)		LMPC	49

CHAPS: 3-(3-cholamidopropyl)dimethylammonio-1-propanesulfonate; C$_{12}$E$_9$: nonaethylene–glycol monododecyl ether; DMPC: dimyristoylphosphatidylcholine; PC: dodecylphosphocholine; OG: octyl-glucoside; DPPG: dipalmitoyl phophatidylglycerol; POPC: 1-palmitoyl-2-oleoyl phosphatidylcholine; NLS: N-lauroyl sarcosine; POPG: 1-palmitoyl-2-oleoyl phosphatidylglycerol; SDS: sodium dodecyl sulfate; DM: dodecyl-β-D-maltoside; LDAO: N,N-dimethyldodecylamine-N-oxide; LMPC: lyso-myristoylphosphatidylcholine.

13.2.3.1 Quasi-Native Membrane Environment

Certain organic solvent mixtures have been used to solubilize complex membrane proteins for NMR studies. Such organic solvent mixtures seem appealing for membrane protein NMR experiments, since they give narrow line-shape and better resolution due to the effective molecular weight of the protein not being increased by detergents or micelle-bound membrane protein preparation.[103] For example, a mixture of methanol/chloroform (1:1) with 0.1 M $LiClO_4$ or $DCOOND_4$ (pH = 5.5) were used to study bacterioopsin fragments.[104,105] The solution NMR structure of the transmembrane H^+-transporting subunit c of the F_1F_0 ATP synthase in a single-phase solution of chloroform-methanol-water (4:4:1) has also been shown to mimic that of the protein in the native complex.[106] The $CD_3OD/CDCl_3$ (1:1) solvent system was also used to study amide H–D exchange of membrane protein fragments. Such a mixed organic solvent system considered as a "quasi-native" membrane environment provided a first approach, which allowed studies on structural and conformational properties of transmembrane segments in high-resolution NMR. The mixed organic solvent systems have been validated by other techniques and have been applied for membrane proteins. Circular dichroism (CD),[107] infrared spectroscopy (FT-IR),[108] I=1$_c$ and NMR[105,107,109–111] approaches have showed that bacteriorhodopsin (BR) retains most of its native secondary structure in methanol–chloroform (1:1) with 0.1 M $LiClO_4$. Arseniev and co-workers[107] also applied ^{19}F NMR showing that BR solubilized in methanol–chloroform retained the secondary structure of the fully active chromophore in the purple membrane and possessed a specific tertiary structure. The data also showed that individual fragments of BR, isolated after polypeptide chain cleavage, retained their "native" conformations in the organic solvent.[107.]

13.2.3.2 Native-Like Membrane Environment

Detergents or lipids used for NMR, such as sodium dodecyl sulfate (SDS), dodecylphosphocholine (DPC), dihexanoylphosphatidylcholine (DHPC), dimyristoylphosphatidylcholine (DMPC), and lysomyristoylphosphatidylcholine (LMPC) (Figure 13.2) are amphiphilic molecules with one or two long hydrocarbon chains and a polar or charged head group. They can self-assemble to form detergent micelles, bicelles, lipid bilayers, or lipid vesicles, which are used as the "native-like" membrane-mimicking environment suitable for solubilization of membrane proteins for high-resolution NMR studies because of their relative small size (usually 10 to 100 kDa) compared with any available lipid bilayered assemblies.

Micelles. Although detergents at concentrations below the critical micelle concentration (CMC) (e.g., 0.1% or 3 mM SDS) are often used to denature recombinant proteins, membrane proteins are particularly stable toward micellar detergents (e.g., 2% SDS).[112,113] Vinogradova and Sanders have demonstrated that the detergent micelles change their shape and aggregation state upon binding a peptide and revealed that there is no direct correlation between the size of the naked micelles and that of the peptide–micelle complexes.[114] In addition, the micelle size also varied upon the peptide bound states,[115] for example, a peptide of ~ 20 residues (M_r ~ 2.3 K), binding on the micellar surface may have NMR spectral line widths approxi-

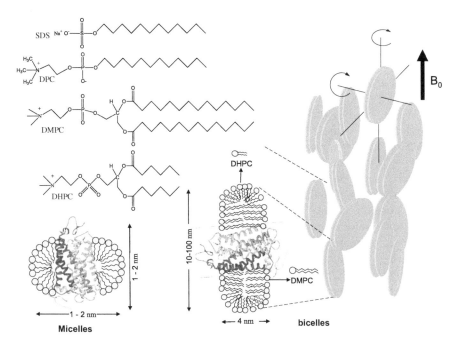

FIGURE 13.2 Molecular structures of the commonly-used detergents and lipids and the graphic representation of the 2D cross-section slice model and sizes of the micelles and bicelles.

mately equal to that of a water-soluble protein of ~ 10 kDa, whereas the value for peptides penetrating the micelle may be ~ 20 kDa.[13] Another example is the 40 kDa diacylglycerol kinase (DAGK) homotrimer forming 100 kDa protein–detergent micelles with octylglucoside, whereas protein-free octylglucoside micelles are only about 20 kDa.[116]

In the absence of bound protein peptide, SDS or DPC forms aggregates or micelles at the concentration above its CMC with a molecular mass of 20 to 30 kDa.[114] SDS is a strong anionic detergent with a CMC of 8 mM, which is affected by salt concentration and temperature, whereas DPC micelles (CMC of 1.1 mM) have a neutral head group and its aggregation state is not affected by salt concentrations.[13] DPC micelle-bound protein peptide samples tend to give a better NMR resonance dispersion and a better ordered structure than SDS.[117] However, the higher costs and the presence of spectral impurities sometimes make DPC less attractive. During the past decades, liquid-state NMR has been used to study micelle-bound membrane proteins such as bacteriorhodopsin fragments, [105,107,109,110,118,119] and the results have been confirmed by high-resolution electron cryomicroscopy using bacteriorhodopsin as a model.[120]

Small bicelles. Bicelles (Figure 13.2) are the smallest bilayered disc-shaped aggregates that self-assemble from mixtures of long-chain phospholipids (e.g., DMPC), short-chain phospholipids (e.g., DHPC), or detergents (e.g., CHAPS).[48] Upon mixing, the long-chain lipids form the extended bilayer portion and the short-chain lipids form the caps at the ends of the discs. The size, shape, and orientation properties of the bicelles can be controlled by the ratio of the two types of lipids: for example, the DMPC to DHPC ratio, ranging from small isotropic bicelles [121] to large bicelles similar to extended bilayers.[122] Typically, when the amount of short-chain phospholipids is increased, or the ratio of long-chain and short-chain lipids is less than 1 and the total concentration of the lipids is 5 to 15% (w/w), bicelles prepared from an unaligned phase are suitable for high-resolution NMR studies.[123–127] Larger bicelles have originally been employed as a medium in which proteins were reconstituted and then characterized by solid-state NMR.[122,128] Only recently have bicelles been recognized to have more uses in introducing small degrees of residual orientation of soluble proteins in order to determine dipolar coupling; this has proven extremely beneficial for structure determination of soluble proteins by high-resolution NMR.[129,130] Small bicelles can be weakly aligned in stressed gels, whereas large bicelles can be completely aligned on the magnetic field; this provides alternative ways of accessing the orientation information of the anisotropic spin system of membrane proteins. However, some proteins have been found incompatible with DMPC-based bicelles. It is known that some membrane proteins interact with bicelles in such a manner that the aggregate morphology is grossly perturbed and also that some bicelles disrupt native folding of the membrane protein as well as their structure and function.[10] In addition, smaller isotopic bicelles are still much too large to be efficiently employed in solution NMR studies of integral membrane protein–bicelles complexes.

Other native-like membrane environments are represented by lipid vesicles and bilayers, which are the most desirable lipid assemblies to accurately mimic the native environment for structural studies of membrane proteins. However, because even the smallest vesicle bilayers are very large from a solution NMR point of view (1000 kDa aggregate molecular weight), the bilayer systems are not suitable for direct solution NMR studies of transmembrane proteins, whereas they are broadly employed in solid-state NMR, as reviewed by Opella[10,48] and others.[131–135] Plasma membranes provide a native milieu for transmembrane proteins. However, due to the slow mobility of the semisolid native plasma membrane samples, solution NMR spectral linewidth is very broad, which limits high-resolution NMR studies on conformation and structure of membrane proteins.[136–138]

13.3 HIGH-RESOLUTION SOLUTION-STATE NMR FOR TRANSMEMBRANE PROTEINS

Once the membrane protein samples are solubilized and renatured using the membrane-mimetic solvent systems described above, advanced solution NMR techniques can be adapted and performed for structural studies on transmembrane proteins as for soluble proteins.[139–141] In general, NMR characterization of recombinant transmembrane proteins can be briefly described in three steps: (1) backbone sequential assignments, (2) conformational parameter measurements, and (3) structural calculations with NMR-measured distance and torsion angle constraints. The choices of

NMR techniques applied to membrane proteins also depend on the effective size of the membrane protein or the protein segments in the membrane-like environment.

Generally, the actively-shielded pulse field gradient techniques are used in all 2D/3D experiments[142–146] for the S/N signal enhancement, water suppression,[147] and diffusion measurement.[148] The recently developed CryoProbe method[142] can generate a three to fourfold increase in signal to noise ratio ([1]H–S/N, 6000:1) by reducing the operating temperature of the NMR coil assembly and the preamplifier. The enhanced sensitivity and resolution are extremely valuable in detecting hydrophobic transmembrane proteins or fragments due to their limited solubility and sample stability.

In addition to the 2D NMR techniques, triple resonance 3D NMR experiments can be useful for the following reasons: (1) 3D NMR overcomes the peak overlap problem by introducing the third dimension and separating overlapping peaks into a number of 2D planes.[149] (2) For moderate-sized proteins (~ 20 kDa), most of the one-bond J couplings are significantly larger than the spectral line width.[150] The many choices of 2D or 3D triple resonance NMR experiments can be used for backbone or side-chain sequential assignments and conformational determination as shown in Table 13.2.

13.3.1 Sequential Assignment

NMR experimental methods and techniques for the protein sequential assignments have been described in other reviews,[139,151–153] and the process does not significantly differ from that for water-soluble proteins and micelle-associated membrane proteins or peptides.[12,154] However, as a consequence of the micellar environment, the effective sizes of membrane proteins or peptides are usually much larger than nonmicellar solvent systems, and the NMR acquisition parameters should therefore be adjusted. In general, the effective size will be increased and the shape will be changed once the peptide is incorporated into a detergent micelle as discussed above. For example, a typical peptide of ~ 20 amino acids (~ 2.3 kDa) that binds to the surface of a micelle may have line widths approximately equal to that of a water-soluble protein of ~ 10 kDa, but approximately 20 kDa when it penetrates the micelle.[13] Therefore, the traditional classification of protein size ranges and corresponding NMR sample requirements and techniques established for water-soluble protein NMR will not be directly applicable to micelle–peptide complex samples. Therefore, the corresponding NMR acquisition parameters should be adjusted to the corresponding effective size of the micelle–peptide complexes as briefly described below.[13]

As reviewed by Baleja, short peptides of ~ 20 to 30 amino acids (~ 2.3 to 3.4 kDa) will have an effective size of less than 10 kDa for peptide–micelle complexes. The micelle-bound peptides can be studied by homonuclear 2D ^{1}H NMR techniques, with or without isotope-enrichment. The pulse gradient 2D NMR techniques include total correlation spectroscopy (TOCSY) (mixing times of 45 and 75 ms for short or long range chemical shift correlations) or double quantum filter correlation spectroscopy (DQF-COSY) experiments for protein resonance assignments, and nuclear Overhauser enhancement spectroscopy (NOESY) experiments (typical mixing times of 100 to 250 ms) for conformational or structural measurements. Practically, several

TABLE 13.2
2D and 3D NMR Experimental Methods and Their Purposes

NMR Experiments		Labelling	Purposes
$^1H^1H$ NOESY	2D	No	NOE constraints and aromatic side chain assignment
$^1H^1H$ TOCSY	2D	No	Backbone and side chain assignment
^{15}N ^1H HSQC	2D	^{15}N	Structural fingerprint studies
^{15}N ^1H TOSCY-HSQC	3D	^{15}N	N^H H^N resonance pair assignment
			N^H H^N pairs to C_αs correlation and assignment for some residues
Backbone and side chain assignments			
HNCO	3D	^{15}N, ^{13}C	Intra-residue for sequential connectivity
HN(CO) CA	3D	^{15}N, ^{13}C	Intra-residue for sequential connectivity
HN(CA) CO	3D	^{15}N, ^{13}C	Intra-residue, also used with HNCO
HNCA	3D	^{15}N, ^{13}C	Intra-residue with weaker inter-residue correlations between C_α
CBCACONH	3D	^{15}N, ^{13}C	Sequential connectivity and assignment
HBHA(CBCACO) NH	3D	^{15}N, ^{13}C	Intra-residue for sequential connectivity
HN(CO) CACB	3D	^{15}N, ^{13}C	Sequential connectivity and assignment
HNCACB	3D	^{15}N, ^{13}C	Intra-residue with weaker inter-residue corelations between C_α and $C\beta$
HN(CA)CB	3D	^{15}N, ^{13}C	Distinct Ser and Thr
HA(CACO) NH	3D	^{15}N, ^{13}C, ^2H	H_α assignment
HCCH-TOCSY	3D	^{15}N, ^{13}C	Side chain assignment
	3D	^{15}N, ^{13}C	Side chain assignment
	3D	^{15}N, ^{13}C	Side chain assignment
	3D	^{15}N, ^{13}C, ^2H	Side chain assignment
	3D	^{15}N, ^{13}C, ^2H	Side chain assignment
Distance NOE measurements and sequential assignments			
^{15}N-edited NOESY-HSQC	3D	^{15}N, ^{13}C, ^2H	NOE constraints and aromatic side chain assignment
^{13}C-edited NOESY-HSQC	3D	^{13}C, ^{15}N, ^2H	NOE constraints and aromatic side chain assignment
15N-edited HSQC-NOESY	3D	^{15}N, ^{13}C, ^2H	NOE constraints, sequential assignment, secondary structure survey

approaches may be used to deal with signal overlapping situations, including D_2O (100%) and temperature variation experiments. Fully deuterated water may be used for deuterium exchange of the amide protons, which helps assign aromatic side-chain signals. Experiments conducted at different temperatures may also be used to resolve the overlapping signal resonances, typically at 10, 25, 37, 45, and 50°C. However, caution must be taken for protein stability and possible loss at NOE cross-peak due to the fast molecule motion at higher temperatures. Homonuclear NMR studies on short peptides of truncated membrane proteins have been conducted for several GPCRs, including cannabinoid receptor CB2 fragments,[155] glycophorin A (GpA) transmembrane fragments,[156] and other membrane protein fragments.[157,158]

However, transmembrane protein fragments can reach a total mass of 10 to 25 kDa, when incorporated into detergent micelles. As a result, the NMR spectral line width substantially exceeds the values of the coupling constant, typically 3 to 12 Hz for ^1H–^1H homonuclear NMR, and the spectral cross-peak intensity (or signal-to-noise ratio S/N) becomes poor. In this case, the peptides have to be isotopically labeled with either single ^{15}N or double ^{13}C and ^{15}N isotopes in order to resolve the signal overlapping and achieve successful correlation spectra. The heteronuclear ^{13}C and ^{15}N-edited 2D/3D NMR experiments take advantage of heteronuclear one-bond coupling constants ($^1J_{H-N} \sim 100$ Hz and $^1J_{H-C} \sim 145$ Hz), which allows collection of multidimensional chemical shift correlation spectra of high sensitivity. The polypeptide backbone assignment and sequential/conformational studies of membrane proteins or fragments require triple resonance (^1H, ^{15}N, and ^{13}C, or ^2H) experiments, which have been extensively reviewed for general applications[150,153,159] and for membrane proteins.[10,11,13,90,160]

3D NMR studies and correlated spin systems are summarized in Table 13.2, in which 3D NMR experiments are named by the spin correlations defined for the specific atoms of the polypeptide backbone. For example, a 3D HN(CO)CA study demonstrated that the backbone resonance of ^{15}N and its attached proton are detected in connection (through the intervening carbonyl) with the $\alpha-$ carbon CA (or Cα) resonance of the preceding amino acid residue. Sensitivity comparisons of commonly used triple resonance 3D NMR studies for backbone resonance assignments are shown in Figure 13.3, including the 3D HNCA, HN(CO)CA, HNCO, NHCACB, and CACB(CO)NH, which have been reviewed extensively elsewhere.[11,13,150,153,159] The sequential assignments may be further verified by the analysis of ^{15}N NOESY-HSQC spectra[161] with a mixing time of 50 to 250 ms, depending on the size of protein samples. Aliphatic side-chain assignments can be made using C(CO)NH,[162,163] H(C)(CO)NH,[162,163] HCCH-COSY,[159,164] and HCCH-TOCSY spectra,[152,153,165] as needed. Aromatic ring proton and carbon resonance will be assigned by analyzing 2D ^1H–^{13}C HSQC and ^1H–^1H NOESY.[152]

For whole integral membrane proteins or helix bundles (30 to 40 kDa), the effective size of peptide–micelle complexes could be very large (over 100 kDa),[116] which results in a slow molecular motion and broad spectral line width so that the 3D spectra will not show adequate spin correlation intensities in triple resonance experiments. Actually, large-sized, hydrophobic, and insufficient amounts of membrane proteins have always posed major challenges to NMR biophysicists.

Recent developments in NMR research has created considerable hope for solving structures of transmembrane proteins with a molecular mass larger than 30 kDa by NMR, including development of a method for full or partial deuteration of carbon-bound hydrogen atoms of the polypeptide and TROSY (transverse relaxation optimized spectroscopy) NMR method[88,166] TROSY NMR presents one of the recent major advances in solution NMR spectroscopy and has made a significant impact on the determination of membrane protein structures in detergent micelles. Basically, the scalar heteronuclear spin–spin couplings are not decoupled in TROSY experiments, but only one of the four peaks in the multiplet is retained, and the chemical

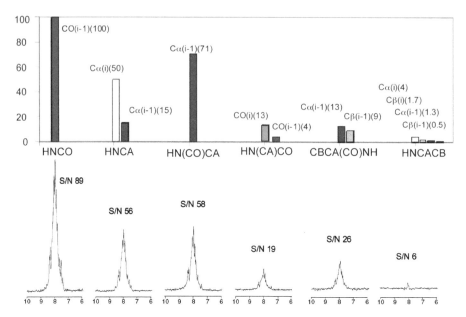

FIGURE 13.3 Basic 3D triple resonance NMR experiments and their S/N sensitivities for backbone assignments. HNCO usually has the highest sensitivity but less information (mainly CO signals), whereas HNCACB and CBCANH provide more information but low sensitivity. (From Xie, X.Q. and Chen, J.Z. *J. Biol. Chem.*, 280, 3605–3612, 2005. With permission.)

shift anisotropy relaxation is used to compensate dipolar relaxation.[167,168] This procedure results in an improved signal–noise ratio.

Another crucial development for resolving and assigning NMR spectra of large complexes, including membrane proteins in detergent micelles, is to selectively and uniformly deuterate residue side-chains.[166] Deuteration essentially dilates ¹H–¹H relaxation, resulting in a substantially narrower resonances line[169–171] and higher sensitivity of through-bond COSY-type experiments.[172–174] The gain in sensitivity comes from the lower gyromagnetic ratio of ²H relative to ¹H and its correspondingly lower effectiveness in changing the rate of transverse relaxation of neighboring heteronuclei.[169] This results in lowering the effect of increasing line broadening with increasing molecular sizes. Many newer studies have taken advantage of TROSY and deuteration experiments to increase the upper limit for the overall size of the protein–detergent complex.[86,175,176]

However, the extensive ²H-labeling TROSY experiment may only be limited in detection distance information between the exchangeable amide protons unless the ²H-labeling is supplemented with selective protonation of methyl groups.[177] In order to overcome the problem of losing too much side-chain information due to the complete deuteration of these moieties, methods have been developed[177,178] to reintroduce methyl protons in a number of aliphatic side chains in order to produce valuable long-range nuclear Overhauser effects (NOEs) with helical membrane proteins.

Thus, combined TROSY and uniform side-chain deuteration approaches are well suited to study large transmembrane proteins incorporated into micelles or bicelles.[11] The TROSY version of the standard triple-resonance experiments used at high field strengths has allowed resonance assignments for proteins larger than 100 kDa.[87,179–181] This approach has been used to determine the backbone structure of the transmembrane domain of the outer membrane protein OmpA (177 residues, 19 kDa) in DPC micelles of 50-kDa molecular mass,[166] and the backbone structure of OmpX (148 residues, 16 kDa) in DHPC micelles of about 60-kDa molecular mass. [167]

13.3.2 CONFORMATIONAL DATA MEASUREMENTS

13.3.2.1 Chemical Shift Index

Chemical shift assignment is the first important step for sequential determination and later conformational studies. However, chemical shifts do not provide particularly useful conformational parameters, although they are also protein folding–dependable. This is because chemical shifts may be affected by many factors, including H-bonding, solvation effects, bound ions, and molecular motion as well as temperature.[13] For example, the inter-residue H-bonding between the amide N–H of one residue and the carbonyl C = O of another one will shift proton N–H downfield (de-shielding effect). A methyl group that lies perpendicular to an aromatic ring will be shifted upfield by as much as –2 ppm (shielding effect).[139]

Nevertheless, large chemical shift changes can be useful qualitatively to evaluate folded and unfolded proteins. Wishart has developed an empirical method, chemical shift index (CSI) approach,[182] based on the fact that the character and nature of protein secondary structure strongly influenced the NMR chemical shifts of the residues. Based on the CSI method, ^{1}H chemical shift of the α-CH proton of all 20 naturally occurring amino acids experienced an upfield shift (with respect to the random coil value) when in a helical configuration and a comparable downfield shift when in a β-strand extended configuration. The CSI can be used to rapidly and quantitatively determine the identity, extent, and location of secondary structural elements in proteins, including α-helices, β-strands, and random coils, on the basis of the α-CH proton chemical shift assignments. For example, Hα chemical shifts that are at higher field (> 0.1 ppm) than that amino acid in a random peptide, are indicative of a α-helices, whereas Hα with lower chemical shifts are indicative that the residue is part of a β sheet.

Practically, the CSI technique is essentially a two-stage digital filtration process described below.[182] First, a CSI (a ternary index having values of –1, 0, and 1) is assigned to all identifiable residues on the basis of their assigned -proton chemical shift values and guided α-proton chemical shift values listed in Table 13.3. The CSI will be marked as a positive value of +1 for the residue if the α-proton chemical shift is greater than the range given in Table 13.3 for that residue, marked as a negative value of –1 if less than the range, and marked as 0 if within the range. Then, secondary structures are delineated and subsequently identified on the basis of the values and "densities" of these CSI. For example, "dense" grouping of four or more "–1's" not interrupted by a "1" is a helix; any "dense" grouping of three

TABLE 13.3
The Chemical Shift Values of α-Protons Used in the Determination of
Secondary Structures

Residue	α-CH range (ppm)	Residue	α-CH range (ppm)
Ala	4.35 ± 0.10	Met	4.52 ± 0.10
Cys	4.65 ± 0.10	Asn	4.75 ± 0.10
Asp	4.76 ± 0.10	Pro	4.44 ± 0.10
Glu	4.29 ± 0.10	Gln	4.37 ± 0.10
Phe	4.66 ± 0.10	Arg	4.38 ± 0.10
Gly	3.97 ± 0.10	Ser	4.50 ± 0.10
His	4.63 ± 0.10	Thr	4.35 ± 0.10
Ile	3.95 ± 0.10	Val	3.95 ± 0.10
Lys	4.36 ± 0.10	Trp	4.70 ± 0.10
Leu	4.17 ± 0.10	Tyr	4.60 ± 0.10

NOEs, dihedral constraints, H-bonding, and RDC constraints for polypeptide tertiary structures.

Source: From Wishart, D.S., Sykes, B.D., and Richards, F.M. *Biochemistry,* 31(6), 1647–1651, 1992.
With permission.

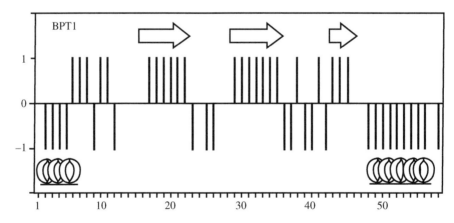

FIGURE 13.4 Chemical shift index plot of α-protons from BPTI proteins, illustrating the
CSI prediction of protein secondary structures (a helix: ∞∞ ; and β-sheet: →). (From Wishart,
D.S., Sykes, B.D., and Richards, F.M. *Biochemistry,* 31(6), 1647–1651, 1992. With permis-
sion.)

or more "1's" not interrupted by a "–1" is a β-strand (Figure 13.4). All other regions
are designated as random coils. The accuracy and consistency of the method has
been demonstrated by comparison of the secondary structures of 12 proteins determined by NMR and x-ray crystallography.[182] The CSI information has been broadly
used to confirm the secondary structure location of polypeptides.[183–185]

13.3.2.2 NOE Distance Constraints

Nuclear Overhauser effect (NOE) is one of the most important NMR parameters used in conformational analysis (Figure 13.5) since the magnitude of the NOE cross-peak volume measured in NOESY is inversely proportional to the sixth power of the inter-proton distance in space ($I_{NOE} \propto r^{-6}$). Similar to water-soluble proteins, ^{13}C- and ^{15}N-separated NOESY experiments are used for the distance constraint measurements.[161,186,187] 3D ^{15}N-edited NOESY–HSQC with 75 to 200 ms mixing time for micelle-bound membrane peptides was used to collect amide (^{15}NH)-based NOEs, including $NH_i–NH_j$, $NH–H_\alpha$, and $NH–H_{side-chain}$. 3D ^{13}C-edited NOESY–HSQC (75 to 200 ms mixing time) was employed to collect various NOEs from side chains using ^{13}C uniformly labeled samples dissolved in 100% D_2O.

NOE-cross-peak volumes measured from the 2D NOESY or 3D ^{15}N- and ^{13}C-resolved NOESY spectra will be used to calculate internuclear distances[188] by programs such as CYANA[189] and MARDIGRAS.[190] These calculated distances are then categorized into boundaries: 1.8 to 2.8 Å (strong), 1.8 to 3.3 Å (medium), or 1.8 to 5.0 Å (weak) according to their integrated cross-peak volumes. NOE-distance restraints are applied with a square-well potential and a force constant of 50 kcal mol^{-1}Å^{-1} for distance geometry calculations [191] and constrained dynamics simulation.[192] The distance restraints will then be linearly released in 10 ps in dynamics structural calculations.

13.3.2.3 Torsional Angle Constraints

Another important parameter is the J-coupling constant measurement that can be used to calculate the torsion angles or dihedral angles as conformational constraints. The peptide dihedral angles ϕ (phi), ψ (psi) and $\chi-1$ (chi-1) rotating around the N-Cα-C, and Cα–Cβ bonds, respectively (Figure 13.6), are used to define the protein secondary structures linked together by these important atoms, namely the alpha carbon (α), carbonyl carbon (C), and amide nitrogen (N) atoms. The dihedral angle ϕ around N–Cα bond is the most commonly measured backbone dihedral angle and is calculated using Karplus relationship [193] from an analysis of the vicinal $^3J_{HN-C\alpha H}$ coupling constants detected from 2D DQF—COSY[194], HMQC—J[195], or 3D HNHA spectra.[190,196] Typically, the torsion angles and their corresponding coupling constants are correlated as follows:[197] $^3J_{H\alpha N-CH}$ ≤ 6 Hz, =-60±30°; 6 Hz< $^3J_{HN-C\alpha H<}$ 8 Hz, =-6±40°; and $^3J_{HN-CH}$ ≥ 8 Hz, =-120±40°.

The torsion angle ψ is the dihedral angle around the C–Cα bond and is often added into the protein secondary structure. Further conformational restraints for ϕ and ψ were usually added on the basis of the consensus chemical shift index [198,199] and NOE patterns characteristics of secondary structures: α-helical residues, –60 ± 40° (ϕ) and -50 ± 50° (ψ); β-strand residues, –120 ± 40° (ϕ) and –130 ± 50° (ψ). The angle restraint should be applied with caution because it may sometimes obscure the local real distortion or the interesting flexible sections in the middle of secondary structure elements.[200–202] Another torsion angle (χ-1) is the dihedral angle around the Cα–Cβ bond of the peptide side chain and can be determined by measuring the $^3J_{HC\alpha–C\beta H}$ coupling constants, or estimated from the Hα–Hβ NOE intensities or the HN–Hβ NOE intensities as reported.[203,204]

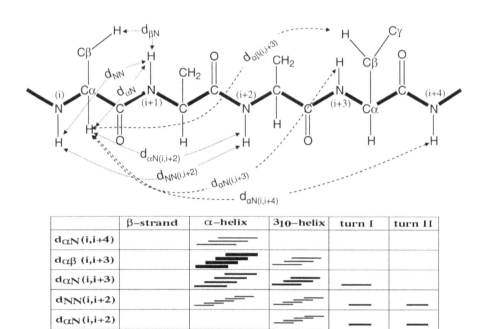

	β–strand	α–helix	3₁₀–helix	turn I	turn II
$d_{\alpha N}$(i,i+4)		═══			
$d_{\alpha\beta}$ (i,i+3)		▬▬▬	═══		
$d_{\alpha N}$(i,i+3)		▬▬▬	═══	───	
d_{NN}(i,i+2)		═══	═══	───	───
$d_{\alpha N}$ (i,i+2)		═══	═══	───	───
d_{NN}	───────	▬▬▬▬▬	▬▬▬▬▬	──▬	▬
$d_{\alpha N}$	▬▬▬▬▬	───────	───────	───	──▬
	1 2 3 4 5 6	1 2 3 4 5 6 7	1 2 3 4 5 6	1 2 3 4	1 2 3 4

FIGURE 13.5 Survey of the sequential and medium range ^{1}H–^{1}H NOEs in the following sequential secondary structures: β-sheet, α-helix, 3₁₀-helix, turns of types I and II. (From Wuthrich, K. *NMR of Proteins and Nucleic Acids.* John Wiley & Sons. Inc., New York, 1986. With permission.)

During calculation, if $^3J_{HN-C\alpha H}$ values are extremely small (< 6.0 Hz) corresponding to more than one φ-angle, the φ-angles that are determined from small coupling constants will be used as restraints only if they appear in regions of α-helical secondary structures as determined from the NOE data and amide exchange patterns.[205] Torsional restraints are applied to φ-angles with a typical force constant of 200 kcal mol^{-1}rad^{-2} during computer simulation.

13.3.2.4 Hydrogen-Bonding Constraints

H-bonds are monitored by measuring amide proton-deuterium (^{1}H–^{2}H) exchange. Amide ^{1}H–^{2}H exchange rates of micelle-bound membrane proteins are usually measured by the rate of decrease of cross-peak volumes in a set of 12 TOCSY spectra as the following protocol.[207] First, DPC micelles containing the peptide were lyophilized from H$_2$O–TFE, and dissolved in D$_2$O–TFE-d$_3$. TFE may be necessary to ensure the complete solubility of membrane protein and assist the proper folding. Then, a series of 12 TOCSY spectra is recorded over 66 hours. The temperature was adjusted to 25°C, and the pH was adjusted to a proper range. The half-exchange

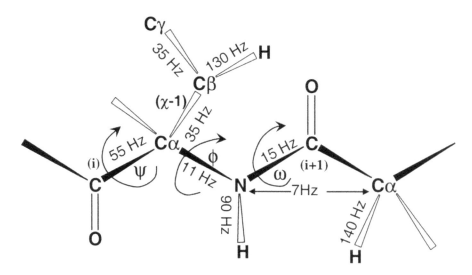

FIGURE 13.6 Nomenclatures for the polypeptide backbone dihedral angles f, ψ, ω, correspondent to the torsion angles rotating about N-Ca, C–Ca bond, and C–N bond and coupling constraints, respectively. The torsional angle (c-1) is the dihedral angle around the side-chain Ca–Cb bond. (From Fersht, A. *Structure and Mechanism in Protein Science: A Guide to Enzyme Catalysis and Protein Folding*. W.H. Freeman, New York, 1999.[206] With permission.)

times of amide protons were found by fitting the cross-peak volume time dependence to the equation:

$$V(t) = V(0) \exp (-\ln2t/t_{1/2})$$

where V(t) is the cross-peak volume (integration limits are the same for all TOCSY spectra) in the 2D spectrum. The recording started at time t once the sample had been placed into the spectrometer. To confirm the results from TOCSY spectra, 30 sets of HMQC (or HSQC) spectra may be collected at 30-minute intervals immediately after the dissolution of the lyophilized samples. It is important that the delay between dissolution and the start of acquisition does not exceed 10 minutes. In this way, a number of slowly exchanging amide protons could be identified in these spectra. H-bond acceptors for many of these slowly exchanging amides can be unambiguously assigned by analysis of the 3D structures of model helices. Then, hydrogen bonds will be included in subsequent calculations with upper and lower bounds of 1.8 to 2.5 Å (H^N–O) and 2.5 to 3.3 Å (H^N–O), typically.

13.3.2.5 Residual Dipolar Coupling Constraints

Residual dipolar coupling (RDC) represents one of the most important emerging NMR techniques. The values can be used to improve the local as well as global accuracy of the geometry and validate the structure of macromolecules in which traditional NOE measurements are limited. RDC is measured from the influence of anisotropic motion on $^{1}J(^{1}H^{-15}N)$ coupling constants[208] by dissolving proteins in dilute liquid crystal media[209-213] or by studying naturally paramagnetic molecules[64,214-217] and has rapidly become a routine method for the investigation of the structure of macromolecules by NMR. The importance is even greater for helical membrane proteins in micelles and small bicelles because of the almost total absence of assignable long-range distance constraints from NOEs in these membrane protein samples.[218] For large membrane proteins or micelle-bound membrane proteins, it usually becomes very difficult to measure the structural/conformational NOEs (particularly long-range NOEs) due to prohibitive relaxation effects, which makes the determination of membrane protein structure beyond 30 kD difficult using classical techniques.

In principle, RDC exists from incomplete averaging of dipole–dipole interactions in solution NMR and provides information of the orientation of internuclear vectors relative to the magnetic field (Figure 13.7). The dipolar coupling between a pair of ^{1}H and ^{15}N nuclei can be simplified as follows:

$$D_{HN} = -\frac{\mu_0 h}{16\pi^3}\gamma_H\gamma_N\left[D_a(3\cos^2\theta-1)+\frac{3}{2}D_r\sin^2\theta\cos 2\phi\right]SA_a \times r_{HN}$$

where θ denotes the fixed polar angle between the H–N inter-atomic vector and the z-axis of the order tensor, and ϕ is the angle that describes the position of projection of the H–N inter-atomic vector on the x-y plane relative to the x-axis. A_a is the

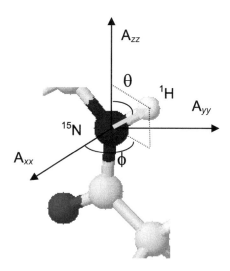

FIGURE 13.7 Orientation of the dipolar couplings vector in a protein N–H bond segment. The vector connects the amide ^{1}H and ^{15}N atoms. As shown, the interaction vector coincides with the chemical bond. The axis system of the alignment tensor is designated as A_{xx}, A_{yy}, and A_{zz}. The angles θ and ϕ define the orientation of the dipolar vector with respect to the alignment tensor. (From de Alba, E. and Tjandra, N. *Prog. Nucl. Magnet. Res. Spectrosc.*, 40(2), 175–197, 2002. With permission.)

unitless axial component of the molecular alignment tensor, that is, $A_{zz} - (A_{zz}+A_{yy})$. S is the generalized order parameter, r_{HN} is the H–N distance, and the square brackets denote the time average due to molecular tumbling and internal motions D_a and D_r (in units of Hz) are the axial and rhombic components of the alignment tensor normalized to the interaction between $^1H^N$ and ^{15}N nuclei or the value observed for NH dipoles. Parameters of alignment tensor (D_a and D_r) can be obtained from the best fitting experimental RDCs to known structure.[219]

The distance dependence of RDC is γ^{-3} (*vs* γ^{-6} for NOE) and therefore has the advantage of allowing long-range (up to 10 Å) internuclear interactions to be monitored, which is particularly useful for long-range NOE measurement in large membrane protein molecules. The RDCs are not limited to 1H–1H interactions and can be monitored for a wide range of nuclei, for example, ^1H–^{13}C, ^{13}C–^{15}N, ^1H–^{15}N, and so on. Thus, RDCs provide additional structural information that is not accessible by the NOEs method.[220]

When a protein is dissolved in the diluted bicelle solutions, the protein weakly aligned in bicelle preparation adopts some of the order of the surrounding bicelles, and the level of order can yield up to a $\pm$ 15 Hz residual dipolar splitting. The actual measurement of residual dipolar coupling was based on 2D IPAP-type ^1H–^{15}N HSQC spectra by suppression of the NH_2 signal on isotropic and weakly aligned samples of protein in SDS or DPC micelles in polyacrylamide gels by following the protocol reported by Opella and co-workers.[218,221]

The structure and topology of the helices are generated by Dipolar waves[222] and polarity index slant angle (PISA) wheels[223] derived from experimental measurements of RDCs and residual chemical shift anisotropies (RCSAs). The ^1H–^{15}N RDC constraints are then introduced with the force constant linearly from 0.01 to 7 kcal Hz^{-1} as the simulated annealing dynamic temperature is lowered to 25 K in steps of 12.5 K over 120 ps. The accepted structures should have violations of the RDC values that are less than 1.05 Hz,[221] and deviations of the helical dihedral angles of less than 10° from those determined by Dipolar Wave fitting.[222]

13.4 HIGH-RESOLUTION SOLID-STATE NMR FOR TRANSMEMBRANE PROTEINS

Solid-state NMR spectroscopy (SSNMR) has been increasingly applied to study orientations and conformations of membrane proteins using both aligned sample and magic angle sample spinning approaches.[10,131,224–226] SSNMR overcomes the membrane protein crystallization difficulties and the size limitation of membrane protein–detergent micelles in solution NMR. The slow tumbling and broad line width, associated with solution NMR study of membrane bilayer–bound large proteins, is solved in SSNMR either by using magic angle spinning (MAS) or uniaxially oriented sample (UOS) NMR techniques.[10] SSNMR has no molecular size limitation because the tumbling is not a band-narrowing mechanism in SSNMR and line width does not depend on the size of the protein and thus is not an intrinsic limitation on resolution.

13.4.1 MAGIC ANGLE SAMPLE SPINNING NMR

Solid-state magic angle spinning (MAS) NMR techniques obtain high resolution and narrowed line width by fast-spinning the sample around an axis that is tilted 54.7°, the magic angle, relative to the external applied magnetic field (Figure 13.8). This method analyzes the orientation in a sample that contains randomly oriented bilayers and allows determination of solution-type high-resolution NMR spectra.[227,228] Although the dipolar interactions that are valuable for analysis of peptide conformation and orientation information relative to the membrane are averaged out due to the fast magic angle sample spinning, the conformational internuclear distances and torsion constraints can be achieved through the developed SSNMR pulse programs and techniques with selective heteronuclear (e.g., ^{13}C, ^{19}F, ^{15}N, or ^{31}P) [229–231] or homonuclear (e.g., ^{13}C and ^{13}C) spins.[135,232]

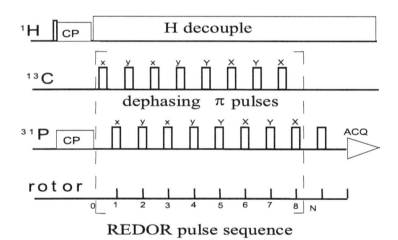

REDOR pulse sequence

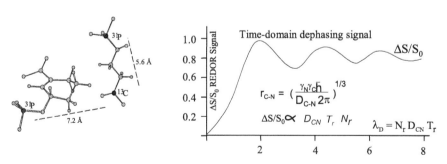

FIGURE 13.8 $^{31}P/^{13}C/^{1}H$ REDOR NMR determined possible structure of the S3P-glycophosphate complex bound to EPSP synthase. (A) REDOR pulse program; (B) The calculated ratio of REDOR difference ΔS (or S_f-S_0) to full-echo intensity S_0 for one dephasing pulse each half rotor cycle was plotted as a function of λ_D, the product of the number of rotor cycles N_r, the C-N dipolar coupling D_{C-N}, and the rotor period T_r. REDOR distances r_{C-N} are calculated from measured dipolar coupling constants D_{C-N}.

For example, triple resonance rotational-echo decoupling resonance (REDOR) NMR[227] is widely applied to structural studies on macromolecules and is also gaining more applications for membrane proteins.[233-236] It provides direct information on internuclear dipolar coupling between isolated pairs of nuclei that can be used to deduce the interhelical distances and the relative orientation of a receptor system. It offers a higher accuracy (0.1 Å) than traditional methods, including x-ray crystallography and solution-state NMR (accuracy of 0.5 –to 2.0 Å). It also allows direct measurement of a receptor system in a nondestructive way, and it is not limited by the molecular weight of the protein-bilayers or ligand complexes or by the availability of high regular crystals for x-ray diffraction.

Basically, as a result of magic angle spinning, the $^{19}F/^{13}C$ or $^{15}N/^{13}C$ dipolar coupling (distance-dependent) is coherently averaged to zero in achieving a high-resolution solid-state spectrum. REDOR[227] employs multiple pulses (refocusing and dephasing pulses) to reintroduce the dipolar coupling in a predictable manner (Figure 13.8). To obtain a spectrum of the dipolar coupling present in the system, a REDOR transform is applied to the dephasing data obtained by calculating the ratio (with or without echo peak intensity differences to the full echo intensity ($\Delta S/S_0$)) as a function of total dephasing time[237,238] (Figure 13.8).

The REDOR technique has been used to infer peptide backbone conformations of melanostatin from $^{13}C-^{15}N$[231]; inter-residue distances of gramicidin A in a model membrane system,[228] distance measurements at the ligand site of a bacterial chemotaxis membrane receptor,[239] protein–catecholamine bond formation in insect cuticle development from $^{13}C-^{15}N$ distance,[226] and the orientation of L–[5–^{13}C] glutamine in the binding site of uniform-^{15}N glutamine-binding *protein.*[237] In addition, a wide variety of pulse sequences that recouple the dipolar interactions (distance constraints) that are averaged out by magic angle sample spinning have been reported for protein structural studies of proteins.[240-243]

13.4.2 UNIAXIALLY ALIGNED SAMPLES NMR

Membrane proteins usually have either transmembrane or in-plane configuration, which are particularly well suited for oriented solid-state NMR approaches. Uniaxially aligned samples (UAS) of membrane proteins in bilayers are usually prepared mechanically or magnetically when exposed to a magnetic field.[244,245] The bilayers would then provide the environment that both immobilizes the protein and aligns it magnetically, thus fulfilling the two prerequisites of structure determination by solid-state NMR of aligned samples. A combination of multiple RF pulses applied as a line width-narrowing mechanism and orientation-dependent measurements provide the basis for protein structure determination. For example, of an ^{15}N-labeled peptide, ^{15}N resonance and ^{15}N-^{1}H dipolar coupling frequencies depend on orientation of the amide N–H bond relative to the external magnetic field.[15] Then, polarization-induced spin exchange at the magic angle (PISEMA) experiments[246] were conducted to determine successive peptide plane orientations of transmembrane helical peptides. Subsequently, PISA wheels were generated by plotting the resonance patterns for all residues in a uniformly ^{15}N-labeled peptide, which can be used to define the average α-helix tilt and even rotational orientation in aid of the 3D structure calculation of membrane protein segments.[247,248] Although the UAS NMR technique is expected to gain importance in membrane protein helical structure studies, the current UAS NMR investigations are limited to single or double transmembrane helices as reported for studying the structure of single α-helical transmembrane segment M2 of the nicotinic acetylcholine receptor [249] and structure determination of the membrane-bound fd coast protein in lipid bilayers[250] and receptor-ligand interactions in membrane bilayers.[244,251,252]

Other solid-state NMR approaches have also been used in structural studies of uniformly ^{15}N-labeled membrane proteins and their fragments, including 2D solid-state heteronuclear chemical shift correlation experiments (HETCOR),[253,254] and the 2D solid-state polarization inversion spin-exchange at the magic angle experiments (PISEMA).[255] These new methods have recently earned increased value for structural studies on membrane proteins. [218,256,257]

13.5 MEMBRANE PROTEIN NMR STRATEGIES AND DEVELOPMENTS

Membrane protein NMR approaches have been developed by Opella and other NMR biophysicists on the basis of their experiences in membrane protein NMR studies and the following strategies are recommended.[10–13,210] For NMR sample preparation, regardless of length or hydrophobicity of the membrane protein peptides, a quick HSQC spectra of uniformly ^{15}N-labeled samples should be evaluated in an initial screen of sample folding conditions in the commonly used NMR detergents and lipids, for example, SDS, DPC, DHPC, and LMPC. For uniaxially aligned NMR samples, it is necessary to simultaneously optimize the conditions for various gel samples, including the protein transfer, into gels. These spectra are highly sensitive monitors of the entire expression, purification, and sample preparation procedures. Any doubling or anomalous broadening of resonance values should be considered an indication of aggregated or improperly folded polypeptide.[258] In this case, no further structural studies should be conducted by solution NMR or even sample preparation in bilayer for solid-state NMR until these problems have been solved. This is a very important checkpoint to prevent further experimental failures or generation of false results. Problems associated with protein aggregation or misfolding can be resolved by the identification of optimum conditions and finding a good combination of detergents, detergent–protein concentrations, ionic strength (not for DPC), pH, and temperature.[10,50,114,225,259–262]

There have been a few high-resolution solution-state and solid-state NMR studies of membrane proteins solubilized in micelles, bicelles, or organic solvents as well as aligned in lipid bilayers for solid-state NMR as previously reviewed.[9–11] In particular, recent developments in solution NMR techniques made it possible to extend NMR studies to structures in the size range 50 to 150 kDa and beyond.[87,263] Using TROSY-based 3D ^{2}H/^{15}N/^{13}C NMR methods, several labs have recently completed the structures of 15- to 20-kDa monomeric β-barrel membrane proteins–micellar complexes,[166,264,265] including OmpX (148 residues) in DHPC micelles (~ 60 kDa),[12] OmpA (177 residues) in DPC micelles (~ 50 kDa),[166] and the outer membrane enzyme PagP (164 residues) in DPC and n-octyl-β-D-glucoside micelles, (50- to 60 kDa).[265] However, NMR structure determination of helical transmembrane proteins has so far yielded less complete results than for the β-barrel membrane proteins.

Examples of NMR studies on α-helical membrane proteins include the dimeric transmembrane domain of human glycoporin A (2 × 40 residues),[156] the light-harvesting β1-subunit of *Rhodobacter sphaeroides* (48 residues),[266] the coat proteins from the filamentous bacteriophages M13 (50 residues)[267] and fd (50 residues),[202] the bacteriorhodopsin fragment comprising residues 1–71,[268] the 81-residue human immunodeficiency virus (HIV) membrane-associated protein Vpu,[269] the 79-residue subunit c of the F_1,F_0–ATPase,[106,270] and the 80-residue bacterial membrane transport protein MerF.[271,272] The most recent NMR research developments are for the 40-kDa homotrimeric protein diacylglycerol kinase (DAGK)(121-residue subunit) in micellar complexes with an overall size larger than 100 kDa,[273] and the 13-kDa MISTIC proteins (110 residues) with ~ 25kDa of detergent-MISTIC micellar complexes.[274] These new findings may soon be envisaged and even be applicable to large and complex membrane proteins such as the G protein-coupled receptors (GPCRs), in the near future. We have illustrated the following examples to represent the recent NMR investigation and development of membrane proteins.

13.5.1 TRUNCATED MEMBRANE PROTEIN FRAGMENTS — CANNABINOID CB2 RECEPTOR

Cannabinoid (CB) receptors, the subtypes CB_1 from the brain[275,276] and CB_2 from the spleen,[277] are classified as members of the rhodopsin-like G protein-coupled transmembrane receptor (GPCR) family of proteins. The receptors are known to have many pharmacological effects.[278–282] The CB_2 receptor (39.7 kDa) is particularly interesting because it is expressed in high quantities in the human spleen and tonsils, is involved in signal transduction processes in the immune system, and can potentially be a drug target for immunotreatments.[277,283,284] Xie and co-workers have engineered CB_2 receptor transmembrane peptide fragments or helix bundles in ^{15}N–^{13}C isotope-enriched M9 media for NMR investigation.[31,44,285]

Despite the rapid development of NMR technology, direct NMR measurement of the seven transmembrane-spanning domain GPCRs still remains a challenge because the spectra are obscured by the broad line width due to the large size and slow molecular correlation time, not to mention the heavy overlapping signal generated by repetitive amino acids. Particularly, the micelle- or bicelle-membrane protein complexes generate a large molecular mass of detergent-bound protein samples that is beyond the NMR detection limits. Xie and co-workers have, instead of tackling the full-length receptor structure at once, systematically divided the target membrane protein into smaller segments or "blocks," determined their structures individually, and later reassembled them to build a model of the 3D structure of the full-length CB_2. 3D high-resolution triple-resonance NMR was applied to characterize recombinant single- or double-helix-bound CB_2 protein fragments in detergent micelles and organic solvent systems. Figure 13.9 illustrates 3D ^{15}N-resolved NOESY and TOCSY NMR experiments for studies of the isotope-labeled recombinant transmembrane protein CB_2 fragment in deuterated DPC detergent micelles solution. The strip plots

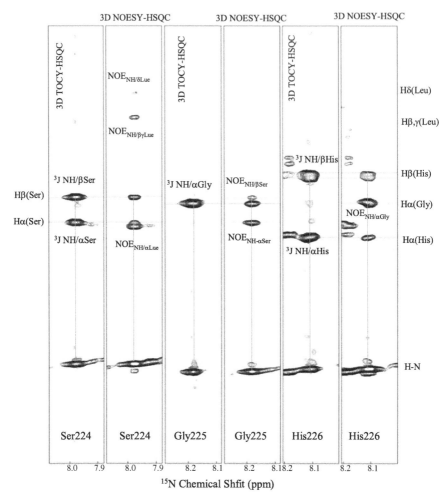

FIGURE 13.9 3D heteronuclear NMR data analysis of CB$_2$ membrane protein fragment CB$_2$180–233 in DPC micelles, 298 K. Strip plots of 3D NMR spectra illustrated the sequential assignments for the Ser-Gly-His section of the CB$_2$ fragment by using ^{15}N-resolved TOCSY–HSQC and NOESY–HSQC experiments.

of 3D ^{15}N-edited HSQC–TOCSY and HSQC–NOESY NMR spectra represent the characteristic sequential assignments of the CB$_2$180–233 protein fragment.[285]

Moreover, extensive high-resolution NMR studies were carried out to investigate the structural role of the cytoplasmic helical domain (helix 8) on the biological activation of the CB$_2$ receptor in comparison with that of CB$_1$ helix 8 fragment as well as their mutants in membrane-mimetic environments.[155] The NMR-based structural calculation first revealed that NMR structural comparisons of the CB$_1$I397–G418 (left) and CB$_2$I298–K319 (right) exhibit a short *helix a* and a long *helix b* with an "L-shape" turn at the Arg residue. As shown, the CB$_1$I397–G418 segment had an extended Arg401 side-chain conformation with a formed salt bridge between Asp404 and His407, whereas $CB_2I298–K319$ only showed a curved side chain of Arg^{302} forming a salt bridge with Glu305. (Figure 13.10, top). Such conformational properties and salt bridge dissimilarity might help in understanding the different structural roles of the fourth cytoplasmic helices in their biological function and regulation of CB_1 and CB_2 receptors.

The combination of NMR techniques with protein expression/purification and peptide chemical synthesis methods allows not only studies of the conformational and dynamic properties of a single helix but also determination of the interhelix distance and relative orientation of helix bundles. These transmembrane or helix bundle fragments might be regarded as natural assembling blocks to be used as a basis for transmembrane receptor spatial structure reconstruction. The "segments to 3D assembly" approach is well supported by existing NMR studies, including rhodopsin[6,158] and bacteriorhodopsin,[109,286–288] which shows that the NMR-determined structures and conformations are highly comparable to the existing x-ray data.[5,289]

13.5.2 TWO-TRANSMEMBRANE-HELICES PROTEIN — MERCURY TRANSPORT PROTEIN MERF

Opella and co-workers[272] have recently reported the 3D backbone structure of the MerF domain of the bacterial mercury detoxification system. MerF has 81 residues, two transmembrane helices, one interhelical loop, and the essential biological function of transporting Hg(II) across membranes. The micelle-bound MerF membrane protein structure was determined using solution NMR methods through the measurement of backbone ^1H–^{15}N RDCs from samples of two different constructs that align differently in stressed polyacrylamide gels. Dipolar wave fitting to the ^1H–^{15}N RDCs determines the helical boundaries based on periodicity and was utilized in the generation of supplemental dihedral restraints for the helical segments. The ^1H–^{15}N RDCs and supplemental dihedral restraints enabled the determination of the structure of the helix–loop–helix core domain of the mercury transport membrane protein MerF with a backbone RMSD of 0.58 Å (Figure 13.11). The structure of MerF in micelles determined by NMR spectroscopy showed that both pairs of cysteine residues are located on the cytoplasmic side of the membrane. The structural model determined is distinctly different from that derived principally from the hydrophobicity plot that located the first pair near the periplasm with only the second pair exposed to the cytoplasm.[272,290] Thus, the structure of the helix–loop–helix core domain of MerF (Figure 13.11) not only demonstrated that NMR can be used to determine the 3D structures of polytopic membrane proteins but also sets the stage for studying the structural biology of heavy metal transport across membranes. Some other double-helix transmembrane peptides have also been studied by high-resolution NMR, including glycophorin A,[156] subunit c of the F_1F_0—ATPase,[291] membrane-bound fd coat protein,[279] synthetic Peptide Fragments from Emr,[280] and the other proteins.[170,202,292]

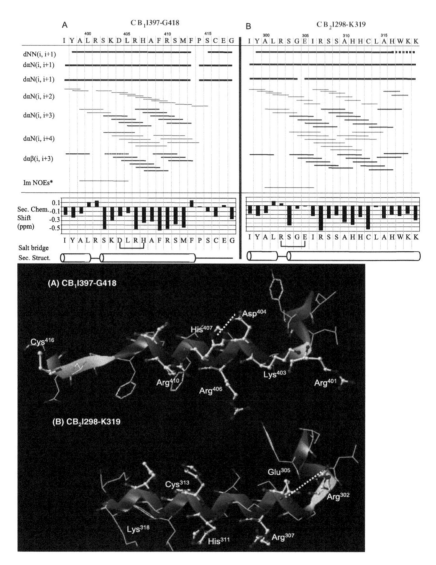

FIGURE 13.10 (See color insert following page 272.) (Top) Schematic summary of the NOE restraints important in the secondary structure determination of CB1I397–G418 (A) and CB2I298–K319 (B) in DPC micelles. *Dashed lines* indicate ambiguities due to the peak overlap. The cylinders showing two helix portions *a* and *b*. * indicate the important NOEs, which are observed between Ala^{399}(αH) and Leu^{405}(NH), Ala^{399}(βH) and Leu^{405}(NH), Ala^{399}(αH) and Leu^{405}(γH) in peptide CB$_1$I397–G418, and NOEs between Ala^{300}(NH) and Ile^{306}(NH), Ala^{300}(NH) and Ile^{306}(αH), and Ala^{300}(NH) and Ile^{306}(γH) in CB$_2$I298–K319, for determination of orientation between helices *a* and *b*. (**Bottom**) NMR structural comparison: (A) CB1I397–G418 shows an extended Arg401 side chain with a formed salt bridge between His407 and Asp404, whereas (B) CB2I298–K319 shows a curved side chain of Arg302 forming a salt bridge with Glu305. (From Xie, X.Q. and Chen, J.Z. *J. Biol. Chem.*, 280, 3605–3612, 2005. With permission.)

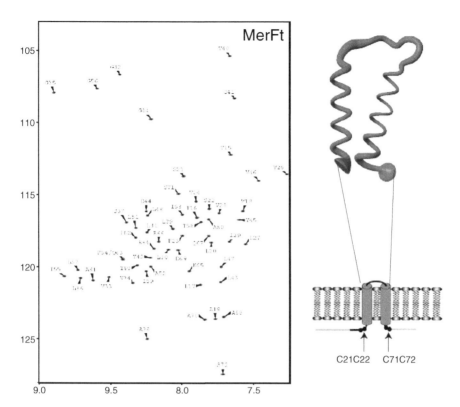

FIGURE 13.11 1H–15N HSQC assigned spectrum of 0.7 mm merft in 500 mM SDS micelles with 10 mM phosphate, pH 6.0, temperature 60°C (td1 = 1024, td2 = 256, d1 = 1.5s) (left). spline representation of an ensemble of structures calculated using the NMR data (residues 24–69), showing the structure (RMSD 0.58) of the helix–loop–helix core of MerF in membrane-mimetic environment. (From Howell, S.C., Mesleh, M.F., and Opella, S.J. *Biochemistry,* 44, 5196–5206, 2005. With permission.)

13.5.3 Three-Helix-Bundle Transmembrane Protein — DAGK

There are few three transmembrane helix bundle proteins that have been studied by NMR methods including diacylglycerol kinase (DAGK), lqE receptor, and synthetic homo-oligomeric peptide trimer.[261] DAGK is a 40 kDa homotrimeric membrane protein from *E. coli*, comprised of 121 residue subunits each having three transmembrane segments (Figure 13.12). Due to the wt-DAGK protein degradation over a period of days at 45°C, a superstable mutant (s-DAGK)[293] was applied to explore the DAGK structure. Sanders and co-workers report that the wt- and s-DAGK retain their homotrimeric oligomeric state in DPC and are catalytically functional showing that V_{max} is similar to that under ideal membrane conditions, but K_m is significantly elevated for both MgATP and diacylglycerol. Although differences in stability exist between wt- and s-DAGK, TROSY spectra were very similar, suggesting little difference in overall structure.

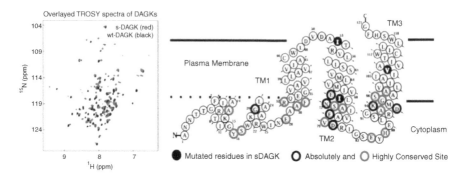

FIGURE 13.12 Secondary structure and membrane topology of DAGK were determined by 3D/4D heteronuclear NMR methods. (From Oxenoid, K., Kim, H.J., Jacob, J., Soennichsen, F.D., and Sanders, C.R.. *J. Am. Chem. Soc.,* 126(16), 5048–5049, 2004. With permission.)

In their studies, backbone NMR assignments were achieved for DAGK in detergent micelles by using TROSY-based pulse sequences including ^{15}N-edited TROSY-HNCA and TROSY-HNCACB experiments, which provided intra-residual and sequential cross-peaks of Cα and Cα/Cβ, respectively. Ambiguities were resolved by ^{15}N-edited TROSY–HN(CO)CA and TROSY–HN(CO)CACB, which provided only sequential cross-peaks.[273] In addition, 4D HNCACO and 4D HNCOCA experiments[180] were also used to confirm assignments. DAGK was found to be an almost exclusively helical protein. This study suggested that it is feasible both to solve the structure of DAGK using solution NMR methods and using NMR as a primary tool in structural studies of other helical integral membrane proteins of similar size and complexity.

13.5.4 FOUR-HELIX-BUNDLE TRANSMEMBRANE PROTEIN — MISTIC

Very few NMR-determined four-helix-bundle transmembrane (4TM) proteins have been reported so far. For instance, the influenza A M2 protein, di-heme four-helix-bundle in cytochrome bc1 and MISTIC protein, have been studied. MISTIC (membrane-integrating sequence for translation of integral membrane protein constructs) is an unusual *Bacillus subtilis* integral membrane protein (110 amino acids, 13 kDa) (Figure 13.13) that folds autonomously into the membrane, bypassing the cellular translocon machinery.[274] As reported by Choe and co-workers, the transmembrane MISTIC protein structure was determined with paramagnetic probes by using TROSY-based 3D NMR experiments, including TROSY–HNCA, TROSY–HNCACO, and TROSY–^{15}N-resolved NOESY (mixing time 200 ms) studies of a ^{2}H, ^{15}N, ^{13}C-labeled MISTIC protein sample as well as electron paramagnetic resonance (EPR) experiments using site-directed spin labels. It was found that MISTIC is a bundle of four α-helices with a polar lipid-facing surface. Additional experiments suggested that MISTIC can be used for high-level production of other membrane proteins in their native conformation, including many eukaryotic proteins that have previously been intractable to bacterial expression. MISTIC has been used to produce nearly 20 different functional eukaryotic membrane proteins in bacteria and demonstrated promising yields for GPCR membrane protein production and NMR structural studies will be carried out in the future.

ACKNOWLEDGMENTS

I would like to thank Dr. Kenneth Lundstrom for inviting me to write this chapter I would also like to acknowledge Jian Wang, Wanyun Sheng, and other members of my research group for putting together some of the figures in this chapter.

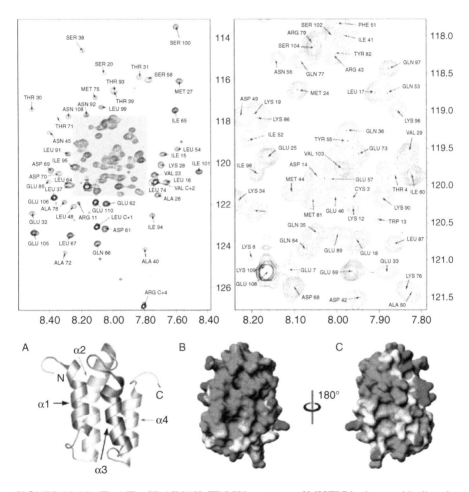

FIGURE 13.13 (**Top**) The 2D 15N/1H–TROSY spectrum of MISTIC is shown with all peaks identified by residue number. The right panel is an enlargement of the region indicated by the gray box in the left panel. (**Bottom**) Ribbon diagram of the lowest energy conformer highlighting the four α-helix bundles determined from NMR (A); surface electrostatic potential is mapped (B and C) (see the text for the details). (From Roosild, T.P., Greenwald, J., Vega, M., Castronovo, S., Riek, R., and Choe, S. *Science,* 307, 1317-1321, 2005. With permission.)

REFERENCES

1. Xie, X.Q., Chen, J.Z., and Billings, E.M. 3D structural model of the G-protein-coupled cannabinoid CB2 receptor. *Prot. Struct. Funct. Genet.,* 53, 307–319, 2003.
2. Ma, P. and Zemmel, R. Value of novelty? *Nat. Rev. Drug Discov.,* 1, 571–572, 2002.
3. Berman, H.M., Bhat, T.N., Bourne, P.E., Feng, Z., Gilliland, G., Weissig, H., and Westbrook, J. The Protein Data Bank and the challenge of structural genomics. *Nat. Struct. Biol.,* 7 Suppl, 957–959, 2000.
4. Tusnady, G.E., Dosztanyi, Z., and Simon, I. Transmembrane proteins in the Protein Data Bank: identification and classification. *Bioinformatics,* 20, 2964–2972, 2004.
5. Palczewski, K., Kumasaka, T., Hori, T., Behnke, C.A., Motoshima, H., Fox, B.A., Le Trong, I., Teller, D.C., Okada, T., Stenkamp, R.E., et al. Crystal structure of rhodopsin: A G protein-coupled receptor. *Science,* 289, 739–745, 2000.
6. Yeagle, P.L., Choi, G., and Albert, A.D. Studies on the structure of the G-protein-coupled receptor rhodopsin including the putative G-protein binding site in unactivated and activated forms. *Biochemistry,* 40, 11932–11937, 2001.
7. Gesell, J., Zasloff, M., and Opella, S.J. Two-dimensional 1H NMR experiments show that the 23-residue magainin antibiotic peptide is an a-helix in dodecylphosphocholine micelles, sodium dodecyl sulfate micelles, and trifluoroethanol/water solution. *J. Biomol. NMR.,* 9, 127–135, 1997.
8. Discotic phospholipid particles (bicelles) revolutionize structural analysis of macromolecules by NMR, http://www.avantilipids.com/BicellePreparation.asp. (accessed 2005)
9. Marcotte, I., Auger, M. Bicelles as model membranes for solid- and solution-state NMR studies of membrane peptides and proteins. *Concepts in Magnetic Resonance, Part A* 2005, 24A, 17-37, 2005.
10. Opella, S.J., Marassi, F.M. Structure determination of membrane proteins by NMR spectroscopy. *Chem. Rev.,* 104, 3587–3606, 2004.
11. Nielsen, N., Malmendal, A., and Vosegaard, T. Techniques and applications of NMR to membrane proteins (Review). *Mol. Membr. Biol.,* 21, 129–141, 2004.
12. Fernandez, C. and Wuthrich, K. NMR solution structure determination of membrane proteins reconstituted in detergent micelles. *FEBS Lett.,* 555, 144–150, 2003.
13. Baleja, J.D. Structure determination of membrane-associated proteins from nuclear magnetic resonance data. *Anal. Biochem.,* 288, 1–15, 2001.
14. Roosild, T.P., Greenwald, J., Vega, M., Castronovo, S., Riek, R., and Choe, S. NMR structure of mistic, a membrane-integrating protein for membrane protein Expression. *Science,* 307, 1317–1321, 2005.
15. Torres, J., Stevens, T.J., and Samso, M. Membrane proteins: the 'Wild West' of structural biology. Erratum to document cited in CA139:2419. *Trends Biochem. Sci.,* 28, 174, 2003.
16. Shire, D., Calandra, B., Rinaldi-Carmona, M., Oustric, D., Pessegue, B., Bonnin-Cabanne, O., Le Fur, G., Caput, D., and Ferrara, P. Molecular cloning, expression and function of the murine CB2 peripheral cannabinoid receptor. *Biochim. Biophys. Acta.,* 1307, 132–136, 1996.
17. Nowell, K.W., Pettit, D.A., Cabral, W.A., Zimmerman, H.W., Jr., Abood, M.E., and Cabral, G.A. High-level expression of the human CB2 cannabinoid receptor using a baculovirus system. *Biochem. Pharmacol.,* 55, 1893–1905, 1998.
18. Calandra, B., Tucker, J., Shire, D., and Grisshammer, R. Expression in *Escherichia coli* and characterization of the human central CB1 and peripheral CB2 cannabinoid receptors. *Biotechnol. Lett.,* 19, 425–428, 1997.

19. Lynch, K.R. *Identification and Expression of G Protein-Coupled Receptors*. John Wiley & Sns, New York, 1998.

20. Sarramegna, V., Talmont, F., Demange, P., and Milon, A. Heterologous expression of G-protein-coupled receptors: Comparison of expression systems from the standpoint of large-scale production and purification. *Cell. Mol. Life Sci.*, 60, 1529–1546, 2003.

21. Bouvier, M., Menard, L., Dennis, M., and Marullo, S. Expression and recovery of functional G-protein-coupled receptors using baculovirus expression systems. *Curr. Opin. Biotechnol.*, 9, 523–527, 1998.

22. Massotte, D. G protein-coupled receptor overexpression with the baculovirus-insect cell system: a tool for structural and functional studies. *Biochim. Biophys. Acta.*, 1610, 77–89, 2003.

23. Sander, P., Grunewald, S., Bach, M., Haase, W., Reilander, H., and Michel, H. Heterologous expression of the human D2S dopamine receptor in protease-deficient Saccharomyces cerevisiae strains. *Eur. J. Biochem.*, 226, 697–705, 1994.

24. King, K., Dohlman, H.G., Thorner, J., Caron, M.G., and Lefkowitz, R.J. Control of yeast mating signal transduction by a mammalian beta 2-adrenergic receptor and Gs alpha subunit. *Science*, 250, 121–123, 1990.

25. Feng, W., Cai, J., Pierce, W.M., and Song, Z-H. Expression of CB2 cannabinoid receptor in Pichia pastoris. *Prot. Exp. Purif.*, 26, 496–505, 2002.

26. Grisshammer, R. Expression of G protein-coupled receptors in *Escherichia coli*. *Identif. Exp. GPCR*, 133–150, 1998.

27. Whitelegge, J.P., Gundersen, C.B., and Faull, K.F. Electrospray-ionization mass spectrometry of intact intrinsic membrane proteins. *Prot. Sci.*, 7, 1423–1430, 1998.

28. Goff, S.A. and Goldberg, A.L. An increased content of protease La, the lon gene product, increases protein degradation and blocks growth in *Escherichia coli*. *J. Biol. Chem.*, 262, 4508–4515, 1987.

29. Strauch, K.L. and Beckwith, J. An *Escherichia coli* mutation preventing degradation of abnormal periplasmic proteins. *Proc. Natl. Acad. Sci. USA*, 85, 1576–1580, 1988.

30. Baneyx, F. and Georgiou, G. Construction and characterization of *Escherichia coli* strains deficient in multiple secreted proteases: protease III degrades high-molecular-weight substrates *in vivo*. *J. Bacteriol.*, 173, 2696–2703, 1991.

31. Zheng, H., Zhao, J., Wang, S., Lin, C.M., Chen, T., Jones, D.H., Ma, C., Opella, S., and Xie, X.Q. Biosynthesis and purification of a hydrophobic peptide from trans-membrane domains of G-protein-coupled CB2 receptor. *Peptide Res.*, 65, 450–458, 2005.

32. Morgan, W.D., Kragt, A., and Feeney, J. Expression of deuterium-isotope-labelled protein in the yeast pichia pastoris for NMR studies. *J. Biomol. NMR*, 17, 337–347, 2000.

33. Yokoyama, S. Protein expression systems for structural genomics and proteomics. *Curr. Opin. Chem. Biol.*, 7, 39–43, 2003.

34. Ong, S.E., Blagoev, B., Kratchmarova, I., Kristensen, D.B., Steen, H., Pandey, A., and Mann, M. Stable isotope labeling by amino acids in cell culture, SILAC, as a simple and accurate approach to expression proteomics. *Mol. Cell. Proteomics*, 1, 376–386, 2002.

35. Baneres, J-L., Martin, A., Hullot, P., Girard, J-P., Rossi, J-C., and Parello, J. Structure-based analysis of GPCR function: conformational adaptation of both agonist and receptor upon leukotriene B4 bBinding to recombinant BLT1. *J. Mol. Biol.*, 329, 801–814, 2003.

36. Freissmuth, M., Selzer, E., Marullo, S., Schutz, W., and Strosberg, A.D. Expression of two human beta-adrenergic receptors in *Escherichia coli*: functional interaction with two forms of the stimulatory G protein. *Proc. Natl. Acad. Sci. USA*, 88, 8548–8552, 1991.

37. Marullo, S., Emorine, L.J., Strosberg, A.D., and Delavier-Klutchko, C. Selective binding of ligands to beta 1, beta 2 or chimeric beta 1/beta 2-adrenergic receptors involves multiple subsites. *EMBO J.*, 9, 1471–1476, 1989.

38. Breyer, R.M., Strosberg, A.D., and Guillet, J.G. Mutational analysis of ligand binding activity of beta 2 adrenergic receptor expressed in *Escherichia coli*. *EMBO J.*, 9, 2679–2684, 1990.

39. Galli, G., Matteoni, R., Bianchi, E., Testa, L., Marazziti, D., Rossi, N., and Tocchini-Valentini, G. Replica filter assay of human beta-adrenergic receptors expressed in *E.coli*. *Biochem, Biophys. Res. Commn.*, 173, 680–688, 1990.

40. Bhave, G., Nadin, B.M., Brasier, D.J., Glauner, K.S., Shah, R.D., Heinemann, S.F., Karim, F., and Gereau, R.W.IV. Membrane topology of a metabotropic glutamate receptor. *J. Biol. Chem.*, 278, 30294–30301, 2003.

41. Baneres, J-L. and Parello, J. Structure-based analysis of GPCR function: evidence for a novel pentameric assembly between the dimeric leukotriene B4 receptor BLT1 and the G-protein. *J. Mol. Biol.*, 329, 815–829, 2003.

42. Lacatena, R.M., Cellini, A., Scavizzi, F., and Tocchini-Valentini, G.P. Topological analysis of the human beta 2-adrenergic receptor expressed in *Escherichia coli*. *Proc. Natl. Acad. Sci. USA*, 91, 10521–10525, 1994.

43. Ishihara, G., Goto, M., Saeki, M., Ito, K., Hori, T., Kigawa, T., Shirouzu, M., and Yokoyama, S. Expression of G protein coupled receptors in a cell-free translational system using detergents and thioredoxin-fusion vectors. *Prot. Exp. Purif.*, 41, 27–37, 2005.

44. Xie, X-Q., Zhao, J., and Zheng, H.A. Expression, purification, and isotope labeling of cannabinoid CB2 receptor fragment, CB2(180-233). *Prot. Exp. Purif.* 38, 61–68, 2004.

45. Tucker, J. and Grisshammer, R. Purification of a rat neurotensin receptor expressed in *Escherichia coli*. *Biochem. J.*, 317, 891–899, 1996.

46. Jockers, R. and Strosberg, A.D. Expression of b-adrenergic receptors in *Escherichia coli*. *Meth. Mol. Biol.*, 126 (Adrenergic Receptor Protocols), 215–220, 2000.

47. Luca, S., White, J.F., Sohal, A.K., Filippov, D.V., van Boom, J.H., Grisshammer, R., and Baldus, M. The conformation of neurotensin bound to its G protein-coupled receptor. *Proc. Natl. Acad. Sci. USA* 100, 10706–10711, 2003.

48. Nevzorov, A.A., Mesleh, M.F., and Opella, S.J. Structure determination of aligned samples of membrane proteins by NMR spectroscopy. *Magnet. Resonan. Chem.*, 42, 162–171, 2004.

49. Tian, C., Breyer, R.M., Kim, H.J., Karra, M.D., Friedman, D.B., Karpay, A., and Sanders, C.R. Solution NMR spectroscopy of the human vasopressin V2 receptor, a G protein-coupled receptor. *J. Am. Chem. Soc.*, 127, 8010–8011, 2005.

50. Ma, C., Marassi, F.M., Jones, D.H., Straus, S.K., Bour, S., Strebel, K., Schubert, U., Oblatt-Montal, M., Montal, M., and Opella, S.J. Expression, purification, and activities of full-length and truncated versions of the integral membrane protein Vpu from HIV-1. *Prot. Sci.*, 11, 546–557, 2002.

51. Yabuki, T., Kigawa, T., Dohmae, N., Takio, K., Terada, T., Ito, Y., Laue, E.D., Cooper, J.A., Kainosho, M., and Yokoyama, S. Dual amino acid-selective and site-directed stable-isotope labeling of the human c-Ha-Ras protein by cell-free synthesis. *J. Biomol. NMR*, 11, 295–306, 1998.

52. Bushweller, J.H., Holmgren, A., and Wuthrich, K. Biosynthetic 15N and 13C isotope labelling of glutathione in the mixed disulfide with *Escherichia coli* glutaredoxin documented by sequence-specific NMR assignments. *Eur. J. Biochem.*, 218, 327–334, 1993.

53. van Gammeren, A.J., Hulsbergen, F.B., Hollander, J.G., and de Groot, H.J.M. Biosynthetic site-specific 13C labeling of the light-harvesting 2 protein complex: a model for solid state NMR structure determination of transmembrane proteins. *J. Biomol. NMR*, 30, 267–274, 2004.

54. Wearne, S.J. Factor Xa cleavage of fusion proteins. Elimination of non-specific cleavage by reversible acylation. *FEBS Lett.*, 263, 23–26, 1990.

55. Owen, W.G., Esmon, C.T., and Jackson, C.M. The conversion of prothrombin to thrombin. I. Characterization of the reaction products formed during the activation of bovine prothrombin. *J. Biol. Chem.*, 249, 594–605, 1974.

56. He, M., Jin, L., and Austen, B. Specificity of factor Xa in the cleavage of fusion proteins. *J. Prot. Chem.*, 12, 1–5, 1993.

57. Nagai, K., Perutz, M.F., and Poyart, C. Oxygen binding properties of human mutant hemoglobins synthesized in *Escherichia coli*. *Proc. Natl. Acad. Sci. USA*, 82, 7252–7255, 1985.

58. Chow, B.K., Morrow, G.W., Ho, M., Pederson, R.A., McIntosh, C.H., Brown, J.C., and MacGillivray, R.T. Expression of recombinant human glucose-dependent insulinotropic polypeptide in *Escherichia coli* by sequence-specific proteolysis of a protein A fusion protein. *Peptides*, 11, 1069–1074, 1990.

59. Chang, J.Y. Thrombin specificity. Requirement for apolar amino acids adjacent to the thrombin cleavage site of polypeptide substrate. *Eur. J. Biochem.*, 151, 217–224, 1985.

60. Smith, D.B. and Johnson, K.S. Single-step purification of polypeptides expressed in *Escherichia coli* as fusions with glutathione S-transferase. *Gene*, 67, 31–40, 1988.

61. Sachdev, D. and Chirgwin, J.M. Fusions to maltose-binding protein: control of folding and solubility in protein purification. *Meth. Enzymol.*, 326, 312–321, 2000.

62. Guan, K.L. and Dixon, J.E. Eukaryotic proteins expressed in *Escherichia coli*: an improved thrombin cleavage and purification procedure of fusion proteins with glutathione S-transferase. *Anal. Biochem.*, 192, 262–267, 1991.

63. Lundblad, R.L. Chemical cleavage of peptide bBonds, in *Techniques in Protien Modification*. CRC Press, pp. 51–61, 1995.

64. Ma, C. and Opella, S.J. Lanthanide ions bind specifically to an added "EF-hand" and orient a membrane protein in micelles for solution NMR spectroscopy. *J. Magnet. Resonan.*, 146, 381–384, 2000.

65. Marassi, F.M., Ma, C., Gesell, J.J., and Opella, S.J. Three-dimensional solid-state NMR spectroscopy is essential for resolution of resonances from in-plane residues in uniformly 15N-labeled helical membrane proteins in oriented lipid bilayers. *J. Magnet. Resonan.*, 144, 156–161, 2000.

66. Rogl, H., Kosemund, K., Kuhlbrandt, W., and Collinson, I. Refolding of *Escherichia coli* produced membrane protein inclusion bodies immobilized by nickel chelating chromatography. *FEBS Lett.*, 432, 21–26, 1998.

67. Smith, M.C., Furman, T.C., Ingolia, T.D., and Pidgeon, C. Chelating peptide-immobilized metal ion affinity chromatography. A new concept in affinity chromatography for recombinant proteins. *J. Biol. Chem.*, 263, 7211–7215, 1988.

68. White, J.F., Trinh, L.B., Shiloach, J., and Grisshammer, R. Automated large-scale purification of a G protein-coupled receptor for neurotensin. *FEBS Lett.*, 564, 289–293, 2004.

69. Liu J. J., Hartman D.S., and Bostwick J.R. An immobilized metal ion affinity adsorption and scintillation proximity assay for receptor-stimulated phosphoinositide hydrolysis. *Anal. Biochem.,* 318, 91–99, 2003.

70. Grisshammer, R. and Tucker, J. Quantitative evaluation of neurotensin receptor purification by immobilized metal affinity chromatography. *Prot. Exp. Purif.,* 11, 53–60, 1997.

71. Shirabe, K., Yamada, M., Merrill, A.R., Cramer, W.A., and Nakazawa, A. Overproduction and purification of the colicin E1 immunity protein. *Plasmid,* 29, 236–240, 1993.

72. Perin, M.S., Fried, V.A., Slaughter, C.A., and Sudhof, T.C. The structure of cytochrome b561, a secretory vesicle-specific electron transport protein. *EMBO J.,* 7, 2697–2703, 1988.

73. Karger, B.L. and Giese, R.W. Reversed phase liquid chromatography and its application to biochemistry. *Anal. Chem.,* 50(12), 1048A–1052A, 1057A–1058A, 1062A, 1064A, 1066A, 1068A, 1070A, 1072A–1073A, 1978.

74. Nestor, J.J., Wilson, C.J., Cantley, L.C., Yaffe, M.B., and Guo, A. Methods for identifying ligands of G protein-coupled receptors (GPCRs) from translated cDNA libraries and therapeutic uses. *Appl. WO,* (Consensus Pharmaceuticals, Inc., USA), pp. 1–80, 2002.

75. Goetz, M., Rusconi, F., Belghazi, M., Schmitter, J.M., and Dufourc, E.J. Purification of the c-erbB2/neu membrane-spanning segment: a hydrophobic challenge. *J. Chromatogr., B: Biomed. Sci. Appl.,* 737, 55–61, 2000.

76. Turner, G.J., Miercke, L.J., Mitra, A.K., Stroud, R.M., Betlach, M.C., and Winter-Vann, A. Expression, purification, and structural characterization of the bacteriorhodopsin-aspartyl transcarbamylase fusion protein. *Prot. Exp. Purif.,* 17, 324–338, 1999.

77. Muccio, D.D. and DeLucas, L.J. Isolation of detergent-solubilized monomers of bacteriorhodopsin by size-exclusion high-performance liquid chromatography. *J. Chromatogr.,* 326, 243–250, 1985.

78. Hooks Shelley, B. and Harden, T.K. Purification and *in vitro* functional analysis of R7 subfamily RGS proteins in complex with Gbeta5. *Meth. Enzymol.,* 390, 163–177, 2004.

79. Burkey, T.H. and Regan, J.W. Activation of mitogen-activated protein kinase by the human prostaglandin EP3A receptor. *Biochem. Biophys. Res. Commn.,* 211, 152–158, 1995.

80. Mahn, A. and Asenjo, J.A. Prediction of protein retention in hydrophobic interaction chromatography. *Biotechnol. Adv.,* 23, 359–368, 2005.

81. Fexby, S. and Buelow, L. Hydrophobic peptide tags as tools in bioseparation. *Trends Biotechnol.,* 22, 511–516, 2004.

82. Jennissen, H.P. Hydrophobic (interaction) chromatography of proteins. *Biochromatography,* 46–71, 2002.

83. Wenk, S.O. and Kruip, J. Novel, rapid purification of the membrane protein photosystem I by high-performance liquid chromatography on porous materials. *J. Chromatogr., B: Biomed. Sci. Appl.,* 737, 131–142, 2000.

84. Liu, T. and Geng, X. New development of hydrophobic interaction chromatography and its applications in biochemical research. *SePu,* 16, 30–34, 1998.

85. Kiefer, H., Vogel, R., and Maier, K. Bacterial expression of G-protein-coupled receptors: prediction of expression levels from sequence. *Recept. Channels,* 7, 109–119, 2000.

86. Riek, R., Pervushin, K., and Wuthrich, K. TROSY and CRINEPT: NMR with large molecular and supramolecular structures in solution. *Trends Biochem. Sci.,* 25, 462–468, 2000.

87. Riek, R., Wider, G., Pervushin, K., and Wüthrich, K. Polarization transfer by cross-correlated relaxation in solution NMR with very large molecules. *Proc. Natl. Acad. Sci. USA,* 96, 4918–4923, 1999.

88. Pervushin, K., Riek, R., Wider, G., and Wuthrich, K. Attenuated T2 relaxation by mutual cancellation of dipole-dipole coupling and chemical shift anisotropy indicates an avenue to NMR structures of very large biological macromolecules in solution. *Proc. Natl. Acad. Sci. USA,* 94, 12366–12371, 1997.

89. Wuthrich, K. *NMR of Proteins and Nucleic Acids,* Wiley-Interscience, New York, 1986.

90. Fernandez, C., Hilty, C., Bonjour, S., Adeishvili, K., Pervushin, K., and Wuthrich, K. Solution NMR studies of the integral membrane proteins OmpX and OmpA from *Escherichia coli. FEBS Lett.,* 504, 173–178, 2001.

91. Vinogradova, O., Sonnichsen, F., and Sanders, C.R., II. On choosing a detergent for solution NMR studies of membrane proteins. *J. Biomol. NMR,* 11(4), 381–386, 1998.

92. Kiefer, H. *In vitro* folding of alpha-helical membrane proteins. *Biochim. Biophys. Acta.,* 1610, 57–62, 2003.

93. Buchanan, S.K. Beta-barrel proteins from bacterial outer membranes: structure, function and refolding. *Curr. Opin. Struct. Biol.,* 9(4), 455–461, 1999.

94. Booth, P.J., Curran, A.R. Membrane protein folding. *Curr. Opin. Struct. Biol.,* 9, 115–121, 1999.

95. Fiermonte, G., Dolce, V., and Palmieri, F. Expression in *Escherichia coli,* functional characterization, and tissue distribution of isoforms A and B of the phosphate carrier from bovine mitochondria. *J. Biol. Chem.,* 273, 22782–22787, 1998.

96. Kiefer, H., Krieger, J., Olszewski, J.D., von Heijne, G., Prestwich, G.D., and Breer, H. Expression of an olfactory receptor in *Escherichia coli*: purification, reconstitution, and ligand binding. *Biochemistry,* 35, 16077–16084, 1996.

97. Braiman, M.S., Stern, L.J., Chao, B.H., and Khorana, H.G. Structure-function studies on bacteriorhodopsin. IV. Purification and renaturation of bacterio-opsin polypeptide expressed in *Escherichia coli. J. Biol. Chem.,*– 262, 9271–9276, 1996.

98. Gorzelle, B.M., Nagy, J.K., Oxenoid, K., Lonzer, W.L., Cafiso, D.S., and Sanders, C.R. Reconstitutive refolding of diacylglycerol kinase, an integral membrane protein. *Biochemistry,* 38, 16373–16382, 1999.

99. Hovius, R., Tairi, A.P., Blasey, H., Bernard, A., Lundstrom, K., and Vogel, H. Characterization of a mouse serotonin 5-HT3 receptor purified from mammalian cells. *J. Neurochem.,* 70, 824–834, 1998.

100. Devesa, F., Chams, V., Dinadayala, P., Stella, A., Ragas, A., Auboiroux, H., Stegmann, T., and Poquet, Y. Functional reconstitution of the HIV receptors CCR5 and CD4 in liposomes. *Eur. J. Biochem.,* 269, 5163–5174, 2002.

101. Manley, D. and O'Neil, J.D. Preparation of glycerol facilitator for protein structure and folding studies in solution. *Meth. Mol. Biol.,* 228, 89–101, 2003.

102. Zheng, L., Kostrewa, D., Berneche, S., Winkler, F.K., and Li, X.D The mechanism of ammonia transport based on the crystal structure of AmtB of *Escherichia coli. Proc. Natl. Acad. Sc.i USA,* 101, 17090–17095, 2004.

103. Sanders, C.R. and Oxenoid, K. Customizing model membranes and samples for NMR spectroscopic studies of complex membrane proteins. *Biochim. Biophys. Acta., (BBA)— Biomembr.,* 1508, 129–145, 2000.

104. Sobol, A.G., Arseniev, A.S., Abdulaeva, G.V., Musina, L.Y., and Bystrov, V.F. Sequence-specific resonance assignment and secondary structure of (1-71) bacterioopsin. *J. Biomol. NMR,* 2, 161–171, 1992.

105. Lomize, A.L., Pervushin, K.V., and Arsen'ev, S. Spatial structure of (34-65)bacterioopsin polypeptide in SDS micelles determined from nuclear magnetic resonance data. *J. Biomol. NMR,* 2, 361–372, 1992.

106. Girvin, M.E., Rastogi, V.K., Abildgaard, F., Markley, J.L., and Fillingame, R.H. Solution structure of the transmembrane H+-transporting subunit c of the F1F0 ATP synthase. *Biochemistry,* 37, 8817–8824, 1998.

107. Arseniev, A.S., Kuryatov, A.B., Tsetlin, V.I., Bystrov, V.F., Ivanov, V.T., and Ovchinnikov, Y.A. Fluorine-19 NMR study of 5-fluorotryptophan-labeled bacteriorhodopsin. *FEBS Lett.,* 213, 283–288, 1987.

108. Torres, J. and Padros, E. The secondary structure of bacteriorhodopsin in organic solution: a Fourier transform infrared study. *FEBS Lett.,* 318, 77–79, 1993.

109. Maslennikov, I.V., Arseniev, A.S., Chikin, L.D., Kozhich, A.T., Bystrov, V.F., and Ivanov, V.T. Conformation of bacteriorhodopsin transmembrane segment D. *Biol. Membr.,* 8, 156–160, 1991.

110. Maslennikov, I.V., Lomize, A.L., and Arsen'ev, A.S. Structure refinement of (34-65) bacterioopsin from NMR data in solution. *Bioorg. Khim.,* 17, 1456–1469, 1991.

111. Barsukov, I.L., Abdulaeva, G.V., Arsen'ev, A.S., and Bystrov, V.F. Sequence-specific proton-NMR assignment and conformation of proteolytic fragment 163-231 of bacterioopsin. *Eur. J. Biochem.,* 192, 321–327, 1990.

112. Kleinschmidt, J.H., Wiener, M.C., and Tamm, L.K. Outer membrane protein A of *E. coli* folds into detergent micelles, but not in the presence of monomeric detergent. *Prot. Sci.,* 8, 2065–2071, 1999.

113. Tanford, C. and Reynolds, J.A. Characterization of membrane proteins in detergent solutions. *Biochim. Biophys. Acta,* 457, 133–170, 1976.

114. Vinogradova, O., Snnichsen, F., and Sanders, C.R., II: On choosing a detergent for solution NMR studies of membrane proteins. *J. Biomol. NMR,* 11, 381–386, 1998.

115. le Maire, M., Champeil, P., and Moller, J.V. Interaction of membrane proteins and lipids with solubilizing detergents. *Biochim. Biophys. Acta. (BBA) — Biomembr.,* 1508, 86–111, 2000.

116. Moller, J.V. and le Maire, M. Detergent binding as a measure of hydrophobic surface area of integral membrane proteins. *J. Biol. Chem.* 268:18659-18672, 1993.

117. Veeraraghavan, S., Baleja, J.D., and Gilbert, G.E. Structure and topography of the membrane-binding C2 domain of factor VIII in the presence of dodecylphosphocholine micelles. *Biochem. J.,* 332, 549–555, 1998.

118. Barsukov, I.L., Nolde, D.E., Lomize, A.L., and Arseniev, A.S. Three-dimensional structure of proteolytic fragment 163-231 of bacterioopsin determined from nuclear magnetic resonance data in solution. *Eur. J. Biochem.,* 206, 665–672, 1992.

119. Pashkov, V.S., Balashova, T.A., Zhemaeva, L.V., Sikilinda, N.N., Kutuzov, M.A., Abdulaev, N.G., and Arseniev, A.S. Conformation of surface exposed N-terminus part of bacteriorhodopsin studied by transferred NOE technique. *FEBS Lett.,* 381, 119–122, 1996.

120. Henderson, R., Baldwin, J.M., Ceska, T.A., Zemlin, F., Beckmann, E., and Downing, K.H. Model for the structure of bacteriorhodopsin based on high-resolution electron cryo-microscopy. *J. Mol. Biol.,* 213, 899–929, 1990.

121. Glover, K.J., Whiles, J.A., Wu, G., Yu, N., Deems, R., Struppe, J.O., Stark, R.E., Komives, E.A., and Vold, R.R. Structural evaluation of phospholipid bicelles for solution-state studies of membrane-associated biomolecules. *Biophys. J.*, 81, 2163–2171, 2001.

122. Sanders, C.R., 2nd. and Landis, G.C. Reconstitution of membrane proteins into lipid-rich bilayered mixed micelles for NMR studies. *Biochemistry*, 34, 4030–4040, 1995.

123. Whiles, J.A., Brasseur, R., Glover, K.J., Melacini, G., Komives, E.A., and Vold, R.R. Orientation and effects of mastoparan X on phospholipid bicelles. *Biophys. J.*, 80, 280–293, 2001.

124. Whiles, J.A., Deems, R., Vold, R.R., and Dennis, E.A. Bicelles in structure-function studies of membrane-associated proteins. *Bioorgic Chem.*, 30, 431–442, 2002.

125. Glover, K.J., Whiles, J.A., Vold, R.R., and Melacini, G. Position of residues in transmembrane peptides with respect to the lipid bilayer: a combined lipid Noes and water chemical exchange approach in phospholipid bicelles. *J. Biomol. NMR*, 22, 57–64, 2002.

126. Glover, K.J., Whiles, J.A., Wood, M.J., Melacini, G., Komives, E.A., and Vold, R.R. Conformational dimorphism and transmembrane orientation of prion protein residues 110-136 in bicelles. *Biochemistry*, 40, 13137–13142, 2001.

127. Vold, R.R., Prosser, R.S., and Deese, A.J. Isotropic solutions of phospholipid bicelles: a new membrane mimetic for high-resolution NMR studies of polypeptides. *J. Biomol. NMR*, 9, 329–335, 1997.

128. Howard, K.P. and Opella, S.J. High-resolution solid-state NMR spectra of integral membrane proteins reconstituted into magnetically oriented phospholipid bilayers. *J. Magnet. Resonan. B.*, 112, 91–94, 1996.

129. Delaglio, F., Kontaxis, G., and Bax, A. Protein structure determination using molecular fragment replacement and NMR dipolar couplings. *J. Am. Chem. Soc.*, 122, 2142–2143, 2000.

130. Tjandra, N. and Bax, A. Direct measurement of distances and angles in biomolecules by NMR in a dilute liquid crystalline medium. *Science*, 278, 1111–1114, 1997.

131. Watts, A., Straus, S.K., Grage, S.L., Kamihira, M., Lam, Y.H., and Zhao, X. Membrane protein structure determination using solid-state NMR. *Meth. Mol. Biol.*, 278, 403–473, 2004.

132. Aisenbrey, C. and Bechinger, B. Investigations of polypeptide rotational diffusion in aligned membranes by 2H and 15N solid-state NMR spectroscopy. *J. Am. Che*, 16676–16683, 2004.

133. Dave, P.C., Tiburu, E.K., Damodaran, K., and Lorigan, G.A. Investigating structural changes in the lipid bilayer upon insertion of the transmembrane domain of the membrane-bound protein phospholamban utilizing 31P and 2H solid-state NMR spectroscopy. *Biophys. J.*, 86, 1564–1573, 2004.

134. Akutsu, H. and Fujiwara, T. Structural analysis of membrane proteins by solid-state NMR. *Tanpakushitsu Kakusan Koso*, 47, 1144–1151, 2002.

135. Smith, S.O. and Peersen, O.B. Solid-state NMR approaches for studying membrane protein structure. *Annu. Rev. Biophys. Biomol. Struct.*, 21, 25–47, 1992.

136. Wieruszeski, J.M., Bohin, A., Bohin, J.P., and Lippens, G. *In vivo* detection of the cyclic osmoregulated periplasmic glucan of Ralstonia solanacearum by high-resolution magic angle spinning NMR. *J. Magnet. Resonan.*, 151, 118–123, 2001.

137. Serber, Z. and Dotsch, V. In-cell NMR spectroscopy. *Biochemistry*, 40(48), 14317–14323, 2001.

138. Moreau, C., Le Floch, M., Segalen, J., Leray, G., Metzinger, L., de Certaines, J.D., and Le Rumeur, E. Static and magic angle spinning (31)P NMR spectroscopy of two natural plasma membranes. *FEBS Lett.*, 461, 258–262, 1999.

139. Wuthrich, K. *NMR of Proteins and Nucleic Acids*. John Wiley & Sons, New York, 1986.

140. Havel, T.F. An evaluation of computational strategies for use in the determination of protein structure from distance constraints obtained by nuclear magnetic resonance. *Prog. Biophys. Mol. Biol.*, 56, 43–78, 1991.

141. Clore, G.M. and Gronenborn, A.M. Determination of three-dimensional structures of proteins and nucleic acids in solution by nuclear magnetic resonance spectroscopy. *Crit.Rev. Biochem. Mol.r Biol.*, 24, 479–564, 1989.

142. Goger, M.J., McDonnell, J.M., and Cowburn, D. Using cryoprobes to decrease acquisition times of triple-resonance experiments used for protein resonance assignments. *Spectroscopy (Amsterdam, Netherlands)*, 17, 161–167, 2003.

143. Peters, A.R., Dekker, N., Van den Berg, L., Boelens, R., Kaptein, R., Slotboom, A.J., and De Haas, G.H. Conformational changes in phospholipase A2 upon binding to micellar interfaces in the absence and presence of competitive inhibitors. A proton and nitrogen-15 NMR study. *Biochemistry*, 31, 10024–10030, 1992.

144. Chapman, B.E., James, G.E., Moore, and W.J. Conformations of P2 protein of peripheral nerve myelin by nuclear magnetic resonance spectroscopy. *J. Neurochem.*, 36, 2032–2036, 1981.

145. Menda, G.L. and Moore, W.J. NMR studies of myelin basic protein. X. Conformation of a determinant encephalitogenic in the rabbit. *Biochim. Biophys. Acta.*, 748, 176–183, 1983.

146. Mammi, S. and Peggion, E. Conformational studies of human 15-2-aminohexanoic acid little gastrin in sodium dodecyl sulfate micelles by proton NMR. *Biochemistry*, 29, 5265–5269, 1990.

147. Piotto, M., Saudek, V., and Sklenar, V. Gradient-tailored excitation for single-quantum NMR spectroscopy of aqueous solutions. *J. Biomol. NMR*, 2, 661–665, 1992.

148. Dekker, N., Peters, A.R., Slotboom, A.J., Boelens, R., Kaptein, R., Dijkman, R., and De Haas, G. Two-dimensional proton-NMR studies of phospholipase-A2-inhibitor complexes bound to a micellar lipid-water inferface. *Eur. J. Biochem.*, 199, 601–607, 1991.

149. Griesinger, C., Soerensen, O.W., and Ernst, R.R. Three-dimensional Fourier spectroscopy: application to high-resolution NMR. *J. Magnet. Resonan.*, 84, 14–63, 1989.

150. Bax, A. and Grzesiek, S. Methodological advances in protein NMR. *Acc. Chem. Res.*, 26, 131–138, 1993.

151. Markley, J.L. Two-dimensional nuclear magnetic resonance spectroscopy of proteins: an overview. *Meth. Enzymol.*, 176, 12-64, 1989.

152. Redfield, C. Using nuclear magnetic resonance spectroscopy to study molten globule states of protiens. *Methods* 34, 121–132, 2004.

153. Kay, L.E. and Gardner, K.H. Solution NMR spectroscopy beyond 25 kDa. *Curr. Opin. Struct. Biol.*, 7, 722–731, 1997.

154. Tamm, L. Structure and function of membrane proteins by solution NMR in detergent micelles, in *Proceedings of the 36th Central Regional Meeting of the American Chemical Society*, American Chemical Society, Indianapolis, IN, 2004.

155. Xie, X.Q. and Chen, J.Z. NMR Structural comparison of the cytoplasmic juxtamembrane domains of G-protein coupled CB1 and CB2 receptors in membrane-mimetic DPC micelles. *J. Biol. Chem.*, 280, 3605–3612, 2005.

156. MacKenzie, K.R., Prestegard, J.H., and Engelman, D.M. A transmembrane helix dimer: structure and implications. *Science*, 276, 131–133, 1997.
157. Yeagle, P.L. and Albert, A.D. Use of nuclear magnetic resonance to study the three-dimensional structure of rhodopsin. *Meth. Enzymol.*, 343, 223–231, 2002.
158. Berlose, J.P., Convert, O., Brunissen, A., Chassaing, G., and Lavielle, S. Three dimensional structure of the highly conserved seventh trans membrane domain of G-protein coupled receptors. *Eur. J. Biochem.*, 225, 827–843, 1994.
159. Cavanagh, J., Fairbrother, W.J., Palmer, A.G., III, and skelton, N.J. *Protein NMR Spectroscopy: Principles and Practice*. Academic Press, San Diego, 1996.
160. Kochendoerfer, G.G., Jones, D.H., Lee, S., Oblatt-Montal, M., Opella, S.J., and Montal, M. Functional characterization and NMR spectroscopy on full-length Vpu from HIV-1 prepared by total chemical synthesis. *J. Am. Chem. Soc.*, 126, 2439–2446, 2004.
161. Hilty, C., Wider, G., Fernandez, C., and Wuthrich, K. Stereospecific assignments of the isopropyl methyl groups of the membrane protein OmpX in DHPC micelles. *J. Biomol. NMR*, 27, 377–382, 2003.
162. Grzesiek, S., Anglister, J., and Bax, A. Correlation of backbone amide and aliphatic side-chain resonances in $^{13}C/^{15}N$-enriched protein by isotropic mixing of ^{13}C magnetization. *J. Magnet. Resonan.*, B101, 114–119, 1993.
163. Muhandiran, D.R. and Kay, L. E. Gradient-enhanced triple-resonance three-dimensional NMR experiments with improved sensitivity. *J. Magnet. Resonan. B.*, 103, 203–216, 1994.
164. Ikura, M., Kay, L.E., and Bax, A. Improved three-dimensional ^1H-^{13}C-^1H correlation spectroscopy of a 13C-labeled protein using constant evolution. *J. Biomol. NMR*, 1, 299–304, 1991.
165. Kay, L.E., Xu, G.Y., Singer, A.U., Muhandiram, D.R., and Forman-Kay, J.D. A gradient-enhanced HCCH-TOCSY experiment for record side-chain ^1H and ^{13}C correlation in H_2O samples of proteins. *J. Magnet. Resonan. B.*, 101, 333–337, 1993.
166. Arora, A., Abildgaard, F., Bushweller, J.H., and Tamm, L.K. Structure of outer membrane protein A transmembrane domain by NMR spectroscopy. *Nat. Struct. Biol.*, 8, 334–338, 2001.
167. Fernandez, C., Adeishvili, K., and Wuthrich, K. Transverse relaxation-optimized NMR spectroscopy with the outer membrane protein OmpX in dihexanoyl phosphatidylcholine micelles. *Proc. Natl. Acad. Sci. USA*, 98, 2358–2363, 2001.
168. Fernandez, C., Hilty, C., Bonjour, S., Adeishvili, K., Pervushin, K., and Wuthrich, K. Solution NMR studies of the integral membrane proteins OmpX and OmpA from *Escherichia coli*. *FEBS Lett.*, 504, 173–178, 2001.
169. Venters, R.A., Thompson, R., and Cavanagh, J. Current approaches for the study of large proteins by NMR. *J. Mol. Struct.*, 602–603, 275–292, 2002.
170. Venters, R.A., Metzler, W.J., Spicer, L.D., Mueller, L., and Farmer, B.T., II. Use of 1HN-1HN NOEs to Determine Protein Global Folds in Perdeuterated Proteins. *J. Am. Chem. Soc.*, 117, 9592–9593, 1995.
171. Nietlispach, D., Clowes, R.T., Broadhurst, R.W., Ito, Y., Keeler, J., Kelly, M., Ashurst, J., Oschkinat, H., Domaille, P.J., and Laue, E.D. An approach to the structure determination of larger proteins using triple resonance NMR experiments in conjunction with random fractional deuteration. *J. Am. Chem. Soc.*, 118, 407–415, 1996.
172. Kushlan, D.M. and LeMaster, D.M. Resolution and sensitivity enhancement of heteronuclear correlation for methylene resonances via 2H enrichment and decoupling. *J. Biomol. NMR*, 3, 701–708, 1993.

173. Yamazaki, T., Lee, W., Arrowsmith, C.H., Muhandiram, D.R., and Kay, L.E. A suite of triple resonance NMR experiments for the backbone assignment of ^{15}N, ^{13}C, ^{2}H labeled proteins with high sensitivity. *J. Am. Chem. Soc.,* 116, 11655–11666, 1994.

174. Nietlispach, D. Suppression of anti-TROSY lines in a sensitivity enhanced gradient selection TROSY scheme. *J. Biomol. NMR,* 31, 161–166, 2005.

175. Nietlispach, D., Mott, H.R., Stott, K.M., Nielsen, P.R., Thiru, A., and Laue, E.D. Structure determination of protein complexes by NMR. *Meth. Mol. Biol. (Totowa, NJ, United States)* 278, 255–288, 2004.

176. Fernandez, C., Adeishvili, K., and Wuthrich, K. Transverse relaxation-optimized NMR spectroscopy with the outer membrane protein OmpX in dihexanoyl phosphatidylcholine micelles. *Proc. Natl. Acad. Sci. USA,* 98, 2358–2363, 2001.

177. Gardner, K.H., Rosen, M.K., and Kay, L.E. Global folds of highly deuterated, methyl-protonated proteins by multidimensional NMR. *Biochemistry,* 36, 1389–1401, 1997.

178. Battiste, J.L. and Wagner, G. Utilization of site-directed spin labeling and high-resolution heteronuclear nuclear magnetic resonance for global fold determination of large proteins with limited nuclear overhauser effect data. *Biochemistry,* 39, 5355–5365, 2000.

179. Salzmann, M., Pervushin, K., Wider, G., Senn, H., and Wüthrich, K. TROSY in triple-resonance experiments: New perspectives for sequential NMR assignment of large proteins. *Proc. Natl. Acad. Sci. USA,* 95, 13585–13590, 1998.

180. Yang, D. and Kay, L.E. TROSY triple-resonance four-dimensional NMR spectroscopy of a 46 ns tumbling protein. *J. Am. Chem. Soc.,* 121, 2571–2575, 1999.

181. Fiaux, J., Bertelsen, E.B., Horwich, A.L., and Wuthrich, K. NMR analysis of a 900K GroEL GroES complex. *Nature,* 418, 207–211, 2002.

182. Wishart, D.S., Sykes, B.D., and Richards, F.M. The chemical shift index: a fast and simple method for the assignment of protein secondary structure through NMR spectroscopy. *Biochemistry,* 31, 1647–1651, 1992.

183. Copie, V., Battles, J.A., Schwab, J.M., and Torchia, D.A. Secondary structure of b-hydroxydecanoyl thiol ester dehydrase, a 39-kDa protein, derived from Ha, Ca, Cb and CO signal assignments and the chemical shift index: comparison with the crystal structure. *J. Biomol. NMR,* 7, 335–340, 1996.

184. Morris Kevin, F., Gao, X., and Wong Tuck, C. The interactions of the HIV gp41 fusion peptides with zwitterionic membrane mimics determined by NMR spectroscopy. *Biochim. Biophys. Acta.,* 1667, 67–81, 2004.

185. Yu, J., Simplaceanu, V., Tjandra, N.L., Cottam, P.F., Lukin, J.A., and Ho, C. 1H, 13C, and 15N NMR backbone assignments and chemical-shift-derived secondary structure of glutamine-binding protein of *Escherichia coli. J. Biomol. NMR,* 9, 167–180, 1997.

186. Sklenar, V., Piotto, M., Leppik, R., and Saudek, V. Gradient-tailored water suppression for proton-nitrogen-15 HSQC experiments optimized to retain full sensitivity. *J. Magnet. Resonan. Ser A.,* 102, 241–245, 1993.

187. Xia, Y., Yee, A., Arrowsmith, C.H., and Gao, X. 1H(C) and 1H(N) total NOE correlations in a single 3D NMR experiment. ^{15}N and ^{13}C time-sharing in t1 and t2 dimensions for simultaneous data acquisition. *J. Biomol. NMR,* 27, 193–203, 2003.

188. Kuszewski, J., Nilgess, M., and Brunger, A.T. Sampling and effiency of metrix distance geometry: a novel partial metrization algorithm. *J. Biomol. NMR,* 2, 33–36, 1992.

189. Guntert, P. Automated NMR structure calculation with CYANA. *Meth. Mol. Biol.,* 278, 353–378, 2004.

190. Grzesiek, S., Vuister, G.W., and Bax, A. A simple and sensitive experiment for measurement of JCC couplings between backbone carbonyl and methyl carbons in isotopically enriched proteins. *J. Biomol. NMR*, 3, 487–493, 1993.

191 Nilges, M., Clore, G.M., and Gronenborn, A.M. Determination of three-dimensional structures of proteins from interproton distance data by dynamical simulated annealing from a random array of atoms. Circumventing problems associated with folding. *FEBS Lett.*, 239, 129-136, 1988.

192. Xie, X-Q., Melvin, L.S., and Makriyannis, A. The conformational properties of the highly selective cannabinoid receptor ligand CP-55,940. *J. Biol. Chem.*, 271(181), 10640–10647, 1996.

193. Karplus, S. and Karplus, M. Nuclear magnetic resonance determination of the angle psi in peptides. *Proc. Natl. Acad. Sci. USA*, 69, 3204–3206, 1972.

194. Rance, M., Soerensen, O.W., Bodenhausen, G., Wagner, G., Ernst, R.R., and Wuethrich, K. Improved spectral resolution in COSY proton NMR spectra of proteins via double quantum filtering. *Biochem. Biophys. Res. Commn.*, 117, 479–485, 1983.

195. Kay, L.E. and Bax, A. New methods for the measurement of NH-CaH coupling constants in nitrogen-15-labeled proteins. *J. Magnet. Resonan.*, 86, 110–126, 1990.

196. Vuister, G.W. and Bax, A. Quantitative J correlation: a new approach for measuring homonuclear three-bond J(HNHa) coupling constants in ^{15}N-enriched proteins. *J. Am. Chem. Soc.*, 115, 7772–7777, 1993.

197. Barnham, K.J., Monks, S.A., Hinds, M.G., Azad, A.A., and Norton, R.S. Solution structure of a polypeptide from the N terminus of the HIV protein Nef. *Biochemistry*, 36, 5970–5980, 1997.

198. Wishart, D.S., Bigam, C.G., Yao, J., Abildgaard, F., Dyson, H.J., Oldfield, E., Markley, J.L., and Sykes, B.D. ^{1}H, ^{13}C and ^{15}N chemical shift referencing in biomolecular NMR. *J. Biomol. NMR*, 6, 135–140, 1995.

199. Wishart, D.S. and Sykes, B.D. Chemical shifts as a tool for structure determination. *Meth. Enzymol.*, 239, 363–392, 1994.

200. Hyberts, S.G., Goldberg, M.S., Havel, T.F., and Wagner, G. The solution structure of eglin c based on measurements of many NOEs and coupling constants and its comparison with X-ray structures. *Prot. Sci.*, 1, 736–751, 1992.

201. Clore, G.M., Bax, A., Ikura, M., and Gronenborn, A.M. Structure of calmodulin-target peptide complexes. *Curr. Opin. Struct. Biol.*, 3, 838–845, 1993.

202. Almeida, F.C. and Opella, S.J. fd coat protein structure in membrane environments: structural dynamics of the loop between the hydrophobic trans-membrane helix and the amphipathic in-plane helix. *J. Mol. Biol.*, 270, 481–495, 1997.

203. Hyberts, S.G., Marki, W., and Wagner, G. Stereospecific assignments of side-chain protons and characterization of torsion angles in Eglin c. *Eur. J. Biochem. /FEBS*, 164, 625–635, 1987.

204. Wang, Y. and Jardetzky, O. Predicting 15N chemical shifts in proteins using the preceding residue-specific individual shielding surfaces from j, yi-1, and c1 torsion angles. *J. Biomol. NMR*, 28, 327–340, 2004.

205. Hansen, A.P., Petros, A.M., Meadows, R.P., Nettesheim, D.G., Mazar, A.P., Olejnic-zak, E.T., Xu, R.X., Pederson, T.M., Henkin, J., and Fesik, S.W. Solution structure of the amino-terminal fragment of urokinase-type plasminogen activator. *Biochemistry*, 33, 4847–4864, 1994.

206. Fersht, A. *Structure and Mechanism in Protein Science: A Guide to Enzyme Catalysis and Protein Folding.* W. H. Freeman, New York, 1999.

207. Pervushin, K.V., Arsen'ev, A.S., Kozhich, A.T., and Ivanov, V.T. Two-dimensional NMR study of the conformation of (34-65)bacterioopsin polypeptide in SDS micelles. *J. Biomol. NMR,* 1, 313–322, 1991.

208. Ottiger, M., Delaglio, F., and Bax, A. Measurement of J and dipolar couplings from simplified two-dimensional NMR spectra. *J. Magnet. Resonan.,* 131, 373–378, 1998.

209. Mesleh, M.F., Lee, S., Veglia, G., Thiriot, D.S., Marassi, F.M., and Opella, S.J. Dipolar waves map the structure and topology of helices in membrane proteins. *J. Am. Chem. Soc.,* 125, 8928–8935, 2003.

210. Lee, S., Mesleh, M.F., and Opella, S.J. Structure and dynamics of a membrane protein in micelles from three solution NMR experiments. *J. Biomol. NMR,* 26, 327–334, 2003.

211. Peti, W., Meiler, J., Brueschweiler, R., and Griesinger, C. Model-free analysis of protein backbone motion from residual dipolar couplings. *J. Am. Chem. Soc.,* 124, 5822–5833, 2002.

212. Fernandes, M.X., Bernado, P., Pons, M., and Garcia de la Torre, J. An analytical solution to the problem of the orientation of rigid particles by planar obstacles. Application to membrane systems and to the calculation of dipolar couplings in protein NMR spectroscopy. *J. Am. Chem. Soc.,* 123, 12037–12047, 2001.

213. Tjandra, N., Grzesiek, S., and Bax, A. Magnetic field dependence of nitrogen-proton J splittings in 15N-enriched human ubiquitin resulting from relaxation interference and residual dipolar coupling. *J. Am. Chem. Soc.,* 118, 6264–6272, 1996.

214. Pintacuda, G., Keniry, M.A., Huber, T., Park, A.Y., Dixon, N.E., and Otting, G. Fast structure-based assignment of ^{15}N HSQC spectra of selectively ^{15}N-labeled paramagnetic proteins. *J. Am. Chem. Soc.,* 126, 2963–2970, 2004.

215. Gaponenko, V., Sarma, S.P., Altieri, A.S., Horita, D.A., Li, J., and Byrd, R.A. Improving the accuracy of NMR structures of large proteins using pseudocontact shifts as long-range restraints. *J. Biomol. NMR,* 28, 205–212, 2004.

216. Assfalg, M., Bertini, I., Turano, P., Mauk, A.G., Winkler, J.R., and Gray, H.B. [15]N-^{1}H residual dipolar coupling analysis of native and alkaline-K79A *Saccharomyces cerevisiae* cytochrome c. *Biophys. J.,* 84, 3917–3923, 2003.

217. Barbieri, R., Bertini, I., Cavallaro, G., Lee, Y-M., Luchinat, C., and Rosato, A. Paramagnetically induced residual dipolar couplings for solution structure determination of lanthanide binding proteins. *J. Am. Chem. Soc.,* 124, 5581–5587, 2002.

218. Opella, S.J. and Marassi, F.M. Structure determination of membrane proteins by NMR spectroscopy. *Chem. Rev.,* 104, 3587–3606, 2004.

219. de Alba, E. and Tjandra, N. NMR dipolar couplings for the structure determination of biopolymers in solution. *Prog. Nucl. Magnet. Resonan. Spectrosc.,* 40, 175–197, 2002.

220. Yan, J. and Zartler, E.R. Application of residual dipolar couplings in organic compounds. *Magnet. Resonan. Chem.,* 43, 53–64, 2005.

221. Lee, S., Mesleh, M.F., and Opella, S.J. Structure and dynamics of a membrane protein in micelles from three solution NMR experiments. . *Biomol. NMR,* 26, 327–334, 2003.

222. Mesleh, M.F., Veglia, G., DeSilva, T.M., Marassi, F.M., and Opella, S.J. Dipolar waves as NMR maps of protein structure. *J. Am. Chem. Soc.,* 124, 4206–4207, 2002.

223. Marassi, F.M. and Opella, S.J. A solid-state NMR index of helical membrane protein structure and topology. *J. Magnet. Resonan.,* 144, 150–155, 2000.

224. Bechinger, B. and Sizun, C. Alignment and structural analysis of membrane polypeptides by 15N and 31P solid-state NMR spectroscopy. *Concepts Magnet. Resonan., Part A,* 18A, 130–145, 2003.

225. Opella, S.J. Membrane protein NMR studies. *Meth. Mol. Biol.,* 227(Membrane Transporters), 307–320, 2003.
226. Marassi, F.M. and Opella, S.J. NMR structural studies of membrane proteins. *Curr. Opin. Struct. Biol.,* 8, 640–648, 1998.
227. Gullion, T. and Schaefer, J. Rotational-echo double-resonance NMR. *J. Magnet. Resonan.,* 81, 196–200, 1989.
228. Hing, A.W. and Schaefer, J. Two-dimensional rotational-echo double resonance of Val1-1-^{13}C]Gly2-^{15}N]Ala3-gramicidin A in multilamellar dimyristoylphosphatidylcholine dispersions. *Biochemistry,* 32, 7593–7604, 1993.
229. Murphy, O.J., III, Kovacs, F.A., Sicard, E.L., and Thompson, L.K. Site-directed solid-state NMR measurement of a ligand-induced conformational change in the serine bacterial chemoreceptor. *Biochemistry,* 40, 1358–1366, 2001.
230. Saito, H., Tuzi, S., Yamaguchi, S., Kimura, S., Tanio, M., Kamihira, M., Nishimura, K., and Naito, A. Conformation and dynamics of membrane proteins and biologically active peptides as studied by high-resolution solid-state ^{13}C NMR. *J. Mol. Struct.,* 44, 137–148, 1998.
231. Garbow, J.R. and McWherter, C.A. Determination of the molecular conformation of melanostatin using carbon-13, nitrogen-15 REDOR NMR spectroscopy. *J. Am. Chem. Soc.,* 115, 238–244, 1993.
232. Verdegem, P.J., Bovee-Geurts, P.H., de Grip, W.J., Lugtenburg, J., and de Groot, H.J. Retinylidene ligand structure in bovine rhodopsin, metarhodopsin-I, and 10-methyl-rhodopsin from internuclear distance measurements using 13C-labeling and 1-D rotational resonance. *Biochemistry,* 38, 11316–11324, 1999.
233. Saito, H., Tuzi, S., Yamaguchi, S., Kimura, S., Tanio, M., Kamihira, M., Nishimura, K., and Naito, A. Conformation and dynamics of membrane proteins and biologically active peptides as studied by high-resolution solid-state 13C NMR. *J. Mol. Struct.,* 441, 137–148, 1998.
234. Nishimura, K., Kim, S., Zhang, L., and Cross, T.A. The closed state of a H+ channel helical bundle combining precise orientational and distance restraints from solid state NMR. *Biochemistry,* 41 13170–13177, 2002.
235. Wang, J., Balazs, Y.S., and Thompson, L.K. Solid-state REDOR NMR distance measurements at the ligand site of a bacterial chemotaxis membrane receptor. *Biochemistry,* 36, 1699–1703, 1997.
236. Smith, S.O. Magic angle spinning NMR as a tool for structural studies of membrane proteins. *Magnet. Resonan. Rev.,* 17, 1–26, 1996.
237. Hing, A.W., Tjandra, N., Cottam, P.F., Schaefer, J., and Ho, C. An investigation of the ligand-binding site of the glutamine-binding protein of *Escherichia coli* using rotational-echo double-resonance NMR. *Biochemistry,* 33, 8651–8661, 1994.
238. Gullion, T. and Schaefer, J. Detection of weak heteronuclear dipolar coupling by rotational-echo double-resaonance nuclear magnetic resonance. *J. Adv. Magnet. Resonan.,* 13, 57–83, 1989.
239. Wang, J., Balazs, Y.S., and Thompson, L.K. Solid-state REDOR NMR distance measurements at the ligand site of a bacterial chemotaxis membrane receptor. *Biochemistry,* 36, 1699–1703, 1997.
240. Schnell, I. Dipolar recoupling in fast-MAS solid-state NMR spectroscopy. *Prog. Nucl. Magnet. Resonan. Spectrosc.,* 45, 145–207, 2004,
241. Schmidt-Rohr, K. and Hong, M. Measurements of carbon to amide-proton distances by C-H dipolar recoupling with 15N NMR detection. *J. Am. Chem. Soc.,* 125, 5648–5649, 2003.

242. Brinkmann, A., Schmedt auf der Guenne, J., and Levitt, M.H. Homonuclear zero-quantum recoupling in fast magic-angle spinning nuclear magnetic resonance. *J. Magnet. Resonan.*, 156, 79–96, 2002.

243. Griffin, R.G. Dipolar recoupling in MAS spectra of biological solids. *Nat. Struct. Biol.*, 5 Suppl, 508–512, 1998.

244. Hallock, K.J., Wildman, K.H., Lee, D-K., and Ramamoorthy, A. An innovative procedure using a sublimable solid to align lipid bilayers for solid-state NMR studies. *Biophys. J.*, 82, 2499–2503, 2002.

245. Cardon, T.B., Tiburu, E.K., and Lorigan, G.A. Magnetically aligned phospholipid bilayers in weak magnetic fields: optimization, mechanism, and advantages for X-band EPR studies. *J. Magnet. Resonan.*, 161, 77–90, 2003.

246. Wu, C.H., Ramamoorthy, A., and Opella, S.J. High-resolution heteronuclear dipolar solid-state NMR spectroscopy. *J. Magnet. Resonan.*, 109, 270–272, 1994.

247. Marassi, F.M. and Opella, S.J. A solid-state NMR index of helical membrane protein structure and topology. *J. Magnet. Resonan.*, 144, 150–155, 2000.

248. Wang, J., Denny, J., Tian, C., Kim, S., Mo, Y., Kovacs, F., Song, Z., Nishimura, K., Gan, Z., Fu, R. et al. Imaging membrane protein helical wheels. *J. Magnet. Resonan.*, 144, 162–167, 2000.

249. Opella, S.J., Marassi, F.M., Gesell, J.J., Valente, A.P., Kim, Y., Oblatt-Montal, M., and Montal, M. Structures of the M2 channel-lining segments from nicotinic acetylcholine and NMDA receptors by NMR spectroscopy. *Nat. Struct. Biol.*, 6, 374–379, 1999.

250. Zeri, A.C., Mesleh, M.F., Nevzorov, A.A., and Opella, S.J. Structure of the coat protein in fd filamentous bacteriophage particles determined by solid-state NMR spectroscopy. *Proc. Natl. Acad. Sci. USA* 100, 6458–6463,, 2003.

251. Nevzorov, A.A., Moltke, S., Heyn, M.P., and Brown, M.F. Solid-state NMR line shapes of uniaxially oriented immobile systems. *J. Am. Chem. Soc.*, 121, 7636–7643, 1999.

252. Lambotte, S., Jasperse, P., and Bechinger, B. Orientational distribution of alpha-helices in the colicin B and E1 channel domains: a one and two dimensional 15N solid-state NMR investigation in uniaxially aligned phospholipid bilayers. *Biochemistry*, 37, 16–22, 1998.

253. Murata, K., Kono, H., Katoh, E., Kuroki, S., and Ando, I. A study of conformational stability of polypeptide blends by solid state two-dimensional ^{13}C-^{1}H heteronuclear correlation NMR spectroscopy. *Polymer*, 44, 4021–4027, 2003.

254. Kamihira, M., Vosegaard, T., Mason, A.J., Straus, S.K., Nielsen, N.C., and Watts, A. Structural and orientational constraints of bacteriorhodopsin in purple membranes determined by oriented-sample solid-state NMR spectroscopy. *J. Struct. Biol.*, 149, 7–16, 2005.

255. Ramamoorthy, A., Wei, Y., and Lee, D-K. PISEMA solid-state NMR spectroscopy. *Annu. Rep. NMR Spectrosc.*, 52, 2–52, 2004.

256. Opella, S.J. Multiple-resonance multi-dimensional solid-state NMR of proteins. *Encyc. NMR*, 9, 427–436, 2002.

257. Bjerring, M., Vosegaard, T., Malmendal, A., and Nielsen, N.C. Methodological development of solid-state NMR for characterization of membrane proteins. *Concepts Magnet. Resonan., Part A*, 18A, 111–129, 2003.

258. McDonnell, P.A. and Opella, S.J. Effect of detergent concentration on multidimensional solution NMR spectra of membrane proteins in micelles. *J. Magnet. Resonan.*, 102, 120–125, 1993.

259. Krueger-Koplin, R.D., Sorgen, P.L., Krueger-Koplin, S.T., Rivera-Torres, I.O., Cahill, S.M., Hicks, D.B., Grinius, L., Krulwich, T.A., and Girvin, M.E. An evaluation of detergents for NMR structural studies of membrane proteins. *J. Biomol. NMR,* 28, 43–57, 2004.

260. Opella, S.J., Ma, C., and Marassi, F.M. Nuclear magnetic resonance of membrane-associated peptides and proteins. *Meth. Enzymol.,* 339, 285–313, 2001.

261. Oxenoid, K., Sonnichsen, F.D., and Sanders, C.R. Topology and secondary structure of the N-terminal domain of diacylglycerol kinase. *Biochemistry,* 41, 12876–12882, 2002.

262. Klein-Seetharaman, J., Reeves, P.J., Loewen, M.C., Getmanova, E.V., Chung, J., Schwalbe, H., Wright, P.E., and Khorana, H.G. Solution NMR spectroscopy of alpha -^{15}N]lysine-labeled rhodopsin: the single peak observed in both conventional and TROSY-type HSQC spectra is ascribed to Lys-339 in the carboxyl-terminal peptide sequence. *Proc. Natl. Acad. Sci. USA,* 99, 3452–3457, 2002.

263. Pervushin, K., Riek, R., Wider, G., and Wuthrich, K. Attenuated T2 relaxation by mutual cancellation of dipole-dipole coupling and chemical shift anisotropy indicates an avenue to NMR structures of very large biological macromolecules in solution. *Proc. Natl. Acad. Sci. USA,* 94, 12366–12371, 1997.

264. Fernandez, C. and Wider, G. TROSY in NMR studies of the structure and function of large biological macromolecules. *Curr. Opin. Struct. Biol.,* 13, 570–580, 2003.

265. Huang, S.M., Bisogno, T., Trevisani, M., Al-Hayani, A., De Petrocellis, L., Fezza, F., Tognetto, M., Petros, T.J., Krey, J.F., Chu, C.J., et al. An endogenous capsaicin-like substance with high potency at recombinant and native vanilloid VR1 receptors. *Proc. Natl. Acad. Sci. USA,* 99, 8400–8405, 2002.

266. Sorgen, P.L., Cahill, S.M., Krueger-Koplin, R.D., Krueger-Koplin, S.T., Schenck, C.C., and Girvin, M.E. Structure of the *Rhodobacter sphaeroides* light-harvesting 1 beta subunit in detergent micelles. *Biochemistry,* 41, 31–41, 2002.

267. Papavoine, C.H., Christiaans, B.E., Folmer, R.H., Konings, R.N., and Hilbers, C.W. Solution structure of the M13 major coat protein in detergent micelles: a basis for a model of phage assembly involving specific residues. *J. Mol. Biol.,* 282, 401–419, 1998.

268. Pervushin, K.V., Orekhov, V., Popov, A.I., Musina, L., and Arseniev, A.S. Three-dimensional structure of (1-71)bacterioopsin solubilized in methanol/chloroform and SDS micelles determined by ^{15}N-^{1}H heteronuclear NMR spectroscopy. *Eur. J. Biochem.,* 219, 571–583, 1994.

269. Ma, C. and Opella, S.J. Lanthanide ions bind specifically to an added "EF-hand" and orient a membrane protein in micelles for solution NMR spectroscopy. *J. Magnet. Resonan.,* 146, 381–384, 2000.

270. Rastogi, V.K. and Girvin, M.E. ^{1}H, ^{13}C, and ^{15}N assignments and secondary structure of the high pH form of subunit c of the F1F0 ATP synthase. *J. Biomol. NMR,* 13, 91–92, 1999.

271. Veglia, G., Porcelli, F., DeSilva, T., Prantner, A., and Opella, S.J. The structure of the metal-binding motif GMTCAAC is similar in an 18-residue linear peptide and the mercury binding protein MerP. *J. Am. Chem. Soc.,* 122, 2389–2390, 2000.

272. Howell, S.C., Mesleh, M.F., and Opella, S.J. NMR structure determination of a membrane protein with two transmembrane helices in micelles: MerF of the bacterial mercury detoxification system. *Biochemistry,* 44, 5196–5206, 2005.

273. Oxenoid, K., Kim, H.J., Jacob, J., Soennichsen, F.D., and Sanders, C.R. NMR Assignments for a helical 40 kDa membrane protein. *J. Am. Chem. Soc.,* 126, 5048–5049, 2004.

274. Roosild, T.P., Greenwald, J., Vega, M., Castronovo, S., Riek, R., and Choe, S. NMR structure of mistic, a membrane-integrating protein for membrane protein expression. *Science,* 307, 1317–1321, 2005.

275. Devane, W.A., Dysarz, F.A., III, Johnson, M.R., Melvin, L.S.,and Howlett, A.C. Determination and characterization of a cannabinoid receptor in rat brain. *Mol. Pharmacol.,* 34, 605–613, 1988.

276. Devane, W.A., Hanus, L., Breuer, A., Pertwee, R.G., Stevenson, L.A., Griffin, G., Gibson, D., Mandelbaum, A., Etinger, A., and Mechoulam, R. Isolation and structure of a brain constituent that binds to the cannabinoid receptor. *Science,* 258, 1946–1949, 1992.

277. Kaminski, N.E., Abood, M.E., Kessler, F.K., Martin, B.R., and Schatz, A.R. Identification of a functionally relevant cannabinoid receptor on mouse spleen cells that is involved in cannabinoid-mediated immune modulation. *Mol. Pharmacol.,* 42, 736–742, 1992.

278. Dewey, W.L. Cannabinoid pharmacology. *Pharmacol. Rev.,* 38, 151–178, 1986.

279. Mechoulam, R. The pharmacohistory of cannabis sativa, in *Cannabinoids as Therapeutics Agents,* Mechoulam, R., ed. CRC Press, Boca Raton, FL, 1986, pp. 1–20.

280. Howlett, A.C., Bidaut-Russell, M., Devane, W.A., Melvin, L.S., Johnson, M.R., and Herkenham, M. The cannabinoid receptor: biochemical, anatomical and behavioral characterization. *Trends Neurosci.,* 13, 420–423, 1990.

281. Matsuda, L.A. Molecular aspects of cannabinoid receptors. *Crit. Rev. Neurobiol.,* 11, 143–166, 1997.

282. Reggio, P.H. Cannabinoid receptors: the relationship between structure and function. *Biol. Marijuana,* 449–490, 2002.

283. Massi, P., Vaccani, A., and Parolaro, D. Cannabinoids and immune system. *Recent Adv. Pharmacol. Physiol. Cannabinoids,* 119–137, 2004.

284. Howlett, A.C., Breivogel, C.S., Childers, S.R., Deadwyler, S.A., Hampson, R.E., and Porrino, L.J. Cannabinoid physiology and pharmacology: 30 years of progress. *Neuropharmacology,* 47(Suppl. 1), 345–358, 2004.

285. Zhao, J., Zheng, H.A., and Xie, X-Q. 3D Heteronuclear NMR characterization of isotope-enriched recombinant CB2 membrane protein segment CB2(180-233). *Proteins,* 2005 (submitted).

286. Oldfield, E., Kinsey, R.A., and Kintanar, A. Recent advances in the study of bacteriorhodopsin dynamic structure using high-field solid-state nuclear magnetic resonance spectroscopy. *Meth. Enzymol.,* 88, 310–325, 1982,.

287. Huang, K.S., Bayley, H., Liao, M.J., London, E., and Khorana, H.G. Refolding of an integral membrane protein. Denaturation, renaturation, and reconstitution of intact bacteriorhodopsin and two proteolytic fragments. *J. Biol. Chem.,* 256(8), 3802–3809, 1981.

288. Katragadda, M., Alderfer, J.L., and Yeagle, P.L. Assembly of a polytopic membrane protein structure from the solution structures of overlapping peptide fragments of bacteriorhodopsin. *Biophys. J.,* 81, 1029–1036, 2001.

289. Kahn, T.W. and Engelman, D.M. Bacteriorhodopsin can be refolded from two independently stable transmembrane helices and the complementary five-helix fragment. *Biochemistry,* 31, 6144–6151, 1992.

290. Wilson, J.R., Leang, C., Morby, A.P., Hobman, J.L., and Brown, N.L. MerF is a mercury transport protein: different structures but a common mechanism for mercuric ion transporters? *FEBS Lett.,* 472, 78–82, 2000.

291. Rastogi, V.K. and Girvin, M.E. Structural changes linked to proton translocation by subunit c of the ATP synthase. *Nature,* 402, 263–268, 1999.

292. Kim, S., Cullis, D.N., Feig, L.A., and Baleja, J.D. Solution structure of the Reps1 EH domain and characterization of its binding to NPF target sequences. *Biochemistry,* 40(23), 6776–6785, 2001.
293. Zhou, F.X., Cocco, M.J., Russ, W.P., Brunger, A.T., and Engelman, D.M. Interhelical hydrogen bonding drives strong interactions in membrane proteins. *Nat. Struct. Biol.,* 7(2), 154–160, 2000.

14 Miniaturization of Structural Biology Technologies — From Expression to Biophysical Analyses

Enrique Abola, Peter Kuhn, and Raymond C. Stevens

CONTENTS

14.1 INTRODUCTION

High-throughput (HT) structural biology is making significant contributions to our attempts at developing a complete understanding of biological systems. Process pipelines designed using HT approaches have now been implemented to explore protein fold/function space (Protein Structure Initiative),[1,2] to accelerate the Structure Based Drug Discovery (SBDD) process,[3,4] to study complete proteomes,[5] and to develop and disseminate methods and technologies for working with integral membrane proteins.[6] Although HT structure determination was initially designed and used by structural genomics efforts, HT tools and protocols are increasingly being used successfully by individual investigator laboratories. This has been made possible by the development of the various gene-to-structure technologies initiated in the late 1990s and by the commercialization and competition of the technologies between 2000 and 2005 that have lowered the price of these instruments significantly.

A central theme in the structural biology technology development area has been automation, integration, and miniaturization of processes in the pipeline. These goals have reduced the cost per structure by decreasing time from gene to structure, material usage, and number of personnel needed to accomplish large numbers of tasks.[4] These factors also shorten feedback loops between processes (e.g., between crystallization and x-ray screening), leading to an almost twofold decrease in time to arrive at a structure. For example, miniaturization of the crystallization process, enabled by the development of the nanovolume crystallization robots,[7] as well as the use of microcoil nuclear magnetic resonance (NMR) probes for routine measurement of NMR spectra,[8] has led to marked reduction in the amount of proteins needed for these studies and has led to significantly improved success rates. Importantly, developments such as nanovolume crystallization have provided new tools and data, greatly facilitating attempts to arrive at a deeper and more thorough understanding of the protein crystallization process. At a minimum, miniaturization efforts and the use of robotics are allowing for establishment of robust best-practice approaches. Although further developments are obviously needed to improve the current gene-to-structure success rates, in which 5 to 10% of protein targets entering the pipeline are structurally characterized, and to address the more challenging problems of working with eukaryotic proteins, membrane proteins, and large protein complexes, the work done in the last five years has introduced a fundamental change in structural biology.

In this chapter, we summarize some of the recent efforts in the miniaturization of processes used in the structural determination pipeline. These processes are currently directed mainly toward soluble proteins but are now being adapted for use with membrane proteins, which are greatly in need of breakthrough technologies to enable membrane protein structure and function studies. Most of these tools are useful not just for structural studies but are also proving to be essential for functional studies and, more recently, are being used to characterize the complete proteome of organisms. Finally, we review several five-year highlights of data obtained through the use of the automation, integration, and miniaturization structural biology processes.

14.2 NEW TOOLS FOR HIGH-THROUGHPUT STRUCTURE DETERMINATION PIPELINES

Notwithstanding the successes of HT genomic and chemical screening efforts, the requirements for structural studies remain particularly demanding in terms of material, personnel, and success rates. Costs for reagents and laboratory disposables continue to limit the number of targets that can be reasonably studied. Recent successful studies in which crystallization success rates[9] have been improved dramatically through the use of various mutational constructs add to the possible number of experiments that could be done. As structural studies generally necessitate a few milligrams of highly purified and homogeneous samples of folded proteins, development of new approaches and tools have been required for sample preparation, characterization, and quality control.

14.2.1 MICROEXPRESSION PROTOCOLS

The large-scale *Escherichia coli* or eukaryotic expression and purification of recombinant proteins needed for structural studies is time-consuming and expensive, especially when costly reagents such as selenomethionine (SeMet)[10] or $^{15}N/^{13}C$-labeled NMR medium are required. The Joint Center for Structural Genomics (JCSG)[2] HT pipeline for processing protein targets for crystallization and NMR studies in *E. coli* uses a 96-well format macroexpression device (GNF-fermentor; Figure 14.1A) that enables cell cultures in enriched media to reach high fermentation absorbance values of 20 to 40 at 600 nm.[11] This device has already resulted in a more than 10-fold reduction in the culture volume required for protein production when compared with conventional expression in shaker flasks. Although in some favorable cases, a single fermentor tube with 65 mL of culture medium yielded sufficient protein with natural isotope distribution for 1D ^1H NMR and crystallization trials, the yield of 4 to 16 fermentor tubes typically needs to be pooled to obtain sufficient SeMet protein for thorough crystallization trials or stable isotope-labeled protein for NMR analysis. Recently, small-scale screens have been developed to identify targets that express at acceptable levels for structural studies and are soluble prior to large-scale expression.[12–15] Although these screens were designed for a parallel expression (e.g., 96-well small-scale format), none adequately predicts the reliable expression behavior with scaled-up milliliter and liter fermentations, which has been a challenge for researchers in the field.

Recently, we adapted a low-cost, high-velocity incubating Glas-Col (Glas-Col, LLC, Terre Haute, IN, USA) Vertiga shaker (Figure 14.1B) to develop an efficient, HT *E. coli* microliter-scale expression screening protocol, which accurately predicts protein behavior expressed in large-scale (milliliter and liter) fermentation conditions.[16] The apparatus shakes cultures in three dimensions at speeds of up to 1000 rpm, allowing small-scale (~ 750 µL) cultures grown in 2-mL, deep-well, 96-well blocks to achieve optical densities (OD_{600}) as high as 10 to 20. This generates sufficient material for analysis of expression, solubility, binding to affinity purification matrices, and initial crystallization or NMR analysis. Moreover, this screening strategy has also been used to identify clones that express and are soluble under SeMet[10] or $^{15}N/^{13}C$-labeled expression conditions that are necessary for the production of labeled recombinant proteins for direct structural analysis.

As proof of principle to larger-scale fermentation behavior, 34 proteins were expressed using the microliter-scale Vertiga expression protocol under native conditions and expressed again using the GNF fermentor; this experiment was carried out in duplicate.[16] Targets expressed in the GNF fermentor were purified as outlined by

A

B

FIGURE 14.1 Genomics Institute of the Novartis Research Foundation (GNF) milliliter expression systems and Glas-Col Vertiga microliter expression system. (A) GNF 96-well milliliter fermenter. A typical run includes the growth of approximately 65 mL of culture, grown to an absorbance (A600) of 20 to 40 using media conditions similar to the microliter-scale expression experiments, with oxygen supplementation. (B) Glas-Col Vertiga microliter-scale expression system. A low-cost vertical shaker apparatus adapted for the development of a scalable high-throughput *Escherichia coli* micro-expression screening. (From Lesley, S.A., Kuhn, P., Godzik, A., Deacon, A.M., Mathews, I., Kreusch, A., Spraggon, G., Klock, H.E., McMullan, D., Shin, T., Vincent, J., Robb, A., Brinen, L.S., Miller, M.D., McPhillips, T.M., Miller, M.A., Scheibe, D., Canaves, J.M., Guda, C., Jaroszewski, L., Selby, T.L., Elsliger, M.A., Wooley, J., Taylor, S.S., Hodgson, K.O., Wilson, I.A., Schultz, P.G., and Stevens, R.C. *Proc. Natl. Acad. Sci. USA*, 99, 11664–11669, 2002; and Page, R., Moy, K., Sims, E.C., Velasquez, J., McManus, B., Grittini, C., Clayton, T.L., and Stevens, R.C. *BioTechniques*, 37, 364–370, 2004. With permission.)

Lesley and co-workers at the JCSG[11] with purity and yield determined by SDS-PAGE (sodium dodecyl sulfate–polyacrylamide gel electrophoresis). Most samples were further analyzed using Western blotting with an anti-His antibody (Sigma-Aldrich, St. Louis, MO, USA) using standard protocols. Out of the 34 proteins tested, 33 consistently expressed (or did not express) in both microliter and larger-scale volumes, illustrating a high level of scalability and correlation in the expression levels of soluble protein between the Vertiga screening trials and large-scale growth conditions. In contrast, without the Vertiga shaker, the correlation between microliter and milliliter (and liter) scale expression is much lower (data not shown). The average protein mass expressed in minimal media and purified per milliliter of culture was also similar between micro- and macroexpressed protein (~ 10 µg/mL). In addition to the prokaryotic proteins that have been studied using the Vertiga instrument, eukaryotic proteins of human, murine, yeast origin, and viral proteins from the SARS (severe acute respiratory syndrome) coronavirus have also been successfully microexpressed.

14.2.2 MICROCOIL NMR PROBE

NMR spectroscopy is a well-established technique for protein structure determination, as well as for screening for the folded state of globular proteins.[17-19] Since NMR spectroscopy has intrinsically low sensitivity, milligram amounts of protein are required for NMR screening and structure determination with conventional equipment. Recently, however, microcoil NMR probes (Figure 14.2) have been developed for use in biomolecular NMR spectroscopy that mitigates this problem.[8,20] Specifically, small-diameter coils enable up to 10-fold (mass-based) sensitivity gain[21] so that microgram amounts of protein are now sufficient for screening by NMR spectroscopy. NMR screening for folded globular proteins is beginning to be widely used in structural proteomic initiatives.[8,22-26] At the JCSG, 1D ^{1}H NMR spectra of newly expressed proteins are recorded, evaluated, and graded. High-graded samples are either forwarded for extensive crystallization trials or assigned for NMR structure determination, while low-graded proteins are only subjected to a coarse-screen crystallization effort.[27]

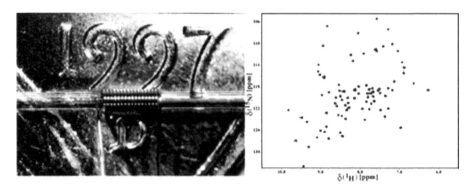

FIGURE 14.2 The microcoil probe shows a mass-based sensitivity increase of a minimal factor of 7.5, allowing for NMR spectroscopy with microgram amounts of proteins. (B) Hypothetical protein TM0979 from *Thermotoga maritime* demonstrating the capabilities of the microcoil NMR probe to completely pursue the sequence specific backbone assignment with less than 500 μg of $^{13}C,^{15}N$ labeled protein. (From Peti, W., Norcross, J., Eldridge, G., and O'Neil-Johnson, M. *J. Am. Chem. Soc.*, 126, 5873–5878, 2004. With permission.)

At this time, miniaturization is primarily aimed at identifying promising targets for structure determination. This methodology effectively guides efforts to focus on targets with a high probability of success and to eliminate poor targets from the target list or replace them with improved constructs. Overall, this process increases the efficiency of the entire pipeline and results in a reduction of the cost per structure. Further developments, including optimized miniaturization, may lead the way directly to structure determination, at least for some proteins.

14.2.3 NANOCALORIMETER OR "ENTHALPY ARRAY"

Understanding the thermodynamics of molecular interactions is central to biology and chemistry. Although a number of methods are available, calorimetry is the only universal assay for the complete thermodynamic characterization of these interactions. Under favorable circumstances, the enthalpy, entropy, free energy, and stoichiometry of a reaction can be determined.[28,29] In addition, calorimetry does not require any labeling or immobilization of the reactants and hence offers a completely generic method for characterizing the interactions. Indeed, titration calorimetry is widely used in both drug discovery and basic science, but its use is severely restricted to a small number of very-high-value measurements due to the large sample requirements and long measurement times. No currently available methods for calorimetric measurements lend themselves to modern approaches in which large libraries of compounds, ranging from small molecules in combinatorial libraries to proteins and other macromolecules, are studied.

Researchers at the Scripps-PARC (Palo Alto Research Center) Institute report a low-cost nanocalorimetry detector (Figure 14.3A) or "enthalpy array" that is used as an HT assay tool to detect enthalpies of binding interactions, enzymatic turnover, and other chemical reactions.[30] The detectors are made by using microscale fabrication technology, resulting in a reduction of nearly three orders of magnitude in

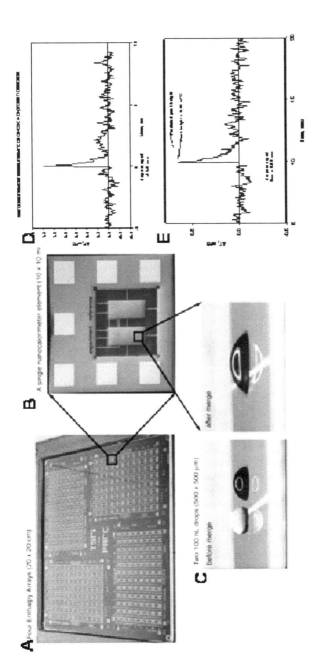

FIGURE 14.3 (See color insert following page 272.) The enthalpy array. (A) The large semiconductor substrate for the parallel manufacturing of four arrays. (B) Single detector set from a 96-format array, photographed and enlarged. The adjacent measurement and reference regions are in the center of the polyimide isolation membrane, surrounded by electrical contact pads supported by an underlying frame. (C) Electrostatic merging/mixing of two 500-nL droplets of water at three different times, starting when a voltage is applied across the gap. One drop has blue coloring to visualize mixing. The noncolored drop is placed asymmetrically across the gap between two electrodes. The mixing started within the first 33 msec, even before surface tension caused the drop to take its final shape. (D) Nanocalorimetric measurements were performed using a membrane preparation containing 0.34 μM G protein-coupled receptor P2Y6 and a buffer containing 0.8 mμ UDP and 2.3 mM GTP, resulting in a temperature rise in 1.2 m°C for a duration of 12 seconds. The estimated total amount of receptor complex used in this reaction was 85 femtomoles; (E) Enzymatic phosphorylation of glucose by hexokinase. Data are plotted as the differential change in temperature versus time. The time at which the hexokinase and glucose drops were merged is indicated. The rms noise at a 1 Hz bandwidth is 50 μK. (From Torres, F.E., Kuhn, P., De Bruyker, D., Bell, A.G., Wolkin, M.V., Peeters, E. Williamson, J.R., Anderson, G.B., Schmitz, G.P., Recht, M.I., Schweizer, S., Scott, L.G., Ho, J.H., Elrod, S.A., Schultz, P.G., Lerner, R.A., and Bruce, R.H. *Proc. Natl. Acad. Sci. USA*, 29 (101), 9517–9522, 2004. With permission.)

both the sample quantity and the measurement time over conventional microcalorimetry. The fabrication technology has a low cost and enables fabrication of 96-detector arrays, which are called enthalpy arrays, on large substrates. Accordingly, the technology will scale to high-volume production of disposable arrays. This increase in performance and reduction in cost promises to enable calorimetry to be used to investigate a substantial number of soluble protein and membrane protein samples. Nanocalorimetry in the enthalpy array format has valuable applications in proteomics for protein interaction and protein chemistry research as well as in HT screening and lead optimization for drug discovery.

As shown in Figure 14.3B, a detector cell consists of two identical adjacent detector regions that provide a differential temperature measurement. Each region contains two thermistors that are combined in an interconnected Wheatstone bridge. Each region also contains an electrostatic merging and mixing mechanism. The detector measures the temperature change arising from a chemical interaction after the mixing of two small (~ 250 nL) drops (Figure 14.3C). The differential measurement enables very precise detection of the temperature rise in a sample under study because the temperature is measured relative to a reference specimen. The reference specimen interacts with the environment in concert with the sample under study, and it also undergoes mixing in concert with the sample under study, effectively subtracting out correlated background drifts in temperature and other common-mode artifacts. Amorphous silicon thermistors with a measured temperature coefficient of resistance of $0.028 \, \text{Å}^\circ\text{C}^{-1}$ are used to detect small temperature changes.

The devices are fabricated on a thin polyimide membrane, which provides thermal isolation, reduces cost, and enables large-scale fabrication. For the current design, the thermal dissipation time is 1.3 seconds, and future improvements are expected to increase this time by a factor of 3 to 5. In a measurement, two drops containing materials of interest are placed on one of the detector regions in Figure 14.3B and Figure 14.3C. Two drops of reference material are placed on the other region. After the drops come to thermal equilibration, they are isothermally merged and mixed in both regions at the same time. If any heat is evolved because of a chemical interaction in the merged drops, the temperature of that region changes relative to the temperature of the reference region, resulting in a change in the voltage output of the bridge.

To cancel the effects of heats of dilution and variations in the environment around the drops as much as possible, the reference drops are chosen to be similar to the measurement drops. To test the level of common mode rejection, a control experiment is performed by using water drops on both sides of the detector. Ideally, the signal should be zero for such a measurement. In the control experiments, the differential temperature is within the noise of the n^+ amorphous silicon thermistors, which is 50 to 100 $\mu^\circ\text{C}$. Thus, these results show that the differential measurement provides successful common mode rejection.

An essential component of the detection system is the electrostatic merging and mixing method. Here, one of the drops is placed asymmetrically across a 50-μm gap between two electrodes on the device surface, as shown in Figure 14.3C. With the application of a voltage (100 V) across the electrodes, an electrostatic force moves the drop until it covers equal amounts of both electrodes. The second drop is placed within the range of this motion and merges with the first drop when they touch each other. The electrostatic energy from the drop-merging device, coupled with the surface tension of the drops, causes the necessary mixing of materials.

Each detector in the array is capable of detecting a 500-μ°C temperature change with a signal-to-noise ratio of 6, resulting from 250 ncal of heat released in a reaction volume of 500 nL. This level of sensitivity permits monitoring a ligand-binding reaction with a nominal binding enthalpy of -10 kcal/mol at a nominal concentration of 50 μM, which corresponds to 25 pmol of material. The sensitivity of the detector is limited by flicker noise in the n^+ amorphous silicon thermistors, which is one order of magnitude higher than the intrinsic thermal (i.e., $k_B T$) noise.

14.2.4 AUTOMATED ELECTRON MICROSCOPY AND TABLE-TOP ELECTRON MICROSCOPY SCREENING

Although electron microscopy has been extremely useful in the study of large macromolecular assemblies,[31] it is also a powerful tool in analyzing complex biological solutions aimed toward higher-resolution structural studies. Conventional electron microscopy (EM) with uranyl acetate staining is occasionally used to observe the interactions of detergents with integral membrane proteins during the solubilization process of these proteins prior to crystallization trials. Key to successful crystallization is the insight into the aggregation state and quality of the detergent extraction of membrane proteins prior to crystallization setup. EM has thus afforded a reliable, simple, and direct method of detergent selection and optimization of detergent concentration based on the protein aggregation state, effectively increasing the likelihood of successful crystallization trials.

Membrane protein samples are difficult to study with traditional homogeneity and monodispersity analyses such as mass spectrometry or dynamic light scattering that are used with soluble proteins, primarily due to domination of the scattering by the micellar species. The JCSG Center for Innovative Membrane Protein Technologies (JCIMPT)[6] has established a routine method of using EM to detect aggregation, or the lack thereof, in the solubilized membrane protein samples using a prototype EM system that is capable of automatically screening 96 grids (negatively stained EM samples) over a weekend time period (Figure 14.4A). This system was developed by the National Resource for Automated Molecular Microscopy (NRAMM) located at The Scripps Research Institute, Center for Integrative Molecular Biosciences (CIMBio). The system includes a sophisticated control software environment that allows for the automated identification of grids and controls the entire data collection. Results from the JCIMPT studies have already provided the preliminary data to specify a technology development program for an automated HT-EM-based approach for selection of novel detergent and lipid combinations that produce optimal membrane protein solution homogeneity.

For the feasibility study, a number of G protein-coupled receptors (GPCRs) were examined with EM. After purification, the concentration of the receptors was approximately 2 mg/mL based on absorbance at 280 nm. An aliquot of the purified and concentrated sample was diluted $10 \times$ in sample buffer (25 mM HEPES pH 7.5, 100 mM NaCl, 0.01% DDM or other detergent, 0.005% phosphatidyl choline). This

A

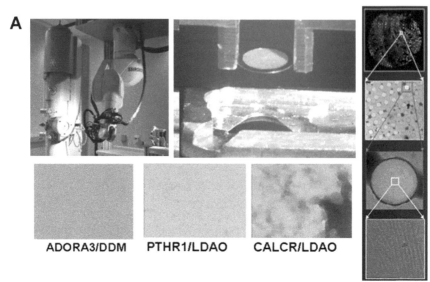

ADORA3/DDM   PTHR1/LDAO   CALCR/LDAO

B

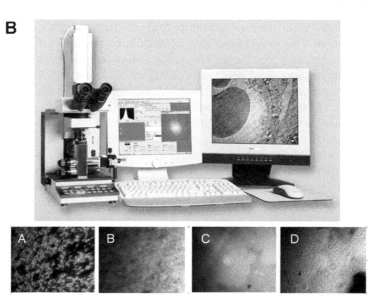

FIGURE 14.4 (See color insert following page 272.) (A) Core research and technology development programs at NRAMM are addressing automation and high-throughput methods for cryo-electron microscopy (see http://nramm.scripps.edu/resource/). Current projects include the engineering of devices for specimen handling and software development for all stages of image acquisition and data processing, as well as software for data information integration. The NRAMM goal is to establish a resource that will serve both as a center for high-throughput molecular imaging as well as for transferring this technology to the research community. Shown in the figure, clockwise from the top left, is the automated sample mounting robot implemented at NRAMM, the robotic manipulation of a sample grid, and the Leginon approach to automated data collection, which enables acquisition of high-quality EM images with minimal supervision. Shown in the bottom panel are three membrane protein–detergents samples illustrating a range of homogeneity in sample combinations. (B) The 5-kV Delong Instruments LVEM does not require staining of the samples and has a very small footprint, which allows it to be used in the laboratory environment. EM images of extracted IMP: Image A: in 1% octyl glucoside (OG) detergent with uranyl acetate stain; image B: with neutral stain. Sample aggregation is readily apparent on image A, probably as a result of the low pH uranyl acetate stain. Image B shows significantly less aggregation probably because a neutral pH of the stain was used in this experiment. Image C: LVEM image of extracted IMP in 1% octyl glucoside (OG) detergent without stain; and image D: a close-up. Both images show very little aggregation, which compares favorably with the stained samples shown in image B).

solution was examined by transmission EM, as shown in Figure 14.4A. The baculovirus-expressed GPCRs adenosine receptor (ADORA) and parathyroid hormone receptor (PTHR1) showed no evidence of vesicles, sheets, or other large-scale structures in detergents assigned as "favorable" solubilizers. The specific adenosine and parathyroid hormone receptor samples are well dispersed by EM measurements, providing suitable samples for crystallization trials. The calcitonin receptor (CALCR) sample showed clear evidence of heterogeneity and was not passed on to crystallization trials.

As a compliment to the traditional EM systems or the automated EM system, an alternate approach for EM data collection is also now available. Evaluation of a table-top low-voltage EM (LVEM) instrument produced excellent data on a number of test samples. These experiments were performed using a 5-kV LVEM (Delong Instruments; see Figure 14.4B), which eliminates the staining step normally required for conventional EM. Sample staining is highly undesirable in the assessment of sample preparations because its acidic pH and relatively high heavy metal concentration could destabilize the protein sample. LVEM experiments carried out on an extracted membrane protein sample that was analyzed with or without the staining procedure show a marked improvement in the appearance of the sample when standard staining is omitted. The results shown in Figure 14.4B illustrate how LVEM could be even more useful than conventional EM for this particular application because it avoids the potentially disrupting staining step, thus providing an undisturbed image of the protein's aggregation state. In addition, the ease of use and ease of accessibility of the instrument make routine screening of many alternative sample preparation conditions feasible. While the current lack of automation on the LVEM is certainly a drawback, the advantages of the ability to examine unstained samples make the LVEM approach advantageous and the path toward automation straightforward.

14.2.5 Miniaturization and Automation of the Crystallization Experiment

When the field of structural genomics was initially proposed (circa 1995), protein crystallization was identified as the most significant bottleneck in the process. It was then a generally held view that once crystals were generated, there was a high probability of a structure solution. Crystallization experiments were laborious, taking considerable time and effort to set up, and there were significant limitations in the amount of crystallization space that could be realistically explored. In addition, experiments were carried out using microliter volumes of protein and hence required large quantities of highly purified protein samples. Now, with the availability and relatively good success of commercial HT crystallization and imaging systems (Figure 14.5 and Figure 14.6; Table 14.1 through Table 14.3), the focus has shifted to the production of protein samples as the perceived bottleneck.

The first of the crystallization systems to be developed in the structural genomics era was the one developed at the University of California at Berkeley and the Lawrence Berkeley National Laboratory (LBNL) called the T2K. This system was capable of doing 10,000 experiments a day and demonstrated the feasibility and the power of using automated systems for crystallization experiments reliably and routinely at 20-nL volumes.[7] By using volumes 50 to 100 times lower than those routinely employed in standard microliter-volume experiments, significantly lesser amounts of sample are required to sample crystallization space. This economy of material allows experimentation with a number of smaller samples to test in parallel for the best construct, tag design, buffer components, and purification method. Traditionally, these experiments would be done in a serial fashion over many weeks, but with the volume load greatly decreased for growth and purification, it is now possible to do tens of experimental protocols per week. The reduced sample requirement also minimizes upstream costs through the reduction in biomass requirements and smaller-scale purifications. The new automated crystallization technologies have incorporated smaller, higher-density drop-plating configurations that allow for a more condensed experimentation scale, minimize plate storage space requirements, and maximize the number of experiments that can be pursued. In addition, efficient tracking of experimental data, a key requirement when working in any HT environment, is facilitated.

Several advantages to the new submicroliter crystallization technology have now been identified. It provides an experimental means of increasing the equilibration rate in vapor diffusion experiments.[7] This often reduces the time required for crystal growth, providing rapid assessment of the experimental parameters that impact crystal growth and structure determination. Faster crystallization also preserves sample integrity and homogeneity by reducing chemical decomposition, an important issue for proteolytically sensitive or oxygen-sensitive samples and SeMet–incorporated proteins that are used for multiple anomalous dispersion (MAD)-based structure determinations.[10] Although nanovolume crystallization produces crystals that are smaller than microliter-volume methods, advances in detector technology, coupled with the availability of intense in-house and synchrotron x-ray sources, allow for routine data collection without significant losses in diffraction resolution. Figure 14.7 compares

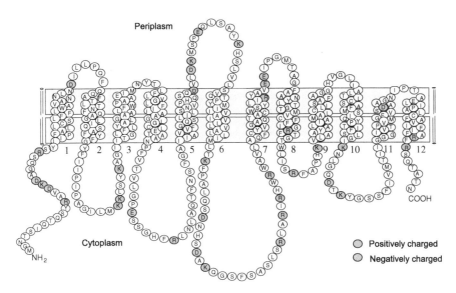

FIGURE 3.2 A two-dimensional model for the folding of the FucP protein of *Escherichia coli* in the cell membrane based on hydropathy plot, positive inside rule, and beta-lactamase fusions. Positive and negative residues are highlighted in gray and black, respectively. (From Gunn, F., Tate, C.G., Sansom, C.E., Henderson, P.J.F. *Molec. Microbiol.*, 15, 771–783, 1995; and Clough, J.L. Ph.D. Thesis, University of Leeds, 2001. With permission.)

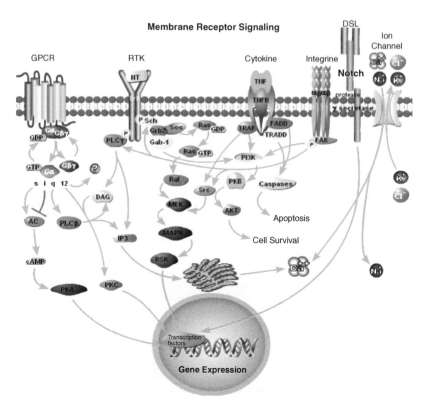

FIGURE 7.2 Signaling through five types of membrane receptors. Activation of a GPCR induces exchange of GDP for GTP on the Gα subunit and dissociation of the G protein complex. GTP-loaded Gα subunits of different families (s, i, q, and 12) act as transducers to activate intracellular effector enzymes leading to the production of second messengers and propagation of the intracellular signal to alter gene transcription or cellular physiology. Ligand binding to single membrane-spanning receptors (receptor tyrosine kinases, cytokine and integrin receptors) induces oligomerization of subunits leading to the creation of binding sites for adapter proteins, which bring effector molecules into association with the receptor for subsequent activation and signal propagation, ultimately affecting cell fate. Activation of notch receptors induces proteolytic cleavage of an intracellular domain that translocates to the nucleus for direct activation of transcription factors. Ion channels, activated by ligands or voltage changes, allow for passage of ions across the membrane to alter intracellular levels of ions and membrane excitability.

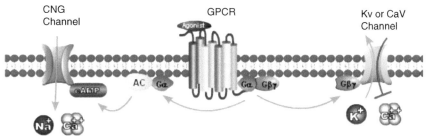

FIGURE 7.3 GPCR regulation of ion channels. Ion channels can function as downstream effectors of GPCR activation through signal transduction mediated by both $G\alpha$ and $G\beta\gamma$ subunits. Cyclic nucleotide–gated channels (CNG) can be activated by cAMP produced through $G\alpha_{s/olf}$ activation of adenylate cyclase (AC) thereby increasing intracellular concentrations of sodium and calcium. $G\beta\gamma$ subunits of $G\alpha_{i/o}$-coupled GPCRs can activate inwardly rectifying voltage-gated potassium channels (Kv) to alter membrane potential or inhibit voltage-gated calcium channels (CaV) to prevent entry of calcium into the cell. (From Reinscheid, R.K., Kim, J., Zeng, J., and Civelli, O. *Eur. J. Pharmacol.* 478(1), 27–34, 2003; Yamada, M., Inanobe, A., and Kurachi, Y. G *Pharmacol. Rev.* 50, 723–757, 1998; Dolphin, A.C. *Pharmacol. Rev.* 55(4), 607–627, 2003. With permission.)

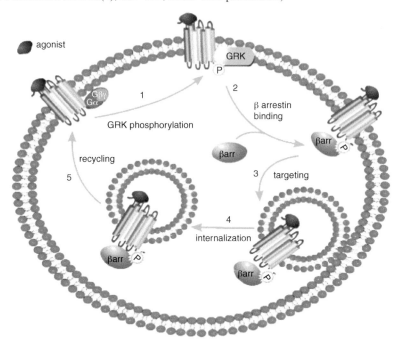

FIGURE 7.5 Internalization of G protein-coupled receptors. (1) The binding of agonist to a GPCR induces signaling through dissociation of the G protein and phosphorylation of the receptor by G protein-coupled receptor kinases (GRKs). (2) Cytosolic β-arrestin interacts with the phosphorylated receptor and inactivates the receptor by preventing reassociation of the G protein complex. (3) The receptor is then targeted to clathrin-coated pits and (4) internalized into endosomes, where β-arrestin is released and the GPCR is dephosphorylated and (5) recycled back to the plasma membrane for another round of stimulation. (From Ferguson, S.S.G. *Pharmacol. Rev.* 53, 1–24, 2001.)

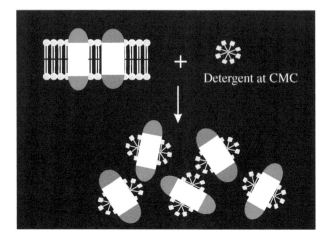

FIGURE 11.4 Solubilization of membrane proteins. Detergents are added to a membrane preparation at or above the CMC value. The detergent disrupts the interactions between the protein and the lipid bilayer at the same time, replacing the lipid molecules to form protein-detergent micelles.

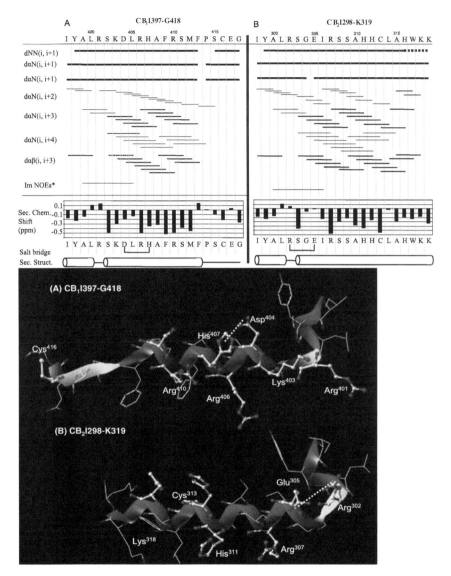

FIGURE 13.10 (**Top**) Schematic summary of the NOE restraints important in the secondary structure determination of CB1I397–G418 (A) and CB2I298–K319 (B) in DPC micelles. *Dashed lines* indicate ambiguities due to the peak overlap. The cylinders showing two helix portions *a* and *b*. * indicate the important NOEs, which are observed between Ala[399](αH) and Leu[405](NH), Ala[399](βH) and Leu[405](NH), Ala[399](αH) and Leu[405](γH) in peptide CB1I397–G418, and NOEs between Ala[300](NH) and Ile[306](NH), Ala[300](NH) and Ile[306](αH), and Ala[300](NH) and Ile[306](γH) in CB2I298–K319, for determination of orientation between helices *a* and *b*. (**Bottom**) NMR structural comparison: (A) CB1I397–G418 shows an extended Arg401 side chain with a formed salt bridge between His407 and Asp404, whereas (B) CB2I298–K319 shows a curved side chain of Arg302 forming a salt bridge with Glu305. (From Xie, X.Q. and Chen, J.Z. *J. Biol. Chem.*, 280, 3605–3612, 2005. With permission.)

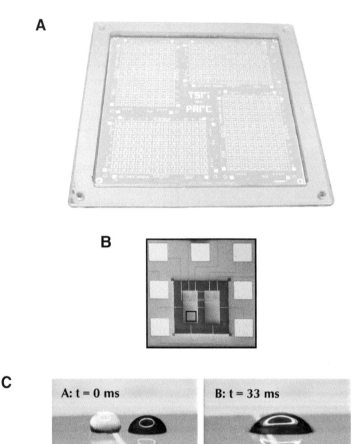

FIGURE 14.3A-C The large semiconductor substrate for the parallel manufacturing of four arrays. (B) Single detector set from a 96-format array, photographed and enlarged. The adjacent measurement and reference regions are in the center of the polyimide isolation membrane, surrounded by electrical contact pads supported by an underlying frame. (C) Electrostatic merging/mixing of two 500-nL droplets of water at three different times, starting when a voltage is applied across the gap. One drop has blue coloring to visualize mixing. The noncolored drop is placed asymmetrically across the gap between two electrodes. The mixing started within the first 33 msec, even before surface tension caused the drop to take its final shape. (From Torres, F.E., Kuhn, P., De Bruyker, D., Bell, A.G., Wolkin, M.V., Peeters, E, Williamson, J.R., Anderson, G.B., Schmitz, G.P., Recht, M.I., Schweizer, S., Scott, L.G., Ho, J.H., Elrod, S.A., Schultz, P.G., Lerner, R.A., and Bruce, R.H. *Proc. Natl. Acad. Sci. USA*, 29 (101), 9517–9522, 2004. With permission.)

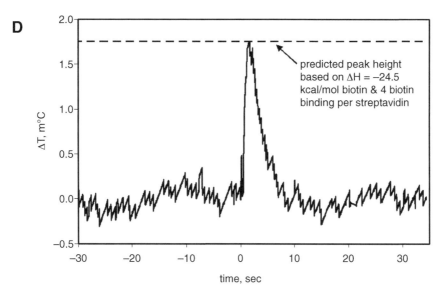

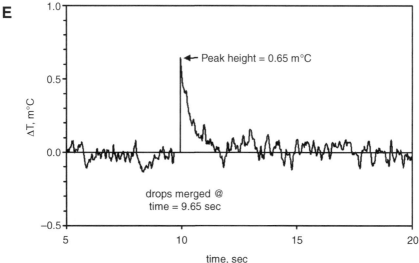

FIGURE 14.3D-E (D) Nanocalorimetric measurements were performed using a membrane preparation containing 0.34 μM G protein-coupled receptor P2Y6 and a buffer containing 0.8 mμ UDP and 2.3 mM GTP, resulting in a temperature rise in 1.2 m°C for a duration of 12 seconds. The estimated total amount of receptor complex used in this reaction was 85 femtomoles; (E) Enzymatic phosphorylation of glucose by hexokinase. Data are plotted as the differential change in temperature versus time. The time at which the hexokinase and glucose drops were merged is indicated. The rms noise at a 1 Hz bandwidth is 50 μK. (From Torres, F.E., Kuhn, P., De Bruyker, D., Bell, A.G., Wolkin, M.V., Peeters, E, Williamson, J.R., Anderson, G.B., Schmitz, G.P., Recht, M.I., Schweizer, S., Scott, L.G., Ho, J.H., Elrod, S.A., Schultz, P.G., Lerner, R.A., and Bruce, R.H. *Proc. Natl. Acad. Sci. USA*, 29 (101), 9517–9522, 2004. With permission.)

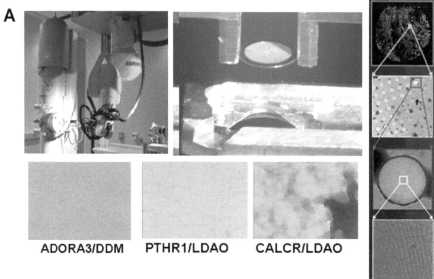

ADORA3/DDM   PTHR1/LDAO   CALCR/LDAO

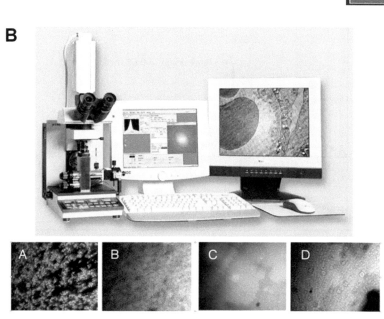

FIGURE 14.4 Core research and technology development programs at NRAMM are addressing automation and high-throughput methods for cryo-electron microscopy (see http://nramm.scripps.edu/resource/). Current projects include the engineering of devices for specimen handling and software development for all stages of image acquisition and data processing, as well as software for data information integration. The NRAMM goal is to establish a resource that will serve both as a center for high-throughput molecular imaging as well as for transferring this technology to the research community. Shown in the figure, clockwise from the top left, is the automated sample mounting robot implemented at NRAMM, the robotic manipulation of a sample grid, and the Leginon approach to automated data collection, which enables acquisition of high-quality EM images with minimal supervision. Shown in the bottom panel are three membrane protein–detergents samples illustrating a range of homogeneity in sample combinations. (B) The 5-kV Delong Instruments LVEM does not require staining of the samples and has a very small footprint, which allows it to be used in the laboratory environment. EM images of extracted IMP: Image A: in 1% octyl glucoside (OG) detergent with uranyl acetate stain; image B: with neutral stain. Sample aggregation is readily apparent on image A, probably as a result of the low pH uranyl acetate stain. Image B shows significantly less aggregation probably because a neutral pH of the stain was used in this experiment. Image C: LVEM image of extracted IMP in 1% octyl glucoside (OG) detergent without stain; and image D: a close-up. Both images show very little aggregation, which compares favorably with the stained samples shown in image B).

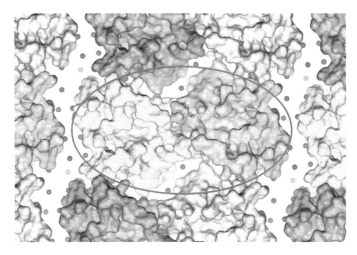

FIGURE 17.1 Rhodopsin oligomer. View from the cytoplasmic side. The structure of a dimer marked by ellipse. Positions of phospholipids are denoted by red (ethanolamine heads) and green (serine heads) spheres.

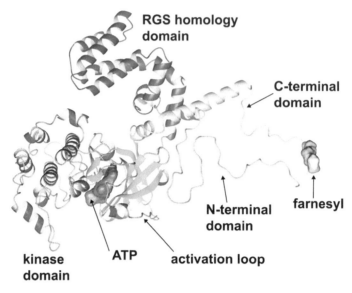

FIGURE 17.2 Homology model of rhodopsin kinase. Three domains are visible: the kinase domain, the RGS homology domain, and the mostly unstructured N- and C-terminal domains

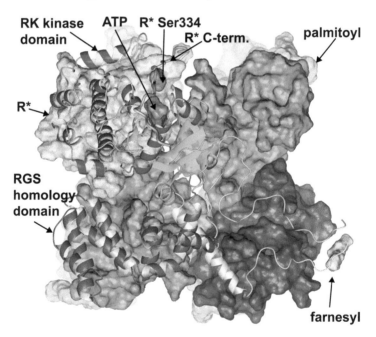

FIGURE 17.3 Complex of RK and rhodopsin tetramer. View from the cytoplasmic side. The kinase domain bound to activated rhodopsin (R* = yellow). RGS homology domain bound to adjacent inactive rhodopsin.

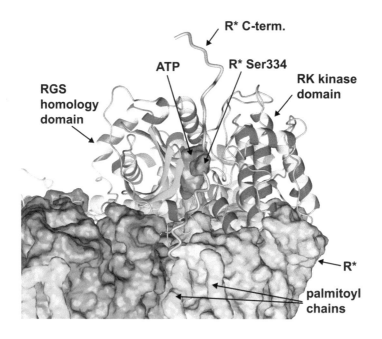

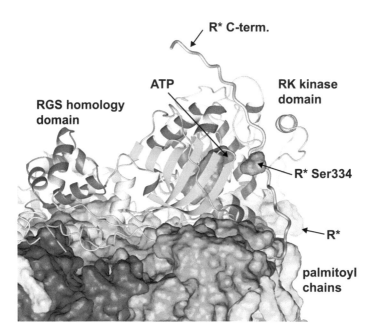

FIGURE 17.4 (A) RK bound to rhodopsin tetramer, side view. Activated rhodopsin shown in yellow. (B) View rotated 90 degrees.

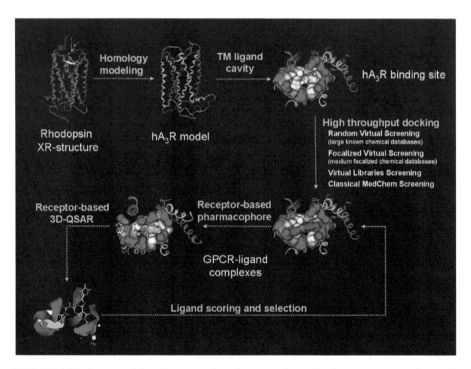

FIGURE 18.3 Integrated ligand-receptor-based anta-gonist design based on molecular modelling.[164] The combined receptor-based and ligand-based drug design approach has been carried out to define a novel pharmacophore model of human adenosine A_3 receptor antagonists. The computational strategy used can be described as follows: (a) GPCR 3D-model(s) are constructed and refined using a conventional rhodopsin-based homology modelling approach; (b) 3D receptor-driven pharmacophores(s) are generated; (c) a 3D receptor-driven quantitative structure-activity relationship (3D-QSAR) is generated; and (d) new adenosine A_3 receptor antagonist candidates are selected, synthesized and characterized pharmacologically. Iterative rounds of receptor-ligand selection are required for better lead design. (With permission from Stefano Moro and Elsevier Press.)

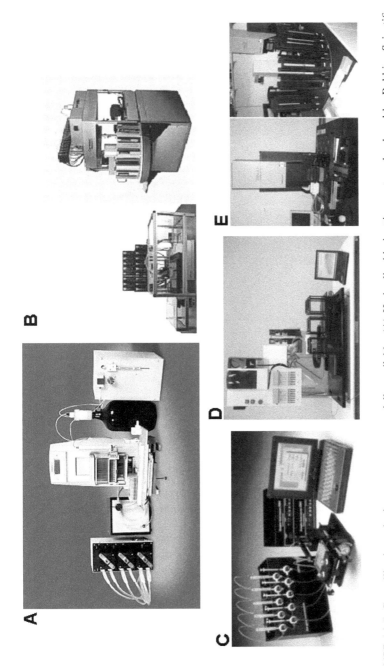

FIGURE 14.5 Crystallization robotics systems commercially available: (A) Hydra liquid pipetting system developed by Robbins Scientific; (B) Cartesian HoneyBee crystallization robot, developed from the initial Syrrx systems; (C) Douglas Instruments IMPAX system, capable of microbatch crystallization; (D) Innovadyne crystallization robot, with a particular strength in handling viscous solutions; (E) Robbins Scientific Tango used to deliver solutions to the 1536-well experiment plates. The main components of the system are exchangeable banks of 1, 12, 96, or 384 syringes and a translation stage that holds up to 12 plates. The robots can simultaneously deliver up to 384 solutions with a volume range of 0.1 to 100 μL. These robots have been operational for over four years in the laboratory of Dr. George DeTitta at the Hauptman-Woodward Institute. (Buffalo, NY).

A
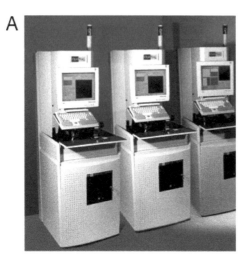

B
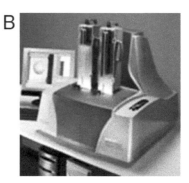

C

D

FIGURE 14.6 Commercially available imaging and plate storage systems (A) Veeco 1700 automated crystallization plate reader capable of inspecting each crystallization experiment within 650 milliseconds; (B) next-generation Oasis LS3 by Veeco; (C) RoboDesign RoboIncubator capable of storing 506 SBS or 394 Linbro plates. This instrument is coupled to a RoboDesign Minstrel III inspection station; (D) Crystal Farm by Discovery Partners Inc.

crystals grown using nanoliter volumes to ones using microliter volumes. Importantly, smaller crystals exhibit an increased surface area–volume ratio that provides easier diffusion of cryoprotectants and small molecule ligands into the crystal lattice. This, as well as the significant reduction in the mass of crystals grown from submicroliter volumes, can often contribute to easier cryopreservation of samples and a reduction in crystal mosaicity.[32] Lastly, it has been shown that a reduction in the volume of a crystallization experiment mimics the effects of microgravity by reducing convective flow (see Section 14.3.3).[33–36] Given the proposal that microgravity experiments improve crystal order and reduce imperfections, crystallization in submicroliter volumes provides an earth-based alternative for producing high-quality crystals.

TABLE 14.1
Summary of Crystallization Systems Built in the Past Few Years*

Manufacturer	Crystallization Method	Web Site Reference
Beckman	Vapor diffusion	http://www.beckman.com
Cartesian Technologies	Vapor diffusion	http://www.cartesiantech.com
Douglas Instruments	Microbatch under oil or vapor diffusion	http://www.douglas.co.uk
Fluidigm (Topaz)	Microfluidic chip-based	http://www.fluidigm.com
Gilson/Cyberlab	Microbatch under oil or vapor diffusion	http://www.gilson.com
Innovadyne Technologies	Vapor diffusion	http://www.innovadyne.com
Matrix Technologies (Hydra PlusOne)	Vapor diffusion	http://www.matrixtechcorp.com/
Tecan	Vapor diffusion	http://www.tecan.com
TTP Labtech	Microbatch under oil or vapror diffusion	http://www.ttplabtech.com

*Many of these systems are now commercially available.

TABLE 14.2
Summary of Commercially Available Fully Integrated Crystallization and Imaging Systems*

Manufacturer	Instrument Type	Web Site Reference
Tritek	Imaging	http://proteincrystalimaging.com/
Veeco (Oasis)	Imaging	http://www.veeco.com
Discovery Partners (Crystal Farm)	Imaging and storage	http://www.discoverypartners.com
The Automation Partnership (HomeBase)	Imaging and storage	http://www.automationpartnership.com

*These systems have recently become available in response to the needs of the various HT efforts, including structural genomics and pharmaceutical companies involved in SBDD programs.

TABLE 14.3
Summary of Commercially Available Fully Integrated Crystallization and Imaging Systems

Manufacturer	Instrument Type	Crystallization Method	Web Site Reference
DataCentric Automation (Rhombix)	Integrated w/ imaging	Vapor diffusion	http://www.dcacorp.com
Diversified Scientific (CrystalScore)	Integrated w/ imaging	Vapor diffusion	http://www.dsitech.com
Emerald BioStructure	Integrated w/ imaging	Vapor diffusion	http://emeraldbiostructures.com
RoboDesign	Integrated w/ imaging	Vapor diffusion	http://robodesign.com
Syrrx-RTS	Integrated w/ imaging	Vapor diffusion	http://www.syrrx.com

Source: From Page, R., Grzechnik, S.K., Canaves, J.M., Spraggon, G., Kreusch, A., Kuhn, P., Stevens, R.C., and Lesley, S.A. *Acta. Crystallogr. D.*, 59, 1028–1037, 2003. With permission.

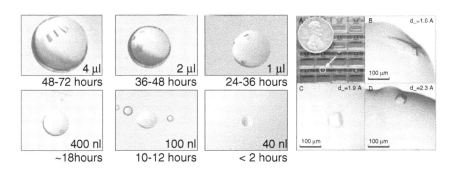

FIGURE 14.7 Convective current experimental study. Examples of crystals of bacterial proteins grown by sitting-drop vapor diffusion under identical conditions in a 4-mL drop (left) and a 100-nL drop (right). (From Carter, D.C., Rhodes, P., McRee, D.E., Tari, L., Dougan, D.R., Snell, G., Abola, E., and Stevens, R.C. *J. Appl. Crystallogr.*, 38, 87–90, 2005. With permission.)

14.2.5.1 Integrated Nanovolume Crystallization System

Beginning in 2001, several groups in the San Diego area (JCSG, GNF [Genomics Institute of the Novartis Research Foundation], the Scripps Research Institute (TSRI), Syrrx) have used the Syrrx fully integrated HT protein crystallization (*Agincourt*) and HT imaging and plate storage system (designed by GNF engineers and the authors of this chapter). Fully integrated second-generation robotic systems consist of various major components, a screen solution maker and dispenser, a protein drop dispenser, and an imaging and plate storage subsystem (Table 14.3). Figure 14.8A shows the *Agincourt* robot in operation at Syrrx. *Agincourt* is capable of dispensing "on the fly" sitting drops, ranging in volume from 20 nL to 1 μL, using a Cartesian Technologies SynQuad dispensing robot.[7] Maximum throughput of the system is on the order of 140,000 crystallization trials per day and has previously set up more than 110 proteins for crystallization trials in a single day. The system sets up a standard sitting-drop vapor diffusion experiment using a commercially available 96-well plate that is sealed with clear tape. Two stations feed plates into the robot; one uses screens that have been premixed offline (coarse-screen), and the other can mix a custom plate from any combination of 96 solutions. The offline station is used to fill the coarse-screen plates by transferring the solution from 10-mL conical tubes into a plate. This is done for repeatedly used screens such as commercially available coarse screens and for premade fine screens for reproducibly growing well-characterized crystals to support co-crystallization and small-molecule soaking experiments.

In practical operation, a mix of fine and coarse screens is often run at the same time, and all the proteins needed for several runs are placed in the dispensing station and held at 4°C. The barcode on each plate instructs the robot about which protein to add to the plate. This allows the robot to be run at maximal efficiency while minimizing operator error. In addition, an additive station allows for dispensing a variable volume from a second 96-well plate into the equivalent wells of the crystallization plate before the well solution is mixed with the protein. To minimize sample evaporation that occurs during reagent mixing, the system has been designed for rapid protein dispensing, precipitant mixing, and plate sealing, taking only 40 seconds to perform these three steps. Under current operating parameters, *Agincourt* can set up a 96-well sitting-drop plate every two minutes, giving it a maximum throughput of 2880 crystallization experiments per hour when setting up 20- to 50-nL crystallization experiments.

Crystallization plate storage using the Syrrx Imaging Forts involves a gantry-configuration crystal image and storage system that can automatically capture and process 138,000 images per day at both 4 and 20°C. Drop imaging uses an OptiMag Veeco-Oasis 1700 system with an "imager" connected over a gigabit network to an image file server. A server process manages the image files and links them to the corresponding experimental data in a relational database. This system architecture decouples imaging and analysis to achieve maximum throughput, imaging one crystallization plate, consisting of 96 individual wells, in less than 60 seconds. Since submicroliter volumes are used, focusing of the crystal drops is not required, since the entire drop depth of field can be observed within one image. Since Syrrx started operations, *Agincourt* has completed more than 5.4 million crystallization experiments, and the crystal imaging stations have collected over 36 million image trials (over 1300 protein targets).

Also shown in Figure 14.8 are two commercially available integrated crystallization and imaging systems with similar major components. One is the Rhombix system (Figure 14.8B), available from Data Centric Automation; the other is CrystalMotion (Figure 14.8C), available from RoboDesign. Both these systems use the vapor diffusion technique for crystallization. An alternative approach to vapor diffusion is the microbatch under the oil method developed by DeTitta and co-workers using 1536-well plates as described in the paper by Luft and colleagues[37] (Figure 14.5e). As a compliment to the crystallization and imaging systems, fine screen

A

B

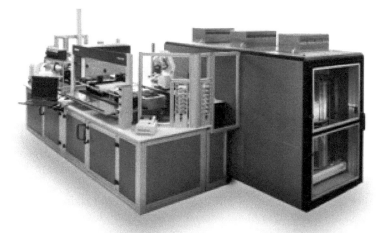

C

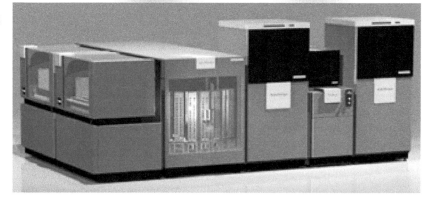

FIGURE 14.8 *Agincourt* robot system represents the first of the new class of integrated systems capable of completing over 14,000 experiments a day using small amounts of proteins. The crystallization system consists of a number of stations arranged around a central robotic arm. The arm moves plates between stations and also serves to position the plates. The stations are a well-filling station that can put any combination of 96 solutions into each well as needed, a mixing station, a well-dispensing station (left rear), a protein-dispensing station (left), a taping station to seal the plates, and a stacker for finished plates (right). (B) Rhombix, Integrated crystallization system from Data Centric Automation. (C) RoboDesign's Crystal-Mation, another example of a commercially available integrated crystallization system.

generators such as the MatrixMaker (Figure 14.9) designed and marketed by Emerald BioSystems, have also emerged recently.

14.3 RESULTS FROM FIVE YEARS OF NANOVOLUME CRYSTALLIZATION EXPERIMENTS

14.3.1 EXPLORING CHEMICAL SPACE

One of the ultimate goals of structural genomics is to provide an understanding of an entire organism at the molecular level. To this end, the JCSG cloned and attempted the expression of the predicted 1877 ORFs from *Thermotoga maritima* (TM).[11] From the original 1877 TM target set, 1851 corresponded to predicted open reading frames (ORFs), 1791 were successfully cloned, 1376 expression clones were generated, 705 proteins expressed in a soluble fashion, and more than 539 have been purified and sent for crystallization analysis. TM proteins were expressed and purified using a single chromatography step (metal chelation; all targets contained an N-terminal tag, MGSDKIHHHHHH, to facilitate expression and purification) for a final purity of 90 to 95%. Each protein was then buffer-exchanged into crystallization screening buffer (10 mM Tris, pH 7.9, 150 mM NaCl, and 0.25 mM Tris [2-carboxyethyl] phosphine hydrochloride [TCEP]), concentrated to »10mg/mL, and screened for crystallization against 480 commercially available crystallization conditions* at 20°C using the nanocrystallization sitting-drop vapor diffusion method[7] with 100 nL drops (50 nL protein plus 50 nL crystallization solution; see Figure 14.10). This effort resulted in 258,720 distinct crystallization experiments and has provided one of the most extensive, systematic datasets of commonly used crystallization conditions for a wide range of proteins to date.[11]

Of the 539 TM proteins screened, 465 (86%) crystallized for a total of 5546 crystal hits, indicating that most of the proteins had an inherent propensity to crystallize. Notably, over half of these proteins crystallized in five or more conditions, 19 crystallized in 50 or more conditions, and one, TM0665, crystallized in over 230 crystallization conditions. The proteins crystallized sampled a wide range of pI and

---

* Conditions compiled from 10 commercially available kits, including Crystal Screen, Crystal Screen 2, Crystal Screen Cryo, PEG/Ion Screen, Grid Screen Ammonium Sulfate, Grid Screen PEG 6000, Grid Screen Peg/LiCl, Grid Screen MPD (Hampton Research, Riverside, CA), Wizard I/II, Cryo I/II (Emerald Biostructures, Bainbridge Island, WA).

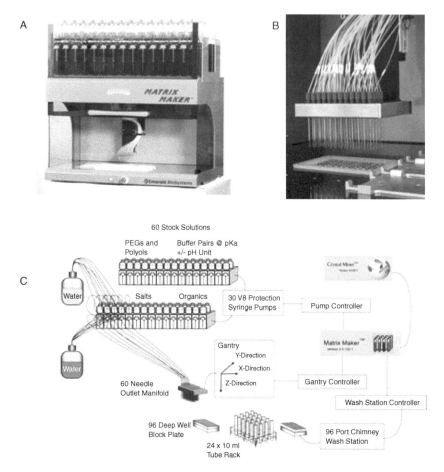

FIGURE 14.9 The Emerald BioSystems Matrix Maker™. (A) An automated liquid handling robot. It can mix any kind of solution from up to 60 different stock solutions such as buffers, salts, and precipitants. (B) A closeup of the liquid dispensing heads. The design, shown schematically in (C) allows the Matrix Maker to handle highly viscous solutions with excellent dispense accuracy, and no cross-contamination between stocks. All of the stock solutions are dispensed by positive-pressure displacement using precision syringe pumps. This design completely eliminates aspiration problems and costly plastic tip waste stream. The Crystal Miner database software controls the Matrix Maker and stores all of the details about each crystallization formulation for cross-reference to the crystallization results.

MW (molecular weight) values, similar to that observed for the proteome as a whole,[11] and there was no observed correlation between protein pI and crystallization pH.[38]

The conditions used for the crystallization trials were also very successful in promoting crystal formation. Of the 480 conditions, 472 (98%) produced crystals for at least one protein, with 37 conditions producing crystals for 25 or more proteins and one, Core Screen #08 (Table 14.4), producing crystals for more than 40 crys-

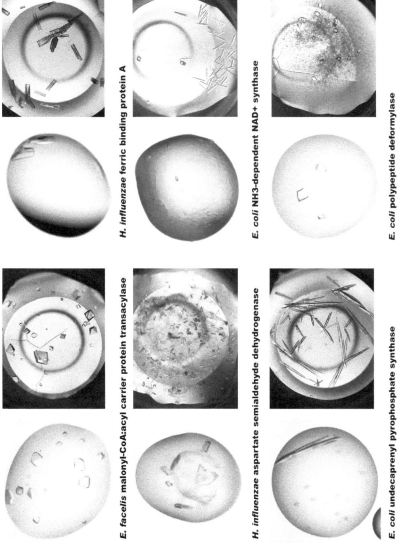

FIGURE 14.10

FIGURE 14.10 (CONTINUED) (Left) Crystals grown with the Agincourt crystallization system. Representative images of lysozyme crystallization trials, obtained at 277 K, with various total drop volumes. Total drop volumes for crystallization contain equal volumes of protein and reservoir solutions (e.g., 4 mL total drop volume equals 2 mL protein plus 2 mL reservoir solution). The time listed after total drop size is the amount of time to see a crystal of ~ 50 mm size: (A) 4-mL drop volume, 48 ± 72 h for initial crystallization; (B) 2 mL, 36 ± 48 h; (C) 1 mL, 24 ± 36 h; (D) 400 nL, 18 h; (E) 100 nL, 10 ± 12 h; (F) 40 nL, 2 h. (G) is a magnifed view of (F) showing crystal formation, roughly 50 mm on the longest edge. (Right) Perspective diagram showing the drop size relative to a U.S. penny. (From Santarsiero, B.D., Yegian, D.T., Lee, C.C., Spraggon, G., Gu, J., Scheibe, D., Uber, D.C., Cornell, E.W., Nordmeyer, R.A., Kolbe, W.F., Jin, J., Jones, A.L., Jaklevic, J.M., Schultz, P.G., and Stevens, R.C. *J. Appl. Crystallogr.*, 35, 278–281, 2002. With permission.)

tallization conditions. Notably, however, the 10 most effective conditions, which account for only 2% of the original conditions used, produced crystals for 196 (42%) different proteins, while the best 108 conditions (23%) produced crystals for all 465. Thus, many of the 480 conditions used for these trials significantly oversampled distinct regions of crystallization space and could be eliminated without impacting the final number of distinct proteins crystallized.

To identify the redundant conditions, an iterative selection algorithm, *Min_Cov*, was used to identify the minimum set of conditions that would have produced crystals for the entire set of crystallized proteins (minimal screens).[38] *Min_Cov* identified 472 minimal screens, one for each condition that successfully crystallized a protein; 415 of these minimal screens were unique. The 67 conditions present in every minimal screen were then identified as those conditions most essential for promoting crystal formation for the most diverse set of proteins (Table 14.4). These conditions are collectively referred to as the Core Screen* and include representatives from all five primary precipitant classes, including high-MW polyethylene glycol (PEG) (31 of the original 171 conditions), low-MW PEGs (8 of 67), ammonium sulfate/salts (10 of 106), polyalcohols (11 of 83), and remaining organics (7 of 53), although high-MW PEG conditions are still the most prevalent (47%). Using only the Core Screen conditions, greater than 84% (392 of 465) of the proteins would have been crystallized. This screen has since been expanded to include the 29 next most effective conditions for promoting crystal formation to make it compatible with 96-well crystallization plates (Table 14.4). This Expanded Core Screen would have crystallized 448 of the total 465 proteins crystallized in this study, for a success rate of 96%. This screen is now regularly used for initial crystallization trials.

A detailed comparison of these results with those published by other SG groups[39] showed some similarities but also a number of significant differences. In particular, the Toronto SG group arrived at a much different set of minimal screens from that of the JCSG. It is well known that several factors such as sample purity, the presence or absence of purification tags, protein buffer, crystallization setup methods, and protein concentration contribute significantly to the protein crystallization

---

* Because each condition in the Core Screen was present in all 472 identified minimal screens, they are all *equally* important for the crystallization of the most diverse set of protein targets in this study and thus their numbered position in the Core Screen Table is arbitrary. See Table 14.4.

TABLE 14.4
JCSG Crystallization Core Screen Developed in Response to the Large Scale Structural Genomics Programs

	Condition	Screen#	Freq.*
1	50%(w/v) PEG 400, 0.2 M Li$_2$SO$_4$, 0.1 M acetate pH 5.1	W1cryo #47	415
2	20%(w/v) PEG 3000, 0.1 M citrate pH 5.5	W1 #06	415
3	20%(w/v) PEG 3350, 0.2 M diammonium hydrogen citrate pH 5.0	PEG/ion #48	415
4	30%(v/v) MPD, 0.02 M CaCl$_2$, 0.1 M NaOAc pH 4.6	H1 #01	415
5	20%(w/v) PEG 3350, 0.2 M magnesium formate pH 5.9	PEG/ion #20	415
6	20%(w/v) PEG 1000, 0.2 M Li$_2$SO$_4$, phosphate-citrate pH 4.2	W1 #39	415
7	20%(w/v) PEG 8000, 0.1 M CHES pH 9.5	W1 #01	415
8	20%(w/v) PEG 3350, 0.2 M ammonium formate pH 6.6	PEG/ion #23	415
9	20%(w/v) PEG 3350, 0.2 M ammonium chloride pH 6.3	PEG/ion #09	415
10	20%(w/v) PEG 3350, 0.2 M potassium formate pH 7.3	PEG/ion #22	415
11	50% MPD, 0.2 M (NH$_4$)H$_2$PO$_4$, 0.1 M Tris pH 8.5	H2 #43	415
12	20%(w/v) PEG 3350, 0.2 M potassium nitrate pH 6.9	PEG/ion #18	415
13	0.8 M (NH$_4$)$_2$SO$_4$, 0.1 M citric acid pH 4.0	AmSO$_4$ #01	415
14	20%(w/v) PEG 3350, 0.2 M sodium thiocyanate pH 6.9	PEG/ion #13	415
15	20%(w/v) PEG 6000, 0.1 M bicine pH 9.0	P6K #18	415
16	10%(w/v) PEG 8000, 8% ethylene glycol, 0.1 M HEPES pH 7.5	H2 #37	415
17	40%(v/v) MPD, 5%(w/v) PEG 8000, 0.1 M cacodylate pH 7.0	W2cryo #01	415
18	40% ethanol, 5%(w/v) PEG 1000, 0.1 M phosphate-citrate pH 5.2	W1cryo #40	415
19	8%(w/v) PEG 4000, 0.1 M NaOAc pH 4.6	H1 #37	415
20	10%(w/v) PEG 8000, 0.2 M MgCl$_2$, 0.1 M Tris pH 7.0	W2 #43	415
21	20%(w/v) PEG 6000, 0.1 M citric acid pH 5.0	P6K #14	415
22	50%(v/v) PEG 200, 0.2 M MgCl$_2$, 0.1 M cacodylate pH 6.6	W2cryo #36	415
23	1.6 M sodium citrate pH 6.5	H2 #28	415
24	20%(w/v) PEG 3350, 0.2 M tripotassium citrate monohydrate pH 8.3	PEG/ion #47	415
25	30% MPD, 0.02 M CaCl$_2$, 0.1 M NaOAc pH 4.6	H1cryo #01	415
26	20%(w/v) PEG 8000, 0.2 M NaCl, 0.1 M phosphate-citrate pH 4.2	W1 #31	415
27	20%(w/v) PEG 6000, 1.0 M LiCl, 0.1 M citric acid pH 4.0	P6K/LiCl #13	415
28	20%(w/v) PEG 3350, 0.2 M ammonium nitrate pH 6.3	PEG/ion #19	415
29	10%(w/v) PEG 6000, 0.1 M HEPES pH 7.0	P6K #10	415
30	0.8 M NaH$_2$PO$_4$/0.8 M KH$_2$PO$_4$, 0.1 M HEPES pH 7.5	H1 #35	415
31	40%(v/v) PEG 300, 0.1 M phosphate-citrate pH 5.2	W2cryo #18	415
32	10%(w/v) PEG 3000, 0.2 M Zn(OAc)$_2$, 0.1 M acetate pH 4.5	W2 #01	415
33	20% ethanol, 0.1 M Tris pH 8.5	H2 #44	415
34	25%(v/v) 1,2-propanediol, 0.1 M Na/K phosphate, 10%(v/v) glycerol pH 6.8	W2cryo #11	415
35	10%(w/v) PEG 20 000, 2% dioxane, 0.1 M bicine pH 9.0	H2 #48	415
36	2.0 M (NH$_4$)$_2$SO$_4$, 0.1 M acetate pH 4.6	H1 #47	415
37	10%(w/v) PEG 1000, 10%(w/v) PEG 8000	H2 #07	415
38	24%(w/v) PEG 1500, 20% glycerol	H1cryo #43	415
39	30%(v/v) PEG 400, 0.2 M MgCl$_2$, 0.1 M HEPES pH 7.5	H1cryo #23	415
40	50%(v/v) PEG 200, 0.2 M NaCl, 0.1 M Na/K phosphate pH 7.2	W2cryo #15	415
41	30%(w/v) PEG 8000, 0.2 M Li$_2$SO$_4$, 0.1 M acetate pH 4.5	W1 #17	415
42	70%(v/v) MPD, 0.2 M MgCl$_2$, 0.1 M HEPES pH 7.5	H2 #35	415
43	20%(w/v) PEG 8000, 0.1 M Tris pH 8.5	W2 #03	415
44	40%(v/v) PEG 400, 0.2 M Li$_2$SO$_4$, 0.1 M Tris pH 8.4	W1cryo #38	415

(continued)

TABLE 14.4 (CONTINUED)
JCSG Crystallization Core Screen Developed in Response to the Large Scale
Structural Genomics Programs

	Condition	Screen[#]	Freq.[*]
45	40%(v/v) MPD, 0.1 M Tris pH 8.0	MPD #17	415
46	25.5%(w/v) PEG 4000, 0.17 M (NH$_4$)$_2$SO$_4$, 15% glycerol	H1cryo #31	415
47	40%(v/v) PEG 300, 0.2 M Ca(OAc)$_2$, 0.1 M cacodylate pH 7.0	W1cryo #37	415
48	14% 2-propanol, 0.14 M CaCl$_2$, 0.07 M acetate pH 4.6, 30% glycerol	H1cryo #24	415
49	16%(w/v) PEG 8000, 0.04 M KH$_2$PO$_4$, 20% glycerol	H1cryo #42	415
50	1.0 M sodium citrate, 0.1 M cacodylate pH 6.5	W1 #14	415
51	2.0 M (NH$_4$)$_2$SO$_4$, 0.2 M NaCl, 0.1 M cacodylate pH 6.5	W2 #04	415
52	10% 2-propanol, 0.2 M NaCl, 0.1 M HEPES pH 7.5	W1 #02	415
53	1.26 M (NH$_4$)$_2$SO$_4$, 0.2 M Li$_2$SO$_4$, 0.1 M Tris pH 8.5	W1 #47	415
54	40%(v/v) MPD, 0.1 M CAPS pH 10.1	W2cryo #25	415
55	20%(w/v) PEG 3000, 0.2 M Zn(OAc)$_2$, 0.1 M imidazole pH 8.0	W2 #40	415
56	10% 2-propanol, 0.2 M Zn(OAc)$_2$, 0.1 M cacodylate pH 6.5	W2 #11	415
57	1.0 M (NH$_4$)$_2$HPO$_4$, 0.1 M acetate pH 4.5	W1 #09	415
58	1.6 M MgSO$_4$, 0.1 M MES pH 6.5	H2 #20	415
59	10%(w/v) PEG 6000, 0.1 M bicine pH 9.0	P6K #12	415
60	14.4%(w/v) PEG 8000, 0.16 M Ca(OAc)$_2$, 0.08 M cacodylate pH 6.5, 20% glycerol	H1cryo #46	415
61	10%(w/v) PEG 8000, 0.1 M imidazole pH 8.0	W2 #34	415
62	30% Jeffamine M-600, 0.05 M CsCl, 0.1 M MES pH 6.5	H2 #24	415
63	3.2 M (NH$_4$)$_2$SO$_4$, 0.1 M citric acid pH 5.0	AmSO4 #20	415
64	20% MPD, 0.1 M Tris pH 8.0	MPD #11	415
65	20% Jeffamine M-600, 0.1 M HEPES pH 6.5	H2 #31	415
66	50%(v/v) ethylene glycol, 0.2 M MgCl$_2$, 0.1 M Tris pH 8.5	W1cryo #43	415
67	10% MPD, 0.1 M bicine pH 9.0	MPD #06	415
68	2.0 M (NH$_4$)H$_2$PO$_4$, 0.1 M Tris, pH 8.5	H1 #41	414
69	3.4 M 1.6 Hexanediol; 0.2 M MgCl$_2$; 0.1 M Tris pH 8.5	H2 #39	414
70	20% (w/v) PEG 6000, 0.1 M Citric Acid pH 4.0	P6K #13	414
71	0.2 M Potassium Chloride, 20% (w/v) PEG 3350, pH 6.9	Peg/Ion #08	414
72	35% (v/v) 2-ethoxyethanol, 0.05 M Ca(OAc)$_2$, 0.1 M Imidazole, pH 7.5	W1cryo #18	414
73	35 % (v/v) MPD; 0.2 M Li$_2$SO$_4$; 0.1 M MES pH 6.0	W2 #02	414
74	1.26 M (NH$_4$)$_2$SO$_4$; 0.2 M NaCl; 0.1 M CHES pH 9.5	W2 #29	414
75	10 % (w/v) PEG 3000; 0.2 M NaCl; 0.1 M phosphate-citrate pH 4.2	W2 #36	414
76	40% (v/v) PEG-600, 0.1 M CHES pH 9.6	W2cryo #31	414
77	40% (v/v) PEG-400, 0.1 M Imidazole pH 7.4	W2cryo #43	414
78	25 % (w/v) PEG 4000, 0.2 M (NH$_4$)$_2$SO$_4$, 0.1 M Acetate pH 4.6	H1 #20	413
79	2.0 M (NH$_4$)$_2$SO$_4$	H1 #32	413
80	8% (w/v) PEG 8000; 0.1 M Tris pH 8.5	H1 #36	413
81	35% (v/v) Dioxane	H2 #04	413
82	1.0 M Hexanediol, 0.01 M CoCl$_2$, 0.1 M Na-acetate pH 4.6	H2 #11	413
83	20% (v/v) PEG 1000, 0.1 M Tris pH 7.0	W1 #19	413
84	2.5 M NaCl, 0.2 M MgCl$_2$, 0.1 M Tris pH 7.0	W2 #17	413
85	20% (w/v) PEG 8000, 0.2 M Ca(Ac)$_2$, 0.1 M MES pH 6.0	W2 #28	413
86	20% (w/v) PEG 3350, 0.2 M di-Ammonium Tartrate, pH 6.6	Peg/ion #38	411
87	40% (v/v) EG, 0.1 M Acetate pH 5.0	W1cryo #02	411
88	20% (w/v) PEG 6000, 1.0 M LiCl, 0.1 M Tris pH 8.0	P6K/LiCl #17	410
89	1.26 M (NH$_4$)$_2$SO$_4$, 0.1 M cacodylate pH 6.5	W1 #13	407

90	12% (w/v) PEG 20000, 0.1 M MES pH 6.5	H2 #22	403
91	20% (w/v) PEG 3350, 0.2 M Lithium Acetate dihydrate, pH 7.8	Peg/ion #24	390
92	20% (w/v) PEG 3350, 0.2 M Sodium Formate, pH 7.2	Peg/ion #21	388
93	20% (v/v) PEG-2000; 0.1 M Tris pH 7.0	W1 #10	381
94	2.0 M (NH$_4$)$_2$SO$_4$; phosphate/citrate pH 4.2	W2 #09	365
95	1.6 M (NH$_4$)H$_2$PO$_4$, 0.08 M Tris pH 8.5, 20 % Glycerol	H1cryo #48	364
96	40% (v/v) ethanol, 0.05 M MgCl$_2$, 0.1 M Tris pH 8.4	W1cryo #07	343

#H1, H2, H1cryo, PEG/Ion, AmSO$_4$, P6K, P6K/LiCl, MPD: Crystal Screen, Crystal Screen 2, Crystal Screen Cryo, PEG/Ion Screen, Grid Screen Ammonium Sulfate, Grid Screen PEG 6000, Grid Screen PEG/LiCl, Grid Screen MPD, respectively (Hampton Research). W1, W2, W1cryo, W2cryo: Wizard I and II and Cryo I and II, respectively (Emerald Biostructures).

*Frequency of the condition was identified in a minimal screen using the Min_Cov algorithm. 473

Source: From Canaves, J.M., Page, R., Wilson, I.A., and Stevens, R.C. J. Mol. Biol., 344, 977–991, 2004. With permission.

process[39–41] and are therefore likely to have contributed to many of the differences observed between these groups. As more studies are carried out, it will become possible to include these factors in the correlation analysis. For now, however, it seems that identified minimal screens are still somewhat correlated to the methods used for sample preparation and screening, and thus additional preparation methods must be taken into consideration when deciding which screen might be appropriate for specific experiments. In addition, however, one must also consider that these screens were developed as initial screens with the purpose of identifying proteins with a natural propensity to crystallize and not necessarily to result in diffraction-quality crystals. Thus, although the exact nature of the crystallization conditions differs between the two minimal screens, both sets of conditions are expected to successfully identify proteins that are amenable to further crystallization studies.

To date, the JCSG has solved more than 150 TM structures out of ~ 200 TM structures in the Protein Data Bank. Examining structural coverage of the TM proteome highlights that there is direct structural coverage of 25% of the expressed soluble proteins and ~ 9% of this organism's proteome (173 unique PDB structures; 267 total). After taking into account structures that can be modeled through homology and fold recognition, this percentage rises to over 70% (88% of predicted crystallizable proteins), which probably represents the highest structural coverage of any organism. In the discussion below, TM protein targets are referred to via their JCSG target number (www.jcsg.org; e.g., TM012). The general approach used by the JCSG in working with TM proteins was to process each target using a single process and consistent quality control. This set of uniformly handled samples can then be used as probes to identify what regions of crystallization and protein property space correlate with crystal formation.

14.3.2 BIOPHYSICAL PROPERTIES OF PROTEINS IN *THERMOTOGA*
MARITIMA THAT CRYSTALLIZED

This internally consistent set of TM proteins, which were expressed, purified, and crystallized under tightly controlled conditions and experimental protocols, can also be analyzed to identify which of the protein properties correlate with crystallization success. An understanding of the space defined by a protein's biophysical properties with respect to the crystallization space defined by the screens is crucial for both target selection and prediction of experimental crystallization conditions. Therefore, to increase protein crystallization success, it is necessary to either select targets whose properties make them likely candidates for crystallization or modify the protein properties so they are compatible with the designated screening space. Several approaches to improve crystallization success include the design of truncations based on deuterium exchange mass spectrometry (DXMS),[42] protein functional domain definitions based on Pfam matches,[43] addition of ligands (substrates, inhibitors, cofactors), target multiplexing by selection of orthologues, surface entropy mutagenesis[44,45] and proteolytic mass spectrometry screening.[46]

To identify useful criteria for future protein target selection and to determine ways to improve current pipeline protocols to increase crystallization success of active targets, the distribution of various parameters in the TM proteome and in the subset of crystallized proteins was analyzed for trends in crystallization success. The parameters analyzed were: (1) biophysical properties, including sequence length, isoelectric point, protein hydropathy, and percentage of charged residues; (2) predicted transmembrane helices and signal peptide sequences; (3) predicted bacterial lipoprotein lipid-binding sites (hydrophobicity pockets); (4) predicted coiled-coils; and (5) predicted low-complexity regions that might lead to disorder. In addition to comparing the differences between the entire proteome and the crystallized subset for each parameter, the distribution of each parameter in the TM proteome was evaluated, since even though a parameter might be a good predictor of crystallization, if it is present in only a very limited number of proteins, the impact on future target selection will be minimal. One such example is the presence or absence of bacterial lipoprotein lipid-binding sites. This trait is 30% less prevalent in crystallized proteins when compared with the proteome, but the population of proteins with this attribute is so small that it makes it irrelevant when the entire proteome is considered. Still, the information is valuable to design filtering strategies for other potential protein target pools that might include a significant number of proteins containing lipid-binding sites. Conversely, the presence of low-complexity regions, as predicted by the SEG program[47] (designated as *SEG regions* throughout this manuscript), or transmembrane helices, as predicted by the TMHMM program,[48] show dramatic distributional differences between both protein populations while also being very prevalent in the proteome, suggesting that they might be good predictors of crystallization success.

Often, in target or construct selection, no structural homologues of a protein target are available to guide the expression, purification, or crystallization experiments, and one can only depend on the protein sequences. In these cases, not only do the selected targets lack significant sequence homology with known structures, but many of the selected targets lack functional domain information (no matches to Pfam or other domain databases), functional annotation (proteins annotated as hypothetical proteins), or even any identifiable sequence homology to any other protein (known as ORFans). Therefore, to predict the likelihood that these proteins will crystallize, it is necessary to use parameters derived from the primary sequence and correlate those parameters with experimental expression and crystallization data to establish empirical rules that can subsequently be used for optimized target selection.

Three possible target filtering strategies are listed in Table 14.5, derived from the JCSG TM experimental work.[49] Seven sequence-derived parameters shown to have a direct effect on protein crystallization were selected for these filtering strategies, including protein length, calculated isoelectric point, percent charged residues, gravy index, the number of SEG residues, the number of predicted transmembrane helices, and the number of predicted signal peptides. The first strategy proposed is based on the absolute maximum and minimum at which crystallization has been observed for each parameter, that is, none of the observed crystals would be lost but would still result in an increase in the ratio of successfully crystallized proteins and selected targets (37.7%; see Table 14.5). The second strategy is based on more stringent cut-offs that tolerate the loss of up to 5% of the crystals per parameter. The goal is to further reduce the pool of potential targets with respect to the first strategy, while further increasing the ratio between successfully crystallized proteins and selected targets (39.5%; see Table 14.5). The loss of a small number of outlier crystallized proteins is tolerated because it allows for a higher success rate for new targets, resulting in an overall increase of successfully crystallized targets. Finally, we propose an even more stringent filtering strategy that uses as limits the area where most of the crystallized proteins cluster in the distribution defined by each protein attribute (maximum clustering strategy, MCS). Whereas the number of lost crystallized proteins and solved structures is higher than in the second strategy, the ratio of crystallized and solved proteins to selected targets is even greater (45.1%; see Table 14.5), indicating that this is a superior target selection or design strategy.

As Table 14.5 shows, filtering based on simple sequence-derived parameters can significantly reduce the time and costs of both crystal and structure production, although each filtering strategy described here has advantages and disadvantages.[49] In all but one strategy, crystals and structures are lost due to filtering. However, the decrease would be more than compensated by the expected increase in crystals and structures from the additional throughput available because of the large number of dubious targets eliminated by the filtering procedures. The last filtering model, MCS, results in a much smaller initial set of targets with a significantly greater chance of crystallizing and even greater chance of being solved when compared with an unfiltered set of initial targets. The MCS model was built using visual data mining, allowing the loss of, at most, 15% of crystallized targets per parameter. This strategy attempted to preserve most of the structures in the regions of maximum clustering while filtering out outlier crystallized and solved structures if their filtration resulted in a significant reduction in the potential target pool. The second constraint used to build the model was a maximal allowed structure loss of 20%. Only 606 TM targets satisfy the MCS criteria, resulting in a loss of 18 out of 86 structures solved. However, the time and effort spent on the complete proteome was three times greater than

TABLE 14.5
Three Different Target Filtering Strategies and Statistics Calculated from Primary Sequence

| | No Filtering | Target Filtering Schema | | |
| | | | 95% Crystal Conservation per Attribute | Maximum Clustering Strategy |
	Proteome Limits	Absolute Limits		
Filtering parameters: Targets not fulfilling these filtering requirements would be discarded as potential targets.	$30<=Length<=1690$ $3.7<=pI<=12.3$ $7.4<=\%Charged<=52.5$ - $1.56<=GRAVY<=1.62$ $0<=SEG<=177$ $0<=TMHMM<=16$ $0<=SignalP<=1$	$41<Length<813$ $4.3<pI<11.2$ $17<\%Charged<47$ - $0.96<GRAVY<0.61$ $SEG<62$ $TMHMM<1$ $SignalP<1$	$65<Length<500$ $4.6<pI<10.2$ $25<\%Charged<41$ - $0.70<GRAVY<0.10$ $SEG<32$ $TMHMM<1$ $SignalP<1$	$90<Length<480$ $4.6<pI<7.4$ $25<\%Charged<40$ - $0.50<GRAVY<0.10$ $SEG<35$ $TMHMM<1$ $SignalP<1$
Final number of targets after filtering (Initial target pool = 1877, i.e., *T. maritima* proteome)	1877	1232	875	606
*Number of proteins that crystallized eliminated (total = 465)	0	0	-118	-191
*Number of protein structures eliminated (total = 86)	0	0	-14	-18
† Chances of Crystallization per target	24.7%	37.7%	39.5%	45.1%
† Chances of Structure solution per target	4.6%	7.0%	8.2%	11.2%
‡Number of proteins crystallized with similar JCSG experimental effort	464 (actual)	708 (theoretical)	741 (theoretical)	847 (theoretical)
‡ Number of protein structures produced with similar JCSG experimental effort	86 (actual)	131 (theoretical)	154 (theoretical)	210 (theoretical)
Theoretical gain in number of structures	n/a	+45	+68	+124
Theoretical increase in JCSG pipeline throughput	n/a	52%	79%	145%

The initial pool of potential targets, the *T. maritima* proteome, contains 1877 ORFs. The analysis shows the predicted effect of different target selection strategies on total pipeline throughput, assuming that the same experimental effort that was devoted to the full shotgun analysis of the *T. maritima* proteome had been used on focused efforts against selected (filtered) sets of targets.

* Number of crystals or structures lost with respect to tier 1, if the proposed filtering schemas had been applied to the *T. maritima* proteome.

† Chances of crystallization and structural solution are calculated as 100* (number of tier1 crystals or structures remaining after filtering or final number of targets after filtering).

‡ Theoretical number of structures gained with similar JCSG resources was calculated extrapolating the chances of structure solution per target for a certain filtering schema to a set of initial targets equivalent in size to the *T. maritima* proteome.

Source: From Canaves, J.M., Page, R., Wilson, I.A., and Stevens, R.C. *J. Mol. Biol.*, 344, 977–991, 2004. With permission.

---

what the MSC filtered strategy would have used. If the same number (1877) of MCS-filtered targets are processed, then a structure output of 210, compared with 86, could be estimated, given the same amount of funding and effort.

The data from the crystallization strategy implemented for the TM proteome have been used successfully to examine parameters that correlate strongly with crystallization success; this has allowed the definition of specific rules that can select proteins with a high propensity for crystal formation. We have identified common characteristics of the TM proteins that were crystallized using a semi-automated structural genomics pipeline and, although some of the observations might be pipe-line-specific, the same methodology can be applied to optimize other HT structural genomics pipelines or to predict which modifications to a given protein are needed to increase the chance of success for other structural biology projects. The calculated parameters described here are derived from primary sequence data, and so they can be applied to any potential targets, even if the protein has no sequence or structural homology to other proteins. With no sequence or structural homology information, we have shown that these easily calculated, sequence-derived parameters can be used to improve the crystallization success rate and substantially reduce the time and cost per structure solved.

14.3.3 Nanovolume Crystallization and Crystal Quality

One of the intriguing results from the analysis of crystals produced from nanoliter-volume experiments has been the frequent anecdotal observation of improvements in the overall x-ray diffraction resolution of the protein structure compared with crystals grown by traditional microliter-scale vapor equilibration methods. These observations from nanoliter experiments bear a striking resemblance to the crystal quality enhancements produced in protein crystal growth under reduced gravity.[50–56] To verify these anecdotal observations, a pilot study was carried out in a collaboration between the groups at New Century Pharmaceuticals, Syrrx, and TSRI.[36] The study included preliminary analysis and computational simulations that were carried out to look for correlations between the observations and experiments done under microgravity conditions with the expectation that these may lead to a better understanding of the nanovolume crystallization experiment.

In a small-scale pilot study, the x-ray diffraction patterns of crystals of six different bacterial proteins in two different drop-volume sizes were analyzed to determine if any trends could be observed relating drop size (for a vapor diffusion crystallization experiment) and crystal quality. In each series of experiments, crystals were grown from the protein samples by sitting-drop vapor diffusion in 100-nL and 4-μL drops, respectively, under identical conditions using the same chemical stock solutions. Proteins that crystallized using distinct precipitating agents and in diverse regions of chemical space were selected. All crystals were frozen in liquid nitrogen using cryoprotectant solutions identical in composition to the reservoir solutions that were used to grow the crystals, supplemented with 25% (v/v) ethylene glycol. Five degree data wedges were collected (on the same beam under identical experimental conditions) on two to three crystals from each drop for each protein sample. The data were integrated and scaled to provide quantitative measures of two parameters that are indicative of crystal quality: (1) diffraction spot mosaicity (the measured width of x-ray peaks at half height), and (2) diffraction resolution of the crystals, which we defined as the resolution shell where the average spot intensity was twofold larger than the background. Examples of crystals grown in these experiments can be seen in Figure 14.10.

Analysis of equations describing convections in nanoliter volumes gives a relation between gravity and crystal size and thus suggests a common chord between these two approaches to solution crystal growth. Using the reliable spacecraft acceleration of $10^{-3}g$ and referencing a 10-mm crystal grown in space, these equations indicate that an equivalent earth-grown crystal would be impractically small; however, if small microcrystals can be analyzed to the same resolution as larger space crystals, then under certain circumstances, the small earth-grown crystal could yield advantages similar to those grown in space. A simple computational simulation implemented to understand the dynamics of buoyancy-induced flows produced in nanoliter volumes of proteins shows that these conditions occur when the drop size in which the crystal is grown approaches the size of the crystal. For a 40-micron crystal, the drop size is varied from 3 μL to 1 nL in volume, with the results shown in Figure 14.11. As the results indicate, the energy decreases gradually from a drop size of 3 μL to about 1 μL, at which time it starts a steep decline. The initial slope defines a condition wherein there is only a limited interaction between the crystal and the surrounding drop fluid–air interface. However, as the size of the drop approaches that of the crystal, the rate of dissipation of vorticity through friction is greatly increased by the confinement of convective flow. This would suggest that the ratio of the crystal size to drop volume is also an important parameter for crystal growth in small drops and may impart an additional order of magnitude attenuation of the disturbance velocities around the growing crystals.[57,58]

The results from the described pilot studies[36] are summarized as follows: (1) Of the six proteins studied, drop size seemed to have little effect on diffraction quality of crystals for two of the proteins (*Eschrichia facelis* malonyl-CoA:acyl carrier protein transacylase and *Haemophilus influenzae* ferric binding protein A); (2) for

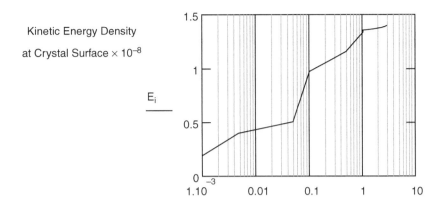

FIGURE 14.11 Kinetic energy density in a 5 micron layer surrounding the crystal surface versus the drop volume in microliters.[36] (From Carter, D.C., Rhodes, P., McRee, D.E., Tari, L., Dougan, D.R., Snell, G., Abola, E., and Stevens, R.C. *J. Appl. Crystallogr.*, 38, 87–90, 2005. With permission.)

E. coli NH_3-dependent NAD+ synthase, the qualities of the diffraction patterns from the nanodrop- and microliter-drop–grown crystals were equivalent, but many of the crystals grown from the larger drop were twinned (unusable), whereas all of the crystals selected from the nanoliter drops were single; (3) with *H. influenzae* aspartate semialdehyde dehydrogenase and *E. coli* undecaprenyl pyrophosphate synthase, crystals grown in the smaller drops yielded superior diffraction such that the average mosaicities of the diffraction patterns from crystals grown from nanodrops were 53% and 70%, respectively, of the mosaicities observed in the crystals grown in the microliter drops. Diffraction to higher resolution was also observed for the nanoliter drop grown crystals; and (4) for *E. coli* polypeptide deformylase, no usable crystals could be obtained from the larger drops. All of the diffraction peaks from the microliter volume crystals were streaked and could not be indexed, integrated, and scaled. Only crystals grown in the nanoliter drops provided usable x-ray diffraction data. A general qualitative trend was also observed, where the effects of drop size were more pronounced in drops with lower viscosity constituents (mainly dictated by the precipitating agent used).

As evidenced by the foregoing study and our experience since the mid-1990s when we introduced nanovolume crystallization approaches, advancements in several experimental methods (e.g., higher brilliance x-ray sources, advanced robotic systems for sample handling and data-collection) have allowed small crystals, in the order of 50 μm, to produce structures comparable with or superior to those using much larger crystals. It has also been postulated that the smaller volume and the faster crystallization times in which there is less time for damage to the protein sample (e.g., oxidation, deamidation, aggregation) are likely to have contributed to the improved crystal quality.[7] What was not clearly understood was whether there were other underlying physical phenomena that could help explain the marked improvement observed.

The analysis and simulation presented above indicate that since convective flows around a growing crystal scale as the cube of the size of the crystal, both microgravity and microcrystal experimental methods yield significant reductions in convective flow velocities. The end result from this change surrounding a growing crystal should lead to improvements in crystal quality. Simulations also show that for nanoliter experiments, the confinement of the crystal by the enclosed drop can yield to further reductions in convective flow velocities, perhaps as much as an additional order of magnitude. In addition, reductions in the flow have previously been implicated in a decrease in incorporation of impurities of growing crystals.[59] Hence, this pilot study, which is suggestive of the influence of convective flow on the quality of protein crystals, raises new questions and suggests new avenues to study the affect of solution and other convective disturbances on crystal growth.

14.4 FUTURE DIRECTIONS

Although robotics systems and the use of nanoliter drops in the crystallization process have allowed us to achieve important milestones, including the complete survey of a microbial proteome, much remains to be done. Our current success rate is 5 to 10% of the targets selected to structure determination, and although this rate significantly improves when multiple processing pathways are used (e.g., through design of several target constructs), it needs to be significantly improved to truly achieve HT rates. In addition, consistently producing high-quality crystals that diffract to resolutions better than 2 Å remains an open issue. Furthermore, most of the experience, to date, has been in working with single-chain globular proteins. Several projects are now underway in laboratories that will use approaches similar to that described here to extend the technology to work with membrane proteins[6] and to work with protein complexes.[5]

On the verge of breakthroughs are three new areas of development — whole gene synthesis, improved microfluidics, and in-house MAD data collection using the Compact Light Source. We are currently attempting to automate, integrate, and miniaturize the process of producing protein constructs primarily through the use of total gene synthesis.[60] With this approach, flexible or unstructured domains that may sterically or entropically impede crystallization can be identified and removed; domain boundaries can be better identified and the genes constructed to remove additional amino acids that may interfere with crystallization.[61] In addition, surface residues that may decrease solubility or hinder crystallization can be substituted.[44] Traditionally, these mutations are introduced using labor-intensive or time-intensive site-directed mutagenesis procedures. A second major advantage of this approach is that the ORFs can be optimized for protein expression in heterologous systems (e.g., *E. coli* and baculovirus-infected insect cells). The genes are synthesized to mimic the codon usage of the heterologous system, rare codons are eliminated, and mRNA secondary structures that can inhibit translation are removed. In addition, cryptic Shine-Dalgarno sequences, promoters, and splice sites are removed to significantly improve the quality of protein production. Finally, "stop" codons are introduced into the second and third reading frames to minimize production of aberrant protein products.

Several tools/instruments/processes will also be better integrated into the HT pipeline to add functionality, improve the success rates, and streamline the process. Microfluidics technology development has radically changed almost every liquid-based instrumentation, from laser printers to DNA/protein/small-molecule analysis. During the late 1990s, companies such as Fluidigm demonstrated the feasibility of microfluidics-based protein crystallization.[62] The current cost of microfluidic chips, however, remains prohibitive for most structural genomics efforts, especially in academia. While microfluidic technology is being established, the breakthrough cost reduction and full implementation into a structural proteomics pipeline has yet to be realized. The problem of aqueous evaporation through poly-(dimethyl siloxane) (PDMS) has been solved by designing a composite PDMS-glass[63] device that forms plugs and then injects them with no dispersion into a glass capillary (Figure 14.12).[64] The filled capillary is cut away from the device, sealed with water-impermeable wax, and incubated. So far, plugs and crystals in these capillaries have been stable.[63] The quality of the crystals grown in plugs in capillaries can be evaluated by on-chip diffraction at room temperature in their original mother liquor (see Figure 14.12). In addition, crystals that form in the plugs inside the capillary can be easily extracted, cryoprotected, and diffracted. Since crystals are grown at a liquid–liquid interface, they do not need to be scraped off a solid surface, and so damage is minimized. Using these devices, it is possible to extend the types of experiments that could be done. Ismagilov and his colleagues have now developed protocols for doing vapor diffusion, microbatch on oil, and free interface diffusion using the new microfluidic chips. Lastly, with almost no technology advancements in actual x-ray generation sources for several decades, we will see the installation of the first compact light source installed in a university laboratory setting.[60] This technology, which combines synchrotron radiation development with laser technology, has the potential to radically change the way in which we solve protein structures using x-ray crystallography.

ACKNOWLEDGMENTS

The authors greatly appreciate the efforts of Angela Walker in manuscript preparation and review. The authors are also very thankful to the collaborative efforts of Syrrx, Genomic Institute of the Novartis Research Foundation, The Scripps Research Institute, and the Joint Center for Structural Genomics, which all contributed to the work described in this chapter and by the many other colleagues who have helped advance the area of high-throughput structural biology. The work was supported in part by The Scripps Research Institute, the NIH PSI-1 Initiative Grant GM062411, and the NIH RoadMap Initiative grant GM073197.

REFERENCES

1. http://www.nigms.nih.gov/psi/ (accessed January 2005).
2. http://www.jcsg.org (accessed January 2005).
3. http://www.syrrx.com (accessed January 2005).

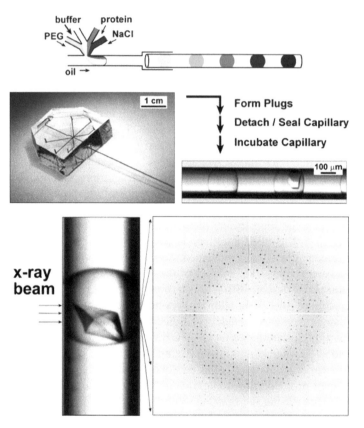

FIGURE 14.12 A) PDMS-glass composite microfluidics device for protein crystallization that does not have evaporation issues as observed with other microfluidics technologies. (B) APS synchrotron diffraction image (right) of a thaumatin crystals (left) grown and diffracted inside a plug on-chip to better than 1.8 Å (I/ (I) = 4.7 at 1.8 Å). (From Zheng, B., Tice, J.D., and Ismagilov, R.F. *Adv. Mater.*, 16, 1365–1368, 2004. With permission.)

4. Hosfield, D., Palan, J., Hilgers, M., Scheibe, D., McRee, D.E., and Stevens, R.C. A fully integrated protein crystallization platform for small molecule drug discovery. *J. Struct. Biol.*, 142, 207–217, 2003.
5. http://sars.scripps.edu (accessed January 2005).
6. http://jcimpt.scripps.edu (accessed January 2005).
7. Santarsiero, B.D., Yegian, D.T., Lee, C.C., Spraggon, G., Gu, J., Scheibe, D., Uber, D.C., Cornell, E.W., Nordmeyer, R.A., Kolbe, W.F., Jin, J., Jones, A.L., Jaklevic, J.M., Schultz, P.G., and Stevens, R.C. An approach to rapid protein crystallization using nanodroplets. *J. Appl. Crystallogr.*, 35, 278–281, 2002.
8. Peti, W., Norcross, J., Eldridge, G., and O'Neil-Johnson, M. Biomolecular NMR using a microcoil NMR probe — new technique for the chemical shift assignment of aromatic side chains in proteins. *J. Am. Chem. Soc.*, 126, 5873–5878, 2004.
9. Derwenda, Z.S. The use of recombinant methods and molecular engineering in protein crystallization. *Methods*, 34, 354–363, 2004.

10. Hendrickson, W.A., Horton, J.R., and LeMaster, D.M. Selenomethionyl proteins produced for analysis by multiwavelength anomalous diffraction (MAD): a vehicle for direct determination of three-dimensional structure. *EMBO. J.*, 9, 1665–1672, 1990.

11. Lesley, S.A., Kuhn, P., Godzik, A., Deacon, A.M., Mathews, I., Kreusch, A., Spraggon, G., Klock, H.E., McMullan, D., Shin, T., Vincent, J., Robb, A., Brinen, L.S., Miller, M.D., McPhillips, T.M., Miller, M.A., Scheibe, D., Canaves, J.M., Guda, C., Jaroszewski, L., Selby, T.L., Elsliger, M.A., Wooley, J., Taylor, S.S., Hodgson, K.O., Wilson, I.A., Schultz, P.G., and Stevens, R.C. Structural genomics of the *Thermotoga maritima* proteome implemented in a high-throughput structure determination pipeline. *Proc. Natl. Acad. Sci. USA*, 99, 11664–11669, 2002.

12. Holz, C., Hesse, O., Bolotina, N., Stahl, U., and Lang, C.. A micro-scale process for high-throughput expression of cDNAs in the yeast *Saccharomyces cerevisiae*. *Prot. Exp. Purif.*, 25, 372–378, 2002.

13. Shih, Y.P., Kung, W.M., Chen, J.C., Yeh, C.H., Wang, A.H.J., and Wang, T.F. High-throughput screening of soluble recombinant proteins. *Prot. Sci.*, 11, 1714–1719, 2002.

14. Adams, M.W.W., Dailey, H.A., DeLucas, L.J., Luo, M., Prestegard, J.H., Rose, J.P., and Wang, B.C. The Southeast Collaboratory for Structural Genomics: a high-throughput gene to structure factory. *Acc. Chem. Res.*, 36, 191–198, 2003.

15. Nguyen, H., Martinez, B., Oganesyan, N., and Rosalind, K. An automated small-scale protein expression and purification screening provides beneficial information for protein production. *J. Struct. Funct. Genom.*, 5, 23–27, 2004.

16. Page, R., Moy, K., Sims, E.C., Velasquez, J., McManus, B., Grittini, C., Clayton, T.L., and Stevens, R.C. Scalable high-throughput micro-expression device for recombinant proteins. *BioTechniques*, 37, 364–370, 2004.

17. Markley, J.L., Ulrich, E.L., Westler, W.M., and Volkman, B.F. Macromolecular structure determination by NMR spectroscopy. *Meth. Biochem. Anal.*, 44, 89–113, 2003.

18. Wüthrich, K. *NMR of Proteins and Nucleic Acids*. John Wiley & Sons, New York, 1986.

19. Wüthrich, K. NMR studies of structure and function of biological macromolecules. *J. Biomol. NMR*, 27, 13–39, 2003.

20. Olson, D.L., Peck, T.L., Webb, A., Magin, R.L., and Sweedler, J.V. High-resolution microcoil ^{1}H-NMR for mass-limited, nanoliter-volume samples. *Science*, 270, 1967–1970, 1995.

21. Peck, T.L., Magin, R.L., and Lauterbur, P.C. Design and analysis of microcoils for NMR microscopy. *J. Magnet. Reson. B.*, 108, 114–124, 1995.

22. Peti, W., Etezady-Esfarjani, T., Herrmann, T., Klock, H.E., Lesley, S.A., and Wuthrich, K. NMR for structural proteomics of *Thermotoga maritima*: screening and structure determination. *J. Struct. Funct. Genom.*, 5, 205–215, 2004.

23. Yokoyama, S. Protein expression systems for structural genomics and proteomics. *Curr. Opin. Chem. Biol.*, 7, 39–43, 2003.

24. Edwards, A.M., Arrowsmith, C.H., Christendat, D., Dharamsi, A., Friesen, J.D., Greenblatt, J.F., and Vedadi, M. Protein production: feeding the crystallographers and NMR spectroscopists. *Nat. Struct. Biol.*, 7 Suppl, 970–972, 2000.

25. Kennedy, M.A., Montelione, G.T., Arrowsmith, C.H., and Markley, J.L. Role for NMR in structural genomics. *J. Struct. Funct. Genomics*, 2, 155–169, 2002.

26. Yee, A., Chang, X., Pineda-Lucena, A., Wu, B., Semesi, A., Le, B., Ramelot, T., Lee, G.M., Bhattacharyya, S., Gutierrez, P., Denisov, A., Lee, C.H., Cort, J.R., Kozlov, G., Liao, J., Finak, G., Chen, L., Wishart, D., Lee, W., McIntosh, L.P., Gehring, K., Kennedy, M.A., Edwards, A.M., and Arrowsmith, C.H. An NMR approach to structural proteomics. *Proc. Natl. Acad. Sci. USA*, 99, 1825–1830, 2002.
27. Page, R., Peti, W., Wilson, I.A., Stevens, R.C., and Wuthrich, K. NMR screening and crystal quality of bacterially expressed prokaryotic and eukaryotic proteins in a structural genomics pipeline. *Proc. Natl. Acad. Sci. USA*, 102, 1901–1905, 2005.
28. Weber, P.C. and Salemme, F.R. Applications of calorimetric methods to drug discovery and the study of protein interactions. *Curr. Opin. Struct. Biol.*, 13, 115–121, 2003.
29. Leavitt, S. and Freire, E. Direct measurement of protein binding energetics by isothermal titration calorimetry. *Curr. Opin. Struct. Biol.*, 11, 560–566, 2001.
30. Torres, F.E., Kuhn, P., De Bruyker, D., Bell, A.G., Wolkin, M.V., Peeters, E., Williamson, J.R., Anderson, G.B., Schmitz, G.P., Recht, M.I., Schweizer, S., Scott, L.G., Ho, J.H., Elrod, S.A., Schultz, P.G., Lerner, R.A., and Bruce, R.H. Enthalpy arrays. *Proc. Natl. Acad. Sci. USA*, 29, 9517–9522, 2004.
31. Vale, R.D. and Milligan, R.A. The way things move: looking under the hood of molecular motor proteins. *Science*, 288, 88–95, 2000.
32. Goodwill, K.E., Tennant, M.G., and Stevens, R.C. High-throughput x-ray crystallography for structure-based drug design. *Drug Disc. Today*, 6, S113–S118, 2001.
33. Carotenuto, L., Cartwright, J.H., Castagnolo, D., Garcia Ruiz, J.M., and Otalora, F. Theory and simulation of buoyancy-driven convection around growing protein crystals in microgravity. *Micrograv. Sci. Technol.*, 13, 14–21, 2002.
34. Ng, J.D., Sauter, C., Lorber, B., Kirkland, N., Arnez, J., and Giege, R. Comparative analysis of space-grown and earth-grown crystals of an aminoacyl-tRNA synthetase: space-grown crystals are more useful for structural determination. *Acta. Crystallogr. D.*, D58, 645–652, 2002.
35. Vekilov, P.G. Protein crystal growth — microgravity aspects. *Adv. Space Res.*, 24, 1231–1240, 1999.
36. Carter, D.C., Rhodes, P., McRee, D.E., Tari, L., Dougan, D.R., Snell, G., Abola, E., and Stevens, R.C. Reduction in diffuse-convective disturbances in nanovolume protein crystallization experiments. *J. Appl. Cryst.*, 38, 87–90, 2005.
37. Luft, J.R., Collins, R.J., Fehrman, N.A., Lauricella, A.M., Veatch, C.K., and DeTitta, G.T. A deliberate approach to screening for initial crystallization conditions of biological macromolecules. *J. Struct. Biol.*, 142, 170–179, 2003.
38. Page, R., Grzechnik, S.K., Canaves, J.M., Spraggon, G., Kreusch, A., Kuhn, P., Stevens, R.C., and Lesley, S.A. Shotgun crystallization strategy for structural genomics: an optimized two-tiered crystallization screen against the *Thermatoga maritima* proteome. *Acta. Crystallogr. D.*, 59, 1028–1037, 2003.
39. Page, R. and Stevens, R.C. Crystallization data mining in structural genomics: using positive and negative results to optimize protein crystallization screens. *Methods*, 34, 373–389, 2004.
40. McPherson, A. Current approaches to macromolecular crystallization. *Eur. J. Biochem.*, 189, 1–23, 1990.
41. McPherson, A. Crystallization of macromolecules: general principles. *Meth. Enzymol.*, 114, 112–120, 1985.
42. Hamuro, Y., Burns, L.L., Canaves, J.M., Hoffman, R.C., Taylor, S.S., and Woods, V.L. Domain organization of D-AKAP2 revealed by enhanced deuterium exchange-mass spectrometry (DXMS). *J. Mol. Biol.*, 321, 703–714, 2002.

43. Bateman, A., Birney, E., Durbin, R., Eddy, S.R., Howe, K.L., and Sonnhammer, E.L.L. The Pfam Protein Families Database. *Nucleic Acids Res.*, 28, 263–266, 2000.
44. Longenecker, K.L., Garrard, S.M., Sheffield, P.J., and Derewenda, Z.S. Protein crystallization by rational mutagenesis of surface residues: Lys to Ala mutations promote crystallization of RhoGDI. *Acta. Crystallogr. D.*, D57, 679–688, 2001.
45. Mateja, A., Devedjiev, Y., Krowarsch, D., Longenecker, K., Dauter, Z., Otlewski, J., and Derewenda, Z.S. The impact of GluAla and GluAsp mutations on the crystallization properties of RhoGDI: the structure of RhoGDI at 1.3 Å resolution. *Acta. Crystallogr. D.*, D58, 1983–1991, 2002.
46. Cohen, S.L., Ferre-D'Amare, A.R., Burley, S.K. and Chait, B.T. Probing the solution structure of the DNA-binding protein MAX by a combination of proteolysis and mass spectrometry. *Prot. Sci.*, 4, 1088–1099, 1995.
47. Wootton, J.C. and Federhen, S. Analysis of compositionally biased regions in sequence databases. *Meth. Enzymol.*, 266, 554–571, 1996.
48. Sonnhammer, E.L.L., von Heijne, G., and Krogh, A. A hidden Markov model for predicting transmembrane helices in protein sequences. *Proc. Int. Conf. Intell. Syst. Mol. Biol.*, 6, 175–182, 1998.
49. Canaves, J.M., Page, R., Wilson, I.A., and Stevens, R.C. Protein biophysical properties that correlate with crystallization success in *Thermotoga maritima*: maximum clustering strategy for structural genomics. *J. Mol. Biol.*, 344, 977–991, 2004.
50. McPherson, A. Effects of a microgravity environment on the crystallization of biological macromolecules. *Micrograv. Sci. Technol.*, 6, 101–109, 1993.
51. Habash, J., Boggon, T.J., Raftery, J., Chayen, N.E., Zagalsky, P.F., and Helliwell, J.R. Apocrustacyanin C1 crystals grown in space and on earth using vapor-diffusion geometry: protein structure refinements and electron-density map comparisons. *Acta. Crystallogr. D.*, D59, 1117–1123, 2003.
52. Wardell, M.R., Skinner, R., Carter, D.C., Twigg, P.D., and Abrahams, J.P. Improved diffraction of antithrombin crystals grown in microgravity. *Acta. Crystallogr. D.*, D53, 622–625, 1997.
53. Cater, C.W. and Yin, Y. Quantitative analysis in the characterization and optimization of protein crystal growth. *Acta. Crystallogr. D.*, D50, 572–590, 1994.
54. Carter, D.C., Wright, B., Miller, T., Chapman, J., Twigg, P., Keeling, K., Moody, K., White, M., Click, J., Ruble, J.R., Ho, J.X., Adcock-Downey, L., Dowling, T., Chang, C.H., Ala, P., Rose, J., Wang, B.C., Declercq, J.P., Evrard, C., Rosenberg, J., Wery, J.P., Clawson, D., Wardell, M., Stallings, W., and Stevens, A. PCAM: a multi-user facility-based protein crystallization apparatus for microgravity. *J. Cryst. Growth*, 196, 610–622, 1999.
55. Declercq, J.P., Evrard, C., Carter, D.C., Wright, B.S., Etienne, G., and Parello, J. A crystal of a typical EF-hand protein grown under microgravity diffracts x-rays beyond 0.9 Å resolution. *J. Cryst. Growth*, 196, 595–601, 1999.
56. Thomas, B.R., Chernov, A.A., Vekilov, P.G., and Carter, D.C. Distribution coefficients of protein impurities in ferritin and lysozyme crystals. Self-purification in microgravity. *J. Cryst. Growth*, 211, 149–156, 2000.
57. Boistelle, R. and Astier, J.P. Crystallization mechanisms in solution. *J. Cryst. Growth*, 90, 14–30, 1988.
58. Luft, J.R., Albright, D.T., Baird, J.K., and DeTitta, G.T. The rate of water equilibration in vapor-diffusion crystallizations: dependence on the distance from the droplet to the reservoir. *Acta. Crystallogr. D.*, D52, 1098–1106, 1996.
59. McPherson, A., Kuznetsov, Y.G., Malkin, A., and Plomp, M. Macromolecular crystal growth as revealed by atomic force microscopy. *J. Struct. Biol.*, 142, 32–46, 2003.

60. http://www.gts10k.org (accessed January 2005).

61. Choi, K.H., Groarke, J.M., Young, D.C., Rossmann, M.G., Pevear, D.C., Kuhn, R.J., and Smith, J.L. Design, expression, and purification of a *Flaviviridae* polymerase using a high-throughput approach to facilitate crystal structure determination. *Prot. Sci.*,13, 2685–2692, 2004.

62. Hansen, C.L., Skordalakes, E., Berger, J.M., and Quake, S.R. A robust and scalable microfluidic metering method that allows protein crystal growth by free interface diffusion. *Proc. Natl. Acad. Sci. USA*, 99, 16531–16536, 2002.

63. Zheng, B., Tice, J.D., Roach, L.S., and Ismagilov, R.F. A droplet-based, composite PDMS/glass capillary microfluidic system for evaluating protein crystallization conditions by microbatch and vapor-diffusion methods with on-chip x-ray diffraction. *Angew Chem-Int. Edit.*, 43, 2508–2511, 2004.

64. Zheng, B., Tice, J.D., and Ismagilov, R.F. Formation of arrayed droplets by soft lithography and two-phase fluid flow, and application in protein crystallization. *Adv. Mater.*, 16, 1365–1368, 2004.

15 Electron Microscopy and Atomic Force Microscopy of Reconstituted Membrane Proteins

Andreas Engel

CONTENTS

Membrane proteins are membrane-embedded nanomachines that fulfill key functions such as energy conversion, solute transport, secretion, and signal transduction. The lack of structural information is related to the instability of membrane proteins in a detergent-solubilized state, making the growth of three-dimensional (3D) crystals difficult. Two-dimensional (2D) crystals of purified membrane proteins reconstituted in the presence of lipids provide an environment close to the native one and allow the structure and function of membrane proteins to be assessed. To this end, electron crystallography is used providing 3D information at the atomic level. Atomic force microscopy allows the surface of membrane proteins to be studied at subnanometer resolution, giving information about their conformational variability that cannot be assessed by crystallographic methods. In addition, atomic force microscopy is the ideal method to directly image the topography of the native membrane. This chapter presents methods to grow 2D crystals, to record them by cryoelectron microscopy, and to extract the structural information by digital image processing. Atomic force microscopy techniques are discussed and relevant results presented. The chapter is intended to provide a compact description of the current possibilities offered by these methods, and their power is demonstrated by the examples selected.

15.1 INTRODUCTION

Membrane proteins are lipid bilayer-embedded nanomachines that fulfill key functions such as energy conversion, solute transport, secretion, and signal transduction. Sequence information indicates that in eukaryotic cells, over 30% of all proteins are membrane bound. As a result of their central roles, membrane proteins are related to many diseases and represent 70% of all drug targets. Progress in membrane protein structure determination is mandatory for understanding the biological membranes and their functions and will provide the basis for the improvement of pharmaceutical therapies.

Major hurdles have hindered progress in membrane protein structure determination, whether achieved by nuclear magnetic resonance (NMR), x-ray, or electron crystallography. The first hurdle is the large-scale expression; membrane proteins cannot often be purified from natural sources in the amounts needed for structural analyses. This is especially true for proteins that are normally expressed at low levels and for proteins of human origin. The second hurdle concerns the purification and stabilization of the purified protein, which must be prepared at a high concentration for both NMR studies and crystallography. Here, the problems are mostly related to the instability of these proteins, which are fragile after extraction from the bilayer, making either the 3D crystallization for x-ray analyses or reaching the concentrations required for solution NMR difficult endeavors.

Reconstitution of membrane proteins into 2D crystals in the presence of lipids offers distinct advantages, as the native environment of membrane proteins as well as their biological activity is restored. However, a significant hurdle is related to the data acquisition and extraction of the structural information. Electron crystallography has not achieved the recent progress made in x-ray crystallography. Nevertheless, atomic resolution information can be extracted by cryoelectron microscopy from 2D crystals of high quality. In addition, atomic force microscopy (AFM) allows native and reconstituted biological membranes to be studied in aqueous solutions and the movement of polypeptide loops to be monitored. Thus, the combined use of electron crystallography and AFM allow atomic structure and function-related dynamics of native membrane proteins to be assessed. In this chapter, the 2D crystallization (2DX) of membrane proteins and their structural analyses by electron crystallography and AFM are discussed.

15.2 TWO-DIMENSIONAL CRYSTALLIZATION

15.2.1 DIFFERENT METHODS FOR 2D CRYSTALLIZATION

Two-dimensional crystals consisting of membrane proteins and lipids can be produced in three different ways.[1] The first method concerns the induction of regular packing of a highly abundant protein in its native membrane. This is achieved by eliminating interspersed lipids using lipases[2] or by extracting lipids with specific detergents.[3] Although this is the most gentle 2DX method, since it does not require solubilization of the membrane protein, it is not generally applicable.

The second method reconstitutes the purified membrane protein into a lipid bilayer at high protein density.[4] The detergent solubilized protein is mixed with solubilized lipids to form a homogeneous solution of mixed protein–detergent and lipid–detergent micelles. Detergent removal then results in the formation of protein aggregates in the worst case, and in the progressive formation of proteoliposomes with large 2D crystalline regions in the best case. Reconstitution begins once the detergent concentration reaches the critical micellar concentration (CMC).[5] The respective affinities between the components of the ternary mixture dictate the progress of the reconstitution process. Ideally, a starting condition should be established where mixed detergent–protein and mixed detergent–lipid micelles have exchanged their constituents to the extent that the mixture consists mainly of ternary detergent–protein–lipid micelles. Assuming that the protein remains in its native, properly folded state during the solubilization and isolation steps, this ideal situation is likely to foster perfect reconstitution and possibly 2DX of a functional membrane protein.

The third method concerns the reconstitution of the membrane proteins at the water–air interface by attaching the solubilized membrane protein to an active lipid monolayer prior to detergent removal.[6] In this process, membrane proteins are concentrated at the monolayer, brought into a planar configuration and finally squeezed together during detergent removal. This approach is useful for membrane proteins that are present in small amounts and are stably solubilized only in low CMC detergents.[7]

Common to all methods (Figure 15.1), the detergent is brought below its CMC to foster assembly of a bilayer, into which the membrane protein should integrate. Generally used methods to bring the detergent concentration below the CMC include dialysis,[5] adsorption of the detergent to Bio-Beads,[8] and dilution of the ternary mixture.[9] Moreover, in all methods, the amount of interspersed lipid must be minimized to ensure regular interactions among the membrane proteins. The pertinent interactions depend on the shape and surface charges of the components. For a given protein, the lipid–detergent mixture, pH, counter ions, and temperature must all be optimized. In addition, the concentration, the ratio of the respective components, and the detergent removal rate are critical. This gives a multidimensional parameter space that needs to be experimentally sampled, a task similar to that carried out in 3D crystallization screens. The difficulty of such experiments is the management of the screens and the assessment of results. With 2DX, the latter is particularly cumbersome because 2D crystals cannot be detected by light microscopy, and screening by electron microscopy is time-consuming. Figure 15.2 illustrates the possible reconstitution results and their interpretation.[10]

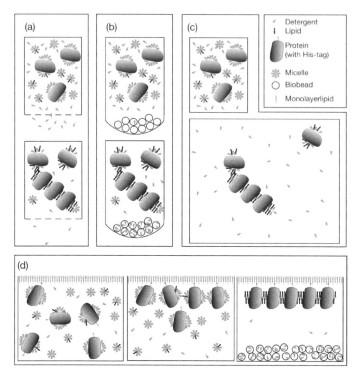

FIGURE 15.1 2DX modes. All modes are based on the principle to bring the detergent concentration in the aqueous phase below the CMC, forcing the detergent in the mixed micelles to partition in the aqueous phase. As result, mixed micelles merge to form larger structures and ultimately 2D crystals. (a) Dialysis can be used to remove the detergent provided its CMC is > 1 mM. (b) Bio-Beads adsorb detergent molecules and can be used for all detergents. Bio-Beads-driven 2D crystallization is particularly successful with low CMC detergents. (c) Dilution is a well-known method for functional reconstitution of membrane proteins. In spite of dilution it is also suitable for 2D crystallization because the protein is highly concentrated after integration in the bilayer. (d) The monolayer technique combines the Bio-Beads method with crystallization at the air–water interface. This method works only with low-CMC detergents because of the necessity to preserve the lipid monolayer. The latter incorporates special lipids having a high affinity for the solubilized protein, for example, by recognition of a specific tag. (By courtesy of Thomas Braun.)

15.2.2 Critical Parameters

The critical moment in the life of a membrane protein during the 2D crystallization process is the time point when the lipid molecules replace the detergent molecules to stabilize the protein.[11] The absorption–desorption kinetics of the detergent molecule is not necessarily the same for the mixed protein–detergent or the lipid–detergent micelles, respectively. As indicated above, the chance for the protein to interact with the appropriate lipids early enough increases when the crystallization cocktail

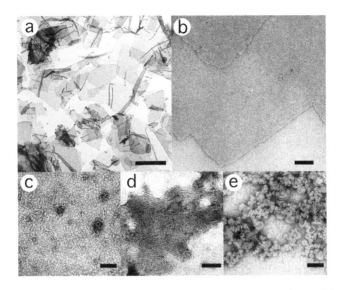

FIGURE 15.2 Electron microscope analysis of 2DX trials by negative staining. (a) Low magnification micrograph (2500 ×) showing large sheets of reconstituted aquaporin from *E. coli* (AqpZ). Scale bar represents 2 μm. (b) Typical micrograph of negatively stained highly ordered 2D crystals of AqpZ. A rectangular lattice can be seen when observed at glancing angle. (c) Micrographs of negatively stained membrane proteins and lipids, which were solubilized with the detergent octylglucoside, reveal mixed detergent–lipid micelles as small dots, while some lipid–protein–detergent structures have elongated (worm-like) shapes (< 10 nm). (d) Partially dialyzed samples often exhibit stacked multilamellar lipid structures. (e) Lipid vesicles of 20–50 nm diameter form when no membrane protein was incorporated (empty vesicles). Such structures appear when the LPR is too high, or when the protein aggregates. Scale bars represent 100 nm in b through e. (By courtesy of Philippe Ringler.)

can be incubated before detergent removal. If the interactions between the detergent–protein and detergent–lipid are not properly balanced during the reconstitution, unspecific aggregation occurs, and the protein is lost.

Many membrane proteins are destabilized upon solubilization,[7] especially when short-chain, high-CMC detergents are used. Smaller but harsher detergents are an advantage for dialysis-driven 2DX. Proteins can be effectively solubilized with low CMC detergents that replace the lipid and keep the hydrophobic surfaces of the protein shielded from water. Unfortunately, such detergents are not easily removed. Thus, the choice of detergent is critical: there is a fine balance between disruption of the membrane to solubilize a membrane protein and preserving its structural integrity. Furthermore, a suitable lipid mixture has to be found for a protein–detergent system to achieve reconstitution and crystallization of the protein.

With an excess of lipid over protein, the protein is mainly incorporated into lipid bilayers, similar to its native state. In an excess of protein over lipid, however, some of the protein aggregates, likely in a denatured form. An important parameter is therefore the lipid–protein ratio (LPR), which should be low enough to promote crystal contacts between protein molecules but not so low that the protein is lost to aggregation.[12] In addition to an appropriate lipid and detergent combination, 2DX of many membrane proteins depend on additives such as divalent ions, among them Mg^{2+}, having a special role, likely as result of its interaction with the lipids.[4]

15.3 ELECTRON MICROSCOPY

15.3.1 IMAGE FORMATION

Electrons interact strongly with matter, making it possible to depict such thin objects as 2D protein crystals. Electrons are elastically scattered by the nuclei of the atoms, which are orders of magnitude heavier than the moving electrons. Electrons are inelastically scattered by the inner- and outer-shell electrons, to which they transmit a fraction of their kinetic energy. Whereas elastic electrons contribute to the coherent axial bright-field image that carries the high-resolution information on the 3D arrangement of the sample atoms, the inelastic electrons carry interesting chemical information. Importantly, however, inelastic scattering is directly related to the beam-induced specimen degradation.

Since only the elastically scattered electrons contribute to a high-resolution image, the coherent-phase contrast image formation is considered. A thin object comprising only light elements (such as 2D membrane protein crystals) is approximately described as a weak-phase object:

$$t(x,y) = 1 + i\ \varphi(x,y), \qquad \varphi(x,y) < \pi/4 \qquad (1)$$

The amplitude distribution in the image plane is the coherent superposition of the unscattered wave and the elastically scattered waves, which is conveniently described as the convolution of the function $t(x,y)$ describing object with the point spread function $h(x,y)$ of the electron optical system:

$$a(x,y) = h(x,y)* (1 + i\varphi(x,y)) \qquad (2)$$

Not amplitudes but intensities $|a(x,y)|^2$ are recorded on the film. Omitting small quadratic terms, the image is thus written as:

$$|a(x,y)|^2 = 1 - 2\ h_i(x,y)* \varphi(x,y) \qquad (3)$$

The imaginary part $h_i(x,y)$ is described by the inverse Fourier transform of the phase contrast transfer function (CTF):

$$h_i(x,y) = FT^{-1}\ [A(p) \sin\ (\pi(C_s\ \lambda^3\ p^4/2 + \Delta f\ \lambda\ p^2))] \qquad (4)$$

where $A(p)$ describes the envelope of the CTF, C_s is the spherical aberration constant, Δf is the defocus, λ is the electron wavelength (about 0.02 Å for a 300-kV electron), and p is the distance from the origin in the reciprocal space. The CTF for weak-phase objects is displayed in Figure 15.3. The contrast is weak when the microscope is operated close to focus because the prominent low-resolution features of the

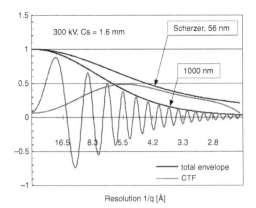

Resolution 1/q [Å]

FIGURE 15.3 CTF of a magnetic lens. The spherical aberration can be partially compensated by operating the lens at Scherzer focus. For a 300-kV instrument having a C_s = 1.6 mm, the Scherzer focus is at Δf = 56 nm (underfocus). Under this condition, all spatial frequencies are transferred with identical sign up to a resolution better than 3 Å. Contrast variations having spatial frequencies below $(15 \text{ Å})^{-1}$ are strongly attenuated, hence micrographs recorded at Scherzer focus exhibit weak contrast. A stronger contrast is obtained by further underfocusing the lens. In this case, the fine structures are transmitted with opposite signs. Source size and electron gun stability dictate the shape of envelope function, which limits contrast transfer at high resolution.

specimen are transferred with small amplitude. However, the contrast can be greatly enhanced moving out of focus because the phase difference between the scattered and unscattered electrons becomes $\pi/2$ or $\pi(n + \sqrt{2})$. The phase shift introduced by the electron optical system has to be corrected for correct image interpretation (see below). The phase shift of electrons scattered elastically by an atom is proportional to the coulomb potential of this atom. Therefore, the ensemble of all electrons singly scattered by a specimen produce a projection of its coulomb potential, which is dominated by the atom's nuclei rather than the electron shells as in x-ray crystallography.

Figure 15.4 shows the power spectrum of a 2D crystal image. This spectrum reveals (1) discrete spots representing the crystal information (see image processing), (2) concentric rings (Thon rings) with gaps in-between, where no structural information is available, and (3) the envelope modulation decreasing the signal intensity at higher resolutions. Thon rings and envelope function reflect the specific nature of the CTF. The decrease of contrast toward high resolution (envelope function) results from the partial incoherence of the electron beam. Modern electron microscopes are equipped with field emission guns exhibiting a high degree of coherence so that the contrast decrease is acceptable even at high resolution. The Thon rings with alternating positive and negative contrast are the result of the phase shift introduced by the electron optical system.

The great advantage to directly acquiring the phase information out to atomic scale resolution in a modern field-emission electron microscope is diluted by several experimental difficulties. First, the instrument has to be stable and

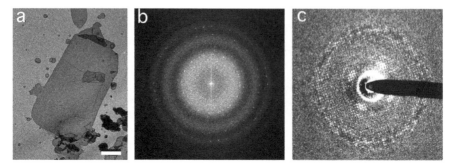

FIGURE 15.4 Electron micrographs and electron diffraction of 2D crystals. (a) Overview image of a vitrified 2D crystal. Scale bar corresponds to 2 μm. (b) Power spectrum of a high-resolution image of a 2D crystal: The periodic arrangement of the protein in the 2D crystal leads to a periodic signal in the recorded image. Therefore, the |modulus |² of the Fourier transform, that is, the power spectrum, shows discrete sharp spots, in which the structural information of the protein is concentrated. Furthermore, noise (all information outside the spots) and the contrast transfer function (CTF, see Figure 15.3) is visible. (c) Electron diffraction of a highly ordered 2D crystal, which is not affected by the CTF but does not carry phase information. Diffraction spots at the periphery of the pattern represent a resolution of $(3 \text{ Å})^{-1}$. (By courtesy of Thomas Braun.)

installed in a field- and vibration-free environment, prerequisites that are routinely reached. Second, beam-induced damage is not only changing the specimen structure, but it also leads to specimen charging. Since these charges act like an electrostatic lens, the focus changes during image recording. If the sample plane is perpendicular to the optical axis, the overall effect is quite small: the focus change occurring during image acquisition may have an influence only at very high resolution. However, when the sample is tilted for collecting the 3D information (see below), the electrostatic lens building up during irradiation introduces an image shift, a problem that cannot always be solved satisfactorily.[13] Third, a further optical defect is related to the changing focus for tilted specimens. In this case, the point-spread function is not space invariant, and the commonly used CTF correction is only an approximate measure to eliminate the phase distortions of the electron optical system. Last but not the least, the depth of focus must be considered, and it limits the thickness of the object to be imaged at high resolution. This limitation is not of concern, if thin samples such as 2D crystals are studied with a 300-kV instrument.

15.3.2 Electron Diffraction

The CTF, the envelope function, and specimen charging do not affect electron diffraction. Therefore, electron diffraction is much more effective for the collection of high-resolution information than is imaging, although the phase information is not retrieved. The directly measured amplitudes can be combined with the phases from the images during image processing. Electron diffraction is not an absolute requirement for determining a structure, but it allows a fast evaluation of the crystal quality, helps correct the CTF, and provides suitable high-resolution information for molecular replacement methods.[14]

15.3.3 SPECIMEN PREPARATION

The major advantage of electron microscopy over x-ray diffraction is its ability to directly record an image and not just a diffraction pattern of the specimen under investigation. However, two major problems have to be overcome to reach high-resolution 3D structures: (1) the strong interaction of electrons with matter leads to rapid sample destruction; and (2) the 2D crystals have to be prepared such that the high-resolution structure is preserved within the vacuum of the electron optical system. Interestingly, the same answer was found for both problems by one technique: cryo electron microscopy.[15] With this technique, crystals are adsorbed onto a thin carbon layer, surplus liquid is blotted away, and the sample is frozen very quickly in liquid ethane. By this preparation method, the crystals are embedded in a thin layer of amorphous ice or in a layer of carbohydrates substituting water; this conserves the fine structures of the 2D crystal. As a result of the low temperature of the specimen (at least liquid nitrogen temperature, 90 K), the sample is less susceptible to beam-induced damage, allowing images to be recorded at electron doses of 5 $e^-/Å^2$. If the sample is cooled down to liquid helium temperature (4.2 K), electron doses of 20 $e^-/Å^2$ are possible without loss of high-resolution structural information ionic.[16]

To efficiently screen the results of 2DX experiments, the quickest preparation method is to embed the sample in a heavy-metal salt solution and let it dry in air. In spite of unphysiological pH and ionic strength, this approach is able to preserve the sample structure to a resolution in the nanometer range that suffices to evaluate the sample quality in a first screen. Negatively stained samples are inspected at room temperature, and the high contrast provided by the heavy-metal salt and beam resistance greatly facilitates the identification of appropriate 2DX conditions (see Figure 15.2).

15.3.4 DATA PROCESSING

Images recorded by cryoelectron microscopy have a very low signal-to-noise ratio. Therefore, the structure of a 2D crystal can only be seen after image processing. In the power spectrum, the crystallinity of the sample is manifested in discrete spots containing the structural information of the periodically arranged protein (Figure 15.5). In contrast to x-ray crystallography, not only the amplitude of the diffraction pattern but also the corresponding phase can be measured, which are both read out from the calculated Fourier transform. A Fourier peak filtering step allows noise reduction (Figure 15.5c). After this simple image-processing step, some structural features can usually be recognized in the filtered image (Figure 15.5d, inset). Crystal defects (distortions, lattice defects, mosaicity) become clearly visible. Before the processing steps for eliminating crystal defects can be applied, the CTF correction must be implemented. This involves the sign changes required to eliminate the phase jump introduced by the optical system. In addition, amplitude attenuation introduced by the CTF is corrected using a Wiener filter. To correct the lattice distortions, the cross-correlation of the image with a small reference (Figure 15.5e) is calculated, and the real position of the unit cells are detected (Figure 15.5f). The reference may be a small area of the original image housing a few unit cells or may be the combined average from other images. The unit-cell positions are compared with the theoretical lattice vectors, and a field of shift vectors describing the crystal distortions can be calculated (Figure 15.5g). This information is used to interpolate the original image along the curved lattice lines to unbend the distorted image of the 2D crystal. By this real-space procedure, the diffraction spots become sharper (see insets in Figure 15.5h). This combination of crystallographic methods in Fourier space and image processing methods in real space allows the resolution to be improved by a factor of 2.[17]

To get a 3D structure, the 2D crystal has to be tilted in the electron microscope so that "side views" of the protein are recorded (Figure 15.6). These images are Fourier peak filtered and unbent in the same way as the untilted ones. However, the CTF correction is complicated by the fact that the point spread function is not space invariant, requiring an additional step to take the effects of the tilt into account.[17] The information from different images is then combined in the Fourier space according to the central section theorem, shifting the phase center of each projection to the origin of the Fourier space. The amplitude and phases of the crystals are concentrated on lattice lines in the z^*-direction (see Figure 15.6b). Since there is no repetition in the z-direction of the crystal, the lines are sampled continuously in the Fourier space and can be interpolated to obtain a regularly spaced 3D sampling (Figure 15.6c). Note that not all of the Fourier space can be sampled as result of the maximal tilt angle (around 60 degrees) of the sample, thus leaving a cone without data.

In theory, zero-degree and high-tilt-angle projections would give all of the information required to calculate a high-resolution 3D representation of the unit cell. Practical limitations necessitate the recording of images over the entire tilt-angle range available. Bending of the carbon support film and the 2D crystals leads to crystal distortions. This has a larger effect at high-tilt angles than around zero-degree tilts. The second problem concerns beam-induced specimen charging, as discussed above. All effects result in a decrease of the resolution perpendicular to the tilt axis, as illustrated in Figure 15.6d. Finally, a continuous set of tilted projections greatly reduces the initial problem of merging the 2D data in 3D space.

Interpreting a cryoelectron microscopy potential map in terms of an atomic protein structure is still a challenging task, even when relatively high-resolution data of better than 4 Å are available. The first step is to determine the macromolecular fold and to trace the backbone through the map, which can be achieved by visual inspection. Structural clues can further be obtained through bioinformatics methods: Extensive sequence alignments and the analysis of correlated mutations have led to valuable insights in the case of aquaporin-1 (AQP1).[18] In the case of α-helical proteins, automated procedures have been developed to determine both the location and the direction of individual helices in the map.[19] When used in conjunction with constraints from the sequence (e.g., the maximal length of a loop), macromolecular folds can be unambiguously derived. Once the

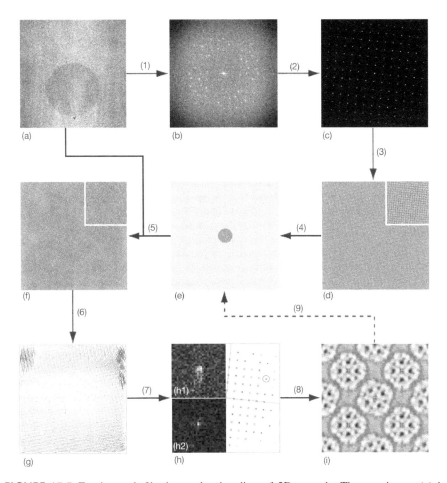

FIGURE 15.5 Fourier peak-filtering and unbending of 2D crystals. The raw image (a) is Fourier transformed (1) and the crystal lattice is indexed in the power spectrum (b) of the raw image. Note that in this case, two crystalline layers of the flattened crystalline vesicle can be separated as they are rotated with respect to each other. For the Fourier peak-filtering (2), the spots containing all the crystal information are cut out, and the amplitudes outside the mask-area (containing the other crystal layer and noise) is set to 0 (c). The inverse Fourier transform (3) reveals the packing of the crystal (d insert). To unbend the 2D crystal, a reference (e) is generated (4) and the cross-correlation function (5) with the raw image is calculated. The cross-correlation (f) reveals the positions of the unit cells. These are compared with the ideal crystal lattice (6) to generate distortion vectors (g). This information is then used to interpolate the raw image along the curved lattice lines to unbend the crystal. This step is only executed after the CTF correction (7). As a result, the spots of the power spectrum are better focused (h): In inset h1, peak 5,3 (indicated with a circle) is depicted before unbending and in h2 after unbending. Amplitudes and phases of the spots are read out and combined with the data of other crystals (8) to yield a final projection map. In this case, the result is the 3.7 Å map of GlpF (i) revealing the typical tetrameric structures of an aquaglyceroporin. This map can also be used to generate a synthetic reference for the unbending procedure (9). (By courtesy of Thomas Braun.)

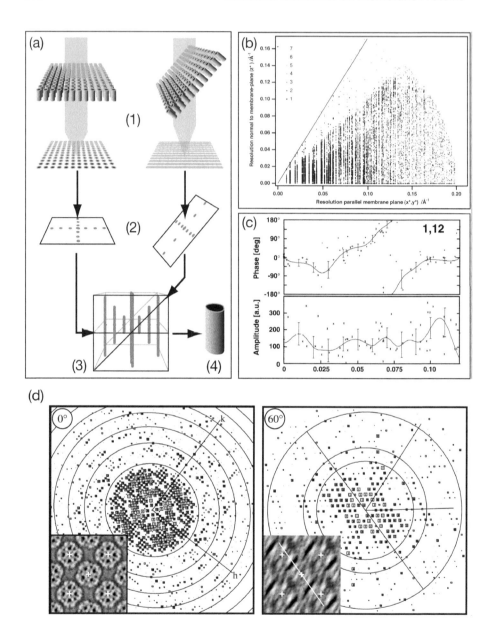

FIGURE 15.6 3D reconstruction of membrane proteins by 2D crystallography. (a) To obtain a 3D reconstruction from 2D crystals, projections are recorded at different tilt angles (1). The images are Fourier filtered and processed as described in Figure 15.5 (2), and the Fourier transforms are combined in the 3D Fourier space according to the central section theorem (3). The discrete orders in the Fourier transform from the crystal are aligned in continuous lattice lines, since the sample is not periodic in the z-direction. The lattice lines are regularly interpolated to sample the 3D Fourier space on a cubic raster. Back-transformation of the combined data finally leads to the representation of the 3D unit cell (4). (b) Azimuthal projection of the sampling in z*-direction shows the sampling of the Fourier space. In this case, the maximal nominal tilt angle was 60 degrees, as indicated by a black line revealing the missing cone. An example of a lattice line is given in panel (c): amplitude and phase values of lattice line 1,12 extend to a z*-resolution of $(7 \text{ Å})^{-1}$. The plotted curve indicates the interpolation of the experimental data. (d) Power spectra of an untitled and 60-degree-tilted 2D crystal. The inset show the Fourier-filtered projection map from an unbent image. Perpendicular to the tilt axis (line in the 60-degree panel) the resolution is reduced as result of support nonflatness and charging. (By courtesy of Thomas Braun.)

fold has been determined, an initial model of the backbone structure is generated. What follows are rounds of manual model (re)building and refinement. The structures of bacteriorhodopsin (bR)[17,20,21] and AQP1[22,23] have been solved by electron crystallography in this way and have been subsequently confirmed by x-ray crystallography. These successes demonstrate that even at a resolution of around 3.5 Å, a cryoelectron microscopy map contains sufficient structural information to uniquely define the atomic structure. The practical challenge is to find this optimal structure in the high-dimensional search space. Since an exhaustive search is not feasible, there is no straightforward general method to arrive at the best possible solution at this level of resolution. The available tools are for a substantial part the same as used in x-ray crystallography. In contrast to x-ray crystallography, experimental phase information is available in the case of electron microscopy. This advantage is not as yet fully exploited and requires the development of novel automated techniques specifically targeted at model building and refinement of electron microscopy data.

15.4 ATOMIC FORCE MICROSCOPY

15.4.1 Image Formation

The only instrument that provides subnanometer resolution and can be operated in solution is the atomic force microscope (AFM).[24,25] In the use of this instrument, a sharp stylus at the end of a flexible cantilever is scanned over the sample surface submerged in buffer solution. Cantilever deflections measured at a resolution of about 1 Å is used to determine the surface contour of the sample, exploiting a sensitive servo system to minimize the force applied to the stylus.

Biomolecular forces lie between a few pN (1 pN = 10^{-12} Newton), as generated, for example, by myosin, and 250 pN, as required to dissociate the streptavidin–biotin complex. It is important to ensure that imaging forces between stylus and sample

are of similar magnitude because higher forces would lead to sample distortion or
even disruption. Electrostatic and van der Waals interactions generate the major
forces interacting between AFM stylus and biological sample when imaging in buffer
solution.[26] Force–distance curves, which are acquired by approaching the sample to
the stylus while measuring its deflection, reveal the nature of these forces. Adjusting
electrolyte concentration and pH of the buffer can minimize these forces. In addition,
AFM feed-back parameters must be optimized to ensure distortion-free imaging,[27]
and cantilevers exhibiting a force constant of about 0.1 N/m must be used. Under
such conditions, topographs of protein surfaces revealing details with a lateral
resolution of 5 Å and a vertical resolution of 1 Å can be recorded routinely.
Importantly, such topographs exhibit an outstanding signal-to-noise ratio, enabling
recognition of single molecule structural features with unprecedented clarity (Figure
15.7 through Figure 15.9).

15.4.2 Observing Membrane Proteins at Work

Because the tip raster scans over the protein surface, the small lateral forces may
displace the biomolecule. Therefore, specimens need to be immobilized on the
solid support such as freshly cleaved mica. Immobilization of membranes requires
small forces that are of the same nature as those interacting between tip and
sample. Thus, pH and ion strength have to be adjusted to allow specimens to
adsorb as a result of van der Waals attractive forces.[28] This procedure is simple
and preserves the biological activity of reconstituted membrane proteins. A relevant example is the structurally and functionally well-characterized porin OmpF,
a trimeric membrane channel, which is a major outer membrane protein of
Escherichia coli. It comprises 16 antiparallel β-strands forming the transmembrane pore. The strands are connected by short turns at the periplasmic (Figure
15.7A) and long loops at the extracellular surface, forming a domain that pro-
trudes by 13 Å above the bilayer (Figure 15.7B). Three conditions induce a
displacement of this domain toward the trimer center, resulting in a structure with
a height of only 6 Å[29]: (1) application of an electric potential > 200 mV across
the membrane, (2) generation of a K^+ gradient > 0.3 M (Figure 15.7d), and (3)
pH 3 (Figure 15.7c). This displacement of the extracellular domain leads to the
closure of the channel, suggesting a protective function at low pH: *E. coli* cells
passing through the acidic milieu of a stomach may survive longer by closing
the outer membrane pores. The first condition, however, is compatible with results
from black lipid membrane experiments, which have demonstrated that porin is
a voltage-gated channel.[30]

15.4.3 Imaging Native Membranes

The imaging power of the AFM has been exploited to reveal the supramolecular
arrangement of membrane proteins in native membranes. Examples concern bacterial
photosynthetic systems comprising antenna proteins as well as reaction centers.[31,32]
Importantly, disc membranes of vertebrate retinal photoreceptor rod outer segments
(ROS) have been imaged by AFM (Figure 15.8). In ROS disc membranes, rhodopsin

FIGURE 15.7 Conformational changes of porin OmpF. (A) The atomic model of the periplasmic surface rendered at 3 Å resolution (left) exhibits features that are recognized in the unprocessed topograph (right). Short β-turns comprising only a few amino acids are sometimes distinct (ellipses and circles). (B) Extracellular surface of OmpF. The comparison between atomic model (left) and topograph (right) illustrates that the loops, which protrude 13 Å from the bilayer, are flexible. The asterisk marks the twofold symmetry axis of the rectangular unit cells housing two porin trimers. (C) pH-dependent conformational change of the extracellular surface. At pH 3, the flexible loops reversibly collapse toward the center of the trimer thereby reducing their height by 6 Å. (D) Conformational change of porin induced by an electrolyte gradient. The monovalent electrolyte gradient across the membrane was > 300 mM. Similar to the pH-dependent conformational change, the extracellular domains reversibly collapsed onto the porin surface. Scale bars are 50 Å, and topographs exhibit a vertical range of 10 Å (A), 15 Å (B), and 12 Å (C, D).

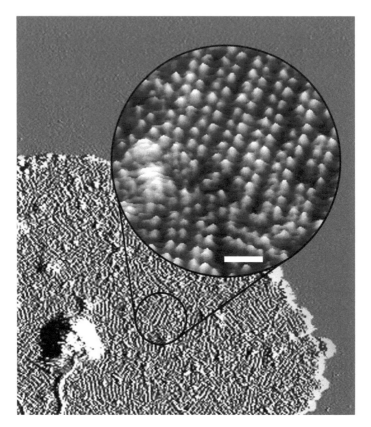

FIGURE 15.8 Topograph of native disc membranes recorded with the AFM reveals higher oligomers of rhodopsin. The overview shows part of a disc membrane that is tightly packed. The inset displays the magnified view of the disk membrane. Rows of rhodopsin dimers separated by 80 to 90 Å are distinct. The scale bar represents 100 Å. (By courtesy of Dimitrios Fotiadis.)

occupies ~ 50% of the space within the discs. Since rhodopsin is the archetypal G protein-coupled receptor (GPCR), these topographs demonstrating the existence of rhodopsin dimers and higher oligomers has had a great impact.[33,34]

How GPCRs operate is one of the most fundamental questions in the field of transmembrane signal transduction. For decades, the concept of how GPCRs, in particular rhodopsin, function in the disc membrane of retinal ROS has been dominated by the hypothesis that in order to encounter the membrane-associated G protein transducin, rhodopsin must rapidly diffuse as a monomeric unit in the fluid disc membrane. This view was supported by biophysical measurements of rhodopsin diffusion and rotation in disc membranes, as well as by low-resolution neutron diffraction, mostly carried out in amphibian photoreceptors. A growing body of pharmacological, biochemical, and biophysical data, however, strongly indicate that these receptors form functional homo- and heterodimers as well as

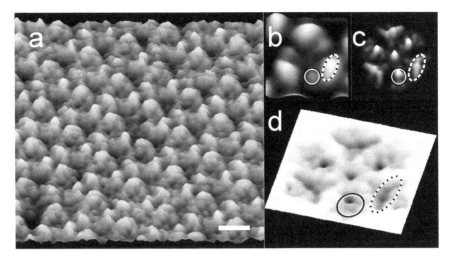

FIGURE 15.9 Atomic force microscope imaging of protein dynamics (for details: see text). (a) Cytoplasmic surface of a purple membrane crystal. (b) Average configuration of 1403 trimers. (c) Map showing the probability p to find a maximum at a position (x, y). (d) The corresponding energy landscape. The scale bar in (a) represents 100 Å and the height represented by the full gray scale is 8 Å. (b)–(d) have frame sizes of 77 Å, and the energy in (d) represented by the full grayscale is $\Delta E = 4.5$ kT. The solid circles and dashed ellipses in (b)–(d) denote the positions of the loops connecting helices A with B and E with F, respectively.

higher-order oligomers and that their oligomeric assembly may have important functional roles.[35–39] The direct visualization of higher oligomers of rhodopsin in the native disc membrane appears to resolve the discrepancy between early biophysical analyses and more recent biochemical observations.

15.4.4 Nanomanipulation

The stylus allows single-protein complexes to be dissected, as demonstrated by the mechanical disassembly of photosystem I complexes, thereby revealing the interface between the membrane resident proteins and the extrinsic proteins.[40] Even without imaging, a wealth of information on the unfolding of proteins and the nature of molecular interactions has been collected by recording force-extension curves. Single proteins can be extracted from a supramolecular structure and the resulting vacancy produced identified by recording a high-resolution topograph after the unzipping event. While initial experiments carried out with the hexagonally packed intermediate (HPI) layer revealed the spokes connecting adjacent hexamer to be the weakest bond,[41] force-extension curves acquired with purple membrane could be interpreted to greater detail based on the atomic structure of bacteriorhodopsin (bR).[42] Here it was observed that transmembrane α-helices unfold in different pathways.[43] Single molecule force spectroscopy has opened an avenue to assess the forces driving the assembly of membrane proteins at the level of single molecules.[44–46]

15.4.5 DATA PROCESSING

As a result of the superb signal-to-noise ratio, images recorded by the AFM do not require averaging to recognize submolecular fine structure. Indeed, averages often are less crisp than original topographs because the AFM is capable of sampling the conformational space of membrane protein surfaces. This has been demonstrated by analyzing topographs of porin OmpF, aquaporin-Z, and bR, all recorded at a lateral resolution of < 7 Å, and a vertical resolution of ~ 1 Å.[47] The example in Figure 15.9 shows the cytoplasmic surface of bR. The major protrusions are at the periphery of the trimers (see average structure in Figure 15.9b) and protrude about 8 Å out of the membrane. This feature results from the loop connecting transmembrane helices E and F, (E–F loop; marked by the ellipses in Figure 15.9b through Figure 15.9d). The smaller protrusions reflect the shorter loop connecting helices A and B (A–B loop; marked with a solid circle in Figure 15.9b through Figure 15.9d). The loop, connecting helices C and D, is hidden by the large E–F loop. To record this image, the force applied to the stylus was approximately 50 pN, preventing a force-induced conformational change of the E–F loop. Interestingly, between different atomic models from x-ray crystallography, the variation of this region is pronounced: the E–F loop is involved in the contacts leading to the 3D crystals, and its conformation is dictated by the 3D packing arrangement of the bR molecules. Although all the bR trimers are identical in the 2D crystal displayed in Figure 15.9a, their surface structure varies significantly. These changes are related to the flexibility of the bR surface rather than any noise introduced by the AFM. The flexibility of the surface lowers the resolution of the ensemble, but structural details of each trimer are still discernable. The variations in the trimer structure can be used to calculate the probability $p(x,y)$ of finding a certain loop at a certain position (x,y) by mapping the corresponding peak positions of all individual bR trimers. The E–F loop exhibiting significant flexibility is more delocalized than is the A–B loop, which occupies a better-defined position. The lipid molecule at the threefold axis is only occasionally visible in the raw data, but if present, it is precisely localized in the center of the trimer. Therefore, a strong signal emerges in the probability map (Figure 15.9c), whereas no signal is present in the average (Figure 15.9b). Precise localization of a surface feature suggests the corresponding structure to be stabilized in a deeper potential well than does a floppy feature, which may exhibit a pronounced thermal motion. The position probability map, $p(x,y)$, is readily converted to a free energy landscape $G(x,y)$ using Boltzmann's law:

$$G(x,y) = -k_B T \ln\left[p(x,y) \right] \tag{5}$$

The energy landscape of the cytoplasmic side of bR is shown in Figure 15.9d. This information relies on the capability of the AFM to address individual molecules with sufficiently high resolution and signal-to-noise ratio and provides valuable information on the nanomechanical properties of a membrane protein surface. It is noteworthy that these properties are expected to change drastically upon binding of a ligand.

15.5 CONCLUSION

Electron microscopy and atomic force microscopy are powerful tools to analyze the structure and function of membrane proteins. These tools allow membrane proteins to be studied in their native environment, where their function is preserved.[48] Technical problems concerning instrumentation, data acquisition technology, and image processing still remain to be solved, but there is no fundamental limitation that would inhibit the achievement of atomic scale insight into functional proteins embedded in a lipid bilayer. Using crystallization robots akin to those now available for 3D crystallization can solve the bottleneck of 2D crystallization. The real bottleneck thus is the necessity to express the membrane protein of interest in sufficient quality and quantity for structural analyses.

ACKNOWLEDGMENTS

This work was supported by the Maurice E. Müller Foundation of Switzerland, the Swiss National Foundation for Research and the National Center of Competence in Structural Biology. The author is indebted to his students who contributed many excellent figures to this review.

REFERENCES

1. Braun, T. and Engel, A. Two-dimensional electron crystallography. *Nat. Encyc. Life Sci.*, A0003044, in press, 2005.
2. Mannella, C.A. Phospholipase-induced crystallization of channels in mitochondrial outer membranes. *Science*, 224, 165–166, 1984.
3. Unger, V.M., Kumar, N.M., Gilula, N.B., and Yeager, M. Expression, two-dimensional crystallization, and electron cryo-crystallography of recombinant gap junction membrane channels. *J. Struct. Biol.*, 128, 98–105, 1999.
4. Jap, B.K., Zulauf, M., Scheybani, T., Hefti, A., Baumeister, W., Aebi, U., and Engel A. 2D crystallization: from art to science. *Ultramicroscopy*, 46, 45–84, 1992.
5. Engel, A., Hoenger, A., Hefti, A., Henn, C., Ford, R.C., Kistler, J., and Zulauf, M. Assembly of 2-D membrane protein crystals — dynamics, crystal order, and fidelity of structure analysis by electron microscopy. *J. Struct. Biol.*, 109, 219–234, 1992.
6. Levy, D., Chami, M., and Rigaud, J.L. Two-dimensional crystallization of membrane proteins: the lipid layer strategy. *FEBS Lett.*, 504, 187–193, 2001.
7. Garavito, R.M. and Ferguson-Miller, S. Detergents as tools in membrane biochemistry. *J. Biol. Chem.*, 276, 32403–32406, 2001.
8. Rigaud, J.L., Mosser, G., Lacapere, J.J., Olofsson, A., Levy, D., and Ranck, J.L. Bio-Beads: an efficient strategy for two-dimensional crystallization of membrane proteins. *J. Struct. Biol.*, 118, 226–235, 1997.
9. Remigy, H.W., Caujolle-Bert, D., Suda, K., Schenk, A., Chami, M., and Engel, A. Membrane protein reconstitution and crystallization by controlled dilution. *FEBS Lett.*, 555, 160–169, 2003.

10. Ringler, P., Heymann, B., and Engel, A. Two-dimensional crystallization of membrane proteins, in: Baldwin, S.A., ed., *Membrane Transport*. Oxford University Press, 2000, pp. 229–268.

11. Dolder, M., Engel, A., and Zulauf, M. The micelle to vesicle transition of lipids and detergents in the presence of a membrane protein: towards a rationale for 2D crystallization. *FEBS Lett.,* 382, 203–208, 1996.

12. Hasler, L., Heymann, J.B., Engel, A., Kistler, J., and Walz, T. 2D crystallization of membrane proteins: rationales and examples. *J. Struct. Biol.,* 121, 162–171, 1998.

13. Gyobu, N., Tani, K., Hiroaki, Y., Kamegawa, A., Mitsuoka, K., and Fujiyoshi, Y. Improved specimen preparation for cryo-electron microscopy using a symmetric carbon sandwich technique. *J. Struct. Biol.,* 146, 325–333, 2004.

14. Gonen, T., Sliz, P., Kistler, J., Cheng, Y., and Walz, T. Aquaporin-0 membrane junctions reveal the structure of a closed water pore. *Nature,* 429, 193–197, 2004.

15. Dubochet, J., Adrian, M., Chang, J.-J., Homo, J.-C., Lepault, J., McDowall, A.W., and Schultz, P. Cryo-electron microscopy of vitrified specimens. *Quart. Rev. Biophys.,* 21, 129–228, 1988.

16. Fujiyoshi, Y. The structural study of membrane proteins by electron crystallography. *Adv. Biophys.,* 35, 25–80, 1998.

17. Henderson, R., Baldwin, J.M., Ceska, T.A., Zemlin, F., Beckmann, E., and Downing, K.H. Model for the structure of bacteriorhodopsin based on high-resolution electron cryo-microscopy. *J. Mol. Biol.,* 213, 899–929, 1990.

18. Heymann, J.B. and Engel, A. Structural clues in the sequences of the aquaporins. *J. Mol. Biol.,* 295, 1039–1053, 2000.

19. de Groot, B.L., Heymann, J.B., Engel, A., Mitsuoka, K., Fujiyoshi, Y., and Grubmüller, H. The fold of human aquaporin 1. *J Mol Biol* 300: 987-94, 2000.

20. Kimura, Y., Vassylyev, D.G., Miyazawa, A., Kidera, A., Matsushima, M., Mitsuoka, K., Murata, K., Hirai, T., and Fujiyoshi, Y. Surface of bacteriorhodopsin revealed by high-resolution electron crystallography. *Nature,* 389, 206–211, 1997.

21. Mitsuoka, K., Hirai, T., Murata, K., Miyazawa, A., Kidera, A., Kimura, Y., and Fujiyoshi, Y. The structure of bacteriorhodopsin at 3.0 Å resolution based on electron crystallography: implication of the charge distribution. *J. Mol. Biol.,* 286, 861–882, 1999.

22. Murata, K., Mitsuoka, K., Hirai, T., Walz, T., Agre, P., Heymann, J.B., Engel, A., and Fujiyoshi, Y. Structural determinants of water permeation through aquaporin-1. *Nature,* 407, 599–605, 2000.

23. de Groot, B.L., Engel, A., and Grubmüller, H. A refined structure of human aquaporin-1. *FEBS Lett.,* 504, 206–211, 2001.

24. Binnig, G., Quate, C.F., and Gerber, C. Atomic force microscopy. *Phys. Rev. Lett.,* 56, 930–933, 1986.

25. Drake, B., Prater, C.B., Weisenhorn, A.L., Gould, S.A., Albrecht, T.R., Quate, C.F., Cannell, D.S., Hansma, H.G., and Hansma, P.K. Imaging crystals, polymers, and processes in water with the atomic force microscope. *Science,* 243, 1586–1589, 1989.

26. Müller, D.J. and Engel, A. The height of biomolecules measured with the atomic force microscope depends on electrostatic interactions. *Biophys. J.,* 73, 1633–1644, 1997.

27. Müller, D.J., Fotiadis, D., Scheuring, S., Müller, S.A., and Engel, A. Electrostatically balanced subnanometer imaging of biological specimens by atomic force microscope. *Biophys. J.,* 76, 1101–1111, 1999.

28. Müller, D.J., Amrein, M., and Engel, A. Adsorption of biological molecules to a solid support for scanning probe microscopy. *J. Struct. Biol.,* 119, 172–188, 1997.

29. Müller, D.J. and Engel, A. Voltage and pH-induced channel closure of porin OmpF visualized by atomic force microscopy. *J. Mol. Biol.,* 285, 1347–1351, 1999.

30. Engel, A., Massalski, A., Schindler, H., Dorset, D.L., and Rosenbusch, J.P. Porin channel triplets merge into single outlets in *Escherichia coli* outer membranes. *Nature,* 317, 643–645, 1985.

31. Scheuring, S., Sturgis, J.N., Prima, V., Bernadac, A., Levy, D., and Rigaud, J.L. Watching the photosynthetic apparatus in native membranes. *Proc. Natl. Acad. Sci. USA,* 101, 11293–11297, 2004.

32. Bahatyrova, S., Frese, R.N., Siebert, C.A., Olsen, J.D., Van Der Werf, K.O., Van Grondelle, R., Niederman, R.A., Bullough, P.A., Otto, C., and Hunter, C.N. The native architecture of a photosynthetic membrane. *Nature,* 430, 1058–1062, 2004.

33. Fotiadis, D., Liang, Y., Filipek, S., Saperstein, D.A., Engel, A., and Palczewski, K. Atomic-force microscopy: rhodopsin dimers in native disc membranes. *Nature,* 421, 127–128, 2003.

34. Liang, Y., Fotiadis, D., Filipek, S., Saperstein, D.A., Palczewski, K., and Engel, A. Organization of the G protein-coupled receptors rhodopsin and opsin in native membranes. *J. Biol. Chem.,* 278, 21655–21662, 2003.

35. Lee, S.P., O'Dowd, B.F., and George, S.R. Homo- and hetero-oligomerization of G protein-coupled receptors. *Life Sci.,* 74, 173–180, 2003.

36. Overton, M.C., Chinault, S.L., and Blumer, K.J. Oligomerization, biogenesis, and signaling is promoted by a glycophorin A-like dimerization motif in transmembrane domain 1 of a yeast G protein-coupled receptor. *J. Biol. Chem.,* 278, 49369–49377, 2003.

37. Stanasila, L., Perez, J.B., Vogel, H., and Cotecchia, S. Oligomerization of the alpha 1a- and alpha 1b-adrenergic receptor subtypes. Potential implications in receptor internalization. *J. Biol. Chem.,* 278, 40239–40251, 2003.

38. Bouvier, M. Oligomerization of G-protein-coupled transmitter receptors. *Nat. Rev. Neurosci.,* 2, 274–286, 2001.

39. Milligan, G. Oligomerisation of G-protein-coupled receptors. *J. Cell Sci.,* 114, 1265–1271, 2001.

40. Fotiadis, D., Müller, D.J., Tsiotis, G., Hasler, L., Tittmann, P., Mini, T., Jenö, P., Gross, H., and Engel, A. Surface analysis of the photosystem I complex by electron and atomic force microscopy. *J. Mol. Biol.,* 283, 83–94, 1998.

41. Müller, D.J., Baumeister, W., and Engel, A. Controlled unzipping of a bacterial surface layer with an AFM. *Proc. Natl. Acad. Sci. USA,* 96, 13170–13174, 1999.

42. Oesterhelt, F., Oesterhelt, D., Pfeiffer, M., Engel, A., Gaub, H.E., and Müller, D.J. Unfolding pathways of individual bacteriorhodopsins. *Science,* 288, 143–146, 2000.

43. Janovjak, H., Kessler, M., Oesterhelt, D., Gaub, H., and Muller, D.J. Unfolding pathways of native bacteriorhodopsin depend on temperature. *EMBO. J.,* 22, 5220–5229, 2003.

44. Möller, C., Fotiadis, D., Suda, K., Engel, A., Kessler, M., and Müller, D.J. Determining molecular forces that stabilize human aquaporin-1. *J. Struct. Biol.,* 142, 369–378, 2003.

45. Janovjak, H., Struckmeier, J., Hubain, M., Kedrov, A., Kessler, M., and Muller, D.J. Probing the energy landscape of the membrane protein bacteriorhodopsin. *Structure (Camb.),* 12, 871–879, 2004.

46. Kedrov, A., Ziegler, C., Janovjak, H., Kuhlbrandt, W., and Muller, D.J. Controlled unfolding and refolding of a single sodium-proton antiporter using atomic force microscopy. *J. Mol. Biol.,* 340, 1143–1152, 2004.

47. Scheuring, S., Müller, D.J., Stahlberg, H., Engel, H.A., and Engel, A. Sampling the conformational space of membrane protein surfaces with the AFM. *Eur. Biophys. J.*, 31, 172–178, 2002.
48. Walz, T., Smith, B.L., Zeidel, M.L., Engel, A., and Agre, P. Biologically active 2-dimensional crystals of aquaporin CHIP. *J. Biol. Chem.*, 269, 1583–1586, 1994.

16 Structural Genomics Networks for Membrane Proteins

Kenneth H. Lundstrom

CONTENTS

Structure determination for membrane proteins has been extremely modest compared with soluble proteins. Among the more than 30,000 structures deposited in public databases, less than 100 have been obtained for membrane proteins. The low success rate is mainly due to the difficulties in expression, purification, and crystallization of membrane proteins. Generally, the quantity and quality of recombinantly expressed membrane proteins have not been sufficient. Moreover, the solubilization and crystallization in the presence of detergents severely affect the yields and reduce the stability of purified proteins. Structural genomics approaches, where a large number of targets are processed in parallel, have provided new means of developing high-throughput (HT) methods for optimization of technologies facilitating structure determination of membrane proteins. The establishment of large national and international networks has enhanced the productivity and, hopefully, will provide the tools for solving many new structures of membrane proteins in the near future.

16.1 INTRODUCTION

The strong interest in membrane proteins has been driven by their large repertoire of functions and also their importance as drug discovery targets. Today more than 70% of the drug targets are based on membrane proteins. The largest family of proteins of medical interest is the G protein-coupled receptors (GPCRs).[1] Additionally, a large number of ion channels, transporters, and efflux pumps are considered interesting drug targets today and perhaps even moreso in the future.

Although there has been such a strong interest in membrane proteins, paradoxically, very little is known about their three-dimensional structure. The main reason for this is their complex composition with generally several transmembrane-spanning regions. This topology has substantially affected their expression pattern resulting in significantly lower yields of recombinant proteins in categorically all expression systems verified. The expressed membrane proteins have been produced in either quantitatively or qualitatively insufficient amounts. To make the situation worse, the purification procedure of membrane proteins is also more difficult than for soluble proteins due to the requirement of strong detergents and the low stability of purified proteins. Finally, the crystallization of membrane proteins, which also has to take place in the presence of detergents, is more complicated than for soluble proteins.

For these reasons, it is not a big surprise that very few high-resolution structures of membrane proteins have been obtained. Today, less than 1% of the more than 30,000 entries deposited for soluble proteins in public databases are on membrane proteins.[2] The most important present development is technology improvement, at various stages, in structural biology, including the expression, purification, and crystallography. Although genuine progress has been achieved in all areas, as described in other chapters, substantial improvement can only be achieved by a systematic approach. This requires the comparison of different expression systems in parallel, studies on a large number of targets, and the introduction of automation and HT methods. The most efficient way to set up such large research programs is through the establishment of networks consisting of research teams with expertise in various fields. A network is commonly composed of a number of academic research teams and small or midsize biotechnology companies. Most importantly, strong emphasis is put on the coordination of the program and free information flow among the laboratories through an active database and good communication skills. Many of the established networks have popularly been named structural genomics initiatives. In this chapter, structural genomics programs studying membrane proteins are described (Table 16.1). They are divided into three types of consortia: (1) networks that concentrate on *prokaryotic targets* uniquely; (2) networks with both *pro- and eukaryotic targets*; and (3) networks with only *eukaryotic targets*.

16.2   NETWORKS ON PROKARYOTIC TARGETS

Prokaryotic targets have clearly been favored in comparison with eukaryotic membrane proteins. The main reason for this is the better success rate of expression and the possibility to use *Escherichia coli* as the host for protein production. A factor that has also favorably influenced certain prokaryotic targets has been their thermoresistant nature, which has facilitated the purification procedure.

The Southeast Collaboratory on Structural Genomics (SECSG) is funded by the National Institutes of Health (NIH) and has, among other programs, set up a structural genomics study on the whole genome of *Pyrococcus furiosus*. It is estimated that the relatively small 1.9 Mb genome of *P. furiosus* contains 2200 ORFs (open reading frames). Many of the ORFs code for proteins with unknown functions.[3] An

TABLE 16.1
Structural Genomics Networks Studying Membrane Proteins

Network	Target Proteins	Expression Host	Web Site
Prokaryotic			
SECSG (USA)	*Pyrococcus furiosus* genome	*E. coli*	www.secsg.org/pfu.htm
TB (USA)	*Mycobacterium tuberculosis* genome	*E. coli*	http://doe-mbi.ucla.edu/TB
JCSG (USA)	*Thermotoga maritima*	*E. coli*	www.jcsq.org
Structurome (Japan)	*Thermus thermophilus* genome	*E. coli* Cell-free	www.thermus.org
ProAMP (Germany)	*Salmonella typhimurium* *Helicobacter pylori*	*E. coli*	www.pst-ag.com
Prokaryotic & Eukaryotic			
E-MeP (Europe)	Pro- and eukaryotic MPs	*E. coli*, *L.lactis* *S. cerevisi ae*, *P. pastoris* Baculovirus, SFV	www.e-mep.org
SweGene (Sweden)	Bacterial, yeast and mammalian MPs	*E. coli* yeast	www.swegene.org
NCCR (Switzerland)	Bacterial and mammalian MPs	*E. coli*	www.structuralbiology.ethz.ch
Eukaryotic			
SGPP (USA)	Pathogenic protozoa	*E. coli* yeast	www.sgpp.org
JCSG (USA)	Mouse genome 182 GPCRs	*E. coli* Baculovirus Adenovirus	www.jcsg.org
SGCE (USA)	*C. elegans* genome	*E. coli*	www.sgce.cbse.uab.edu
MePNet	Mammalian GPCRs	*E. coli* *P. pastoris* SFV	www.mepnet.org

JCSG, Joint Center for Structural Genomics, SECSG, Southeast Collaboratory on Structural Genomics; ProAMP, Protein-wide Analysis of Membrane Proteins; NCCR, National Center of Competence in Research; SGPP, Structural Genomics on Pathogenic Protozoa; JCSG, Joint Center for Structural Genomics; SGCE, Structural Genomics on *C. elegans*; MePNet, Membrane Protein Network.

HT strategy for cloning, expression, and purification has been planned. The expression is carried out in conventional *E. coli* vectors harboring N-terminal hexa-histidine tags. Today, 86% of the genome has been cloned, more than 500 proteins expressed, and 22 crystal structures obtained. Currently, activities are in progress to initiate studies on membrane proteins, for which further developed prokaryotic and eukaryotic expression vectors might be used.

Another NIH-funded structural genomics initiative aims at studying the genome of *Mycobacterium tuberculosis*.[4] Of the 3924 ORFs, 1165 genes are suggested to code for membrane proteins. The network consists of 60 laboratories in nine countries. The expression of the target proteins will take place in *E. coli*. In addition to the importance of *M. tuberculosis* in human infectious disease, this organism has eukaryotic homologues such as serine/threonine kinases[5] and adenylate cyclases.[6] Today, some 30 structures of soluble proteins have been solved. The Joint Center for Structural Genomics (JCSG) has initiated a program on the 1877 ORFs at *Thermotoga maritima*, which is decribed in more detail in Chapter 14.

Within the RIKEN Genomics Science Center (GSC) the Structurome Project has focused on the small genome of *Thermus thermophilus*. More than 600 of the 1000 to 2000 postulated genes have been amplified by polymerase chain reaction (PCR) and expression is currently in progress in *E. coli* and a cell-free translation system. The advantage of studying *T. thermophilus* proteins is their thermostability and good crystallizability.[7]

A German network named Protein-wide Analysis of Membrane Proteins (Pro-AMP), funded by the German Research Ministry (BMBF) was recently established to study bacterial membrane proteins of pathogenic interest. In this context, selected membrane proteins from *Salmonella typhimurium* and *Helicobacter pylori* are subjected to x-ray crystallography, nuclear magnetic resonance (NMR) spectroscopy, and electron microscopy. The targets are expressed in *E. coli* and, when possible, in other suitable prokaryotic hosts. The structural information collected from the program will be used for the design of novel anti-infective drugs.

16.3 NETWORKS ON PROKARYOTIC AND EUKARYOTIC TARGETS

Quite a few networks have decided to include both prokaryotic and eukaryotic targets in their programs. The advantage of this approach is that homologous proteins exist among prokaryotes and eukaryotes, and generally, the success rate for expression, purification, and crystallography of bacterial targets has been higher. It is therefore expected that structure determination will be faster for prokaryotic targets and that empirically accumulated information from work on bacterial targets can be transferred to mammalian and other eukaryotic membrane proteins.

SweGene, a national network established in South Western Sweden, has a Membrane Protein Platform (MPP), which studies mainly bacterial and yeast membrane proteins. Much emphasis is placed on method development, particularly in the areas of host cell engineering, fermentation, protein production, and purification. Moreover, attention is given to crystallization methods applying lipid cubic phase crys-

tallography. The targets in SweGene are bacterial reaction centers, trans-hydroge-
nases, aquaporins, and glycerol facilitator proteins. Additionally, a few mammalian
GPCRs are included in the program. The recombinant membrane proteins are
expressed in *E. coli* and yeast cells.

Another national network, the National Center for Competence in Research
(NCCR), has been established in Switzerland among research teams from the Swiss
Federal Institute of Technology (ETH) in Zurich, the Universities of Zurich and
Basel, and the Paul Sherrer Institute (PSI) in Villigen. The program concentrates its
efforts on recombinant protein expression, protein engineering, computational biology and structure determination by x-ray crystallography, NMR spectroscopy, and
electron microscopy. Among the targets are both prokaryotic and eukaryotic membrane proteins such as adenosine triphosphate (ATP)–binding cassette (ABC) transporters and the *E. coli* DsbD system. Expression is mainly carried out in bacterial
systems, but mammalian expression is also performed. Scientists within the NCCR
have generated libraries containing naturally occurring ankyrin repeat proteins with
highly specific binding to a number of target proteins, which could lead to complex
formation as a means of facilitating crystallization.

Recently the five-year European Union (EU)–funded E-MeP (European Membrane Protein) network was established by bringing together some 20 research
teams from six European countries. This consortium will study 100 prokaryotic
and 200 eukaryotic membrane proteins. Among the targets are GPCRs, ion channels, transporters, and other membrane proteins. Six well-established expression
systems will be applied for recombinant protein expression. These are based on
E. coli,[5] *Lactococcus lactis*,[8] *Saccharomyces cerevisiae*,[9] *Pichia pastoris*,[10] baculovirus,[11] and Semliki Forest virus (SFV)[12] vectors. Furthermore, cell-free translation will be verified as an exploratory activity within the program. Evidently,
the prokaryotic targets will be subjected to bacterial expression, and the success
rate, in general, is anticipated to be much higher than for the eukaryotic membrane
proteins. However, some eukaryotic membrane proteins such as the six-transmem-
brane (6-TM) yeast mitochondrial carrier protein and the human KDEL (Lys-Asp-
Glu-Leu) receptor with a 7-TM topology have given at least moderate yields in
L. lactis.[9] Likewise, the neurotensin[13] and the adenosine A_{2a} receptor —[14] two
human GPCRs — have been expressed at relatively high levels in *E. coli*. However,
expression of mammalian GPCRs in *E. coli* membranes has required massive
construct engineering to find the appropriate fusion protein partner and purification
tags, and it is therefore doubtful whether this approach is compatible for a large
number of targets. Therefore, expression in insect and mammalian cells will be
of prime interest.

16.4 NETWORKS ON EUKARYOTIC TARGETS

There are a number of networks that uniquely concentrate on eukaryotic targets.
The main reason for this is that whole genomes are targeted. In these structural
genomics initiatives, obviously only a part of the studied targets are membrane

proteins. Alternatively, some networks have decided to target specific families of eukaryotic membrane proteins.

The Joint Center for Structural Genomics (JCSG) was established in California by The Scripps Research Institute (TSRI), the Genomics Institute for the Novartis Research Foundation (GNF), the San Diego Supercomputer Center (SDSC) at the University of California, San Diego (UCSD), and the Stanford Synchrotron Radiation Laboratory (SSRL, a Division of the Stanford Linear Accelerator Center, SLAC). Additional collaborators from other entities such as the Salk Institute and Syrrx Inc. are part of the consortium. The aim of this network is to study the whole mouse proteome. Much effort is dedicated to HT expression, purification, and crystallography methods. Expression studies will be mainly carried out in *E. coli*, but alternative expression vectors such as baculovirus will be applied for membrane proteins. Additionally, 182 GPCRs will be expressed from baculovirus and adenovirus vectors within the JCSG.

A network called Structural Genomics of Pathogenic Protozoa (SGPP) also established in the United States has developed HT methods for cloning, expression, and purification of membrane proteins. To facilitate the cloning procedure, *E. coli* and *Pichia pastoris* vectors have been engineered for ligation-independent cloning. Rhinovirus 3C protease cleavable C-terminal His-6 and calmodulin tags were introduced for affinity column chromatography purification. Currently, some 100 ORFs from *Leishmania major* with predicted 2- or 4-TM topology have been expressed in *E. coli*. Half of the targets showed positive signals by immunoblotting, and 20% could be detected by Coomassie blue staining. Introduction of the bacterial pelB signal sequence significantly enhanced the expression levels for most of the targets tested. Expression from *P. pastoris* vectors resulted in success rates higher than 50%, as measured by immunoblotting with expression levels in the range of 20 mg/L. The presence of a yeast signal sequence did not affect the expression levels. A number of targets are currently subjected to solubilization and purification attempts.

The NIH-funded Center for the Southeast Collaboratory on Structural Genomics (SECSG) has on its program an initiative called Structural Genomics of *Caenerabdhitis elegans* (SGCE), in which the aim is to study the whole genome of this worm, considered to be a valuable model system. So far, only soluble targets have been expressed in *E. coli*, purified, and crystallized. However, the genome encodes a total of 19,092 proteins, including a large number of membrane proteins, for many of which the function is unknown. Previously *C. elegans* olfactory receptors belonging to the superfamily of GPCRs have been expressed from SFV vectors in various mammalian host cell lines.[15] Within the SGCE, there are plans to overexpress a selected number of GPCRs and other *C. elegans* membrane proteins in insect and mammalian cells from baculovirus and lentivirus vectors.

As Japan has pioneered the efforts in structural genomics, several programs have also been dedicated to membrane proteins. The RIKEN Structural Genomics Initiative has a program with the aim to study the whole mouse genome, from which cDNAs will be selected for expression evaluation and further purification efforts. It is too early to say whether any GPCRs will be chosen within this program. However, the Biological Information Research Center (AIST) in Tokyo and the Japan Biological Informatics Consortium (JBIC) have established a joint-venture structural genomics program on membrane proteins. Bacteriorhodopsin was chosen as a model for GPCRs, and two-dimensional crystallography and electron crystallography approaches are in progress; mammalian GPCRs are also included. For instance, mutant forms of the β_2-adrenergic receptor have been expressed from baculovirus vectors at high levels in large-scale production to support the purification efforts.

16.5 MePNet

The MePNet consortium was established in 2001 with the financial support of more than 30 pharmaceutical and biotechnology companies. The execution of the practical experiments has taken place in three academic laboratories (Frankfurt, Marseilles, and Strasbourg) and in a biotechnology company (BioXtal, Switzerland). The selection of 100 GPCRs was based on ligand availability, representation of all GPCR families, and a large number of receptor subtypes. Moreover, the link to human disease was one of the selection criteria. Three well-established systems have been applied for expression evaluation. In this context, conventional pET15 *E. coli* vectors and Gateway vectors with various fusion partners were applied for expression in bacterial inclusion bodies. The bacterial expression was strongly dependent on the expression vector, the host strain, and culture temperature. The success rate evaluated by immunoblotting was relatively low (46%), although several targets could be expressed at relatively high levels. Several GPCRs have been scaled up for fermentor production to support the refolding attempts from inclusion bodies. *Pichia pastoris* was chosen as a lower eukaryotic expression system.[10] After clonal selection, 94% of the GPCRs showed a signal when verified by immunoblots. Interestingly, the expression levels could be substantially enhanced by lowering the growth temperature and supplementation of the medium with additives and receptor-specific ligands. The expression optimization resulted in more than 100-fold increase in specific binding activity with the highest B_{max} values of 180 pmol/mg protein. The mammalian expression system of choice was the SFV system[12], for which previously extreme levels of expression and binding activity up to176 pmol/mg protein were obtained.[16] The success rate for SFV-based GPCR expression was also very high (95%) and a large number of the receptors were expressed at levels of more than 1 mg/L. The highest binding activity obtained for a GPCR in SFV-infected[17] Chinese hamster ovary (CHO) cells was 287 pmol/mg.[17]

Recently, some 10 GPCRs expressed from *P. pastoris* vectors have been solubilized and purified from yeast cell membranes, and currently, a few targets have been subjected to crystallography attempts. Likewise, GPCRs from SFV-infected cells are solubilized and purified. In parallel, several GPCRs expressed in bacterial inclusion bodies are purified and subjected to refolding and crystallization studies. As the strategy of MePNet now switches to crystallography, a selected number of well-expressed targets from the three expression systems are subjected to crystallography attempts with the aim of obtaining high-resolution structures of several GPCRs within the next three years.

16.6 CONCLUSION

Structural biology of membrane proteins is far from the almost routine process to solve high-resolution structures of soluble proteins. Despite some good progress made in several areas of structural biology, some bottlenecks still exist. Although improvement in both yields and quality of recombinantly expressed membrane proteins has been achieved, there is still concern about the crystallizability in comparison with, for instance, receptors isolated from native tissue. There is great concern that after successful crystallization of the bovine rhodopsin isolated from cow retina,[18] no success of recombinantly expressed and purified GPCRs have been obtained so far. The most important questions are whether the properties of the recombinantly expressed material are different from those of native receptors and whether they present an obstacle for obtaining high-quality crystals with sufficient diffraction to generate high-resolution structures. Although excellent pioneering work under long-term commitments has been carried out on individual membrane proteins, perhaps the most significant contribution to progress will come from large structural genomics initiatives, where parallel expression, purification, and crystallization of a large number of targets will provide the expected success.

REFERENCES

1. Vanti, W.B., Swaminathan, S., Blevins, R., Bonini, J.A., O'Dowd, B.F., George, S.R., Weinshank, R.L., Smith, K.E., and Bailey, W.J. Patent status of the therapeutically important G protein-coupled receptors. *Exp. Opin. Ther. Patents*, 11, 1861–1887, 2001.
2. Michel, H. Membrane proteins of known structure. Max Planck Institute for Biophysics, Frankfurt, Germany. http://www.mpibp-frankfurt.mpg.de/michel/public/membrotstruct.html (accessed 2005).
3. Robb, F.T., Maeder, D.L., Brown, J.R., DiRuggiero, J., Stump, M.D., Yeh, R.K., Weiss, R.B., and Dunn, D.M. Genomic sequence of hyperthermophile, *Pyrococcus furiosus*: implications for physiology and enzymology. *Meth. Enzymol.*, 330, 134–157, 2001.
4. Goulding, C.W., Apostol, M., Anderson, D.H., Gill, H.S., Smith, C.V., Kuo, M.R., Yang, J.K., Waldo, G.S., Suh, S.W., Chauhan, R., et al. The TB structural genomics consortium: providing a structural foundation for drug discovery. *Curr. Drug Targets Infect. Disord.*, 2, 121–141, 2002.
5. Cole, S.T., Brosch, R., Parkhill, J., Garnier, T., Churcher, C., Harris, D., Gordon, S.V., Eiglmeier, K., Gas, S., Barry, III, C.E. et al. Deciphering the biology of *Mycobacterium tuberculosis* from the complete genome sequence. *Nature*, 393, 537–544, 1998.
6. McCue, L.A., McDonough, K.A., and Lawrence, C.E. Functional classification of cNMP-binding proteins and nucleotide cyclases with implications for novel regulatory pathways in *Mycobacterium tuberculosis*. *Genome Res.*, 10, 204–219, 2000.
7. Yokoyama, S., Hirota, H., Kigawa, T., Yabuki, T., Shirouzu, M., Terada, T., Ito, Y., Matsuo, Y., Kuroda, Y., Nishimura, Y., et al. Structural genomics projects in Japan. *Nat. Struct. Biol.*, Suppl: 943–945, 2000.

8. Kunji, E.R., Slotboom, D.J., and Poolman, B. *Lactococcus lactis* as host for over-production of functional membrane proteins. *Biochim. Biophys. Acta.*, 1610, 97–108, 2003.

9. Andersen, B. and Stevens, R.C. The human D1A dopamine receptor: heterologous expression in *Saccharomyces cerevisiae* and purification of the functional receptor. *Prot. Exp. Purif.*, 13, 111–119, 1998.

10. Weiss, H.M., Haase, W., Michel, H., and Reilander, H. Comparative biochemical and pharmacological characterization of the mouse 5HT5A 5-hydroxytryptamine receptor and the human beta$_2$adrenergic receptor produced in the methylotrophic yeast *Pichia pastoris*. *Biochem. J.*, 330, 1137–1147, 1998.

11. Possee, R.D. Baculoviruses as expression vectors. *Curr. Opin. Biotechnol.*, 8, 569–572, 1997.

12. Lundstrom, K. Semliki Forest virus vectors for rapid and high-level expression of integral membrane proteins. *Biochim. Biophys. Acta.*, 1610, 90–96, 2003.

13. Luca, S., White, J.F., Sohal, A.K., Filippov, D.V., van Boom, J.H., Grisshammer, R., and Baldus, M. The conformation of neurotensin bound to its G protein-coupled receptor. *Proc. Natl. Acad. Sci. USA*, 100, 10706–10711, 2003.

14. Weiss, H.M. and Grisshammer, R. Purification and characterization of the human adenosine A$_{2a}$ receptor functionally expressed in *Escherichia coli. Eur. J. Biochem.*, 269, 82–92, 2002.

15. Monastyrskaia, K., Goepfert, F., Hochstrasser, R., Acuna, G., Leighton, J., Pink, J.R. and Lundstrom, K. Expression and intracellular localisation of odorant receptors in mammalian cell lines using Semliki Forest virus vectors. *J. Receptor Signal Transduction Res.*, 19, 687–701, 1999.

16. Sen, S., Jaakola, V.P., Heimo, H., Engstrom, M., Larjomaa, P., Scheinin, M., Lundstrom, K., and Goldman, A. Functional expression and direct visualization of the human alpha 2B-adrenergic receptor and alpha 2B-AR-green fluorescent fusion protein in mammalian cell using Semliki Forest virus vectors. *Prot. Exp. Purif.*, 32, 265–275, 2003.

17. Hassane, G., Wagner, R., Kemph, J., Cherouati, N., Hassane, N., Prual, C., André, N. Reinhart, C., Pattis, F. and Lundstrom, K. Semliki Forest virus vectors for over-expression at 101 g proteins - coupled receptors in mammalian host cells. *Prot. Expr. Purif.*, in press, 2005.

18. Palczewski, K., Kumasaka, T., Hori, T., Behnke, C.A., Motoshima, H., Fox, B.A., Le Trong, I., Teller, D.C., Okada, T., Stenkamp, R.E. et al. Crystal structure of rhodopsin: a G protein-coupled receptor. *Science*, 289, 739–745, 2000.

17 Molecular Modelling of Membrane Proteins

Slawomir Filipek and Anna Modzelewska

CONTENTS

17.1 INTRODUCTION

Molecular modelling is a robust methodology for creating and analyzing the structures of biological molecules and their complexes. On the one hand, it is an integral part of crystallographic and nuclear magnetic resonance (NMR) structure determination packages; on the other hand, the protein structures can be created from sequences using only *in silico* computational methods. The second approach is especially important now in the genomics era when genome sequencing projects are producing plenty of valuable data. Additionally, in the case of membrane proteins, experimental determination of protein structures and the structures of their complexes is extremely difficult (see Chapter 13 and Chapter 14 on NMR and crystallography, respectively).

Nowadays, as computer power increases, many *in silico* methods have been developed and used with success to predict the structures of proteins and also of protein–ligand and protein–protein complexes. In addition to the prediction of still structures, they are also able to trace the motion of particular domains of proteins, and so potentially they are able to reveal the mechanisms of protein action. However, to obtain the correct structure or mechanism, detailed examination is required on every step. Smart methods such as inverse protein folding that calculates the compatibility of the protein model structure with its own sequence have been developed to evaluate predicted models. Furthermore, energy minimization and simulation methods provide corrections to rough protein models and facilitate our understanding of the motion of particular parts of protein and also reveal details of the formation of the complex. In the case of membrane proteins, molecular modelling methods are also able to check *in silico* hypotheses about the collective behavior of the proteins involved in raft superstructures.

17.2 PROTEIN STRUCTURE PREDICTION

Proteins whose structures have not yet been obtained by experimental methods can be characterized by theoretical structure prediction algorithms based on genome sequences. There are two main types of such methods based on two distinct sets of principles: the laws of physics and the rules of evolution. The first approach — *ab initio* methods — predicts structure from sequence alone, without relying on similarity of the folds between the modelled sequence and any of the known structures. They assume that the native structure corresponds to the global free-energy minimum of the protein, and so they use computational procedures for the exploration of the protein conformational space. The other factor influencing search and solution is free-energy function approximation used for evaluation of possible conformations. These methods are still more popular because with them full automation is possible and they are also more successful in prediction.[1,2] For 35% of examined proteins that have lengths shorter than 150 amino acids, the generated models had sufficient global similarity to the true structure. However, the accuracy of even the "correct" models was only ~ 4 Å RMSD (root mean square deviation) over ~ 80 residues.[3] This is too low for applications requiring high-resolution structures, for instance, drug design, but such low-resolution structures may be sufficient for the functional description of protein sequences on a genome-wide scale.[4]

The second group of methods includes homology (comparative) modelling and threading. These methods rely on detectable similarity among aligned sequences, mostly from similar families of proteins, and at least one known protein structure. The templates for modelling can be found by sequence comparisons or by sequence-structure threading that can sometimes reveal more distant relationships than can purely sequence-based methods. In threading, fold assignment and alignment are achieved by threading the sequence through each of the structures in a library of all known folds. Each sequence-structure alignment is assessed by the energy of a corresponding model of the resulted fold and not by sequence similarity as in the sequence comparison methods. Next, all predicted folds are linked together to form a final structure. High-accuracy comparative models based on more than 50% sequence identity to the template have approximately 1 Å RMSD error for the main chain atoms, which is comparable with a medium-resolution NMR or low-resolution x-ray structure. Low-accuracy comparative models are based on less than 30% of sequence identity and roughly contain less than 70% of residues within 3.5 Å of their correct positions. It is currently possible to model domains in ~ 60% of all known protein sequences,[3] but usually, one domain per protein is predicted (on the average, proteins have slightly more than two domains). Furthermore, two-thirds of the models are based on < 30% sequence identity to the closest template.

In order to construct a homology model for a query protein sequence, the query must be first aligned with one or more reference proteins of known structure. However, when sequence identity between two proteins is < 30%, the alignment becomes unreliable. The consequence of this will be incorrectly folded regions in the protein model. Fragment based homology procedures use the alignment between proteins to identify a number of structurally conserved regions (SCRs). They are regions of the template protein structure with no insertions or deletions in the corresponding parts of aligned sequences. Usually, the sequence conservation is highest in these regions, and they have well-defined secondary structures (helices or strands). The regions between the SCRs are called variable regions (VRs) or loops. The SCRs provide a consistent framework between the known and the unknown proteins. Hence, the coordinates of the protein backbone in the query protein can be copied from one of the known proteins. Loops are more difficult to predict, and typically, they are modelled by searching a database of structures for regions of suitable length and geometry at the interface with the SCRs.[5]

The restraint-based homology modelling does not generally break the model into two distinct phases, that is, building conserved regions and finding loop structures. Instead, the alignment is used to derive geometrical restraints such as distances between pairs of backbone atoms, ranges of backbone, and side-chain dihedral angles. The restraints come to an overall scoring function that defines how well the model structure matches the set of geometric criteria. The structure generation procedure is then used to create models that best satisfy the restraints.[6] Both homology and threading methods require an expert user to achieve the best results. However, the fully automated methods of protein structure prediction are still more powerful, and they even compete in a contest called CAFASP — Critical Assessment of Fully Automated Structure Prediction.[7]

17.3 EVALUATION OF PREDICTED MODELS

Difficult cases in homology/threading modelling correspond to protein sequences that only possess distant homology to known structures, and the level of sequence identity is low. In such cases, incorrect alignment can lead to significant structural errors in modelled structures. To select the right model or to identify the erroneous regions, some tools for predicting the quality of the model have been developed. They range from simple checks on geometry of the model bonds, bond angles, torsions, and so on[8] to knowledge-based methods to identify residues located in unfavorable environments. The accuracy of the model protein structures is most strongly restricted by the accuracy of protein loop regions. Attempts to refine this area of the protein models are subjected to active research. The methods can be tested within the series of blind prediction contests known as the Critical Assessment of Structure Prediction (CASP).[9,10] Protein modellers use sequences of proteins that will soon be solved by x-ray or NMR methods, but these are not yet available. Such a competition is also valuable for providing a common set of benchmark proteins for comparison of all accessible protein structure prediction methods.

Early modern methods on the compatibility of protein sequences with their own three-dimensional (3D) structures were based on 3D profiles.[11,12] In this approach, a scoring scheme is based on the tendency of each amino acid for its structural surroundings (helix, strand, or other) as well as on the degree of solvent exposure and the polarity of that environment. This methodology can also be used to search a database of proteins structures to find a fold that the protein can adopt — this is called the fold-recognition method. A related approach uses a known protein structure to search a database of protein sequences to find which structure is most likely to be implemented by a given sequence.[12] This approach is known as the inverse protein folding. The overall goal of assessing protein sequence–structure compatibility is identified as protein threading. The profile-3D methodology can be used to assess potentially misfolded regions in a protein model structure.[13] This is achieved by calculating the compatibility of the protein model structure with its own sequence. While applying other methods,[14,15] the scoring function is obtained by performing statistical analysis of known protein structures and deriving pairwise contact, solvation, and hydrophobic energy terms. One important application of threading methods is high-throughput (HT) screening of translated protein sequences obtained by genome sequencing. As for the prediction of the protein fold and the functional class, it is possible to use methods related to protein threading, even when the level of sequence similarity is too low for global sequence alignment and local sequence motif identification to work.[16,17] The method looks at the structural conservation and variation in the active site for a set of known structures. This can be used to derive a 3D motif that identifies the essential requirements for the protein (usually an enzyme) to function.

17.4 STRUCTURE OPTIMIZATION AND ANALYSIS

Predicted protein models require optimization. However, detailed quantum-mechanical description of such complex systems as proteins is not possible now. Molecular mechanics (MM) is based on several assumptions such as the Born-Oppenheimer approximation leading to the possibility of writing potential energy as a function of positions of nuclei. Such a function is called a force-field, and there are dozens of them.[18] Yet even the simplest ones, the four-component force fields, can lead to valuable results, when properly parametrized. One possible realization of such a function is the following, implemented in the AMBER program[19] (see scheme below), while recent developments in force-fields and simulations are described by Wang and co-workers.[20]

$$E_{pot} = \sum_{i}^{bonds} \frac{k_i}{2}(l_i - l_{i,0})^2 + \sum_{i}^{angles} \frac{k_i}{2}(\theta_i - \theta_{i,0})^2 + \sum_{i}^{torsions} \sum_{n}^{3} \frac{V_n}{2}(1 + \cos(n\omega_i - \gamma))$$

$$+ \sum_{i}^{N} \sum_{j}^{N} \left(4\varepsilon_{ij} \left[\left(\frac{\sigma_{ij}}{r_{ij}} \right)^{12} - \left(\frac{\sigma_{ij}}{r_{ij}} \right)^6 \right] + \frac{q_i q_j}{4\pi\varepsilon_0 r_{ij}} \right)$$

The first and second terms are based on Hooke's law used for stretching bonds (l) and bending angles between bonds (θ). The third term approximates shape of energy barriers during changing of torsion angles (ω), and the last term involves nonbonded interactions between every pair of N atoms in the system and is composed of van der Waals and Coulombic potentials, respectively. Such a simple expression with separated internal coordinates can simplify the parametrization process and facilitate the understanding of the influence of particular parameters on final results.

The most fundamental concept in force field is the atom type. It is based not only on the atomic number of the atom but also on its hybridization, formal charge, spin, and character of bonds to other atoms. The number of possible atom types is infinite, and the more the atom types, the more detailed (and complicated) the force field is. As the power of computers increases, modifications of the function are introduced like higher-order polynomials in bond and angle terms, improper torsions to confine *out-of-plane* bending motions, and finally *cross terms* to couple different kinds of motion, for example, stretch–stretch for adjacent bonds, stretch–bend (two bonds and angle), or bend–torsion (two angles and torsion). Many such parameters are determined quantum-mechanically.[21]

Minimal energy of terms composed of bonds, angles, and torsions are always zero because they are constructed as penalty terms for the ideal structure. The situation is different for nonbonded interactions: the van der Waals energy is approximated by the Lennard-Jones potential with a shallow minimum; the Coulomb potential has no minimum at all. The final structure of minimal energy is therefore a result of opposing forces and is also very sensitive to parametrization. MM methods are still called *empirical* methods, although many parameters are calculated using *ab initio* quantum procedures. This is especially true for partial atomic charges, since there is no empirical method to estimate them.

The next step is the minimization of the calculated energy of the protein to obtain relaxed — and hence more reliable — protein structure. However, the problem arises because proteins cannot take the proper shape in vacuum; for membrane proteins, additionally, a membrane can influence the shape of the protein. For less detailed simulations, the surrounding solvents and membranes can be simplified by proper dielectric constants (ε), but much better results are obtained by explicit inclusion of molecules of water and phospholipids (sometimes hydrocarbons are used instead) in the calculations. To avoid surface-tension effects and evaporation of solvent molecules from the surface, a periodic box containing a whole analyzed system should be included. The periodic box is like a unit cell in a crystal, in which the central cell is surrounded by identical cells that are the exact images of the source cell. It prevents the escape of molecules and introduces pressure into the calculations. The periodic box is the only proper way to simulate membranes because without it, the membranes turn into micelles. Additionally, some important domains of proteins such as active sites (especially with ligand bound) are modelled with greater accuracy using quantum methods, while the rest of the protein is treated classically. This approach is called QM/MM and has been successfully applied to retinal isomerization of bacteriorhodopsin[22] and is reviewed by Gogonea and co-workers.[23]

Analysis of protein stability and motion of particular parts can be done by molecular dynamics (MD) or Monte Carlo (MC) methods. MD is based on Newton's equations of motion, which are calculated by finite difference procedures; MC is a stochastic method, and the proper sampling of structures with low energy is provided by the Metropolis algorithm. Both methods can calculate the same thermodynamic quantities of the analyzed system. MD is usually used for time variability of protein.[24] However, to investigate longer-lasting motions such as ligand diffusion to and from an active site or the movement of domains, some force should be applied to different parts of the analyzed system. Such methods are implemented, for instance, in Steered Molecular Dynamics.[25] MC is very effective in the sampling of low-energy structures, since there are no energy barriers to overcome in this method. Furthermore, MC is frequently used for studying lattice models of biopolymers with simplified representation of amino acids. This is usually the case in protein folding studies because of very long simulation times.[26,27]

Simulations conducted in a membrane require powerful computers and, first of all, an understanding of the different behaviors of membrane proteins compared with cytosolic ones and the influence of lipids on protein structure and dynamics. In a review on bacterial outer membrane proteins,[28] the authors discuss the application of specific simulation techniques for ion conduction in porins, channel gating of *scavenger* transporters, and even catalytic mechanisms of enzymes (outer membrane phospholipase, OMPLA) that degrade phospholipids in perturbed cell membranes. Only limited data on lipid-protein interactions may be obtained from crystal structures where some lipid molecules are present.[29] Early simulations have provided a glimpse at lipid–protein interactions;[30] more detailed analysis was not possible due to relatively short simulation times (~ 1 ns). It has been suggested that only longer simulations (10 ns or more) will reveal more reliable protein–lipid interactions.

Apart from longer simulation times, proper treatment of long-range electrostatic interactions or the type of pressure coupling have important consequences for the equilibrium properties observed in membranes. Anezo and co-workers reported a series of long (up to 150 ns) MD simulations of dipalmitoylphosphatidylcholine bilayers, in which the methodology of simulation is systematically varied. They compared simulations with truncation schemes, Ewald summations, and modified Coulomb interactions.[31] A recent review of membrane protein targets investigated by computer simulations was done by Ash and co-workers.[32] They discuss bacteriorhodopsin, G protein-coupled receptors (GPCRs), aquaglyceroporins, ion channels, adenosine triphosphatases (ATPases), and outer membrane proteins.

17.5 MODELLING PROTEIN–LIGAND COMPLEXES

One of the most important and useful areas of application of molecular modelling is fitting (docking) to a protein a second molecule, typically a small molecule ligand. Such a model of possible interactions between the protein and the ligand may facilitate the understanding of biologically important protein–ligand complexes. Also, drug discovery processes could be more efficient in the case of effective docking procedures for eliminating poor ligands at an early stage of the drug development. The first stage of such a process is to identify the ligand binding site and the general geometry of the complex. Then, using the series of distinct ligands, their relative binding affinities are calculated, sometimes with decomposition of calculated affinities to specific parts of liquids provided the docking and other calculations are sufficiently rapid. This topic has been discussed in several reviews,[33–35] and the usefulness of homology modelling in drug design process has been reviewed by Hillisch and co-workers.[36]

In order to perform computational docking, experimental or modelled 3D structures of both the protein and the ligand molecules are required. Electrostatic contribution is among the most crucial components of protein–ligand interactions, therefore it is important to estimate the charge distribution for both molecules. For amino acids, these parameters are often available from standard molecular mechanics force fields in the form of atom-centered partial atomic charges. The same situation goes for peptidic ligands as well. For unusual amino acids or nonpeptidic ligands, the common procedure is to calculate such charges from quantum-mechanics methods by fitting the charge values to the computed electron distribution at the molecular surface.[37] If the ligand molecule is too large for direct quantum calculations, it is divided into smaller parts; and after calculations, the charges are tweaked together. The most desired situation is when a protein structure is available from experimental data, since homology or *ab initio* models are less detailed.

There are several programs available to perform docking calculations. One of the best programs is Dock.[38] The algorithm used by Dock has several steps. In the first step, a set of overlapping spheres is used to construct a negative image of a specified site on a protein. In the second step, this negative image of the active site is matched against the structures of potential ligands. The matches can be scored by the quality of the geometric fit as well as by the molecular mechanics interaction energy. The program Grid[39] utilizes a 3D grid around a protein. At each grid point, the program computes the MM interaction energy between proteins and a series of chemical functional groups called probes. This allows identification of likely protein binding sites for ligands. The AutoDock program[40] uses the concept of a grid-based scheme for energies of individual atoms, which is why the interaction energy of the protein–ligand complex can be computed more rapidly as an interaction between the ligand and the grid. In AutoDock, the likely ligand-binding site is then identified by a simulated annealing algorithm or genetic algorithm search procedure.[41] The program also tries to identify the binding affinity of a particular ligand by estimating the free energy of binding.

A very useful comparison of different docking techniques and scoring algorithms is presented by Bissantz and co-workers[42]; in their study, homology models of several GPCRs (the human dopamine D3, the muscarinic M1, the vasopressin V1a, the β_2-adrenergic, and the δ-opioid receptors) were used for virtual screening. While antagonists could be distinguished from other molecules in a test database, receptor models were not precise enough for agonists. Only after conformational rearrangement of the active site normally occurring during receptor activation were such modified receptors suitable for agonist screening.[43] The application of rhodopsin for modelling of GPCRs was also strongly confirmed by Ballesteros and co-workers.[44] They proposed that the overall structures of rhodopsin and amine receptors are very similar and that several of the highly unusual structural features of rhodopsin are also present in amine GPCRs. Thus, different amino acids or dissimilar microdomains can support analogous deviations from the regular helical structure, thereby resulting in similar 3D structure. This feature called *structural mimicry* is a mechanism by which a common ancestor of GPCRs could diverge to develop the selectivity necessary to interact with diverse signals while preserving the overall fold.

17.6 MODELLING PROTEIN–PROTEIN COMPLEXES

Modelling and simulation methods are frequently used to study protein–protein complexes. The topic is extremely important because molecular recognition processes affect many areas of biology. While there are many crystal or NMR structures less than 1% represent membrane proteins.[45] However are only a few and fragmented data on protein–protein complexes exist. Thus, the reliable modelling methods that could describe the nature of the protein–protein interactions would be highly valuable. Many of the computational methods used for modelling protein–protein complexes are similar to those used to model protein–ligand complexes. For instance, the Dock program, discussed above, has been applied in studies conducted by Brooijmans and co-workers[46] and Shoichet and Kuntz.[47] Due to the size of the computational task involved in the docking of two large protein structures, an approximation is often used to treat them as rigid bodies, which means that only rotational and translational degrees of freedom are necessary. The program FTDock (Fourier Transform Dock)[48] can effectively search for optimum interactions between two rigid proteins. This program was tested on various systems, including enzyme–inhibitor and antibody–antigen complexes. Addition of an electrostatic component greatly improved the ability to find final solutions in FTDock. Side-chain flexibility and the effect of solvation[49] were also introduced to this program.

Another approach is to use low-resolution structures of interacting proteins.[50,51] In this method, simplified atom–atom and residue–residue potential terms are used. This neglects small structural details on a scale below 7 Å and smoothes out many of the local minima present in the full atom representation. The search procedure is then more rapid. One drawback of rigid body docking is that it will not be able to identify changes in the conformations of the two proteins when the complex is formed. Flexible docking procedures are designed to accommodate such rearrangements by allowing for protein side-chain and backbone movements during the computations.[49] Protein–protein interface can also be modelled by molecular surface fitting, with surface flexibility implicitly addressed through liberal intermolecular penetration.[52] Pattern recognition fuzzy logic algorithms were also implemented in solving the surface complementarity problem of interacting proteins.[53] All such methods may now be tested in a communitywide contest called CAPRI (Critical Assessment of PRedicted Interactions).[54] The predictions are compared with unpublished x-ray structures of protein–protein complexes. The results obtained so far underline the need for new scoring functions and for methods for properly handling the conformation changes that were observed in some of the target complexes.

Nearly all tests for the above mentioned methods were conducted for soluble complexes of small proteins. This is understandable, since in the development phase, such algorithms should be tested on simple cases. Secondly, there are not so many membrane targets with resolved 3D structures. Nevertheless, molecular modelling techniques are used also for membrane proteins. Cai and co-workers identified caveolin-binding sites in several proteins by docking simulations.[55] Caveolin-1 may serve as a possible receptor of the SARS-associated coronavirus (SARS CoV) proteins. Simon and co-workers modelled major histocompatibility complex (MHC) class II molecules complexed with various peptides by fitting them into the binding groove and analyzed them with the help of experimental data.[56] Modelling was also used for aggregation of subunits of ion channels into a cylindrical configuration of a pore,[57] for conformational preference of cell surface monosialogangliosides to bind cholera toxin[58] (which might help in designing cholera toxin inhibitors), and for an ExbB–ExbD–TonB protein complex that serves to energize transport of iron sidero-phores and vitamin B_{12} across the outer membrane of Gram-negative bacteria.[59]

17.7 APPLICATION TO COMPLEX OF RHODOPSIN AND RHODOPSIN KINASE

17.7.1 MOLECULAR MODEL OF THE RHODOPSIN OLIGOMER

Bovine rhodopsin is an important membrane protein target, since it is the only GPCR with a resolved 3D structure. It can serve as a template for GPCR family A receptors. The recently shown oligomerization of rhodopsin[60] may also be an important feature of other GPCRs and affect their functional states. Our current model of the rhodopsin dimer[61,62] is based on the 1N3M model of the rhodopsin oligomer[60] deposited in the Protein Data Bank (PDB). The basis for this model was the crystal structure of rhodopsin[63,64] deposited under the 1HZX identifier in the PDB. Based on atomic force microscopy (AFM) measurements of distances between rhodopsins in the paracrystal, we assembled a model, where helices IV and V form an interface between rhodopsin monomers. Energetic considerations, together with geometrical constraints obtained from AFM, excluded other models of the oligomeric structures. The oligomers in our model (1N3M) are built from separate dimers and linked together by a long loop between helices V and VI. A dimer is a repetitive motif in the oligomer (double row), and thus tetramers and higher structures are connected in an identical manner. Optimization of structure and molecular dynamics were carried out using the consistent valence force field (CVFF) in the Discover program (InsightII 2000, Accelrys Inc.). Atomic charges were determined by minimizing the electrostatic energy of the system, while charges were variable. A series of short molecular dynamics simulations, up to 100 ps in a single run, was used to build a reliable system of interacting proteins. After each molecular dynamics run, optimi-zation of the whole structure was performed (maintaining frozen parts, when nec-essary).

The 1N3M model was improved by the addition of phospholipids. Specifically, three types of phospholipids were used with phosphatidylcholine headgroups on the intradiscal side and phosphatidylethanolamine and phosphatidylserine headgroups (three times more phosphatidylethanolamine headgroups than phosphatidylserine) on the cytoplasmic side.[65,66] All three types of phospholipids contain the saturated stearoyl chain (18:0) in the *sn*1 position and the polyunsaturated docosahexaenoyl chain (22:6n-3) in the *sn*2 position. Phospholipids were inserted between rhodopsin monomers, and the complex was optimized by molecular dynamics, followed by energy minimization with the rhodopsin monomers frozen in their initial positions. Next, the complex was subjected to several steps of short molecular dynamics simulations, followed by energy minimization to remove disallowed contacts. Favorable interactions among the inserted or modified subunits were created, usually during the first 10 ps of molecular dynamics simulation. The distances between the rhodopsin monomers in the paracrystal remained unchanged after addition of the phospholipids and optimization of the model without any constraints. A longer molecular dynamics run, up to 500 ps in the periodic box, was used to validate that the oligomer model was stable. The cytoplasmic view of the rhodopsin oligomer is shown in Figure 17.1.

17.7.2 MODELLING RHODOPSIN KINASE AND ITS COMPLEX WITH RHODOPSIN OLIGOMER

Rhodopsin kinase (RK) was built by homology modelling from the crystal structure of GPCR kinase 2 (PDB code 1OMW)[67] using the Modeller module.[68] RK is composed of three domains: the kinase domain, the RGS homology domain, and the floppy domain assembled from N- and C-termini. GPCR kinase 2 possesses additionally a Pleckstrin Homology (PH) domain responsible for binding the G$\alpha\beta\gamma$ subunits from the trimeric G protein. This domain was removed from RK, since it does not bind G$\beta\gamma$. Additionally, GPCR kinase 2 was crystallized in an open (inactive) state of kinase domain without a substrate bound. To have a closed (active)–state structure of RK, we performed homology modelling of the kinase domain of RK based on the crystal structure of the activated AKT–protein kinase B with GSK3 peptide bound (PDB code 1O6L).[69] The resulting structure of RK lacked the N terminus and the activation loop (residues 1-31 and 475-492, respectively), since they were not visible in the crystal structure of GPCR kinase 2. They were modelled without a template with the same software Modeller. Verification of the whole

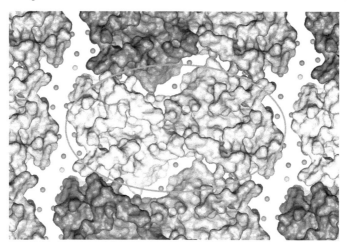

FIGURE 17.1 (See color insert following page 272.) Rhodopsin oligomer. View from the cytoplasmic side. The structure of a dimer marked by ellipse. Positions of phospholipids are denoted by red (ethanolamine heads) and green (serine heads) spheres.

structure was done in the Profile-3D[12] module of InsightII by evaluating the compatibility between sequence and structure. The MolMol program[70] was used to analyze the modelled macromolecular structures and for creating graphics. RK with all domains modelled is presented in Figure 17.2.

RK was farnesylated at position Cys558, and the resulting structure was optimized by short molecular dynamics and minimization. Then, RK was docked onto the structure of activated rhodopsin. Since there is no experimental structure of activated rhodopsin, we followed the commonly accepted mechanism, where the cytoplasmic part of transmembrane helix VI (from Trp265) is rotated about 90 degrees and the cytoplasmic end of this helix is moved about 5 Å out of the center. A loop between helix V and VI should accommodate to this motion. Since this loop also bridges rhodopsin dimers together, we carefully designed an altered loop and selected the one having the most reliable interactions with adjacent rhodopsin dimers. Retinal was changed to the all-*trans* form, and the neighboring residues (6 Å from retinal) of rhodopsin were subjected to minimization and molecular dynamics to accommodate the changes. Other rhodopsin molecules in the oligomer were left in inactive conformations. For docking with RK, we decided to use a tetramer of rhodopsin. We previously found that the tetramer structure was enough to bind the trimeric rhodopsin G protein transducin (Gt)[71] and that all four rhodopsins were busy interacting with Gt. After Gt is activated and the rhodopsin surface freed, RK phosphorylates rhodopsin, thus preparing it to be turned off by arrestin. Only the C-terminus of rhodopsin is phosphorylated. Several residues can be phosphorylated, but the most important ones are Ser334, Ser338, and Ser343, since phosphorylation at these sites is sufficient to bind arrestin.[72]

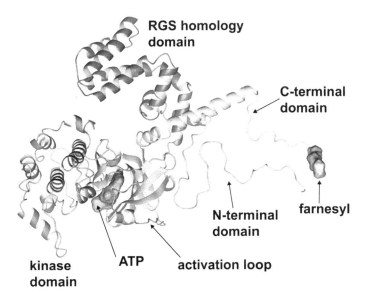

FIGURE 17.2 (**See color insert following page 272.**) Homology model of rhodopsin kinase. Three domains are visible: the kinase domain, the RGS homology domain, and the mostly unstructured N- and C-terminal domains.

It is unclear how RK can bind to rhodopsin. There is almost no experimental data on this topic. We used the constraints from the structure itself. RK is farnesylated in the C-terminus, and so this part should be soaked in a membrane. The C-terminus of RK is mostly unstructured, and the domain following the last helix is freely movable. If RK would bind to rhodopsin at the position shown in Figure 17.1 (the imaginary rhodopsin is below RK), the C-terminus of RK would have been elongated to full length to enable farnesyl soaking. Such an interface between interacting proteins prevents RK from complete probing of an active site of activated rhodopsin. Furthermore, the activation loop would be in contact with rhodopsin and eclipsed from interactions of other proteins. Having this in mind, we decided to lay down RK and examine the interactions of the flat side of RK with the rhodopsin tetramer. Short molecular dynamics runs, ~ 10 ps each, were applied to different parts of the model to secure proper interactions between molecules in the complex and to tweak a structure. A longer run of 100 ps was applied to the interface between rhodopsin and RK.

The resulting structure is presented in Figure 17.3 (top view), while side views are shown in Figures 17.4A and 17.4B. The activation loop is now facing up and is broadly accessible, and most interestingly, the RGS homology domain is interacting with the adjacent inactive rhodopsin. Such an interaction can facilitate the binding of RK to rhodopsin and keep close to the rhodopsin raft, while adenosine triphosphate (ATP) and adenosine diphosphate (ADP) are diffusing in and out. The C-terminus of rhodopsin is now pointing up and fits into the active site of RK. Figure

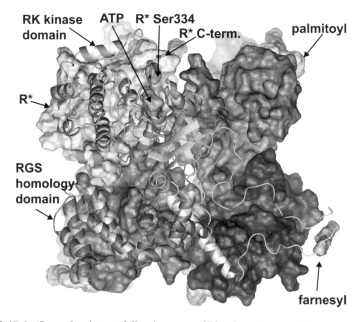

FIGURE 17.3 (**See color insert following page 272.**) Complex of RK and rhodopsin tetramer. View from the cytoplasmic side. The kinase domain bound to activated rhodopsin (R* = yellow). RGS homology domain bound to adjacent inactive rhodopsin.

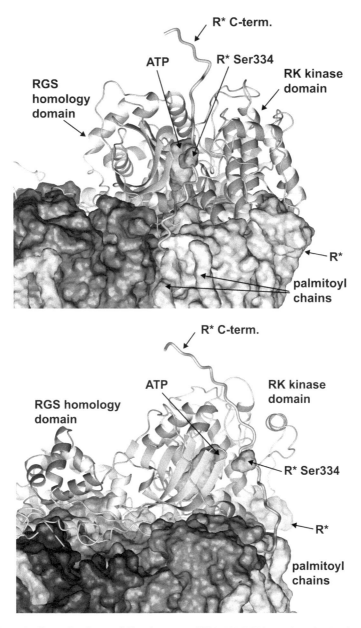

FIGURE 17.4 (See color insert following page 272.) (A) RK bound to rhodopsin tetramer, side view. Activated rhodopsin shown in yellow. (B) View rotated 90 degrees.

17.3 and Figure 17.4 show the complex ready for phosphorylation at Ser334. All other residues for phosphorylation are located further in sequence, which makes this process feasible. The cytoplasmic active site of rhodopsin (formed by the movement of helix VI) is recognized by an RK loop 279–287 (not shown). It moves deeply into a cleft between helices VI and VII of rhodopsin. Our model of the complex is in agreement with disulfide cross-linking experiments.[73] The artificial disulfide bond in rhodopsin, between Cys245 (end of helix VI) and Cys338 (C-terminus), enhanced the activation of Gt but abolished phosphorylation by RK. Enhancement of activation is understandable because helix VI was moved out and made room for establishing an activation site for rhodopsin. However, RK is not able to phosphorylate the rest of the C-terminus of rhodopsin. We modelled a complex of rhodopsin and RK in the vertical position and found that such an arrangement would fulfill requirements for appropriate positions of ATP and Ser343 of rhodopsin because this phosphorylation site is close to the rhodopsin surface. Lack of phosphorylation may come from the inability of RK in a vertical position to bind closely to the activated rhodopsin.

17.8 FUTURE PROSPECTS

Despite many experimental and computational efforts, the area of membrane proteins is still the "Wild West" of structural biology.[74] This territory is mostly unexplored, but new experimental techniques, together with sophisticated modelling methods (especially in synergistic approach), are making it possible to explore this area to an increasing extent. Simulations of pure lipid membranes investigate phenomena on length of 10 nm and time scale of 100 ns.[32] It is now possible to compare results of such simulations with those of experiments carried out at mesoscopic scales. Additionally, the development of computational methods, especially the sampling procedures in simulation methods, may increase both the simulation time and the accuracy of predictions. Simulations can provide dynamic views of proteins and can be used to build models of other conformational states (such as open and closed conformations of channels or substates during receptor activation) and to explore different modes of binding in protein–ligand and protein–protein complexes. The mostly unexplored area of protein–lipid interface (because of lack of experimental structures and inefficient sampling in simulation methods) is an exciting field for investigation, provided efficient hardware and software are available. Binary and higher mixtures of phospholipids in membranes, the effect of cholesterol on the stability of membranes, and the formation of rafts are new areas of membrane protein research that need to be addressed in the near future.

REFERENCES

1. Aloy, P., Stark, A., Hadley, S., and Russell, R.B. Predictions without templates: new folds, secondary structure, and contacts in CASP5. *Proteins*, 53, 436–456, 2003.
2. Bonneau, R. and Baker, D. *Ab initio* protein structure prediction: progress and prospects. *Annu. Rev. Biophys. Biomol. Struct.*, 30, 173–189, 2001.
3. Sali, A., Glaeser, R., Earnest, T., and Baumeister, W. From words to literature in structural proteomics. *Nature*, 422, 216–225, 2003.
4. Hardin, C., Pogorelov, T.V., and Luthey-Schulten, Z. *Ab initio* protein structure prediction. *Curr. Opin. Struct. Biol.*, 12, 176–181, 2002.
5. Forster, M.J. Molecular modelling in structural biology. *Micron*, 33, 365–384, 2002.

6. Sali, A. and Blundell, T.L. Comparative protein modelling by satisfaction of spatial restraints. *J. Mol. Biol.*, 234, 779–815, 1993.
7. Fischer, D., Rychlewski, L., Dunbrack, R.L., Ortiz, A.R., and Elofsson, A. CAFASP3: The third critical assessment of fully automated structure prediction methods. *Proteins*, 53, 503–516, 2003.
8. Laskowski, R.A., MacArthur, M.W., Moss, D.S., and Thornton, J.M. PROCHECK: a program to check the stereochemical quality of protein structures. *J. Appl. Crystallogr.*, 26, 283–291, 1993.
9. Moult, J., Fidelis, K., Zemla, A., and Hubbard, T. Critical assessment of methods of protein structure prediction (CASP)-round V. *Proteins*, 53, 334–339, 2003.
10. Venclovas, C., Zemla, A., Fidelis, K., and Moult, J. Assessment of progress over the CASP experiments. *Proteins*, 53, 585–595, 2003.
11. Luthy, R., McLachlan, A.D., and Eisenberg, D. Secondary structure-based profiles: use of structure-conserving scoring table in searching protein sequence databases for structural similarities. *Proteins*, 10, 229–239, 1991.
12. Bowie, J.U., Luthy, R., and Eisenberg, D. A method to identify protein sequences that fold into a known three-dimensional structure. *Science*, 253, 164–170, 1991.
13. Luthy, R., Bowie, J.U., and Eisenberg, D. Assessment of protein models with three-dimensional profiles. *Nature*, 356, 83–85, 1992.
14. Jones, D.T., Taylor, W.R., and Thornton, J.M. A new approach to protein fold recognition. *Nature*, 358, 86–89, 1992.
15. Bryant, S.H. and Lawrence, C.E. An empirical energy function for threading protein sequence through the folding motif. *Proteins*, 16, 92–112, 1993.
16. Rychlewski, L., Zhang, B.H., and Godzik, A. Functional insights from structural predictions: analysis of the *Escherichia coli* genome. *Prot. Sci.*, 8, 614–624, 1999.
17. Zhang, L., Godzik, A., Skolnick, J., and Fetrow, J.S. Functional analysis of the *Escherichia coli* genome for members of the alpha/beta hydrolase family. *Fold Des.*, 3, 535–548, 1998.
18. Ponder, J.W. and Case, D.A. Force fields for protein simulations. *Adv. Prot. Chem.*, 66, 27–85, 2003.
19. Pearlman, D.A., Case, D.A., Caldwell, J.W., et al. AMBER, a computer program for applying molecular mechanics, normal mode analysis, molecular dynamics and free energy calculations to elucidate the structures and energies of molecules. *Comput. Phys. Commn.*, 91, 1–41, 1995.
20. Wang, W., Donini, O., Reyes, C.M., and Kollman, P.A. Biomolecular simulations: recent developments in force fields, simulations of enzyme catalysis, protein-ligand, protein-protein, and protein-nucleic acid noncovalent interactions. *Annu. Rev. Biophys. Biomol. Struct.*, 30, 211–243, 2001.
21. Ewig, C.S., Berry, R., Dinur, U., et al. Derivation of class II force fields. VIII. Derivation of a general quantum mechanical force field for organic compounds. J. *Comput. Chem.*, 22, 1782–1800, 2001.
22. Hayashi, S., Tajkhorshid, E., and Schulten, K. Structural changes during the formation of early intermediates in the bacteriorhodopsin photocycle. *Biophys. J.*, 83, 1281–1297, 2002.
23. Gogonea, V., Suarez, D., van der Vaart, A., and Merz, K.W. New developments in applying quantum mechanics to proteins. *Curr. Opin. Struct. Biol.*, 11, 217–223, 2001.
24. Zhu, F.Q., Tajkhorshid, E., and Schulten, K. Molecular dynamics study of aquaporin-1 water channel in a lipid bilayer. *FEBS Lett.*, 504, 212–218, 2001.

25. Saam, J., Tajkhorshid, E., Hayashi, S., and Schulten, K. Molecular dynamics investigation of primary photoinduced events in the activation of rhodopsin. *Biophys. J.*, 83, 3097–3112, 2002.

26. Kolinski, A. and Skolnick, J. Reduced models of proteins and their applications. *Polymer*, 45, 511–524, 2004.

27. Kolinski, A., Klein, P., Romiszowski, P., and Skolnick, J. Unfolding of globular proteins: Monte Carlo dynamics of a realistic reduced model. *Biophys. J.*, 85, 3271–3278, 2003.

28. Bond, P.J. and Sansom, M.S.P. The simulation approach to bacterial outer membrane proteins (Review). *Mol. Membr. Biol.*, 21, 151–161, 2004.

29. Lee, A.G. Lipid-protein interactions in biological membranes: a structural perspective. *Biochim. Biophys. Acta-Biomembr.*, 1612, 1–40, 2003.

30. Tieleman, D.P., Marrink, S.J., and Berendsen, H.J.C. A computer perspective of membranes: molecular dynamics studies of lipid bilayer systems. *Biochim. Biophys. Acta-Rev Biomembr.*, 1331, 235–270, 1997.

31. Anezo, C., de Vries, A.H., Holtje, H.D., Tieleman, D.P., and Marrink, S.J. Methodological issues in lipid bilayer simulations. *J. Phys. Chem. B.*, 107, 9424–9433, 2003.

32. Ash, W.L., Zlomislic, M.R., Oloo, E.O., and Tieleman, D.P. Computer simulations of membrane proteins. *Biochim. Biophys. Acta-Biomembr.*, 1666, 158–189, 2004.

33. Joseph-McCarthy, D. Computational approaches to structure-based ligand design. *Pharmacol. Ther.*, 84, 179–191, 1999.

34. Lengauer, T. and Rarey, M. Computational methods for biomolecular docking. *Curr. Opin. Struct. Biol.*, 6, 402–406, 1996.

35. Anderson, A.C. The process of structure-based drug design. *Chem. Biol.*, 10, 787–797, 2003.

36. Hillisch, A., Pineda, L.F., and Hilgenfeld, R. Utility of homology models in the drug discovery process. *Drug Discov. Today*, 9, 659–669, 2004.

37. Merz K.M. Analysis of large data-base of electrostatic potential derived atomic charges. *J. Comput. Chem.* 13, 749-767, 1992.

38. Kuntz, I.D. Structure-based strategies for drug design and discovery. *Science*, 257, 1078–1082, 1992.

39. Goodford, P.J. A computational procedure for determining energetically favorable binding sites on biologically important macromolecules. *J. Med. Chem.*, 28, 849–857, 1985.

40. Morris, G.M., Goodsell, D.S., Huey, R., and Olson, A.J. Distributed automated docking of flexible ligands to proteins: parallel applications of AutoDock 2.4. *J. Comput. Aided Mol. Des.*, 10, 293–304, 1996.

41. Morris, G.M., Goodsell, D.S., Halliday, R.S., et al. Automated docking using a Lamarckian genetic algorithm and an empirical binding free energy function. *J. Comput. Chem.*, 19, 1639–1662, 1998.

42. Bissantz, C., Bernard, P., Hibert, M., and Rognan, D. Protein-based virtual screening of chemical databases. II. Are homology models of G-protein coupled receptors suitable targets? *Proteins*, 50, 5–25, 2003.

43. Bissantz, C. Conformational changes of G protein-coupled receptors during their activation by agonist binding. *J. Recept. Sig. Transd.*, 23, 123–153, 2003.

44. Ballesteros, J.A., Shi, L., and Javitch, J.A. Structural mimicry in G protein-coupled receptors: implications of the high-resolution structure of rhodopsin for structure-function analysis of rhodopsin-like receptors. *Mol. Pharmacol.*, 60, 1–19, 2001.

45. Lundstrom, K. Structural genomics on membrane proteins: mini review. *Comb. Chem. High Throughput Screen.*, 7, 431–439, 2004.

46. Brooijmans, N., Sharp, K.A., and Kuntz, I.D. Stability of macromolecular complexes. *Proteins*, 48, 645–653, 2002.
47. Shoichet, B.K. and Kuntz, I.D. Predicting the structure of protein complexes: a step in the right direction. *Chem. Biol.*, 3, 151–156, 1996.
48. Gabb, H.A., Jackson, R.M., and Sternberg, M.J.E. Modelling protein docking using shape complementarity, electrostatics and biochemical information. *J. Mol. Biol.*, 272, 106–120, 1997.
49. Jackson, R.M., Gabb, H.A., and Sternberg, M.J.E. Rapid refinement of protein interfaces incorporating solvation: application to the docking problem. *J. Mol. Biol.*, 276, 265–285, 1998.
50. Tovchigrechko, A., Wells, C.A., and Vakser, I.A. Docking of protein models. *Prot. Sci.*, 11, 1888–1896, 2002.
51. Vakser, I.A. Low-resolution docking: prediction of complexes for underdetermined structures. *Biopolymers*, 39, 455–464, 1996.
52. Duhovny, D., Nussinov, R., and Wolfson, H.J. Efficient unbound docking of rigid molecules. *Algorithms Bioinform. Proc.*, 185–200, 2002.
53. Exner, T.E., Keil, M., and Brickmann, J. Pattern recognition strategies for molecular surfaces. II. Surface complementarity. *J. Comput. Chem.*, 23, 1188–1197, 2002.
54. Janin, J., Henrick, K., Moult, J., et al. CAPRI: a critical assessment of PRedicted interactions. *Proteins*, 52, 2–9, 2003.
55. Cai, Q.C., Jaing, Q.W., Zhao, G.M., Guo, Q., Cao, G.W., and Chen, T. Putative caveolin-binding sites in SARS-CoV proteins. *Acta. Pharmacol. Sin.*, 24, 1051–1059, 2003.
56. Simon, A., Simon, I., and Rajnavolgyi, E. Modelling MHC class II molecules and their bound peptides as expressed at the cell surface. *Mol. Immunol.*, 38, 681–687, 2002.
57. Chou, K.C. Insights from modelling three-dimensional structures of the human potassium and sodium channels. *J. Proteome Res.*, 3, 856–861, 2004.
58. Sharmila, D.J.S. and Veluraja, K. Monosialogangliosides and their interaction with cholera toxin — Investigation by molecular modelling and molecular mechanics. *J. Biomol. Struct. Dyn.*, 21, 591–613, 2004.
59. Zhai, Y.F., Heijne, W., and Saier, M.H. Molecular modelling of the bacterial outer membrane receptor energizer, ExbBD/TonB, based on homology with the flagellar motor, MotAB. *Biochim. Biophys. Acta-Biomembr.*, 1614, 201–210, 2003.
60. Fotiadis, D., Liang, Y., Filipek, S., Saperstein, D.A., Engel, A., and Palczewski, K. Atomic-force microscopy: rhodopsin dimers in native disc membranes. *Nature*, 421, 127–128, 2003.
61. Liang, Y., Fotiadis, D., Filipek, S., Saperstein, D.A., Palczewski, K., and Engel, A. Organization of the G protein-coupled receptors rhodopsin and opsin in native membranes. *J. Biol. Chem.*, 278, 21655–21662, 2003.
62. Fotiadis, D., Liang, Y., Filipek, S., Saperstein, D.A., Engel, A., and Palczewski, K. The G protein-coupled receptor rhodopsin in the native membrane. *FEBS Lett.*, 564, 281–288, 2004.
63. Teller, D.C., Okada, T., Behnke, C.A., Palczewski, K., and Stenkamp, R.E. Advances in determination of a high-resolution three-dimensional structure of rhodopsin, a model of G protein-coupled receptors (GPCRs). *Biochemistry*, 40, 7761–7772, 2001.
64. Palczewski, K., Kumasaka, T., Hori, T., et al. Crystal structure of rhodopsin: a G protein-coupled receptor. *Science*, 289, 739–745, 2000.

65. Giusto, N.M., Pasquare, S.J., Salvador, G.A., Castagnet, P.I., Roque, M.E., and Ilincheta de Boschero, M.G. Lipid metabolism in vertebrate retinal rod outer segments. *Prog. Lipid Res.*, 39, 315–391, 2000.

66. Saiz, L. and Klein, M.L. Structural properties of a highly polyunsaturated lipid bilayer from molecular dynamics simulations. *Biophys. J.*, 81, 204–216, 2001.

67. Lodowski, D.T., Pitcher, J.A., Capel, W.D., Lefkowitz, R.J., and Tesmer, J.J.G. Keeping G proteins at bay: a complex between G protein-coupled receptor kinase 2 and G beta gamma. *Science*, 300, 1256–1262, 2003.

68. Sali, A., Potterton, L., Yuan, F., van Vlijmen, H., and Karplus, M. Evaluation of comparative protein structure modelling by MODELLER. *Proteins*, 23, 318–326, 1995.

69. Yang, J., Cron, P., Good, V.M., Thompson, V., Hemmings, B.A., and Barford, D. Crystal structure of an activated Akt/protein kinase B ternary complex with GSK3-peptide and AMP-PNP. *Nat. Struct. Biol.*, 9, 940–944, 2002.

70. Koradi, R., Billeter, M., and Wuthrich, K. MOLMOL: A program for display and analysis of macromolecular structures. *J. Mol. Graph.*, 14, 51–55, 1996.

71. Filipek, S., Krzysko, K.A., Fotiadis, D., et al. A concept for G protein activation by G protein-coupled receptor dimers: the transducin/rhodopsin interface. *Photochem. Photobiol. Sci.*, 3, 628–638, 2004.

72. Maeda, T., Imanishi, Y., and Palczewski, K. Rhodopsin phosphorylation: 30 years later. *Prog. Retin. Eye Res.*, 22, 417–434, 2003.

73. Cai, K., Klein-Seetharaman, J., Hwa, J., Hubbell, W.L., and Khorana, H.G. Structure and function in rhodopsin: effects of disulfide cross-links in the cytoplasmic face of rhodopsin on transducin activation and phosphorylation by rhodopsin kinase. *Biochemistry*, 38, 12893–12898, 1999.

74. Torres, J., Stevens, T.J., and Samso, M. Membrane proteins: the "Wild West" of structural biology. *Trends Biochem. Sci.*, 28, 137–144, 2003.

18 Toward Structural Bases for GPCR Ligand Binding: A Path for Drug Discovery

Sabrina A. Beretta, Alla Korepanova, Mulugeta Mamo, Geoffrey F. Stamper, and Mark L. Chiu

CONTENTS

18.1 INTRODUCTION

G protein-coupled receptors (GPCRs) comprise a family of seven-transmembrane (7TM) receptors that mediate most of the cell–cell communication in humans via a wide variety of extracellular activators such as hormones, light, neurotransmitters, ions, odorants, and amino acids.[1,2] Because of their physiological importance in the metabolic, endocrine, neuromuscular, and central nervous systems, many pharmaceutical companies have made significant efforts to develop drug therapies that act on this family of proteins.[3,4] More than 50% of marketed drugs treat diseases by targeting more than 20 GPCRs, with worldwide sales surpassing $25 billion.[5–7]

Surveys of genomic data suggest that there are about 1000 GPCRs in the human genome: 614 annotated entries and the remainder classified as orphan receptors.[8–10] These as-yet-untargeted GPCRs may have important physiological roles and unique mechanisms of interaction, making GPCRs a class of proteins with great drug discovery potential.[11]

All members of the GPCR superfamily share a common structural framework: putative seven-transmembrane (7TM), domains connected by three intracellular loops and three extracellular loops with an extracellular amino-terminus and an intracellular carboxyl-terminus.[12,13] Superimposed on the basic structure of GPCRs are a number of variations that provide specificity in ligand binding, G protein coupling, and interactions with other proteins. Sequence alignment of the putative 7TM domains divides the superfamily into several subfamilies (Figure 18.1). Family 1 (Class A) is the largest group and includes the rhodopsins and biogenic amine GPCRs. Family 2 (Class B) shows essentially little sequence homology to Family 1, even within the 7TM segments. Members of this family include receptors for peptide hormones such as secretin, parathyroid hormone, parathyroid-hormone-related protein, and calcitonin. Family 3 (Class C) members include the gamma amino butyric acid (GABA), metabotropic glutamate, and extracellular Ca^{2+} sensing receptors. In addition to these three family members, there are the olfactory, taste, and pheromone receptors.

Each GPCR family has different post-translational characteristics that could be important for activity. Family 3 GPCRs dimerize via interactions between the ligand-binding domains. While the ligand-binding properties of these domains in monomers are similar in dimers, the regulation of both ligand-binding and G protein-activation in Family 3 GPCRs differ.[14] Homo- and heterodimerization is possible with this and other GPCR families as well. The consequences of heterodimerization may extend the repertoire of GPCR pharmacology.[15]

Protein X-ray crystallography and nuclear magnetic resonance (NMR) are recognized as powerful techniques for lead discovery.[16–18] However, their most common and potent role continues to be played during lead optimization. While approaches in structure-based lead optimization vary, the predominant methodology employs an iterative process by which structural information provides a rationale for chemical modifications to a lead molecule; this improves target–ligand interactions until the desired potency is achieved as measured by a simple *in vitro* activity assay. The use of protein X-ray crystallography in structure-based lead optimization is the preferred technique in the pharmaceutical industry, and its successful use in this purpose was reviewed recently.[19] Though it is not without its limitations,[20] there are now several examples of commercially available drugs, whose discovery was driven largely by structure-based lead discovery and optimization.[21] Thus far, the successes are limited to soluble proteins, whose overexpression, isolation, crystallization, and structure determination are more commonplace.

However, the ever increasing knowledge of the medical importance of membrane bound proteins, specifically GPCRs, continues to steer the focus of structure-based lead discovery and optimization efforts in the pharmaceutical industry toward these lucrative targets. To date, there is no experimentally derived structural information available for a therapeutically relevant GPCR. However, the recent determination of

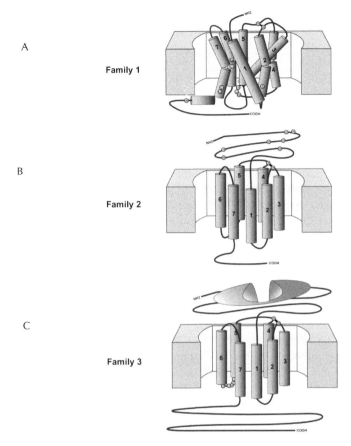

FIGURE 18.1 Schematic representations of GPCR families.[182] Monomeric GPCRs are depicted with structural aspects of the three main GPCR families. **(A)** Family 1 (Class A) GPCRs are characterized by highly conserved amino acids indicated by circles around single letter amino acid codes and a disulfide bridge that connects the first and second extracellular loops. The 7TM domains are numbered after the TM positions of rhodopsin. Most of these receptors have a palmitoylated-cysteine in the carboxyl terminal tail. Threading of the amino acid sequences of Family 1 GPCRs to the crystal structure of rhodopsin suggest that the transmembrane domains can be tilted and kinked. **(B)** Family 2 (Class B) GPCRs are characterized by having a relatively long amino-terminal domain that contains several cysteines that form a network of disulfide bridges. Little is known about the orientation of the transmembrane domains because of the divergence in amino acid sequences from Family 1 GPCRs. Hence, only schematic representations of the TM domains are shown. **(C)** Family 3 (Class C) GPCRs are characterized by having a long amino terminus and carboxyl terminus. The crystal structure of the amino-terminal ligand-binding domain with the Venus flytrap (VFT) motif has a disulfide linker. Except for two cysteines in extracellular loops 1 and 2 that form a putative disulfide bridge, Family 3 receptors do not have any features that characterize Family 1 and 2 receptors. A unique characteristic of these receptors is that the third intracellular loop is short and highly conserved. Again, only the schematic representations of the TM domains are shown. (With permission from Susan George and Nature Reviews Publishing Co.)

the crystal structure of rhodopsin[22] has brought the potential for rational lead discovery and optimization of GPCR ligands to the forefront of structural biology. Indeed, there are now several academic, commercial, and combined efforts designed to develop general methods for GPCR structure determination and lead discovery.[23,24]

The determination of the crystal structure of rhodopsin represents the first three-dimensional (3D) view of a GPCR, including its covalently bound ligand, 11-*cis*-retinal.[22] Recent improvement in the diffraction quality of the rhodopsin crystals yielded a structure that resolves the entire protein chain to a much higher resolution, allowing for a more complete analysis of the retinal-receptor interactions at an atomic level.[25] However, this crystal structure represents the inactive state of rhodopsin. Although not a drug target itself, this structure may be useful in the design of antagonists to therapeutic members of Family 1 GPCRs.

The aforementioned diversity of GPCR states requires methods that can characterize the aspects of GPCR function. In this chapter, we dissect aspects of the multifactorial nature of GPCR ligand-binding by addressing the characterization of ligand-binding and functional assay screening; the localization, characterization, and structure determination of GPCR ligand-binding domains; the use of NMR to elucidate structures of GPCRs and their ligands; and the prospective use of molecular modeling to generate lead compounds for GPCRs.

18.2 DRUG DISCOVERY VIA GPCR LIGAND BINDING AND FUNCTIONAL ASSAY SCREENING

Presently, most approaches to GPCR drug discovery focus on primary screens to identify the binding affinities of lead compounds to the receptor followed by screening for biochemical functional responses that look for the effect of the compounds on the signal pathway. The methods used to determine the binding affinities of membrane preparations of GPCRs can be homogeneous or heterogeneous, exploiting either radiometric or nonradioactive processes.[2]

One of the most common radioactive high-throughput (HT) platforms for GPCR binding is based on the scintillation proximity assays.[26] The receptor is immobilized on a protein-coated scintillant surface found on either beads or microplate plastic. When a radioisotope-labeled ligand binds to the receptor, it is brought in close proximity to the scintillant, and the energy transfer produces an emission that is read by a counter. These assays are robust and precise, even though the assays provide information only on the compound binding, not on the intrinsic method of G protein activation. Classification of agonism or antagonism depends on whether the compounds enhance or inhibit reference compound binding.

The most effective nonradioactive GPCR binding assays are based on the principle of fluorescence polarization. Fluorescence intensity and resonant energy transfer signals can also be exploited. The platforms available are highly automated with a very fast readout. With these methods, there is no need to immobilize the receptor. Either a fluorescent ligand or a fluorescent dye–labeled ligand can be used. However, the fluorescent properties of the ligand are important for the assay such as having an emission spectrum without background interference, a long lifetime of emission for sensitivity of detection, and a good quantum yield for detection efficiency. When a fluorescent ligand lacks such qualities, attaching an optimal fluorescent dye to a ligand is required. When a fluorescent dye-labeled ligand binds to the receptor, there will be a change in the fluorescent signal, proportional to the degree of binding in solution, for example, the bound compound will slow down the rate of molecular motion, producing a shift in the fluorescence polarization signal. Fluorescent-labeled ligands can be stable but have lower affinity than the unlabeled ligand due to steric hindrance or quenching.

Considerable interest and effort are applied in screening GPCR functional responses. The assays target well-characterized signal pathways such as GTP binding,[27] cAMP production,[28,29] or calcium ion flux.[30] All these functional platforms are highly compatible with high throughput (HT) screening approaches, and the various formats can investigate the affinity and efficacy of compounds and distinguish among antagonists, full agonists, and partial agonists. The transient nature of the functional response, however, can complicate the evaluation of the compound potency.

In spite of the benefits and the great importance of cell-based screens in industry, some caveats emerge. Transfected cells are artificial systems, and the results may not reflect events in native systems. Often, there are background problems due to drug interactions with related intrinsic receptors present in transfected cells that cause difficulty in resolving either potency or selectivity. Moreover, cell-based assay systems for GPCRs monitor signals at very distal points and may not tell anything about the mechanism of receptor-ligand-binding. Hence, it can be difficult to distinguish GPCR–ligand, ligand–GPCR–G protein, and ligand–GPCR–G protein–secondary messenger interactions. Likewise, the relative amounts of GPCR, G proteins, and secondary messenger proteins greatly affect the kinetic responses. Finally, assays and models do not take into account the possible physical heterogeneity of the receptor that can function as hetero-oligomers.

18.3 THE VALUE OF STRUCTURE-BASED DRUG DESIGN OF GPCRs

One of the great challenges in the GPCR field is to elucidate the conformational changes associated with the binding of agonists, antagonists, and allosteric modulators.[31] Little is known about the structure–function relationships that govern the specificity of ligand binding and activation in the GPCR–G protein systems as well as GPCR–G protein–effector protein interactions.[32] This ignorance is due, in part, to the multifactorial nature of signal transduction regulation. Notwithstanding this, the molecular details of GPCR ligand binding and activation could provide diagnostic tools that prevent progression of receptor-mediated disease and generate new strategies for more specific and effective therapeutic intervention.[33]

Engineering ligand specificity and potency can best be obtained with structure-based drug design. In the absence of GPCR structures, extensive efforts are made to model GPCRs based on the high-resolution structures of the 7TM proteins bovine rhodopsin and bacteriorhodopsin.[8,34,35] Although current GPCR models attempt to accommodate related mutagenesis data in a 3D context, much is to be learned from experimentally derived 3D structures. The structure determination of rhodopsin and bacterial transport proteins reveal kinks in the TM helices and orientations of the interhelical loops that are not resolved by modeling and site-directed mutagenesis experiments.[22,36,37] Moreover, there are many GPCRs that cannot be modeled because of their divergence in sequence similarity and functional properties from bovine rhodopsin.[38] Knowledge of a GPCR structure would elucidate the molecular basis of ligand binding and receptor conformation in normal and abnormal conditions.[39]

To approach structure-based drug design of GPCRs, we need to obtain high-resolution membrane protein structures. Four different methods are currently used for the elucidation of high-resolution structures of membrane proteins: X-ray crystallography,[40–42] electron crystallography,[40,43,44] atomic force microscopy,[45] and NMR spectroscopy.[46–49]

Presently, most membrane protein structures have been determined via x-ray crystallography because data collection and processing are routine and allow for a large number of structures to be solved rapidly to high-resolution without much concern for protein size. Although the other methods can provide complementary structural information, there are still many barriers to proper sample preparation that prevent routine use of electron crystallography, atomic force microscopy, and NMR spectroscopy. Detailed mechanistic insights of proteins require structure resolution better than 4 Å, particularly if small conformational changes and the positions of structurally relevant water molecules are to be observed. Of the high-resolution integral membrane protein structures posted on two Web sites put together by the groups of Stephen White and Hartmut Michel,[50,51] bovine rhodopsin is the only example of a high-resolution membrane protein structure in the GPCR family.

Bovine rhodopsin is highly overexpressed in bovine retinae and can form crystals when purified to homogeneity with detergents that retain native lipids.[22] From this example, we can generalize that the first challenge in getting structures is finding the best method to overexpress GPCRs. Since most GPCRs are not overexpressed in native cells, assessment of diverse heterologous expression systems is required. The next challenge would be to purify the active form of the protein in a state that is amenable to the structural methods noted above. More information can be obtained from excellent reviews on overexpression and purification of membrane proteins.[52–55] Also, methods to crystallize membrane proteins have been extensively reviewed.[56–60] Because of the logistical complexity of getting membrane protein structures, several consortia and structural genomic centers have been formed to develop methods to obtain GPCR and other membrane protein structures.[23,24]

18.4 LOCALIZATION OF GPCR LIGAND-BINDING DOMAINS

Multiple experimental approaches using tools in molecular biology, biochemistry, and biophysics have been used to identify the structural determinants of ligand binding and functional activation of the different GPCRs.[61,62] The endogenous ligands that bind to the three major classes of GPCRs are exceptionally diverse, varying from Ca^{2+} ions and small molecules such as amines and amino acids to larger molecules such as peptides and glycoproteins. The chemical diversity of the ligands translates into the potential diversity of the ligand interactions with GPCRs. Likewise, the main targets for drug design focus on the discovery of small non-endogenous ligands that can act as agonists, antagonists, and allosteric regulators.

In the absence of being able to obtain the structure of the intact full-length protein, significant efforts have been made to extricate the conserved residues and modular domains of GPCRs that have distinct roles in the structure–function relationships in different classes of receptors. In fact, there is a number of amino acid residues conserved among all GPCRs in the 7TM domains. For example, a conserved cysteine at the extracellular end of the third transmembrane domain (TM3) forms a disulfide bond with a conserved cysteine in the second extracellular loop in most GPCRs.[63] The TM3 domain is centrally positioned in rhodopsin[22] and contains a highly-conserved Asp-Arg-Tyr (DRY) motif found in Family 1 GPCRs.[63] Interactions between TM3 and TM6 control the equilibrium between the active and inactive conformations of the receptor.[64] Since the conserved Arg in the DRY motif is present only in a subset of Family 3 receptors and is absent in Family 2 GPCRs, the TM3 domain may have different roles in the activation mechanism of these receptors.[63]

Often, site-directed mutagenesis of the conserved amino acids can be used to assess their roles in establishing ligand potency, G protein activation, and effector protein interactions. Several experimental observations reveal that conserved amino acids are involved in important movements such as changes in the kinks and tilts of the helices or mutual movements of one domain over the other.[65,66] The characterization of the relationships among ligand-binding residues can help decide which changes in ligand design can be useful in achieving greater ligand specificity.

18.4.1 Ligand-Binding Domains of Family 1 GPCRs

The ligand-binding sites for small-molecule agonist ligands in Family 1 GPCRs are conserved.[67–69] Site-directed mutagenesis and spectroscopic analysis of the fluorescent antagonist carazolol binding to the β$_2$ adrenergic receptor (β2AR) has mapped the binding site to be residues in TM3, TM5, TM6, and TM7 located deep inside the receptor.[70] Asp 113 in TM3 of β2AR is conserved among all biogenic amine receptors.[61,62] The region that defines the ligand binding of these receptors is approximately near the region for the ligand–chromophore 11-cis-retinal that is covalently bound to rhodopsin.[22,71,72]

Unlike rhodopsins, the majority of GPCRs interact noncovalently with their endogenous and exogenous ligands. Family 1 receptors, such as rhodopsin, δ-opoid[73] and CXCR4[74] have similar ligand binding features that involve putative TM2, TM3, TM5, TM6, and TM7 with their respective ligands (Figure 18.2). At the same time, peptide agonists interact with the residues in the amino-termini and extracellular loops of their cognate receptor.[61] Hence, the loop regions are not just linkers holding together TM helices but serve as important functional elements of the GPCRs.[75]

The agonist-binding site of Family 1 GPCRs involves more that 30 conserved residues in the TM3, TM5, TM6, and TM7 domains of the receptors.[76] Small-

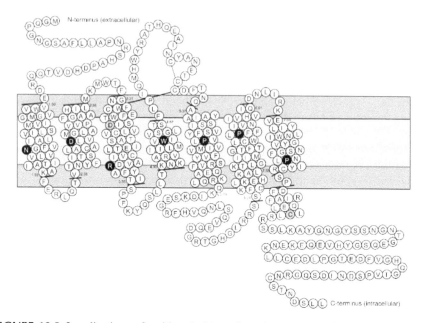

FIGURE 18.2 Localizations of residues in biogenic amine receptors that are important for ligand binding (gray color) within a membrane topology model of the rat $5HT_{2A}$ receptor (amino terminus and most of the carboxyl terminus are not shown).[62] The localizations of TM segments in the $5HT_{2A}$ receptor are indicated (black lines). The core and water-lipid interface regions of the lipid membrane are indicated with light gray and dark gray colors on the background. The protonated amine group of agonists and antagonists interacts with the carboxylate group of Asp3.32 (gray). The convention 3.32 refers to Asp32 on the TM3 domain. Polar moieties of agonists can interact with polar residues at positions 3.36, 5.42, 5.43, 5.46, 6.55, and 7.36 (gray). Aromatic residues Trp3.28, Phe5.47, Phe5.48, Trp6.48, Phe6.51, Phe6.52, Trp6.40, and Tyr6.43 (gray) in biogenic amine receptors play important roles in ligand binding and agonist-induced receptor activation. The highly conserved reference residue within each TM helix is indicated with black circles and white text. (With permission from Kurt Kristiansen and Elsevier Press.)

molecule ligands and peptides are also shown to interact with at least one extracellular loop.[61,77] Residues in the extracellular side of the binding pocket determine the specificity of binding. In particular, the TM5 domain, the least conserved domain in Family 1 GPCRs, is important for the interaction with different ligands.[78] The putative buried parts of the binding pocket consist of conserved hydrophobic and aromatic residues that mediate key binding interactions for several receptors.[34] The predicted structural features of ligand binding suggest the existence of privileged structures of the ligands that can be accommodated in the buried region of the binding pocket of Family 1 GPCRs.[34] Ideally, the agonist ligand should interact with the conservative and variable parts of the binding pocket providing the bases of ligand specificity and activation of signal transduction.

Models based on the rhodopsin structure do not provide a satisfactory explanation for stereoselective binding of β2AR antagonists.[79] It is likely that antagonists bind preferentially to a receptor conformation that differs from that of agonists. Movements of TM3 and TM6 at the cytoplasmic side of the membrane play an important role in G protein coupling.[80] To identify potential regions responsible for spontaneous receptor activation, chimeric or mutated GPCRs have demonstrated that the third intracellular loop that connects TM5 and TM6 is an important determinant for G protein coupling.[81,82]

An elegant method to elucidate ligand specific conformational changes is to label purified β2AR with a fluorescent dye in specific locations. Changes in fluorescence indicate how agonists and antagonists bind to β2AR.[83,84] This method is also used to show the enhancement of protein stability in the presence of ligands.[85] Alternatively, the swapping of domains of β2AR and its cognate G protein permits localization of domains that are important for ligand-induced conformational changes.[86] The combination of site-directed mutagenesis and fluorescence experiments have been used to elucidate sequential mechanisms for the binding of agonists to β2AR.[87] The resolution of slow and rapid conformational states also introduces a novel method to distinguish the receptor states for better drug design.

18.4.2 Ligand-Binding Domains of Family 2 GPCRs

The prominent feature of Family 2 GPCRs is a long amino-terminal domain (100 to 200 amino acids) with six conserved cysteine residues. The endogenous agonistic ligands are peptides of around 30 residues that bind to the amino-terminal domain and then dip into the pocket formed by TM1, TM2, and TM7.[88–94] The primary amino acid sequences of peptides of the glucagon receptor superfamily are 50% identical and more than 70% homologous.[95] Although the peptide ligands of the Family 2 glucagon receptor superfamily are highly homologous to each other, they are able to activate their respective receptors specifically to elicit the appropriate physiological responses.[9,91] The most conservative residues are located at the amino-terminus of the peptides with the amino-terminal histidine being strictly conserved within the glucagon family. The negatively charged side chain of Asp9 of the glucagon peptide is required more for the receptor activation than for binding. Substitution of Asp9 of the glucagon peptide with a different amino acid residue did not abolish the binding to the receptor but resulted in the generation of antagonist activity.[96] Identified sequence homology of the peptide ligands suggests that the amino-terminal part of the ligand is more important for signal transduction and the carboxyl-terminal part of the ligand is more involved in receptor recognition.[95]

The individual isolated amino-terminal domains of the glucagon-like peptide 1 (GLP-1), parathyroid hormone (PTH) and pituitary adenylate cyclase activating polypeptide (PACAP) receptors bind their respective ligands with lower binding affinities than that determined for the full-length receptors.[97–100] The juxtamembrane domain (extracellular region of the 7TM bundle) and the extracellular loops of the PTH receptor have been shown to form contacts with the ligand.[101,102] Polyclonal antibodies against epitopes based on the sequences of the first extracellular loop of the rat glucagon receptor blocked glucagon binding, indicating that this loop is part of the ligand-binding site.[91,103]

The NMR structure of the amino-terminal domain of the corticotropin-releasing factor (CRF) receptor uncovers the details of the hormone binding site and supports the two-site model of ligand binding for Family 2 GPCRs.[104] The structure of the extracellular domain is identified as a short consensus repeat, a common protein interaction module containing two anti-parallel β-sheets flanked by disordered regions. The structure is stabilized by three disulfide bonds and includes several key residues identified as highly conserved in Family 2 receptors. Amino acid sequence similarity of Family 2 GPCRs suggests that the identified fold of the extracellular domain from the CRF should be similar in all extracellular domains of Family 2 receptors.[104]

A clearer view of the mechanism of ligand-binding domains is obtained only after a structure is obtained. For example, many questions about the ligand-binding domains of two nuclear receptors, erythropoietin (EpoR) and glucocorticoid receptors, could be resolved only after high-resolution structure determinations.[105,106] The crystal structure of the ligand-binding domain of EpoR in its unliganded form revealed a dimer in which the individual membrane-spanning and intracellular domains would be too far apart to permit phosphorylation. The unliganded dimer is formed from self-association of the same key binding site residues that interact with EpoR ligands. The structures of agonist and antagonist peptide complexes of EpoR have shown that the actual dimer configuration is critical for the biological response and signal efficiency. In the glucocorticoid receptor, structures of the antagonist and agonist with co-activator peptide reveal different conformations, suggesting that there can be several ways to antagonize the glucocorticoid receptor. All these structures together reveal an extremely flexible protein capable of adapting its interactions with other proteins in a number of different ways. Hence, the most productive generation of ligand SAR (structure–activity relationship) would involve comparing different structures of receptor ligand complexes.

18.4.3 LIGAND-BINDING DOMAINS OF FAMILY 3 GPCRS

In Family 1 and Family 2 GPCRs, the 7TM domain is involved in ligand binding and conformational changes. In contrast to Family 2 GPCRs, Family 3 GPCRs possess a 500- to 600-amino acid residue amino-terminal extracellular domain solely responsible for ligand binding.[63] The extracellular domain of Family 3 receptors is divided into two subdomains: the Venus Flytrap (VFT) module and the cysteine-rich domain positioned between the VFT and the 7TM domains.[107,108] The agonist-binding site of the VFT domain lies on the surface of the cleft between two lobes of the VFT domain. The 2.2 Å resolution crystal structure of the extracellular domain of the metabotropic glutamate type 1 receptor (mGluR1) with and without glutamate shows bi-lobe architecture of the VFT domain.[109] The mGluR1 structures with and without glutamate demonstrate that the dimeric ligand-binding domains of mGluR1 adopt "open" or "closed" conformations that exist in dynamic equilibrium. Lobe one of the VFT has a buried surface area of 168 Å^2 that mediates glutamate interactions via direct hydrogen bonds to the conserved residues Ser165, Thr188, Tyr74, and Lys409. Lobe two has a 135 $Å^2$ buried surface area that provides water-mediated hydrogen bonds and salt bridges that link glutamate to the conserved residues Asp208, Asp318, and Tyr236. The structural differences in ligand binding imply that glutamate interacts with one lobe of the VFT cleft in the "open" form and then stabilizes a "closed" form by making additional contacts to the second lobe. The suggested role of glutamate is to stabilize the closed conformation of the VFT, and the closure of the VFT is indicated as a key step in the activation of the glutamate receptor.[63] The crystal structure of the extracellular ligand-binding region of mGluR1 with (S)-(α)-methyl-4-carboxyphenylglycine, confirms that this antagonist maintains the open conformation of the VFT.[110] Unveiled molecular details of the receptor interaction with the agonist and antagonist ligands provide structural bases for the design of new drugs.

Family 3 GPCRs also form homo- and heterodimers.[107,108,111,112] The mGluR5, as in the case of all other mGluR- and Ca^{2+}-sensing receptors, form homodimers.[111,113,114] The $GABA_B$ receptor is a heterodimer constituted of the homologous $GABA_{B1}$ and $GABA_{B2}$ subunits,[108] whereas $GABA_{B1}$ is responsible for GABA recognition,[115] and $GABA_{B2}$ is required for the correct trafficking of $GABA_{B1}$ to the cell surface.[116] The dimeric state of the receptors has been recently implicated in intracellular G protein activation. Differences in crystal structures, with or without ligand bound to the dimeric extracellular domain of mGluR1, suggest that the movements of the extracellular ligand-binding domains in the dimers should induce a structural rearrangement in the membrane-spanning portion of the receptor.[109] Fluorescent proteins inserted into the cytosolic loops of the receptor served as probes to detect changes in the orientation of the two protomers that comprise the dimer in intact cells. The fluorescence Förster resonance energy transfer (FRET) analysis of the ligand binding to the mGluR1 demonstrate that the first intracellular loops of each protomer in the mGluR1 dimer move apart from each other, while the second intracellular loops move closer.[117] These results support a model of glutamate action; glutamate binding causes a rearrangement of dimeric extracellular domains, which, in turn, leads to the conformational changes of the 7TM domain and the intracellular domains that mediate the corresponding signal transduction pathway. Whether the molecular rearrangements involve the 7TM domain conformational changes or just changes in relative orientation of the transmembrane helices remains to be determined.

An active area of research is to elucidate the link between ligand–receptor stoichiometry and receptor activation. In the case of heterodimeric $GABA_B$ receptors, the agonist binds only to the $GABA_{B1}$ VFT domain to activate the receptor.[115] In the case of homodimeric mGlu5R receptors, partial activation is observed when only one of two glutamate binding sites is occupied.[118] Although two bound agonists are required for full activation of homodimeric mGluR5 receptor,[118] partial receptor activation upon binding of a single agonist molecule may increase the probability of receptor activation at low ligand concentrations thereby providing an important advantage for the fast agonist response of mGlu receptors in modulating synaptic transmission.[118] Hence, the kinetics of GPCR regulation and activation can depend on the GPCR–ligand occupancy.

18.5 NMR OF GPCRs

NMR presents many great opportunities to obtain structural and dynamic details of membrane proteins. NMR-based structure determination requires sequence specific resonance assignments and other parameters that provide distance and torsion angle constraints. Because of the size and complexity of GPCR solutions, novel solution and solid-state NMR methods are required to elucidate the structures of GPCRs with their ligands.

18.5.1 SOLUTION NMR STUDIES OF LIGAND INTERACTIONS WITH GPCRs

Significant efforts have been made to use NMR to characterize ligand interactions with GPCRs. Initial efforts focused on covalent ligand interactions found in the 7TM proteins — bacteriorhodopsin and rhodopsin — where the covalent ligands are the isomeric forms of the chromophore retinal.[120, 141] While solution NMR was used in some of the earliest studies of rhodopsin–chromophore interactions, overall resolution was very poor for the sonicated membrane suspensions, and so NMR efforts, for the most part, have shifted to solid-state methods. However, with the advent of higher field instruments, novel pulse sequences and recombinant expression technology, solution NMR has, of late, been applied to structural analyses of rhodopsin in a number of interesting ways.

Solution NMR was initially applied to study the interaction between rhodopsin and the G protein alpha subunit transducin.[119] A transferred nuclear Overhauser effect (Tr-NOESY) experiment characterized binding of the fragment peptide of transducin (residues 340 to 350 with mutation Lys341Arg) to rhodopsin.[119] The observed NOEs provided information on distances between various protons in the bound peptide, which, along with distance geometry constraints, NOE-constrained molecular dynamics, simulated annealing, and grid searches, produced a series of structures of the transducin peptide in association with the ground state rhodopsin and the excited Meta II intermediate. Although a number of problems, including a low signal-to-noise ratio and assignment errors, were later identified in this innovative work,[120] a similar analysis using Tr-NOESY and total correlation spectroscopy (TOCSY) experiments comparing free transducin peptide (amino acid residues 340 to 350) to the Meta II intermediate bound rhodopsin form has been successful.[119] Disordered peptide is observed in solution, which, upon light activation, undergoes a dramatic shift in conformation, including a helical turn followed by an open reverse turn along the peptide chain. Comparison of the structures of the free transducin peptide to the corresponding domain in the intact transducin suggest that G protein activation involves a series of reversible conformational changes.[119,121–123] This model was later confirmed by measurement of residual dipolar couplings.[124,125] Magnetically oriented bilayer ^{31}P NMR studies indicate that rod outer segment disc membranes — which by themselves lack the cooperativity in alignment that exists in the intact rod outer segment — actually have a sufficiently large total magnetic susceptibility anisotropy by themselves to align their membrane normal parallel to a magnetic field.[125]

In the example of the C5a receptor, a series of agonist and antagonist peptides have been used to establish the putative interaction sites not only of the intact C5a receptor but of the isolated carboxyl-terminal domain.[126] By comparing the solution structures of these peptides, common elements of the structures of agonist and antagonist molecules can be resolved. Having peptides simplifies the process of obtaining isotopically labeled ligands that can be characterized using NMR methods. Alternatively, engineering of cysteine residues near a proposed binding site in the C5a receptor has been used to trap short cysteine-containing 3-mer peptides derived from the C5a peptide by disulfide bond formation.[127] Ligand-binding experiments using agonists and antagonists are able to identify where such molecules could bind to the C5a receptor.[128] Fragment-based drug discovery opens the possibility of linking these bound fragments to make novel ligands.

The functional roles of GPCR domains can be determined by NMR character-ization of the isolated domains. In cannabinoid CB1 and CB2 GPCRs, isolated carboxyl-terminal juxtamembrane segments in dodecylphosphocholine micelles have been shown to directly activate their corresponding G proteins.[129–131] The conformational specificities of these two domains have been defined by NMR struc-ture determination.[132] Similar studies have examined the structures of the TM7 domain of the tachykinin NK1[133] and the bradykinin[134] receptors in a variety of organic solvents and detergents. By coupling functional mutagenesis data to struc-tural data of the TM domains, more details of the structural framework of G protein activation can be learned. The redox state of the isolated ligand-binding domain has been useful in defining the role of oxidation on the intact receptor.[135] NMR structure determination as with other methods can be hampered by the nature of the receptor domain or ligands when incorporated into organic solvents, micelles, bicelles, and vesicles.

NMR structures of the amino-terminal extracellular domain of the cortico-trophin-releasing-factor (CRF) receptor helped provide a structural rationale for the mechanism of activation upon hormone peptide binding.[104] Similar work has been conducted with the secretin receptor.[136] The similarity of equilibrium dissociation constants for hor-mone binding to the intact receptor and amino-terminal domain suggest that the mechanism of ligand–receptor–G protein activation can be modular in nature.

18.5.2 SOLID-STATE NMR STUDIES OF LIGAND INTERACTIONS WITH GPCRS

GPCRs typically exist in a membrane bilayer that is an intrinsically anisotropic environment. These molecules are partially or fully immobilized and can become intransigent to structural analyses by standard techniques such as X-ray crystallo-graphy or solution-state NMR. Further, detergent-solubilized GPCRs have high correlation times that make solution NMR methods problematic. Instead, solid-state NMR (ssNMR) techniques specifically designed for the study of slowly tumbling or solid-phase systems can offer unique possibilities to elucidate structural or dynamic parameters at atomic resolution.

For more than three decades, ssNMR has been applied to the structural study of membranes and proteins associated with them.[137,138] Unlike in solution, the spectral resolution and the overall sensitivity of ssNMR are influenced by the size and orientation dependence of the nuclear spin interactions such as the chemical shielding and the homo- and heteronuclear dipolar spin–spin couplings. As a result, spectral resolution and overall sensitivity can be compromised; earlier biophysical applications of ssNMR concentrated on the determination of individual structural and dynamic parameters in insoluble and noncrystalline systems. Combined with site-specific single or pair-wise isotope labeling, powerful applications ranging from the determination of intermolecular distances and torsion angles in membrane proteins and their natural ligands[139–141] to the determination of a complete 3D structure of a membrane peptide[142] are reported. In addition to rhodopsin and other retinal proteins, a variety of membrane embedded systems have been the subject of ssNMR-based structural studies, often involving macroscopically oriented systems,[143] and the interested reader is referred to a series of review articles[144–147] for further information.

Application of state-of-the-art ssNMR technology permits the spectroscopic study of 25 nmol of labeled GPCR agonist and the corresponding nonlabeled purified receptor. In most cases, GPCRs must be solubilized from cell membranes and purified in the presence of detergents. Exceptions are rhodopsin and bacteriorhodopsin, which can be isolated and purified in membranes and can be directly centrifuged into magic-angle spinning rotors, frozen, and stored at 190 K.[148,149]

Sophisticated ssNMR experiments, which can be employed to determine the 3D conformation of a protein at atomic resolution, are, by a factor of one half to threefold, less sensitive than experiments used for NMR resonance assignments.[150,151] Hence, their application requires samples containing at least 60 nmol of labeled material. Considering the size of MAS rotors and an average molecular weight of 50 kDa, 25 nmol of GPCR would correspond to 1.25 mg in 100 μL (12.5 mg/mL). Obtaining greater than or equal to 12.5 mg/mL GPCR concentration can be difficult to achieve with detergent-solubilized GPCR samples. Typically, an increase in the concentration of the solubilized receptor leads to concomitant concentration of detergent micelles. The resulting high detergent concentration can seriously destabilize the GPCRs, since detergents compete not only with lipid–lipid and lipid–protein interactions but also with protein–protein interactions.[152]

Several groups have shown that solubilized and purified GPCRs can be functionally reconstituted into lipid vesicles. A frequently used strategy for proteoliposome preparation involves saturation of preformed liposomal vesicles with detergents, followed by an incubation with the solubilized membrane protein and detergent removal by various methods.[153,154] Applications to GPCRs include the functional reconstitution of the amino-terminally truncated form of rat neurotensin receptor (NTS-1)[155] and the chemokine CCR5 receptor with the CD4 peptide.[156] In these studies, the detergent was removed with hydrophobic polystyrene beads. A different approach is based on selective extraction of detergents from mixed detergent–lipid–protein micelles using cyclodextrin-based compounds. This method has been successfully applied to the functional reconstitution of bovine rhodopsin[157] and the human histamine H$_1$ receptor.[158] For ssNMR studies, the recovered proteoliposomes are incubated with ligand and can subsequently be centrifuged into magic-angle spinning rotors.

An elegant description of the structure of an agonist, neurotensin, bound to its cognate GPCR, NTS-1, has been obtained via ssNMR.[155] [13]C, ^{15}N-labeled 13-amino-acid neurotensin has been shown to complex to functional neurotensin receptors expressed in *Escherichia coli* and reconstituted into lipid vesicles. The neurotensin peptide is disordered in the absence of a receptor and undergoes a β-strand rearrangement upon binding. The structure provides direct experimental evidence for a distinct conformation of a neuropeptide bound with high affinity to its cognate receptor. The structure of the neuropeptide can provide clues to novel ligand structures that can possess similar geometry and electrostatic properties.

18.6 MOLECULAR MODELLING OF GPCRS FROM THE RHODOPSIN STRUCTURE

Comparative or homology protein structure modelling is a critical tool for providing structural information of a protein that is related by sequence identity to a protein of known (experimentally determined) structure. Accurate comparative models of GPCRs based on the crystal structure of rhodopsin would provide a huge boost to the success of structure-based lead discovery. High-accuracy models are typically based on proteins that have 50% sequence identity to the template structure. The resulting model can have a root mean square (RMS) error for main-chain atoms of approximately 1 Å or be comparable in quality with a low-resolution x-ray structure.[159] However, the most accurate comparative models produce an all-atom model of the unknown structure based on the sequence alignment of multiple crystal structures. The low sequence identity, less than 35% in some cases, of medically relevant Family 1 GPCRs to rhodopsin,[64] combined with the existence of only one crystal structure, makes the task of generating models accurate enough for meaningful structure-based drug design a challenge. In fact, experience using the rhodopsin structure to generate a homology model of a Family 1 receptor suggests that producing meaningful 3D models remains an arduous task.[160] In contrast, a recent comparison of the rhodopsin structure with homology models of several GPCRs that incorporated experimental data on the respective GPCRs concluded that the overall structure of rhodopsin is likely similar to the biogenic amine receptors in this family.[64] There have been several reports of models of biogenic amine receptors that help corroborate mutagenic[161] and pharmacological data,[162] including some species-specific selectivity differences.[163] For use in structure-based lead discovery or optimization efforts, however, the utility of these models is less clear.

The use of rhodopsin-derived models of the dopamine D3, muscarinic M1, and vasopressin V1a receptors in virtual screening methods successfully identified known antagonists seeded into a chemical library; however, these same models were not suitable for finding known agonists by the same methodology.[164] The limited utility of the rhodopsin-derived models of therapeutic targets has led to the development of several combined target-based and pharmacophore-based approaches for lead discovery of GPCR targets.

In the absence of structural information about a particular target of interest, pharmacophore-based screening can be employed,[165] though its success often requires several pharmacophore descriptions.[166] By combining the two methodologies, these hybrid methods have had some success in identifying submicromolar lead compounds; two recent examples include the discovery of ligands for the NK1[167] and the D3 dopamine[168] receptors. The latter effort included extensive use of experimental data based on the SARs of known ligands, receptor mutational data, and the results of the substituted cysteine accessibility method (SCAM). This work helped validate the homology model prior to the virtual screening methods. This led to the discovery of compounds with experimentally determined equilibrium dissociation inhibition constant (K_i) values of approximately 1 μM.[168]

In a similar approach, HT molecular docking and comparative molecular field analysis (CoMFA) were used to build a pharmacophore model for the adenosine A_3 receptor. Combined with a rhodopsin-derived model of the receptor, several new antagonists were identified.[169] This is an iterative approach (Figure 18.3), akin to experimental structure-based lead optimization, which allows for the identification of ligand–receptor interactions and is used to build novel real and virtual compound libraries.[170] As expected, this approach is highly dependent on experimentally derived SAR.

Though comparative molecular modelling remains the most reliable method of predicting protein structure,[159] the limitations of applying this technique to GPCRs have been discussed. Recently, a new *ab initio* method was developed that is capable of generating molecular models of GPCRs from primary sequence only.[171] These authors report the identification of several lead compounds with K_i values in the nM range for the serotonin 5-HT1A and 5-HT4, tachykinin NK1, dopamine D2, and chemokine CCR3 receptors.[172] Focusing on preclinical and early-stage-clinical candidates that target GPCRs and ion channels, this methodology is being used to guide medicinal chemistry efforts in both lead discovery and optimization.[173]

The crystal structure of rhodopsin has opened the door for structural biology to impact the discovery and optimization of compounds that target medically relevant GPCRs. However, the efforts are currently limited to computational methods that, while able to provide important guidance and corroboration of experimental data, are not strong enough to replace the more traditional methods of lead identification when available (e.g., high throughput screening). Further, the *in silico* methods are not yet rigorous enough to address some of the more important questions about the complexities of these molecules. Issues of activation, selectivity, antagonist-versus-agonist recognition, structural changes upon ligand binding, and how ligand binding affects signal transduction will all likely require further experimental verification. In the near term, it is conceivable that some of these questions will be answered by X-ray crystallography and NMR spectroscopy. The ability to overexpress using heterologous systems and larger-scale purification efforts of GPCRs are improving,[23,24,174] suggesting that the experimental structure determination of a human GPCR is inevitable. As this information emerges, more traditional structure-based lead discovery and optimization methods will likely be employed in the pharmaceutical industry. And the larger community will begin to address some of the critical biochemical and biological questions in light of this new experimentally derived structural information.

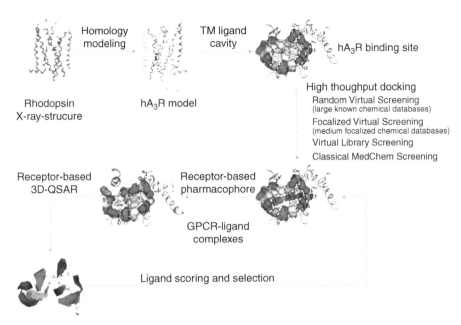

FIGURE 18.3 **(See color insert following page 272.)** Integrated ligand-receptor-based antagonist design based on molecular modelling.[164] The combined receptor-based and ligand-based drug design approach has been carried out to define a novel pharmacophore model of human adenosine A_3 receptor antagonists. The computational strategy used can be described as follows: (a) GPCR 3D-model(s) are constructed and refined using a conventional rhodopsin-based homology modelling approach; (b) 3D receptor-driven pharmacophores(s) are generated; (c) a 3D receptor-driven quantitative structure-activity relationship (3D-QSAR) is generated; and (d) new adenosine A_3 receptor antagonist candidates are selected, synthesized and characterized pharmacologically. Iterative rounds of receptor-ligand selection are required for better lead design. (With permission from Stefano Moro and Elsevier Press.)

18.7 CONCLUSION

In this chapter, we have reviewed methods that elucidate the nature of GPCR–ligand binding. We discussed the advantages and disadvantages of using screening methods for drug design and the roles of conserved regions of GPCRs in ligand binding. To enhance the potential of structure-based drug design, characterization of the GPCR ligand-binding domain–ligand interactions helps define potency. NMR is used to elaborate on details of the protein–ligand interactions that can identify the elements of ligand specificity. In addition, we reviewed the use of molecular modelling to help make 3D models of GPCR–ligand interactions.

The kinds of ligand interactions with GPCRs are particularly diverse. While agonists bind to certain regions, antagonists may bind in conformationally distinct modes. Probably, the most competitive antagonists and inverse agonists occupy the agonist site, albeit in a different orientation. Noncompetitive antagonists may bind at other sites. While the exact locations of the alternative sites are as yet unknown, it is likely that TM domains other than the set lining the agonist-binding pocket are involved, since such ligands do not compete with agonists. There is contention that the active form is somehow conformationally more flexible than the antagonist bound form.

On the contrary, it is usually claimed that proteins are stabilized by their endogenous ligands. GPCR multistate models have been invoked to explain their complex behavior in the presence of agonists, antagonists, and other binding partners.[84] The occurrence of intermediate receptor conformational states during GPCR activation and antagonist binding has been shown in a variety of fluorescence measurements with constitutively active receptor mutants and antagonists.

Structural modification and exogenous and endogenous allosteric modulators also affect ligand–GPCR interaction and receptor activation. This is where structure-based drug design can best help design more selective therapeutic drugs. At present, it is difficult to model conformational transitions from the rhodopsin crystal structure conformation (i.e., the inactive conformation) to the proposed active conformation due to the complexity of potentially numerous coupled local conformational changes that may occur. It is unlikely that this conformational transition would occur spontaneously in equilibrium molecular dynamics simulations.

GPCRs were initially considered to adopt just inactive and activate conformations and to activate a single type of G protein. However, studies with recombinant cell systems have led to a more complex picture in that GPCRs have been shown to activate several G proteins. There have been suggestions to rename GPCRs 7TM proteins, since they mediate many other regulatory elements besides G proteins.[1,12] For instance, long-term GPCR desensitization due to prolonged time of agonist exposure invokes multifactorial mechanisms that involve: endocytosis through β-arrestin; interactions of kinases for phosphorylation; expression of receptor splice variants that exhibit different sensitivities to ligands and G proteins; and receptor coupling to more than one G protein class and cross-talk between GPCRs.[162,175] Notwithstanding the potential of differential interactions with GPCR isoforms, there is also the possibility of receptor oligomerization, which can play a role in subtype-specific receptor regulation.[162,176] Future designs of assays that probe the interactions of GPCRs to other accessory proteins may elucidate another exciting dimension of GPCR drug discovery.

The access to GPCR structures alone may not be sufficient to develop a mechanistic view of all of these interactions. Therefore, additional experimental data will be needed to help define plausible pathways for the conformational changes. Since there is no simple consistent explanation for ligand stereoselectivity, a single static 3D model is not adequate to explain such ligand binding. This is quite consistent with the idea that GPCRs undergo facile transitions between multiple conformational states under normal conditions[177–179] and that agonists, neutral antagonists, and inverse agonists preferentially bind and stabilize different receptor conformations.[180,181] If this is true, then many independent ligand–receptor complex structures must be solved to achieve a more complete understanding of the details of receptor subtype-selective ligand binding. The future holds exciting opportunities for efforts that include structure determination of GPCRs with their respective ligands and effector proteins.

REFERENCES

1. Marinissen, M.J. and Gutkind, J.S. G protein-coupled receptors and signalling networks: emerging paradigms. *Trends Pharmacol. Sci.*, 22, 368–376, 2001.
2. Cooper, M.A. Advances in membrane receptor screening and analysis. *J. Mol. Recognit.*, 17, 286–315, 2004.
3. Chalmers, D.T. and Behan, D.P. The use of Constitutively Active GPCRs in Drug Discovery and Functional Genomics. *Nat. Rev. Drug Disc.*, 1, 599–608, 2002.
4. Sautel, M. and Milligan, G. Molecular Manipulation of G Protein coupled Receptors: A New Avenue into Drug Discovery. *Curr. Med. Chem.*, 7, 889–896, 2000.
5. Klabunde, T. and Hessler, G. Drug design strategies for targeting G protein-coupled receptors. *Chem. Biochem.*, 3, 928–944, 2002.
6. Drews, J. Drug discovery: a historical perspective. *Science*, 287, 1960–1964, 2000.
7. Intelligence, E.B. *The Pharma R&D Fact Book*, in *The Pharma R&D Fact Book*. Espicom Business Intelligence Ltd., London, 2004.
8. Bywater, R.P. Location and nature of the residues important for ligand recognition in G protein coupled receptors. *J. Mol. Recogn.*, 2004.
9. Horn, F., Bettler, E., Oliveira, L., Campagne, F., Cohen, F.E., and Vriend, G. GPCRDB information system for G protein-coupled receptors. *Nucleic Acids Res.*, 31, 294–297, 2003.
10. Civelli, O., Reinscheid, R.K., and Nothacker, H.P. Orphan receptors, novel neuropeptides and reverse pharmaceutical research. *Brain Res.*, 848, 63–65, 1999.
11. Scussa, F. World's best selling drugs. *Med. Adv. News*, 21, 1–46, 2002.
12. Pierce, K.L., Premont, R.T., and Lefkowitz, R.J. Seven-transmembrane receptors. *Nat. Rev. Mol. Cell Biol.*, 3, 639–650, 2002.
13. Vassilatis, D.K., Hohmann, J.G., Zeng, H., Li, F., Ranchalis, J.E., Mortrud, M.T., Brown, A., Rodriguez, S.S., Weller, J.R., Wright, A.C., Bergmann, J.E., and Gaitanaris, G.A. The G protein-coupled receptor repertoires of human and mouse. *Proc. Natl. Acad. Sci. USA*, 100, 4903–4908, 2003.
14. Bywater, R.P., Sorensen, A., Rogen, P., and Hjorth, P.G. Construction of the simplest model to explain complex receptor activation kinetics. *J. Theoret. Biol.*, 218, 139–147, 2002.
15. Gouldson, P.R., Snell, C.R., Bywater, R.P., Higgs, C., and Reynolds, C.A. Domain swapping: a mechanism for functional rescue and activation in G protein coupled receptors. *Prot. Eng.*, 11, 1181–1193, 1998.
16. Nienaber, V.L., Richardson, P.L., Klighofer, V., Bouska, J.J., Giranda, V.L., and Greer, J. Discovering novel ligands for macromolecules using X-ray crystallographic screening. *Nat. Biotechnol.*, 18, 1105–1108, 2000.
17. Muchmore, S.W. and Hajduk, P.J. Crystallography, NMR and virtual screening: integrated tools for drug discovery. *Curr. Opin. Drug Discov. Dev.*, 6, 544–549, 2003.
18. Shuker, S.B., Hajduk, P.J., Meadows, R.P., and Fesik, S.W. Discovering high-affinity ligands for proteins: SAR by NMR. *Science*, 274, 1531–1534, 1996.
19. Blundell, T.L. and Patel, S. High-throughput X-ray crystallography for drug discovery. *Curr. Opin. Pharmacol.*, 4, 490–496, 2004.
20. Davis, A.M., Teague, S.J., and Kleywegt, G.J. Application and limitations of X-ray crystallographic data in structure-based ligand and drug design. *Angew Chem. Int. Ed. Engl.*, 42, 2718–2736, 2003.
21. Blundell, T.L., Jhoti, H., and Abell, C. High-throughput crystallography for lead discovery in drug design. *Nat. Rev. Drug Discov.*, 1, 45–54, 2002.

22. Palczewski, K., Kumasaka, T., Hori, T., Behnke, C.A., Motoshima, H., Fox, B.A., Le Trong, I., Teller, D.C., Okada, T., Stenkamp, R.E., Yamamoto, M., and Miyano, M. Crystal structure of rhodopsin: A G protein-coupled receptor. *Science*, 289, 739–745, 2000.

23. Lundstrom, K. Structural genomics of GPCRs. *Trends Biotechnol.*, 23, 103–108, 2005.

24. Laible, P.D., Scott, H.N., Henry, L., and Hanson, D.K. Towards higher-throughput membrane protein production for structural genomics initiatives. *J. Struct. Funct. Genomics*, 5, 167–172, 2004.

25. Okada, T., Sugihara, M., Bondar, A.N., Elstner, M., Entel, P., and Buss, V. The retinal conformation and its environment in rhodopsin in light of a new 2.2 A crystal structure. *J. Mol. Biol.*, 342, 571–583, 2004.

26. Carpenter, J.W., Laethem, C., Hubbard, F.R., Eckols, T.K., Baez, M., McClure, D., Nelson, D.L., and Johnston, P.A. Configuring radioligand receptor binding assays for HTS using scintillation proximity assay technology. *Meth. Mol. Biol.*, 190, 31–49, 2002.

27. Bidlack, J.M. and Parkhill, A.L. Assay of G protein-coupled receptor activation of G proteins in native cell membranes using 35SGTP gamma S binding. *Meth. Mol. Biol.*, 237, 135–143, 2004.

28. Lad, P.M., Nielsen T.B., and Rodbell, M. A probe for the organization of the beta-adrenergic receptor-regulated adenylate cyclase system in turkey erythrocyte membranes by the use of a complementation assay. *FEBS Lett.*, 122, 179 183, 1980.

29. Schutz, W., Steurer, G., Tuisl, E., and Plass, H. Phosphorylated adenosine derivatives as low-affinity adenosine-receptor agonists. Methodological implications for the adenylate cyclase assay. *Biochem. J.*, 220, 207–212, 1984.

30. Zhang, Y., Kowal, D., Kramer, A., and Dunlop, J. Evaluation of FLIPR Calcium 3 Assay Kit — a new no-wash fluorescence calcium indicator reagent. *J. Biomol. Screen.*, 8, 571–577, 2003.

31. Behan, D.P. and Chalmers, D.T. The use of constitutively active receptors for drug discovery at the G protein-coupled receptor gene pool. *Curr. Opin. Drug Discov. Dev.*, 4, 548–560, 2001.

32. Ellis, C.R. State of GPCR Research in 2004. *Nat. Rev. Drug Discov.*, 3, 577–626, 2004.

33. Ulloa-Aguirre, A., Stanislaus, D., Janovick, J.A., and Conn, P.M. Structure-activity relationships of G protein coupled receptors. *Arch. Med. Res.*, 30, 420–435, 1999.

34. Bondensgaard, K., Ankersen, M., Thogersen, H., Hansen, B.S., Wulff, B.S., and Bywater, R.P. Recognition of privileged structures by G protein coupled receptors. *J. Med. Chem.*, 47, 888–899, 2004.

35. Muller, G. Towards 3D structures of GPCRs: a multidisciplinary approach. *Curr. Med. Chem.*, 7, 861–888, 2000.

36. Abramson, J., Smirnova, I., Kasho, V., Verner, G., Kaback, H.R., and Iwata, S. Structure and mechanism of the lactose permease of *Escherichia coli. Science*, 301, 610–615, 2003.

37. Huang, Y., Lemieux, M.J., Song, J., Auer, M., and Wang, D.N. Structure and mechanism of the glycerol-3-phosphate transporter from *Escherichia coli. Science*, 301, 616–620, 2003.

38. Foord, S.M. Receptor classification: post genome. *Curr. Opin. Pharmacol.*, 2, 561–566, 2002.

39. Spiegel, A.M. and Weinstein, L.S. Inherited Diseases involving G proteins and G protein-coupled receptors. *Annu. Rev. Med.*, 55, 27–39, 2004.

40. Kuhlbrandt, W. and Gouaux, E. Membrane proteins. *Curr. Opin. Struct. Biol.*, 9, 445–447, 1999.

41. Loll, P.J. Membrane protein structural biology: the high throughput challenge. *J. Struct Biol.*, 142, 144–153, 2003.
42. Michel, H. Three-dimensional crystals of a membrane protein complex. The photosynthetic reaction centre from *Rhodopseudomonas viridis. J. Mol. Biol.*, 158, 567–572, 1982.
43. Engel, A., Hoenger, A., Hefti, A., Henn, C., Ford, R.C., Kistler, J., and Zulauf, M. Assembly of 2-D membrane protein crystals: dynamics, crystal order, and fidelity of structure analysis by electron microscopy. *J. Struct. Biol.*, 109, 219–234, 1992.
44. Werten, P.J., Remigy, H.W., de Groot B.L., Fotiadis, D., Philippsen, A., Stahlberg, H., Grubmuller, H., and Engel, A. Progress in the analysis of membrane protein structure and function. *FEBS Lett.*, 529, 65–72, 2002.
45. Liang, Y., Fotiadis, D., Filipek, S., Saperstein, D.A., Palczewski, K., and Engel, A. Organization of the G protein-coupled receptors rhodopsin and opsin in native membranes. *J. Biol. Chem.*, 278, 21655–21662, 2003.
46. Fernandez, C. and Wuthrich, K. *NMR solution structure determination of membrane proteins reconstituted in detergent micelles.* FEBS Lett, 2003. 555: 144-50.
47. Nielsen, N., Malmendal, A., and Vosegaard, T. Techniques and applications of NMR to membrane proteins. *Mol. Membr. Biol.*, 21, 129–141, 2004.
48. Opella, S.J. Membrane protein NMR studies. *Meth. Mol. Biol.*, 227, 307–320, 2003.
49. Watts, A., Straus, S.K., Grage, S.L., Kamihira, M., Lam, Y.H., and Zhao, X. Membrane protein structure determination using solid-state NMR. *Meth. Mol. Biol.*, 278, 403–473, 2004.
50. White, S.H. Blanco.biomol.uci.edu/Membrane_Proteins_xtal.html., (accessed 2005).
51. Michel, H. www.mpibp-frankfurt.mpg.de/michel/public/memprotstruct.html., 2004, (accessed 2005).
52. Tate, C.G. Overexpression of mammalian integral membrane proteins for structural studies. *FEBS Lett.*, 504, 94–98, 2001.
53. Grisshammer, R. and Tate, C.G. Overexpression of integral membrane proteins for structural studies. *Q. Rev. Biophys.*, 28, 315–422, 1995.
54. Sarramegna, V., Talmont, F., Demange, P., and Milon, A. Heterologous expression of G-protein-coupled receptors: comparison of expression systems fron the standpoint of large-scale production and purification. *Cell Mol. Life Sci.*, 60, 1529–1546, 2003.
55. Kashino, Y. Separation mehtods in the analysis of protein membrane complexes. *J. Chromatogr. B.*, 797, 191–216, 2003.
56. Caffrey, M. Membrane protein crystallization. *J. Struct. Biol.*, 142, 108–132, 2003.
57. Caffrey, M. A lipid's eye view of membrane protein crystallization in mesophases. *Curr. Opin. Struct. Biol.*, 10, 486–497, 2000.
58. Michel, H. *Crystallization of Membrane Proteins.* CRC Press, Boca Raton, 1990.
59. Ostermeier, C. and Michel, H. Crystallization of membrane proteins. *Curr. Opin. Struct. Biol.*, 7, 697–701, 1997.
60. Iwata, S. *Methods and Results in Crystallization of Membrane Proteins.* IUL Biotechnology Series, ed. I.F. Tsigelny. Vol. 4., International University Line, La Jolla, CA, 2003.
61. Gether, U. Uncovering molecular mechanisms involved in activation of G protein-coupled receptors. *Endocrinol. Rev.*, 21, 90–113, 2000.
62. Kristiansen, K. Molecular mechanisms of ligand binding, signaling, and regulation within the superfamily of G-protein-coupled receptors: molecular modeling and mutagenesis approaches to receptor structure and function. *Pharmacol. Ther.*, 103, 21–80, 2004.

63. Pin, J.P., Galvez, T., and Prezeau, L. Evolution, structure, and activation mechanism of family 3/C G-protein-coupled receptors. *Pharmacol. Ther.*, 98, 325–354, 2003.

64. Ballesteros, J.A., Shi, L., and Javitch, J.A. Structural mimicry in G protein-coupled receptors: implications of the high-resolution structure of rhodopsin for structure-function analysis of rhodopsin-like receptors. *Mol. Pharmacol.*, 60, 1–19, 2001.

65. Spalding, T.A., Burstein, E.S., Henderson, S.C., Ducote, K.R., and Brann, M.R. Identification of a ligand-dependent switch within a muscarinic receptor. *J. Biol. Chem.*, 273, 21563–21568, 1998.

66. Ghanouni, P., Steenhuis, J.J., Farrens, D.L., and Kobilka, B.K. Agonist-induced conformational changes in the G-protein-coupling domain of the beta 2 adrenergic receptor. *Proc. Natl. Acad. Sci. USA*, 98, 5997–6002, 2001.

67. Strader, C.D., Sigal, I.S., Register, R.B., Candelore, M.R., Rands, E., and Dixon, R.A. Identification of residues required for ligand binding to the beta-adrenergic receptor. *Proc. Natl. Acad. Sci. USA*, 84, 4384–4388, 1987.

68. Bywater, R.P. Location and nature of the residues important for ligand recognition in G-protein coupled receptors. *J. Mol. Recogn.*, 18, 60–72, 2005.

69. Sheikh, S.P., Vilardarga, J.P., Baranski, T.J., Lichtarge, O., Iiri, T., Meng, E.C., Nissenson, R.A., and Bourne, H.R. Similar structures and shared switch mechanisms of the beta2-adrenoceptor and the parathyroid hormone receptor. Zn(II) bridges between helices III and VI block activation. *J. Biol. Chem.*, 274, 17033–17041, 1999.

70. Tota, R.T. and Strader, C.D. Characterization of the binding domain of the beta 2 adrenergic receptor with the fluorescent antagonist carazolol. *J. Biol. Chem.*, 265, 16891–16897, 1990.

71. Teller, D.C., Okada, T., Behnke, C.A., Palczewski, K., and Stenkamp, R.E. Advances in determination of a high-resolution three-dimensional structure of rhodopsin, a model of G-protein-coupled receptors (GPCRs). *Biochemistry*, 40, 7761–7772, 2001.

72. Filipek, S., Stenkamp, R.E., Teller, D.C., and Palczewski, K. G protein-coupled receptor rhodopsin: a prospectus. *Annu. Rev. Physiol.*, 65, 851–879, 2003.

73. Valiquette, M., Vu, H.K., Yue, S.Y., Wahlestedt, C., and Walker, P. Involvement of Trp-284, Val-296, and Val-297 of the human delta-opioid receptor in binding of delta-selective ligands. *J. Biol. Chem.*, 271, 18789–18796, 1996.

74. Brelot, A., Heveker, N., Montes, M., and Alizon, M. Identification of residues of CXCR4 critical for human immunodeficiency virus coreceptor and chemokine receptor activities. *J. Biol. Chem.*, 275, 23736–23744, 2000.

75. Lawson, Z. and Wheatley, M. The third extracellular loop of G protein-coupled receptors: more than just a linker between two important transmembrane helices. *Biochem. Soc. Transactions*, 32, 1048–1050, 2004.

76. Kristiansen, K., Kroeze, W.K., Willins, D.L., Gelber, E.I., Savage, J.E., Glennon, R.A., and Roth, B.L. A highly conserved aspartic acid (Asp-155) anchors the terminal amine moiety of tryptamines and is involved in membrane targeting of the 5-HT(2A) serotonin receptor but does not participate in activation via a "salt-bridge disruption" mechanism. *J. Pharmacol. Exp. Ther.*, 293, 735–746, 2000.

77. Shi, L. and Javitch, J.A. The second extracellular loop of the dopamine D2 receptor lines the binding-site crevice. *Proc. Natl. Acad. Sci. USA*, 101, 440–445, 2004.

78. Bywater, R.P. Location and nature of the residues important for ligand recognition in G-protein coupled receptors. *J. Molecular Recogn.*, 18, 60–72, 2005.

79. Furse, K.E. and Lybrand, T.P. Three-dimensional models for beta-adrenergic receptor complexes with agonists and antagonists. *J. Med. Chem.*, 46, 4450–4462, 2003.

80. Ballesteros, J.A., Jensen, A.D., Liapakis, G., Rasmussen, S.G.F., Shi, L., Gether, U., and Javitch, J.A. Activation of beta 2-adrenergic receptor involves disruption of an ionic lock between the cytoplasmic ends of transmembrane segments 3 and 6. *J. Biol. Chem.*, 276, 29171–29177, 2001.

81. O'Dowd, B.F., Hnatowich, M., Regan, J.W., Leader, W.M., Caron, M.G., and Lefkowitz, R.J. Site-directed mutagenesis of the cytoplasmic domains of the human beta 2-adrenergic receptor. Localization of regions involved in G protein-receptor coupling. *J. Biol. Chem.*, 263, 15985–15992, 1988.

82. Chakir, K., Xiang, Y., Yang, D., Zhang, S.J., Cheng, H., Kobilka, B.K., and Xiao, R.P. The third intracellular loop and the carboxyl terminus of beta2-adrenergic receptor confer spontaneous activity of the receptor. *Mol. Pharmacol.*, 64, 1048–1058, 2003.

83. Gether, U., Lin, S., and Kobilka, B.K. Fluorescent labeling of purified beta 2 adrenergic receptor. Evidence for ligand-specific conformational changes. *J. Biol. Chem.*, 270, 28268–28275, 1995.

84. Kobilka, B., Gether, U., Seifert, R., Lin, S., and Ghanouni, P. Characterization of ligand-induced conformational states in the beta 2 adrenergic receptor. *J. Recept. Signal. Transduct. Res.*, 19, 293–300, 1999.

85. Lin, S., Gether, U., and Kobilka, B.K. Ligand stabilization of the beta 2 adrenergic receptor: effect of DTT on receptor conformation monitored by circular dichroism and fluorescence spectroscopy. *Biochemistry*, 35, 14445–14451, 1996.

86. Seifert, R., Wenzel-Seifert, K., Gether, U., and Kobilka, B.K. Functional differences between full and partial agonists: evidence for ligand-specific receptor conformations. *J. Pharmacol. Exp. Ther.*, 297, 1218–12126, 2001.

87. Swaminath, G., Xiang, Y., Lee, T.W., Steenhuis, J., Parnot, C., and Kobilka, B.K. Sequential binding of agonists to the beta2 adrenoceptor. Kinetic evidence for intermediate conformational states. *J. Biol. Chem.*, 279, 686–691, 2004.

88. Vilardaga, J.P., di Paolo, E., de Neef, P., Waelbroeck, M., Bollen, A., and Robberecht, P. Lysine 173 residue within the first exoloop of rat secretin receptor is involved in carboxylate moiety recognition of Asp 3 in secretin. *Biochem. Biophys. Res. Commn.*, 218, 842–846, 1996.

89. Gardella, T.J., Luck, M.D., Fan, M.H., and Lee, C. Transmembrane residues of the parathyroid hormone (PTH)/PTH-related peptide receptor that specifically affect binding and signaling by agonist ligands. *J. Biol. Chem.*, 271, 12820–12825.

90. Frimurer, T.M. and Bywater, R.P. Structure of the integral membrane domain of the GLP1 receptor. *Proteins*, 35, 375–386, 1999.

91. Unson, C.G. Molecular determinants of glucagon receptor signaling. *Biopolymers*, 66, 218–235, 2002.

92. Lopez de Maturana, R., Willshaw, A., Kuntzsch, A., Rudolph, R., and Donnelly, D. The isolated N-terminal domain of the glucagon-like peptide-1 (GLP-1) receptor binds exendin peptides with much higher affinity than GLP-1. *J. Biol. Chem.*, 278, 10195–10200, 2003.

93. Al-Sabah, S. and Donnelly, D. The positive charge at Lys-288 of the glucagon-like peptide-1 (GLP-1) receptor is important for binding the N-terminus of peptide agonists. *FEBS Lett.*, 553, 342–346, 2003.

94. Grace, C.R., Perrin, M.H., DiGruccio, M.R., Miller, C.L., Rivier, J.E., Vale, W.W., and Riek, R. NMR structure and peptide hormone binding site of the first extracellular domain of a type B1 G protein-coupled receptor. *Proc. Natl. Acad. Sci. USA*, 101, 12836–12841, 2004.

95. Kieffer, T.J. and Habener, J.F. The glucagon-like peptides. *Endocrinol. Rev.*, 20, 876–913, 1999.

96. Unson, C.G., Macdonald, D., Ray, K., Durrah, T.L., and Merrifield, R.B. Position 9
    replacement analogs of glucagon uncouple biological activity and receptor binding.
    J. Biol. Chem., 266, 2763–2766, 1991.
97. Wilmen, A., Goke, B., and Goke, R. The isolated N-terminal extracellular domain of
    the glucagon-like peptide-1 (GLP)-1 receptor has intrinsic binding activity. *FEBS
    Lett.*, 398, 43–47, 1996.
98. Cao, Y.J., Gimpl, G., and Fahrenholz, F. The amino-terminal fragment of the adenylate
    cyclase activating polypeptide (PACAP) receptor functions as a high affinity PACAP
    binding domain. *Biochem. Biophys. Res. Commn.*, 212, 673–680, 1995.
99. Xiao, Q., Giguere, J., Parisien, M., Jeng, W., St. Pierre, S.A., Brubaker, P.L., and
    Wheeler, M.B. Biological activities of glucagon-like peptide-1 analogues *in vitro* and
    in vivo. Biochemistry, 40, 2860–2869, 2001.
100. Bazarsuren, A., Grauschopf, U., Wozny, M., Reusch, D., Hoffmann, E., Schaefer, W.,
    Panzner, S., and Rudolph, R. *In vitro* folding, functional characterization, and disulfide
    pattern of the extracellular domain of human GLP-1 receptor. *Biophys. Chem.*, 96,
    305–318, 2002.
101. Lee, C., Luck, M.D., Juppner, H., Potts, J.T., Kronenberg, Jr., H.M., and Gardella,
    T.J. Homolog-scanning mutagenesis of the parathyroid hormone (PTH) receptor
    reveals PTH-(1-34) binding determinants in the third extracellular loop. *Mol. Endo-
    crinol.*, 9, 1269–1278, 1995.
102. Gensure, R.C., Shimizu, N., Tsang, J., and Gardella, T.J. Identification of a contact
    site for residue 19 of parathyroid hormone (PTH) and PTH-related protein analogs
    in transmembrane domain two of the type 1 PTH receptor. *Mol. Endocrinol.*, 17,
    2647–2658, 2003.
103. Unson, C.G., Cypess, A.M., Wu, C.R., Goldsmith, P.K., Merrifield, R.B., and Sakmar,
    T.P. Antibodies against specific extracellular epitopes of the glucagon receptor block
    glucagon binding. *Proc. Natl. Acad. Sci. USA*, 93, 310–315, 1996.
104. Grace, C.R.R., Perrin, M.H., DiGruccio, M.R., Miller, C.L., Rivier, J.E., Vale, W.W.,
    and Riek, R. NMR structure and peptide hormone binding site of the first extracellular
    domain of a type B1 G protein-coupled receptor. *Proc. Natl. Acad. Sci. USA*, 101,
    12836–12841, 2004.
105. Livnah, O., Stura, E.A., Middleton, S.A., Johnson, D.L., Jolliffe, L.K., and Wilson,
    I.A. Crystallographic evidence for Preformed Dimers of Erythropoietin Receptors
    before Ligand Activation. *Science*, 283, 987–991, 1999.
106. Kauppi, B., Jakob, C., Farnegardh, M., Yang, J., Ahola, H., Alarcon, M., Calles, K.,
    Engstrom, O., Harlan, J., Muchmore, S., Ramqvist, A.K., Thorell, S., Ohman, L.,
    Greer, J., Gustafsson, J.A., Carlstedt-Duke, J., and Carlquist, M. The three-dimen-
    sional structures of antagonistic and agonistic forms of the glucocorticoid receptor
    ligand-binding domain. *J. Biol. Chem.*, 278, 22748–22754, 2003.
107. Kaupmann, K., Huggel, K., Heid, J., Flor, P.J., Biscoff, S., Mickel, S.J., McMaster,
    G., Angst, C., Bittiger, H., Froestl, W., and Bettler, B. Expression cloning of GABA(B)
    receptors uncovers similarity to metabotropic glutamate receptors. *Nature*, 386,
    239–246, 1997.
108. Kaupmann, K., Malitschek, B., Schuler, V., Heid, J., Froestl, W., Beck, P., Mosbacher,
    J., Bischoff, S., Kulik, A., Shigemoto, R., Karschin, A., and Bettler, B. GABA (B)-
    receptor subtypes assemble into functional heteromeric complexes. *Nature*, 396,
    683–687, 1998.
109. Kunishima, N., Shimada, Y., Tsuji, Y., Sato, T., Yamamoto, M., Kumasaka, T., Nakan-
    ishi, S., Jingami, H., and Morikawa, K. Structural basis of glutamate recognition by
    a dimeric metabotropic glutamate receptor. *Nature*, 407, 971–977, 2000.

110. Tsuchiya, D., Kunishima, N., Kamiya, N., Jingami, H., and Morikawa, K. Structural views of the ligand-binding cores of a metabotropic glutamate receptor complexed with an antagonist and both glutamate and Gd^{3+}. *Proc. Natl. Acad. Sci. USA*, 99, 2660–2665, 2002.

111. Romano, C., Yang, W.L., and O'Malley, K.L. Metabotropic glutamate receptor 5 is a disulfide-linked dimer. *J. Biol. Chem.*, 271, 28612–28616, 1996.

112. Marshall, F.H., Jones, K.A., Kaupmann, K., and Bettler, B. GABA(B) receptors — the first 7TM heterodimers. *Trends Pharmacol. Sci.*, 20, 396–399, 1999.

113. Bai, M., Trivedi, S., and Brown, E.M. Dimerization of the extracellular calcium-sensing receptor (CaR) on the cell surface of CaR-transfected HEK293 cells. *J. Biol. Chem.*, 273, 23605–23610, 1998.

114. Robbins, M.J., Ciruela, F., Rhodes, A., and McIlhinney, R.A. Characterization of the dimerization of metabotropic glutamate receptors using an N-terminal truncation of mGluR1alpha. *J. Neurochem.*, 72, 2539–2547, 1999.

115. Galvez, T., Prezeau, L., Milioti, G., Franek, M., Joly, C., Froestl, W., Bettler, B., Bertrand, H.O., Blahos, J., and Pin, J.P. Mapping the agonist-binding site of GABA(B) type 1 subunit sheds light on the activation process of GABA(B) receptors. *J. Biol. Chem.*, 275, 41166–41174, 2000.

116. Margeta-Mitrovic, M., Jan, Y.N., and Jan, L.Y. A trafficking checkpoint controls GABA(B) receptor heterodimerization. *Neuron*, 27, 97–106, 2000.

117. Tateyama, M., Abe, H., Nakata, H., Saito, O., and Kubo, Y. Ligand-induced rearrangement of the dimeric metabotropic glutamate receptor 1alpha. *Nat. Struct. Mol. Biol.*, 11, 637–642, 2004.

118. Kniazeff, J., Bessis, A.S., Maurel, D., Ansanay, H., Prezeau, L., and Pin, J.P. Closed state of both binding domains of homodimeric mGlu receptors is required for full activity. *Nat. Struct. Mol. Biol.*, 11, 706–713, 2004.

119. Harbison, G.S., Smith, S.O., Pardoen, J.A., Courtin, J.M., Lugtenburg, J., Herzfeld, J., Mathies, R.A., and Griffin, R.G. Solid-state ^{13}C NMR detection of a perturbed 6-s-trans chromophore in bacteriorhodopsin. *Biochemistry*, 24, 6955–6962, 1985.

120. Smith, S.O., Palings, I., Copie, V., Raleigh, D.P., Courtin, J., Pardoen, J.A., Lugtenburg, J., Mathies, R.A., and Griffin, R.G. Low-temperature solid-state ^{13}C NMR studies of the retinal chromophore in rhodopsin. *Biochemistry*, 26, 1606–1611, 1987.

121. Mollevanger, L.C., Kentgens, A.P., Pardoen, J.A., Courtin, J.M., Veeman, W.S., Lugtenburg, J., and de Grip, W.J. High-resolution solid-state 13C-NMR study of carbons C-5 and C-12 of the chromophore of bovine rhodopsin. Evidence for a 6-S-cis conformation with negative-charge perturbation near C-12. *Eur. J. Biochem.*, 163, 9–14, 1987.

122. Kakitani, H., Kakitani, T., Rodman, H., and Honig, B. On the mechanism of wavelength regulation in visual pigments. *Photochem. Photobiol.*, 41, 471–479, 1985.

123. Han, M., DeDecker, B.S., and Smith, S.O. Localization of the retinal protonated Schiff base counterion in rhodopsin. *Biophys. J.*, 65, 899–906, 1993.

124. Birge, R.R., Einterz, C.M., Knapp, H.M., and Murray, L.P. The nature of the primary photochemical events in rhodopsin and isorhodopsin. *Biophys. J.*, 53, 367–385, 1988.

125. Birge, R.R., Murray, L.P., Pierce, B.M., Akita, H., Balogh-Nair, V., Findsen, L.A., and Nakanishi, K. Two-photon spectroscopy of locked-11-cis-rhodopsin: evidence for a protonated Schiff base in a neutral protein binding site. *Proc. Natl. Acad. Sci. USA*, 82, 4117–4121, 1985.

126. Higginbottom, A., Cain, S.A., Woodruff, T.M., Proctor, L.M., Madala, P.K., Tyndall, J.D., Taylor, S.M., Fairlie, D.P., and Monk, P.N. Comparative agonist/antagonist responses in mutant human C5a receptors define the ligand binding site. *J. Biol. Chem.*, 280, 17831–17840, 2005.

127. Buck, E., Bourne, H., and Wells, J.A. Site-specific disulfide capture of agonist and antagonist peptides on the C5a receptor. *J. Biol. Chem.*, 280, 4009–4012, 2005.

128. Buck, E. and Wells, J.A. Disulfide trapping to localize small-molecule agonists and antagonists for a G protein-coupled receptor. *Proc. Natl. Acad. Sci. USA*, 102, 2719–2724, 2005.

129. Mukhopadhyay, S., Cowsik, S.M., Lynn, A.M., Welsh, W.J., and Howlett, A.C. Regulation of Gi by the CB1 cannabinoid receptor C-terminal juxtamembrane region: structural requirements determined by peptide analysis. *Biochemistry*, 38, 3447–3455, 1999.

130. Mukhopadhyay, S. and Howlett, A.C. CB1 receptor-G protein association. Subtype selectivity is determined by distinct intracellular domains. *Eur. J. Biochem.*, 268, 499–505, 2001.

131. Mukhopadhyay, S., McIntosh, H.H., Houston, D.B., and Howlett, A.C. The CB(1) cannabinoid receptor juxtamembrane C-terminal peptide confers activation to specific G proteins in brain. *Mol. Pharmacol.*, 57, 162–170, 2000.

132. Xie, X.Q. and Chen, J.Z. NMR structural comparison of the cytoplasmic juxtamembrane domains of G protein-coupled CB1 and CB2 receptors in membrane mimetic dodecylphosphocholine micelles. *J. Biol. Chem.*, 280, 3605–3612, 2005.

133. Berlose, J.P., Convert, O., Brunisson, A., Chassaing, G., and Lavielle, S. Three-dimensional structure of the highly conserved seventh transmembrane domain of G protein coupled receptors. *Eur. J. Biochem.*, 225, 827–843, 1994.

134. Piserchio, A., Zelesky, V., Yu, J., Taylor, L., Polgar, P., and Mierke, D.F. Bradykinin B2 receptor signaling: structural and functional characterization of the C-terminus. *Biopolymers*, 80, 367–373, 2005.

135. Ruan, K.H., Wu, J.X., So, S.P., Jenkins, L.A., and Ruan, C.H. NMR Structure of the thromboxane A2 receptor ligand recognition pocket. *Eur. J. Biochem.*, 271, 3006–3016, 2004.

136. Lisenbee, C.S., Dong, M.Q., and Miller, L.J. Paired Cysteine Mutagenesis to Establish the Pattern of Disulfide Bonds in the functional intact Secretin Receptor. *J. Biol. Chem.*, 280, 12330–12338, 2005.

137. Watts, A., Ulrich, A.S., and Middleton, D.A. Membrane protein structure: the contribution and potential of novel solid state NMR approaches. *Mol. Membr. Biol.*, 12, 233–246, 1995.

138. Smith, S.O. and Peersen, O.B. Solid-state NMR approaches for studying membrane protein structure. *Annu. Rev. Biophys. Biomol. Struct.*, 21, 25–47, 1992.

139. Feng, X., Lee, Y.K., Sandstrom, D., Eden, M., Maisel, H., Sebald, A., and Levitt, M.H. Direct determination of a molecular torsional angle by solid-state NMR. *Chem. Phys. Lett.*, 257, 314–320, 199.

140. Creuzet, F., McDermott, A., Gebhard, R., Vanderhoef, K., Spijkerassink, M.B., Herzfeld, J., Lugtenburg, J., Levitt, M.H., and Griffin, R.G. Determination of Membrane-Protein Structure by Rotational Resonance NMR — Bacteriorhodopsin. *Science*, 21, 783–786, 199.

141. Han, M. and Smith, S.O. NMR Constraints on the Location of the Retinal Chromophore in Rhodopsin and Bathorhodopsin. *Biochemistry*, 34, 1425–1432, 1995.

142. Ketchem, R.R., Hu, W., and Cross, T.A. High-resolution conformation of gramicidin A in a lipid bilayer by solid-state NMR. *Science*, 261, 1457–1460, 1993.

143. Cross, T.A.O. and Opella, S.J. Solid-State NMR Structural Studies of Peptides and Proteins in Membranes. *Curr. Opin. Struct. Biol*, 4, 640–648, 1994.

144. Marassi, F.M. and S.J. Opella. NMR structural studies of membrane proteins. *Curr. Opin. Struct. Biol.*, 8, 640–648, 1998.

145. Engelhard, M.B., Application of NMR-spectroscopy to retinal proteins. *Isr. J. Chem*, 35, 273–288, 1995.

146. Eilers, M., Ying, W., Reeves, P.J., Khorana, H.G., and Smith, S.O. Magic angle spinning nuclear magnetic resonance of isotopically labeled rhodopsin. *Meth. Enzymol.*, 343, 212–222, 2002.

147. Luca, S., Heise, H., and Baldus, M. High-resolution Solid-State NMR Applied to Polypeptides and Membrane Proteins. *Acc. Chem. Res.*, 36, 858–865, 2003.

148. Carravetta, M., Zhao, X., Johannessen, O.G., Lai, W.C., Verhoeven, M.A., Bovee-Geurts, P.H., Verdegem, P.J., Kiihne, S., Luthman, H., de Groot, H.J., de Grip, W.J., Lugtenburg, J., and Levitt, M.H. Protein-induced bonding perturbation of the rhodopsin chromophore detected by double-quantum solid-state NMR. *J. Am. Chem. Soc.*, 126, 3948–3953, 2004.

149. Petkova, A.T., Hu, J.G., Bizounok, M., Simpson, M., Griffin, R.G., and Herzfeld, J. Arginine activity in the proton-motive photocycle of bacteriorhodopsin: solid-state NMR studies of the wild-type and D85N proteins. *Biochemistry*, 38, 1562–1572, 1999.

150. Lange, A., Luca, S., and Baldus, M. Structural constraints from proton-mediated rare-spin correlation spectroscopy in rotating solids. *J. Am. Chem. Soc.*, 124, 9704–9705, 2002.

151. Etzkorn, M., Bockmann, A., Lange, A., and Baldus, M. Probing molecular interfaces using 2D magic-angle-spinning NMR on protein mixtures with different uniform labeling. *J. Am. Chem. Soc.*, 126, 14746–14751, 2004.

152. Gohon, Y.P. and Popot, J.L. Membrane protein-surfactant complexes. *Curr. Opin. Colloid & Interface Sci.*, 8, 15–22, 2003.

153. Rigaud, J.L., Pitard, B., and Levy, D. Reconstitution of membrane proteins into liposomes: application to energy-transducing membrane proteins. *Biochim. Biophys. Acta.*, 1231, 223–246, 1995.

154. Rigaud, J.L., Mosser, G., and Lambert, O. Detergent removal by non-polar polystyrene beads — Applications to membrane protein reconstitution and two-dimensional crystallization. *Eur. Biophys. J. Biophys. Lett.*, 27, 305–319, 1998.

155. Luca, S., Heise, H., and Baldus, M. High-resolution solid-state NMR applied to polypeptides and membrane proteins. *Acc. Chem. Res.*, 36, 858–865, 2003.

156. Devesa, F., Chams, V., Dinadayala, P., Stella, A., Ragas, A., Auboiroux, H., Stegmann, T., and Poquet, Y. Functional reconstitution of the HIV receptors CCR5 and CD4 in liposomes. *Eur. J. Biochem.*, 269, 5163–5174, 2002.

157. De Grip, W.J., Vanoostrum, J., and Bovee-Geurts, P.H. Selective detergent-extraction from mixed detergent/lipid/protein micelles, using cyclodextrin inclusion compounds: a novel generic approach for the preparation of proteoliposomes. *Biochem. J.*, 330 (Pt 2), 667–674, 1998.

158. Prasad Ratnala, V.R., Hulsbbergen, F.B., de Groot, H.J., and de Grip, W.J. Analysis of histamine and modeling of ligand-receptor interactions in the histamine H(1) receptor for Magic Angle Spinning NMR studies. *Inflamm. Res.*, 52, 417–423, 2003.

159. Jacobson, M. and Sali, A. Comparative Protein Structure Modeling and its Applications to Drug Discovery. *Annu. Rep. Med. Chem.*, 39, 259–276, 2004.

160. Archer, E., Maigret, B., Escrieut, C., Pradayrol, L., and Fourmy, D. Rhodopsin crystal: new template yielding realistic models of G-protein-coupled receptors? *Trends Pharmacol. Sci.*, 24, 36–40, 2003.

161. Shin, N., Coates, E., Murgolo, N.J., Morse, K.L., Bayne, M., Strader, C.D., and Monsma, F.J., Jr. Molecular modeling and site-specific mutagenesis of the histamine-binding site of the histamine H4 receptor. *Mol. Pharmacol.*, 62, 38–47, 2002.

162. Bockaert, J. and Pin, J.P. Molecular tinkering of G protein-coupled receptors: an evolutionary success. *EMBO. J.*, 18, 1723–1729, 1999.

163. Yao, B.B., Hutchins, C.W., Carr, T.L., Cassar, S., Masters, J.N., Bennani, Y.L., Esbenshade, T.A., and Hancock, A.A. Molecular modeling and pharmacological analysis of species-related histamine H(3) receptor heterogeneity. *Neuropharmacology*, 44, 773–786, 2003.

164. Bissantz, C., Bernard, P., Hibert, M., and Rognan, D. Protein-based virtual screening of chemical databases. II. Are homology models of G Protein Coupled Receptors suitable targets? *Proteins*, 50, 5–25, 2003.

165. Mason, J.S. and Hermsmeier, M.A. Diversity assessment. *Curr. Opin. Chem. Biol.*, 3, 342–349, 1999.

166. Mason, J.S., Morize, I., Menard, P.R., Cheney, D.L., Hulme, C., and Labaudiniere, R.F. New 4-point pharmacophore method for molecular similarity and diversity applications: overview of the method and applications, including a novel approach to the design of combinatorial libraries containing privileged substructures. *J. Med. Chem.*, 42, 3251–3264, 1999.

167. Evers, A. and Klebe, G. Successful virtual screening for a submicromolar antagonist of the neurokinin-1 receptor based on a ligand-supported homology model. *J. Med. Chem.*, 47, 5381–5392, 2004.

168. Varady, J., Wu, X., Fang, X., Min, J., Hu, Z., Levant, B., and Wang, S. Molecular modeling of the three-dimensional structure of dopamine 3 (D3) subtype receptor: discovery of novel and potent D3 ligands through a hybrid pharmacophore- and structure-based database searching approach. *J. Med. Chem.*, 46, 4377–4392, 2003.

169. Moro, S., Braiuca, P., Deflorian, F., Ferrari, C., Pastorin, G., Cacciari, B., Baraldi, P.G., Varani, K., Borea, P.A., and Spalluto, G. Combined target-based and ligand-based drug design approach as a tool to define a novel 3D-pharmacophore model of human A3 adenosine receptor antagonists: pyrazolo4,3-e1,2,4-triazolo1,5-cpyrimidine derivatives as a key study. *J. Med. Chem.*, 48, 152–162, 2005.

170. Moro, S., Spalluto, G., and Jacobson, K.A. Techniques: Recent developments in computer-aided engineering of GPCR ligands using the human adenosine A3 receptor as an example. *Trends Pharmacol. Sci.*, 26, 44–51, 2005.

171. Shacham, S., Marantz, Y., Bar-Haim, S., Kalid, O., Warshaviak, D., Avisar, N., Inbal, B., Heifetz, A., Fichman, M., Topf, M., Naor, Z., Noiman, S., and Becker, O.M. PREDICT modeling and in-silico screening for G-protein coupled receptors. *Proteins*, 57, 51–86, 2004.

172. Becker, O.M., Marantz, Y., Shacham, S., Inbal, B., Heifetz, A., Kalid, O., Bar-Haim, S., Warshaviak, D., Fichman, M., and Noiman, S. G protein-coupled receptors: in silico drug discovery in 3D. *Proc. Natl. Acad. Sci. USA*, 101, 11304–11309, 2004.

173. PrediX Pharmaceuticals, in www.predixpharm.com. 2005, (accessed 2005).

174. Lundstrom, K. Structural genomics on membrane proteins: The MePNet approach. *Curr. Opin. Drug Disc. Devt.*, 7, 342–346, 2004.

175. Ferguson, S.S., Zhang, J., Barak, L.S., and Caron, M.G. Molecular mechanisms of G protein-coupled receptor desensitization and resensitization. *Life Sci.*, 62, 1561–1565, 1998.

176. Park, P.S., Filipek, S., Wells, J.W., and Palczewski, K. Oligomerization of G protein-coupled receptors: past, present, and future. *Biochemistry*, 43, 15643–15656, 2004.
177. Kenakin, T.P. and Beek, D. The effects on Schild regressions of antagonist removal from the receptor compartment by a saturable process. *Naunyn Schmiedebergs Arch. Pharmacol.*, 335, 103–108, 1987.
178. Kenakin, T. Differences between natural and recombinant G protein-coupled receptor systems with varying receptor/G protein stoichiometry. *Trends Pharmacol. Sci.*, 18, 456–464, 1997.
179. Kenakin, T. Agonist-specific receptor conformations. *Trends Pharmacol. Sci.*, 18, 416–417, 1997.
180. Peleg, G., Ghanouni, P., Kobilka, B.K., and Zare, R.N. Single-molecule spectroscopy of the beta(2) adrenergic receptor: observation of conformational substates in a membrane protein. *Proc. Natl. Acad. Sci. USA*, 98: 8469–8474, 2001.
181. Salamon, Z., Hruby, V.J., Tollin, G., and Cowell, S. Binding of agonists, antagonists and inverse agonists to the human delta-opioid receptor produces distinctly different conformational states distinguishable by plasmon-waveguide resonance spectroscopy. *J. Pept. Res.*, 60, 322–328, 2002.
182. George, S.R., O'Dowd, B.F., and Lee, S.P. G protein-coupled receptor oligomerization and its potential for drug discovery. *Nat. Rev. Drug Discov.*, 1, 808–820, 2002.

Index

A

ABC transporter, 22-24
Adenovirus, 172-173, 326
Affinity tags
    FLAG-tag, 47-48
    His-tag, 47-48
    MBP-tag, 47-48
Alcohol dehydrogenase (ADH) promoter, 119
Alcohol oxidase (AOX) promoter, 118
Allosteric modulation, 100-102
Alphaviruses, see Semliki Forest virus, 174-175, 327
Aquaporin, 128
Aradopsis thaliana, 127
Arrestins, β-arrestins, 103-104
Atomic force microscopy (AFM), 311-317
    data processing, 316-317
    image formation, 311-312
    imaging native membranes, 314-315
    observation of membrane proteins, 312-314
    modelling, 339
    nanomanipulation, 315-316
    porin OmpF, 313
    rhodopsin, 314

B

Bacillus stearothermophilus, 29
Bacillus subtilis, 24, 25, 26
Bacteriorhodopsin
    refolding, 63-64
    atomic force microscopy, 315-317
Baculovirus, 155-162
Bioinformatics, 5-16
beta-barrel finder (BBF), 13
    BPROMPT, 12
    consensus techniques, 11-12
    Dense Alignment Surface (DAS), 8, 11, 12
    Hidden Markov Model (HMM), 8, 10, 11, 13, 15, 16
    HMMTOP, 10, 11, 12, 13, 14
    hydrophobicity analysis, 7-8
    model recognition approaches, 10-11
    multiple sequence alignments, 9
    partial predictions with accuracy, 14
    PHD-htm, 9, 11, 13
    positive inside rule, 9
    prediction confidence, 13-14
    prediction of beta-barrel proteins, 12-13
    prediction of membrane-spanning regions of proteins, 6-13
    SignalP-HMM, 10
    support vector machines, 11, 12-13
    THUMBUP, 8
    TMAP, 9, 11
    TMFinder, 8
    TMHMM, 10, 11, 13, 14
    TMMOD, 10
    TMPDB, 15
    TMpred, 12
    TopPred, 9, 11, 12, 13
Brain-derived neurothrophic factor (BDNF), 99, 106
Brucella abortus, 24, 25
Brucella melitensis, 24, 25

C

Campylobacter jejuni, 24, 25
Caenorhabditis elegans, 324, 326
Cannabinoid CB2 receptor, 235-237
Chaperon proteins, 160-161
Circular dichroism, 34-35
Clathrin-coated pit, 104
COSY (correlation spectroscopy), 221, 223-224, 227
Critical micelle concentration, 32-33, 63, 76, 219, 301
Crystallization strategies, 80-84, 301-304
    co-crystallization, 82-83
    lipidic cubic phase, 83-84
    microdialysis, 81-82
    systems, 275-279
    two-dimensional crystallization, 301-304
    type III crystallization method, 84
    vapour diffusion, 81
Crystallography
    bicelles, 219-220
    crystallization methods, 76-84
    crystallization of membrane proteins, 73-85,
    micelles, 218-219
    nanovolume crystallization, 276-285, 289-292
Cystic fibrosis transmembrane regulator (CFTR), 93
Cytokine receptors, 91, 96-97

T - #0349 - 071024 - C10 - 234/156/18 - PB - 9780367391102 - Gloss Lamination